STUDENT SOLUTIONS MANUAL

GENERAL CHEMISTRY

EIGHTH EDITION

Ebbing/Gammon

DAVID BOOKIN
MT. SAN JACINTO COLLEGE

DARRELL D. EBBING
WAYNE STATE UNIVERSITY

STEVEN D. GAMMON
WESTERN WASHINGTON UNIVERSITY

HOUGHTON MIFFLIN COMPANY ■ BOSTON ■ NEW YORK

Vice President and Publisher: Charles Hartford
Executive Editor: Richard Stratton
Development Editor: Danielle Richardson
Editorial Associate: Rosemary Mack
Senior Project Editor: Nancy Blodget
Senior Manufacturing Coordinator: Priscilla Bailey
Executive Marketing Manager: Katherine Greig
Marketing Associate: Alexandra Shaw

Printed in the U.S.A.

ISBN: 0-618-399453

9-MP-08 07

CONTENTS

Preface / vii

1. **CHEMISTRY AND MEASUREMENT**   3
   Solutions to Exercises 3 / Answers to Review Questions 6 /
   Solutions to Practice Problems 8 / Solutions to General Problems 12 /
   Solutions to Cumulative-Skills Problems 16

2. **ATOMS, MOLECULES, AND IONS**   20
   Solutions to Exercises 20 / Answers to Review Questions 22 /
   Solutions to Practice Problems 25 / Solutions to General Problems 30 /
   Solutions to Cumulative-Skills Problems 33

3. **CALCULATIONS WITH CHEMICAL FORMULAS AND EQUATIONS**   34
   Solutions to Exercises 34 / Answers to Review Questions 40 /
   Solutions to Practice Problems 42 / Solutions to General Problems 57 /
   Solutions to Cumulative-Skills Problems 61

4. **CHEMICAL REACTIONS**   63
   Solutions to Exercises 63 / Answers to Review Questions 69 /
   Solutions to Practice Problems 71 / Solutions to General Problems 80 /
   Solutions to Cumulative-Skills Problems 86

5. **THE GASEOUS STATE**   91
   Solutions to Exercises 91 / Answers to Review Questions 98 /
   Solutions to Practice Problems 102 / Solutions to General Problems 111 /
   Solutions to Cumulative-Skills Problems 115

6. **THERMOCHEMISTRY**   118
   Solutions to Exercises 118 / Answers to Review Questions 121 /
   Solutions to Practice Problems 124 / Solutions to General Problems 129 /
   Solutions to Cumulative-Skills Problems 134

7. **QUANTUM THEORY OF THE ATOM**   139
   Solutions to Exercises 139 / Answers to Review Questions 141 /
   Solutions to Practice Problems 143 / Solutions to General Problems 147 /
   Solutions to Cumulative-Skills Problems 151

8. **ELECTRON CONFIGURATIONS AND PERIODICITY**   153
   Solutions to Exercises 153 / Answers to Review Questions 154 /
   Solutions to Practice Problems 158 / Solutions to General Problems 160 /
   Solutions to Cumulative-Skills Problems 161

9. **IONIC AND COVALENT BONDING**   **163**
Solutions to Exercises 163 / Answers to Review Questions 167 /
Solutions to Practice Problems 169 / Solutions to General Problems 180 /
Solutions to Cumulative-Skills Problems 186

10. **MOLECULAR GEOMETRY AND CHEMICAL BONDING THEORY**   **190**
Solutions to Exercises 190 / Answers to Review Questions 194 /
Solutions to Practice Problems 197 / Solutions to General Problems 203 /
Solutions to Cumulative-Skills Problems 207

11. **STATES OF MATTER; LIQUIDS AND SOLIDS**   **210**
Solutions to Exercises 210 / Answers to Review Questions 213 /
Solutions to Practice Problems 216 / Solutions to General Problems 223 /
Solutions to Cumulative-Skills Problems 227

12. **SOLUTIONS**   **229**
Solutions to Exercises 229 / Answers to Review Questions 234 /
Solutions to Practice Problems 237 / Solutions to General Problems 244 /
Solutions to Cumulative-Skills Problems 250

13. **MATERIALS OF TECHNOLOGY**   **253**
Answers to Review Questions 253 / Solutions to Practice Problems 258 /

14. **RATES OF REACTION**   **262**
Solutions to Exercises 262 / Answers to Review Questions 265 /
Solutions to Practice Problems 269 / Solutions to General Problems 278 /
Solutions to Cumulative-Skills Problems 285

15. **CHEMICAL EQUILIBRIUM**   **287**
Solutions to Exercises 287 / Answers to Review Questions 292 /
Solutions to Practice Problems 295 / Solutions to General Problems 304 /
Solutions to Cumulative-Skills Problems 313

16. **ACIDS AND BASES**   **315**
Solutions to Exercises 315 / Answers to Review Questions 317 /
Solutions to Practice Problems 319 / Solutions to General Problems 325 /
Solutions to Cumulative-Skills Problems 329

17. **ACID-BASE EQUILIBRIA**   **330**
Solutions to Exercises 330 / Answers to Review Questions 342 /
Solutions to Practice Problems 345 / Solutions to General Problems 365 /
Solutions to Cumulative-Skills Problems 380

18. **SOLUBILITY AND COMPLEX-ION EQUILIBRIA**   **382**
Solutions to Exercises 382 / Answers to Review Questions 389 /
Solutions to Practice Problems 390 / Solutions to General Problems 401 /
Solutions to Cumulative-Skills Problems 411

19. **THERMODYNAMICS AND EQUILIBRIUM**   **414**
Solutions to Exercises 414 / Answers to Review Questions 419 /
Solutions to Practice Problems 422 / Solutions to General Problems 430 /
Solutions to Cumulative-Skills Problems 436

20. ELECTROCHEMISTRY   **440**
Solutions to Exercises 440 / Answers to Review Questions 447 /
Solutions to Practice Problems 450 / Solutions to General Problems 468 /
Solutions to Cumulative-Skills Problems 477

21. NUCLEAR CHEMISTRY   **480**
Solutions to Exercises 480 / Answers to Review Questions 484 /
Solutions to Practice Problems 487 / Solutions to General Problems 494 /
Solutions to Cumulative-Skills Problems 499

22. CHEMISTRY OF THE MAIN-GROUP ELEMENTS   **502**
Answers to Review Questions 502 / Solutions to Practice Problems 509 /
Solutions to General Problems 519

23. THE TRANSITION ELEMENTS AND COORDINATION COMPOUNDS   **524**
Solutions to Exercises 524 / Answers to Review Questions 526 /
Solutions to Practice Problems 530 / Solutions to General Problems 536 /

24. ORGANIC CHEMISTRY   **539**
Solutions to Exercises 539 / Answers to Review Questions 543 /
Solutions to Practice Problems 546 / Solutions to General Problems 550 /

25. POLYMERS: SYNTHETIC AND BIOLOGICAL   **553**
Solutions to Exercises 553 / Answers to Review Questions 553 /
Solutions to Practice Problems 556

APPENDIX A   **561**
Solutions to Exercises 561

PREFACE

This student solutions manual provides worked-out answers to the problems that appear in *General Chemistry, 8th Edition*, by Darrell D. Ebbing and Steven D. Gammon. This includes detailed, step-by-step solutions for all in-chapter Exercises as well as for the odd-numbered Practice Problems, General Problems, and Cumulative-Skills Problems, that appear at the end of the chapters. Also provided are answers to all the Review Questions. No answers are given for the Concept Checks and Conceptual Problems.

Please note the following:

Significant figures: The answer is first shown with 1 to 2 nonsignificant figures and no units, and the least significant digit is underlined. The answer is then rounded off to the correct number of significant figures, and the units are added. No attempt has been made to round off intermediate answers, but the least significant digit has been underlined wherever possible.

Great effort and care have gone into the preparation of this manual. The solutions have been checked and rechecked for accuracy and completeness several times. I would like to express my thanks to Conrad Bergo of East Stroudsburg University for his assistance in accuracy reviewing and for his careful work page proofing this manuscript.

D. B.

GENERAL CHEMISTRY

STUDENT SOLUTIONS MANUAL

1. CHEMISTRY AND MEASUREMENT

■ Solutions to Exercises

Note on significant figures: If the final answer to a solution needs to be rounded off, it is given first with one nonsignificant figure, and the last significant figure is underlined. The final answer is then rounded to the correct number of significant figures. In multiple-step problems, intermediate answers are given with at least one nonsignificant figure; however, only the final answer has been rounded off.

1.1   From the law of conservation of mass,

Mass of wood + mass of air = mass of ash + mass of gases

Substituting, you obtain

1.85 grams + 9.45 grams = 0.28 grams + mass of gases

or,

Mass of gases = (1.85 + 9.45 - 0.28) grams = 11.02 grams

Thus, the mass of gases in the vessel at the end of the experiment is 11.02 grams.

1.2   Physical properties: soft, silvery-colored metal; melts at 64°C.

Chemical properties: reacts vigorously with water; reacts with oxygen; reacts with chlorine.

1.3   a. The factor 9.1 has the fewest significant figures, so the answer should be reported to two significant figures.

$$\frac{5.61 \times 7.891}{9.1} = 4.\underline{8}6 = 4.9$$

b. The number with the least number of decimal places is 8.91. Therefore, round the answer to two decimal places.

$$8.91 - 6.435 = 2.4\underline{7}5 = 2.48$$

c. The number with the least number of decimal places is 6.81. Therefore, round the answer to two decimal places.

$$6.81 - 6.730 = 0.0\underline{8}0 = 0.08$$

d. You first do the subtraction within parentheses. In this step, the number with the least number of decimal places is 6.81, so the result of the subtraction has two decimal places. The least significant figure for this step is underlined.

$$38.91 \times (6.81 - 6.730) = 38.91 \times 0.0\underline{8}0$$

Next, perform the multiplication. In this step, the factor 0.0$\underline{8}$0 has the fewest significant figures, so round the answer to one significant figure.

$$38.91 \times 0.0\underline{8}0 = \underline{3}.11 = 3$$

1.4   a. 1.84×10^{-9} m = 1.84 nm   b. 5.67×10^{-12} s = 5.67 ps

c. 7.85×10^{-3} g = 7.85 mg   d. 9.7×10^{3} m = 9.7 km

e. 0.000732 s = 0.732 ms, or 732 μs   f. 0.000000000154 m = 0.154 nm, or 154 pm

1.5   a. Substituting, we find that

$$t_C = \frac{5°C}{9°F} \times (t_F - 32°F) = \frac{5°C}{9°F} \times (102.5°F - 32°F) = 39.\underline{1}67°C = 39.2°C$$

b. Substituting, we find that

$$T_K = \left(t_C \times \frac{1K}{1°C}\right) + 273.15K = \left(-78°C \times \frac{1K}{1°C}\right) + 273.15K = 195.\underline{1}5K = 195K$$

1.6   Recall that density equals mass divided by volume. You substitute 159 g for the mass and 20.2 g/cm^3 for the volume.

$$d = \frac{m}{V} = \frac{159\ g}{20.2\ cm^3} = 7.8\underline{7}1\ g/cm^3 = 7.87\ g/cm^3$$

The density of the metal equals that of iron.

1.7   Rearrange the formula defining the density to obtain the volume.

$$V = \frac{m}{d}$$

Substitute 30.3 g for the mass and 0.789 g/cm^3 for the density.

$$V = \frac{30.3\ g}{0.789\ g/cm^3} = 38.\underline{4}0\ cm^3 = 38.4\ cm^3$$

1.8   Since one pm = 10^{-12} m, and the prefix milli- means 10^{-3}, you can write

$$121\ pm\ \times\ \frac{10^{-12}\ m}{1\ pm}\ \times\ \frac{1\ mm}{10^{-3}\ m} = 1.21 \times 10^{-7}\ mm$$

1.9 $67.6\ \text{Å}^3\ \times\ \left(\dfrac{10^{-10}\ m}{1\ \text{Å}}\right)^3\ \times\ \left(\dfrac{1\ dm}{10^{-1}\ m}\right)^3 = 6.76 \times 10^{-26}\ dm^3$

1.10   From the definitions, you obtain the following conversion factors:

$$1 = \frac{36\ in}{1\ yd} \qquad 1 = \frac{2.54\ cm}{1\ in} \qquad 1 = \frac{10^{-2}\ m}{1\ cm}$$

Then,

$$3.54\ yd\ \times\ \frac{36\ in}{1\ yd}\ \times\ \frac{2.54\ cm}{1\ in}\ \times\ \frac{10^{-2}\ m}{1\ cm} = 3.2\underline{3}6\ m = 3.24\ m$$

■ Answers to Review Questions

1.1   One area of technology that chemistry has changed is the characteristics of materials. In devices such as watches and calculators, the liquid-crystal displays (LCDs) are materials made of molecules designed by chemists. Electronics and communications have been transformed by the development of optical fibers to replace copper wires. In biology, chemistry has changed the way scientists view life. Biochemists have found that all forms of life share many of the same molecules and molecular processes.

1.2   An experiment is an observation of natural phenomena carried out in a controlled manner so the results can be duplicated and rational conclusions obtained. A theory is a tested explanation of basic natural phenomena. They are related in that a theory is based on the results of many experiments and is fruitful in suggesting other new experiments. Also, an experiment can disprove a theory but never prove it absolutely. A hypothesis is a tentative explanation of some regularity of nature.

1.3   Rosenberg conducted controlled experiments and noted a basic relationship that could be stated as a hypothesis; that is, that certain platinum compounds inhibit cell division. This led him to do new experiments on the anti-cancer activity of these compounds.

1.4   Matter is the general term for the material things around us. It is whatever occupies space and can be perceived by our senses. Mass is the quantity of matter in a material. The difference between mass and weight is that mass remains the same wherever it is measured, but weight is proportional to the mass of the object divided by the square of the distance between the center of mass of the object and that of the earth.

1.5   The law of conservation of mass states that the total mass remains constant during a chemical change (chemical reaction). To demonstrate this law, place a sample of wood in a sealed vessel with air, and weigh it. Heat the vessel to burn the wood, and weigh the vessel after the experiment. The weight before and after the experiment should be the same.

1.6   Mercury metal, which is a liquid, reacts with oxygen gas to form solid mercury(II) oxide. The color changes from that of metallic mercury (silvery) to a color that varies from red to yellow depending on the particle size of the oxide.

1.7   Gases are easily compressible fluids. Liquids are relatively incompressible fluids. Solids are relatively incompressible and rigid. A gas fits into a container of any size and shape. A liquid has a fixed volume but no fixed shape. A solid has fixed shape and volume.

1.8   An example of a substance is the element sodium. Among its physical properties: It is a solid, and it melts at 98°C. Among its chemical properties: It reacts vigorously with water, and it burns in chlorine gas to form sodium chloride.

1.9   An example of an element: sodium; of a compound: sodium chloride, or table salt; of a heterogeneous mixture: salt and sugar; of a homogeneous mixture: sodium chloride dissolved in water to form a solution.

1.10   A glass of bubbling carbonated beverage with ice cubes contains three phases, gas, liquid, and solid.

1.11   A compound may be decomposed by chemical reactions into elements. An element cannot be decomposed by any chemical reaction. Thus, a compound cannot also be an element in any case.

1.12   The precision refers to the closeness of the set of values obtained from identical measurements of a quantity. The number of digits reported for the value of a measured or calculated quantity (significant figures) indicates the precision of the value.

1.13   Multiplication and division rule: In performing the calculation 100.0 x 0.0634 ÷ 25.31, the calculator display shows 0.2504398. We would report the answer as 0.250 because the factor 0.0634 has the least number of significant figures (three).

    Addition and subtraction rule: In performing the calculation 184.2 + 2.324, the calculator display shows 186.524. Because the quantity 184.2 has the least number of decimal places (one), the answer is reported as 186.5.

1.14   An exact number is a number that arises when you count items or sometimes when you define a unit. For example, a foot is defined to be 12 inches. A measured number is the result of a comparison of a physical quantity with a fixed standard of measurement. For example, a steel rod measures 9.12 centimeters or 9.12 times the standard centimeter unit of measurement.

1.15   For a given unit, the SI system uses prefixes to obtain units of different sizes. Units for all other possible quantities are obtained by deriving them from any of the seven base units. You do this by using the base units in equations that define other physical quantities.

1.16   An absolute temperature scale is a scale in which the lowest temperature that can be attained theoretically is zero. Degrees Celsius and kelvins have equal size units and are related by the formula

$$t_C = (T_K - 273.15K) \times \frac{1°C}{1K}$$

1.17   The density of an object is its mass per unit volume. Because the density is characteristic of a substance, it can be helpful in identifying it. Density can also be useful in determining whether a substance is pure. It also provides a useful relationship between mass and volume.

1.18   Units should be carried along because first, the units for the answers will come out in the calculations, and second, if you make an error in arranging factors in the calculation, this will become apparent because the final units will be nonsense.

■ Solutions to Practice Problems

Note on significant figures: If the final answer to a solution needs to be rounded off, it is given first with one nonsignificant figure, and the last significant figure is underlined. The final answer is then rounded to the correct number of significant figures. In multiple-step problems, intermediate answers are given with at least one nonsignificant figure; however, only the final answer has been rounded off.

1.31   By the law of conservation of mass:

Mass of sodium carbonate + mass of acetic acid solution

= mass of contents of reaction vessel + mass of carbon dioxide

Plugging in gives

15.9 g + 20.0 g = 29.3 g + mass of carbon dioxide

Mass of carbon dioxide = 15.9 g + 20.0 g - 29.3 g = 6.6 g

1.33   By the law of conservation of mass:

Mass of zinc + mass of sulfur = mass of zinc sulfide

Rearranging and plugging in gives

Mass of zinc sulfide = 65.4 g + 32.1 g = 97.5 g

For the second part, let x = mass of zinc sulfide that could be produced. By the law of conservation of mass:

20.0 g + mass of sulfur = x

Write a proportion that relates the mass of zinc reacted to the mass of zinc sulfide formed, which should be the same for both cases.

$$\frac{\text{mass zinc}}{\text{mass zinc sulfide}} = \frac{65.4 \text{ g}}{97.5 \text{ g}} = \frac{20.0 \text{ g}}{x}$$

Solving gives x = 29.8̲1 g = 29.8 g

1.35 a. Solid   b. Liquid   c. Gas   d. Solid

1.37 a. Physical change   b. Physical change   c. Chemical change   d. Physical change

1.39   Physical change: Liquid mercury is cooled to solid mercury.

Chemical changes: (1) Solid mercury oxide forms liquid mercury metal and gaseous oxygen; (2) glowing wood and oxygen form burning wood (form ash and gaseous products).

1.41 a. Physical property   b. Chemical property   c. Physical property
     d. Physical property   e. Chemical property

1.43   Physical properties: (1) Iodine is solid; (2) the solid has lustrous blue-black crystals; (3) the crystals vaporize readily to a violet-colored gas.

Chemical properties: (1) Iodine combines with many metals such as with aluminum to give aluminum iodide.

1.45 a. Physical process   b. Chemical reaction   c. Physical process
     d. Chemical reaction   e. Physical process

1.47 a. Solution   b. Substance   c. Substance   d. Heterogeneous mixture

1.49 a. A pure substance with two phases present, liquid and gas.

b. A mixture with two phases present, solid and liquid.

c. A pure substance with two phases present, solid and liquid.

d. A mixture with two phases present, solid and solid.

1.51 a. six   b. three   c. four   d. five   e. three   f. four

1.53   40,000 km = 4.0×10^4 km

1.55 a. $\dfrac{8.71 \times 0.0301}{0.031}$ = 8.457 = 8.5

b. 0.71 + 92.2 = 92.91 = 92.9

c. 934 x 0.00435 + 107 = 4.0629 + 107 = 111.06 = 111

d. (847.89 - 847.73) x 14673 = 0.16 x 14673 = 2347 = 2.3×10^3

1.57   The volume of the first sphere is

$$V_1 = (4/3)\pi r^3 = (4/3) \times 3.1416 \times (5.10\ cm)^3 = 55\underline{5}.64\ cm^3$$

The volume of the second sphere is

$$V_2 = (4/3)\pi r^3 = (4/3) \times 3.1416 \times (5.00\ cm)^3 = 52\underline{3}.60\ cm^3$$

The difference in volume is

$$V_1 - V_2 = 55\underline{5}.64\ cm^3 - 52\underline{3}.60\ cm^3 = 3\underline{2}.04\ cm^3 = 32\ cm^3$$

1.59 a. $5.89 \times 10^{-12}\ s = 5.89\ ps$ b. $0.2010\ m = 20.1\ cm$

   c. $2.56 \times 10^{-9}\ g = 2.560\ ng$ d. $6.05 \times 10^3\ m = 6.05\ km$

1.61 a. $6.15\ ps = 6.15 \times 10^{-12}\ s$ b. $3.781\ \mu m = 3.781 \times 10^{-6}\ m$

   c. $1.546\ Å = 1.546 \times 10^{-10}\ m$ d. $9.7\ mg = 9.7 \times 10^{-3}\ g$

1.63 a. $t_C = \dfrac{5°C}{9°F} \times (t_F - 32°F) = \dfrac{5°C}{9°F} \times (68°F - 32°F) = 2\underline{0}.0°C = 20.°C$

   b. $t_C = \dfrac{5°C}{9°F} \times (t_F - 32°F) = \dfrac{5°C}{9°F} \times (-23°F - 32°F) = -3\underline{0}.55°C = -31°C$

   c. $t_F = (t_C \times \dfrac{9°F}{5°C}) + 32°F = (26°C \times \dfrac{9°F}{5°C}) + 32°F = 7\underline{8}.8°F = 79°F$

   d. $t_F = (t_C \times \dfrac{9°F}{5°C}) + 32°F = (-70°C \times \dfrac{9°F}{5°C}) + 32°F = -9\underline{4}.0°F = -94°F$

1.65 $t_F = (t_C \times \dfrac{9°F}{5°C}) + 32°F = (-21.1°C \times \dfrac{9°F}{5°C}) + 32°F = -5.\underline{9}8°F = -6.0°F$

1.67 $d = \dfrac{m}{V} = \dfrac{12.4\ g}{1.64\ cm^3} = 7.5\underline{6}0\ g/cm^3 = 7.56\ g/cm^3$

1.69   First, determine the density of the liquid.

$$d = \frac{m}{V} = \frac{6.71\ g}{8.5\ mL} = 0.7\underline{8}94 = 0.79\ g/mL$$

The density is closest to ethanol (0.789 g/cm^3).

1.71   The mass of platinum is obtained as follows.

$$Mass = d \times V = 21.4\ g/cm^3 \times 5.9\ cm^3 = 12\underline{6}\ g = 1.3 \times 10^2\ g$$

1.73   The volume of ethanol is obtained as follows. Recall that 1 mL = 1 cm^3.

$$Volume = \frac{m}{d} = \frac{19.8\ g}{0.789\ g/cm^3} = 25.\underline{0}9\ cm^3 = 25.1\ cm^3 = 25.1\ mL$$

1.75   Since 1 kg = 10^3 g, and 1 mg = 10^{-3} g, you can write

$$0.480\ kg \times \frac{10^3\ g}{1\ kg} \times \frac{1\ mg}{10^{-3}\ g} = 4.80 \times 10^5\ mg$$

1.77   Since 1 nm = 10^{-9} m, and 1 cm = 10^{-2} m, you can write

$$555\ nm \times \frac{10^{-9}\ m}{1\ nm} \times \frac{1\ cm}{10^{-2}\ m} = 5.55 \times 10^{-5}\ cm$$

1.79   Since 1 km = 10^3 m, you can write

$$3.73 \times 10^8\ km^3 \times \left(\frac{10^3\ m}{1\ km}\right)^3 = 3.73 \times 10^{17}\ m^3$$

Now, 1 dm = 10^{-1} m. Also, note that 1 dm^3 = 1 L. Therefore, you can write

$$3.73 \times 10^{17}\ m^3 \times \left(\frac{1\ dm}{10^{-1}\ m}\right)^3 = 3.73 \times 10^{20}\ dm^3 = 3.73 \times 10^{20}\ L$$

1.81 3.58 short ton $\times \dfrac{2000 \text{ lb}}{1 \text{ short ton}} \times \dfrac{16 \text{ oz}}{1 \text{ lb}} \times \dfrac{1 \text{ g}}{0.03527 \text{ oz}}$

$$= 3.2\underline{4}8 \times 10^6 \text{ g} = 3.25 \times 10^6 \text{ g}$$

1.83 2425 fathoms $\times \dfrac{6 \text{ ft}}{1 \text{ fathom}} \times \dfrac{12 \text{ in}}{1 \text{ ft}} \times \dfrac{2.54 \times 10^{-2} \text{ m}}{1 \text{ in}} = 4434.8 \text{ m} = 4435 \text{ m}$

1.85 (20.0 in)(20.0 in)(10.0 in) $\times \left(\dfrac{2.54 \text{ cm}}{1 \text{ in}}\right)^3 \times \dfrac{1 \text{ L}}{1000 \text{ cm}^3} = 65.\underline{5}4 \text{ L} = 65.5 \text{ L}$

■ Solutions to General Problems

1.87 From the law of conservation of mass,

Mass of sodium + mass of water = mass of hydrogen + mass of solution

Substituting, you obtain

19.70 g + 126.22 g = mass of hydrogen + 145.06 g

or,

Mass of hydrogen = 19.70 g + 126.22 g - 145.06 g = 0.8$\underline{6}$ g

Thus, the mass of hydrogen produced was 0.86 g.

1.89 From the law of conservation of mass,

Mass of aluminum + mass of iron(III) oxide

= mass of iron + mass of aluminum oxide + mass of unreacted iron(III) oxide

5.40 g + 18.50 g = 11.17 g + 10.20 g + mass of iron(III) oxide unreacted

mass of iron(III) oxide unreacted = 5.40 g + 18.50 g - 11.17 g - 10.20 g

= 2.5$\underline{3}$ g

Thus, the mass of unreacted iron(III) oxide is 2.53 g.

1.91 53.10 g + 5.348 g + 56.1 g = 114.$\underline{5}$4 g = 114.5 g total

1.93 a. Bromine   b. Phosphorus   c. Gold   d. Carbon (as graphite)

1.95   Compounds always contain the same proportions of the elements by mass. Thus, if we let X be the proportion of iron in a sample, we can calculate the proportion of iron in each sample as follows.

Sample A: $X = \dfrac{\text{mass of iron}}{\text{mass of sample}} = \dfrac{1.094 \text{ g}}{1.518 \text{ g}} = 0.72068 = 0.7207$

Sample B: $X = \dfrac{\text{mass of iron}}{\text{mass of sample}} = \dfrac{1.449 \text{ g}}{2.056 \text{ g}} = 0.70476 = 0.7048$

Sample C: $X = \dfrac{\text{mass of iron}}{\text{mass of sample}} = \dfrac{1.335 \text{ g}}{1.873 \text{ g}} = 0.71276 = 0.7128$

Since each sample has a different proportion of iron by mass, the material is not a compound.

1.97 $V = (\text{edge})^3 = (39.3 \text{ cm})^3 = 6.069 \times 10^4 \text{ cm}^3 = 6.07 \times 10^4 \text{ cm}^3$

1.99 $V = LWH = 47.8 \text{ in} \times 12.5 \text{ in} \times 19.5 \text{ in} \times \dfrac{1 \text{ gal}}{231 \text{ in}^3} = 50.43 \text{ gal} = 50.4 \text{ gal}$

1.101   The volume of the first sphere is given by

$V_1 = (4/3)\pi r^3 = (4/3) \times 3.1416 \times (5.61 \text{ cm})^3 = 739.5 \text{ cm}^3$

The volume of the second sphere is given by

$V_2 = (4/3)\pi r^3 = (4/3) \times 3.1416 \times (5.85 \text{ cm})^3 = 838.6 \text{ cm}^3$

The difference in volume between the two spheres is given by

$V = V_2 - V_1 = 838.6 \text{ cm}^3 - 739.5 \text{ cm}^3 = 9.91 \times 10^1 = 9.9 \times 10^1 \text{ cm}^3$

1.103 a. $\dfrac{56.1 - 51.1}{6.58} = 7.59 \times 10^{-1} = 7.6 \times 10^{-1}$

b. $\dfrac{56.1 + 51.1}{6.58} = 1.629 \times 10^1 = 1.63 \times 10^1$

c. $(9.1 + 8.6) \times 26.91 = 4.763 \times 10^2 = 4.76 \times 10^2$

d. $0.0065 \times 3.21 + 0.0911 = 1.119 \times 10^{-1} = 1.12 \times 10^{-1}$

1.105 a. 9.12 cg/mL   b. 66 pm   c. 7.1 μm   d. 56 nm

1.107 a. 1.07×10^{-12} s   b. 5.8×10^{-6} m   c. 3.19×10^{-7} m   d. 1.53×10^{-2} s

1.109 $t_F = (t_C \times \dfrac{9°F}{5°C}) + 32°F = (3410°C \times \dfrac{9°F}{5°C}) + 32°F = 617\underline{0}°F = 6170°F$

1.111 $t_F = (t_C \times \dfrac{9°F}{5°C}) + 32°F = (825°C \times \dfrac{9°F}{5°C}) + 32°F = 15\underline{1}7°F = 1.52 \times 10^3 \ °F$

1.113 The temperature in kelvin units is

$$T_K = (t_C \times \frac{1K}{1°C}) + 273.15K = (29.8°C \times \frac{1K}{1°C}) + 273.15K$$

$$= 302.\underline{9}5K = 303.0K$$

The temperature in degrees Fahrenheit is

$$t_F = (t_C \times \frac{9°F}{5°C}) + 32°F = (29.8°C \times \frac{9°F}{5°C}) + 32°F = 85.\underline{6}4°F = 85.6°F$$

1.115 The temperature in degrees Celsius is

$$t_C = \frac{5°C}{9°F} \times (t_F - 32°F) = \frac{5°C}{9°F} \times (1666°F - 32°F) = 907.\underline{7}7°C = 907.8°C$$

The temperature in kelvin units is

$$T_K = (t_C \times \frac{1K}{1°C}) + 273.15K = (907.\underline{7}7°C \times \frac{1K}{1°C}) + 273.15K$$

$$= 118\underline{0}.92K = 1180.9K$$

1.117 $\text{Density} = \dfrac{1.74 \text{ g}}{1 \text{ cm}^3} \times \dfrac{1 \text{ kg}}{10^3 \text{ g}} \times \left(\dfrac{1 \text{ cm}}{10^{-2} \text{ m}}\right)^3 = 1.74 \times 10^3 \text{ kg/m}^3$

1.119 The volume of the quartz is 65.7 mL - 51.2 mL = 14.5 mL. Then, the density is

$$\text{Density} = \frac{\text{mass}}{\text{volume}} = \frac{38.4 \text{ g}}{14.5 \text{ mL}} = 2.6\underline{4}8 \text{ g/mL} = 2.65 \text{ g/mL} = 2.65 \text{ g/cm}^3$$

1.121   First, determine the density of the liquid sample.

$$\text{Density} = \frac{\text{mass}}{\text{volume}} = \frac{22.3 \text{ g}}{15.0 \text{ mL}} = 1.4\underline{8}6 \text{ g/mL} = 1.49 \text{ g/mL} = 1.49 \text{ g/cm}^3$$

This density is closest to that of chloroform (1.489 g/cm^3), so the unknown liquid is chloroform.

1.123   First, determine the volume of the cube of platinum.

$$V = (\text{edge})^3 = (4.40 \text{ cm})^3 = 85.\underline{1}8 \text{ cm}^3$$

Now, use the density to determine the mass of the platinum.

$$\text{Mass} = d \times V = 21.4 \text{ g/cm}^3 \times 85.\underline{1}8 \text{ cm}^3 = 18\underline{2}2.9 \text{ g} = 1.82 \times 10^3 \text{ g}$$

1.125 $$\text{Volume} = \frac{\text{mass}}{\text{density}} = \frac{35.00 \text{ g}}{1.053 \text{ g/ mL}} = 33.2\underline{3}8 \text{ mL} = 33.24 \text{ mL}$$

1.127 a. $8.45 \text{ kg} \times \dfrac{10^3 \text{ g}}{1 \text{ kg}} \times \dfrac{1 \text{ μg}}{10^{-6} \text{ g}} = 8.45 \times 10^9 \text{ μg}$

b. $318 \text{ μs} \times \dfrac{10^{-6} \text{ s}}{1 \text{ μs}} \times \dfrac{1 \text{ ms}}{10^{-3} \text{ s}} = 3.18 \times 10^{-1} \text{ ms}$

c. $93 \text{ km} \times \dfrac{10^3 \text{ m}}{1 \text{ km}} \times \dfrac{1 \text{ nm}}{10^{-9} \text{ m}} = 9.3 \times 10^{13} \text{ nm}$

d. $37.1 \text{ mm} \times \dfrac{10^{-3} \text{ m}}{1 \text{ mm}} \times \dfrac{1 \text{ cm}}{10^{-2} \text{ m}} = 3.71 \text{ cm}$

1.129 a. $5.91 \text{ kg} \times \dfrac{10^3 \text{ g}}{1 \text{ kg}} \times \dfrac{1 \text{ mg}}{10^{-3} \text{ g}} = 5.91 \times 10^6 \text{ mg}$

b. $753 \text{ mg} \times \dfrac{10^{-3} \text{ g}}{1 \text{ mg}} \times \dfrac{1 \text{ μg}}{10^{-6} \text{ g}} = 7.53 \times 10^5 \text{ μg}$

c. $90.1 \text{ MHz} \times \dfrac{10^6 \text{ Hz}}{1 \text{ MHz}} \times \dfrac{1 \text{ kHz}}{10^3 \text{ Hz}} = 9.01 \times 10^4 \text{ kHz}$

(continued)

d. $498 \text{ mJ} \times \dfrac{10^{-3} \text{ J}}{1 \text{ mJ}} \times \dfrac{1 \text{ kJ}}{10^3 \text{ J}} = 4.98 \times 10^{-4} \text{ kJ}$

1.131 Volume $= 12{,}230 \text{ km}^3 \times \left(\dfrac{10^3 \text{ m}}{1 \text{ km}}\right)^3 \times \left(\dfrac{1 \text{ dm}}{10^{-1} \text{ m}}\right)^3 \times \dfrac{1 \text{ L}}{1 \text{ dm}^3} = 1.2230 \times 10^{16} \text{ L}$

1.133 First, calculate the volume of the room in cubic feet.

Volume $= \text{LWH} = 10.0 \text{ ft} \times 11.0 \text{ ft} \times 9.0 \text{ ft} = 9\underline{9}0 \text{ ft}^3$

Next, convert the volume to liters.

$V = 9\underline{9}0 \text{ ft}^3 \times \left(\dfrac{12 \text{ in}}{1 \text{ ft}}\right)^3 \times \left(\dfrac{2.54 \text{ cm}}{1 \text{ in}}\right)^3 \times \dfrac{1 \text{ L}}{10^3 \text{ cm}^3}$

$= 2.\underline{8}0 \times 10^4 \text{ L} = 2.8 \times 10^4 \text{ L}$

1.135 Mass $= 275 \text{ carats} \times \dfrac{200 \text{ mg}}{1 \text{ carat}} \times \dfrac{10^{-3} \text{ g}}{1 \text{ mg}} = 55.\underline{0}0 \text{ g} = 55.0 \text{ g}$

■ Solutions to Cumulative-Skills Problems

1.137 The mass of hydrochloric acid is obtained from the density and the volume.

Mass $=$ density $\times$ volume $= 1.096 \text{ g/mL} \times 50.0 \text{ mL} = 54.\underline{8}0 \text{ g}$

Next, from the law of conservation of mass,

Mass of marble $+$ mass of acid

$=$ mass of solution $+$ mass of carbon dioxide gas

Plugging in gives

$10.0 \text{ g} + 54.\underline{8}0 \text{ g} = 60.4 \text{ g} +$ mass of carbon dioxide gas

mass of carbon dioxide gas $= 10.0 \text{ g} + 54.\underline{8}0 \text{ g} - 60.4 \text{ g} = 4.\underline{4}0 \text{ g}$

(continued)

Finally, use the density to convert the mass of carbon dioxide gas to volume.

$$\text{Volume} = \frac{\text{mass}}{\text{density}} = \frac{4.40 \text{ g}}{1.798 \text{ g/ L}} = 2.\underline{4}47 \text{ L} = 2.4 \text{ L}$$

1.139   First, calculate the volume of the steel sphere.

$$V = (4/3) \pi r^3 = (4/3) \times 3.1416 \times (1.58 \text{ in})^3 \times \left(\frac{2.54 \text{ cm}}{1 \text{ in}}\right)^3 = 27\underline{0}.7 \text{ cm}^3$$

Next, determine the mass of the sphere using the density.

$$\text{Mass} = \text{density} \times \text{volume} = 7.88 \text{ g/cm}^3 \times 27\underline{0}.7 \text{ cm}^3$$

$$= 21\underline{3}3 \text{ g} = 2.13 \times 10^3 \text{ g}$$

1.141   The area of the ice is 840,000 mi^2 - 132,000 mi^2 = 708,000 mi^2. Now, determine the volume of this ice.

$$\text{Volume} = \text{area} \times \text{thickness}$$

$$= 708,000 \text{ mi}^2 \times 5000 \text{ ft} \times \left(\frac{5280 \text{ ft}}{1 \text{ mi}}\right)^2 \times \left(\frac{12 \text{ in}}{1 \text{ ft}}\right)^3 \times \left(\frac{2.54 \text{ cm}}{1 \text{ in}}\right)^3$$

$$= 2.794 \times 10^{21} \text{ cm}^3$$

Now use the density to determine the mass of the ice.

$$\text{mass} = \text{density} \times \text{volume} = 0.917 \text{ g/cm}^3 \times 2.794 \times 10^{21} \text{ cm}^3 = 2.\underline{5}6 \times 10^{21} \text{ g}$$

$$= 2.6 \times 10^{21} \text{ g}$$

1.143   Let x = mass of ethanol and y = mass of water. Then, use the total mass to write x + y = 49.6 g, or y = 49.6 g - x. So, the mass of water is 49.6 g - x. Next,

$$\text{Total volume} = \text{volume of ethanol} + \text{volume of water}$$

Since the volume is equal to the mass divided by density, you can write

$$\text{Total volume} = \frac{\text{mass of ethanol}}{\text{density of ethanol}} + \frac{\text{mass of water}}{\text{density of water}}$$

(continued)

Substitute in the known and unknown values to get an equation for x.

$$54.2 \text{ cm}^3 = \frac{x}{0.789 \text{ g/ cm}^3} + \frac{49.6 \text{ g} - x}{0.998 \text{ g/ cm}^3}$$

Multiply both sides of this equation by (0.789)(0.998). Also, multiply both sides by g/cm³ to simplify the units. This gives the following equation to solve for x.

$$(0.789)(0.998)(54.2) \text{ g} = (0.998) x + (0.789)(49.6 \text{ g} - x)$$

$$42.678 \text{ g} = 0.998 x + 39.134 \text{ g} - 0.789 x$$

$$0.209 x = 3.544 \text{ g}$$

$$x = \text{mass of ethanol} = 16.95 \text{ g}$$

The percentage of ethanol (by mass) in the solution can now be calculated.

$$\text{Percent (mass)} = \frac{\text{mass of ethanol}}{\text{mass of solution}} \times 100\% = \frac{16.95 \text{ g}}{49.6 \text{ g}} \times 100\% = 34.1\% = 34\%$$

To determine the proof, you must first find the percentage by volume of ethanol in the solution. The volume of ethanol is obtained using the mass and the density.

$$\text{Volume} = \frac{\text{mass of ethanol}}{\text{density of ethanol}} = \frac{16.95 \text{ g}}{0.789 \text{ g/ cm}^3} = 21.48 \text{ cm}^3$$

The percentage of ethanol (by volume) in the solution can now be calculated.

$$\text{Percent (volume)} = \frac{\text{volume of ethanol}}{\text{volume of solution}} \times 100\% = \frac{21.48 \text{ cm}^3}{54.2 \text{ cm}^3} \times 100\% = 39.63\%$$

The proof can now be calculated.

$$\text{Proof} = 2 \times \text{Percent (volume)} = 2 \times 39.63 = 79.27 = 79 \text{ proof}$$

1.145　The volume of the mineral can be obtained from the mass of the water displaced and the density of water.

$$\text{Mass of water displaced} = 18.49 \text{ g} - 16.21 \text{ g} = 2.28 \text{ g}$$

$$\text{Volume of mineral} = \frac{\text{mass of water}}{\text{density of water}} = \frac{2.28 \text{ g}}{0.9982 \text{ g/ cm}^3} = 2.284 \text{ cm}^3$$

(continued)

The mass of the mineral is equal to its mass in air plus the weight of the displaced air. The weight of the displaced air is obtained from the volume of the mineral and the density of air.

Mass of displaced air = density x volume

$$= 1.205 \text{ g/L} \times 2.2\underline{8}4 \text{ cm}^3 \times \frac{1 \text{ L}}{10^3 \text{ cm}^3} = 2.7\underline{5}2 \times 10^{-3} \text{ g}$$

Mass of mineral = 18.49 g + 2.7$\underline{5}$2 x 10^{-3} g = 18.4$\underline{9}$27 g

The density of the mineral can now be calculated.

$$\text{Density} = \frac{\text{mass}}{\text{volume}} = \frac{18.4927 \text{ g}}{2.2\underline{8}4 \text{ cm}^3} = 8.0\underline{9}6 \text{ g/cm}^3 = 8.10 \text{ g/cm}^3$$

1.147 The volume of the object can be obtained from the mass of the ethanol displaced and the density of ethanol.

Mass of ethanol displaced = 15.8 g - 10.5 g = 5.3 g

$$\text{Volume of object} = \frac{\text{mass of ethanol}}{\text{density of ethanol}} = \frac{5.3 \text{ g}}{0.789 \text{ g/ cm}^3} = 6.\underline{7}17 \text{ cm}^3$$

The density of the object can now be calculated.

$$\text{Density} = \frac{\text{mass}}{\text{volume}} = \frac{15.8 \text{ g}}{6.\underline{7}17 \text{ cm}^3} = 2.\underline{3}52 \text{ g/cm}^3 = 2.4 \text{ g/cm}^3$$

2. ATOMS, MOLECULES, AND IONS

■ Solutions to Exercises

Note on significant figures: If the final answer to a solution needs to be rounded off, it is given first with one nonsignificant figure, and the last significant figure is underlined. The final answer is then rounded to the correct number of significant figures. In multiple-step problems, intermediate answers are given with at least one nonsignificant figure; however, only the final answer has been rounded off.

2.1 The element with atomic number 17 (the number of protons in the nucleus) is chlorine, symbol Cl. The mass number is 17 + 18 = 35. The symbol is: $^{35}_{17}Cl$.

2.2 Multiply each isotopic mass by its fractional abundance; then sum:

$$
\begin{array}{lll}
34.96885 \text{ amu} \times 0.75771 & = & 26.49\underline{6}247 \\
36.96590 \text{ amu} \times 0.24229 & = & \underline{8.956467} \\
& & 35.45\underline{2}714 = 35.453 \text{ amu}
\end{array}
$$

The atomic weight of chlorine is 35.453 amu.

2.3 a. Se: Group VIA, Period 4; nonmetal   b. Cs: Group IA, Period 6; metal

     c. Fe: Group VIIIB, Period 4; metal   d. Cu: Group IB, Period 4; metal

     e. Br: Group VIIA, Period 4; nonmetal

2.4 Take as many cations as there are units of charge on the anion and as many anions as there are units of charge on the cation. Two K^+ ions have a total charge of 2+, and one CrO_4^{2-} ion has a charge of 2-, giving a net charge of zero. The simplest ratio of K^+ to CrO_4^{2-} is 1:2, and the formula is K_2CrO_4.

2.5 a. CaO: Calcium, a group IIA metal, is expected to form only a 2+ ion (Ca^{2+}, the calcium ion). Oxygen (Group VIA) is expected to form an anion of charge equal to the group number minus eight (O^{2-}, the oxide ion). The name of the compound is calcium oxide.

   b. $PbCrO_4$: Lead has more than one monatomic ion. You can find the charge on the Pb ion if you know the formula of the anion. From Table 2.6, the CrO_4 refers to the anion CrO_4^{2-} (the chromate ion). Therefore, the Pb cation must be Pb^{2+} to give electrical neutrality. The name of Pb^{2+} is lead(II) ion, so the name of the compound is lead(II) chromate.

2.6   Thallium(III) nitrate contains the thallium(III) ion, Tl^{3+}, and the nitrate ion, NO_3^-. The formula is $Tl(NO_3)_3$.

2.7 a. Dichlorine hexoxide   b. Phosphorus trichloride   c. Phosphorus pentachloride

2.8 a. CS_2 b. SO_3

2.9 a. Boron trifluoride   b. Hydrogen selenide

2.10 When you remove one H^+ ion from $HBrO_4$, you obtain the BrO_4^- ion. You name the ion from the acid by replacing -ic with -ate. The anion is called the perbromate ion.

2.11 Sodium carbonate decahydrate

2.12 Sodium thiosulfate is composed of sodium ions (Na^+) and thiosulfate ions ($S_2O_3^{2-}$), so the formula of the anhydrous compound is $Na_2S_2O_3$. Since the material is a pentahydrate, the formula of the compound is $Na_2S_2O_3 \cdot 5H_2O$.

2.13 Balance O first in parts (a) and (b) because it occurs in only one product. Balance S first in part (c) because it appears in only one product. Balance H first in part (d) because it appears in just one reactant as well as in the product.

   a. Write a two in front of $POCl_3$ for O; this requires a two in front of PCl_3 for final balance:

$$2PCl_3 + O_2 \rightarrow 2POCl_3$$

   b. Write a six in front of N_2O to balance O; this requires a six in front of N_2 for final balance:

$$6N_2O + P_4 \rightarrow 6N_2 + P_4O_6$$

c. Write $2As_2S_3$ and $6SO_2$ to achieve ultimately an even number of oxygens on the right to balance what will always be an even number of oxygens on the left. The $2As_2S_3$ then requires $2As_2O_3$. Finally, to balance (6 + 12) O's on the right, write $9O_2$.

$$2As_2S_3 + 9O_2 \rightarrow 2As_2O_3 + 6SO_2$$

d. Write a four in front of H_3PO_4; this requires a three in front of $Ca(H_2PO_4)_2$ for twelve H's.

$$4H_3PO_4 + Ca_3(PO_4)_2 \rightarrow 3Ca(H_2PO_4)_2$$

■ Answers to Review Questions

2.1 Atomic theory is an explanation of the structure of matter in terms of different combinations of very small particles called atoms. Since compounds are composed of atoms of two or more elements, there is no limit to the number of ways in which the elements can be combined. Each compound has its own unique properties. A chemical reaction consists of the rearrangement of the atoms present in the reacting substances to give new chemical combinations present in the substances formed by the reaction.

2.2 Divide each amount of chlorine, 1.270 g and 1.904 g, by the lower amount, 1.270 g. This gives 1.000 and 1.499, respectively. Convert these to whole numbers by multiplying by 2, giving 2.000 and 2.998. The ratio of these amounts of chlorine is essentially 2:3. This is consistent with the law of multiple proportions because, for a fixed mass of iron (one gram), the masses of chlorine in the other two compounds are in the ratio of small whole numbers.

2.3 A cathode-ray tube consists of a negative electrode, or cathode, and a positive electrode, or anode, in an evacuated tube. Cathode rays travel from the cathode to the anode when a high voltage is turned on. Some of the rays pass through the hole in the anode to form a beam, which is then bent toward positively charged electric plates in the tube. This implies that a cathode ray consists of a beam of negatively charged particles (or electrons) and that electrons are constituents of all matter.

2.4 Millikan performed a series of experiments in which he obtained the charge on the electron by observing how a charged drop of oil falls in the presence and in the absence of an electric field. An atomizer introduces a fine mist of oil drops into the top chamber (Figure 2.6). Several drops happen to fall through a small hole into the lower chamber where the experimenter follows the motion of one drop with a microscope. Some of these drops have picked up one or more electrons as a result of friction in the atomizer and have become negatively charged. A negatively charged drop will be attracted upwards when the experimenter turns on a current to the electric plates. The drop's upward speed (obtained by timing its rise) is related to its mass-to-charge ratio from which you can calculate the charge on the electron.

2.5   The nuclear model of the atom is based on experiments of Geiger, Marsden, and Rutherford. Rutherford stated that most of the mass of an atom is concentrated in a positively charged center called the nucleus around which negatively charged electrons move. The nucleus, although containing most of the mass, occupies only a very small portion of the space of the atom. Most of the alpha particles passed through the metal atoms of the foil undeflected by the lightweight electrons. When an alpha particle does happen to hit a metal-atom nucleus, it is scattered at a wide angle because it is deflected by the massive, positively charged nucleus (Figure 2.8).

2.6   The atomic nucleus consists of two kinds of particles, protons and neutrons. The mass of each is about the same, on the order of 1.67×10^{-27} kg, and about 1800 times that of the electron. An electron has a much smaller mass, on the order of 9.11×10^{-31} kg. The neutron is electrically neutral, but the proton is positively charged. An electron is negatively charged. The magnitude of the charges on the proton and the electron are equal.

2.7   Protons (hydrogen nuclei) were discovered as products of experiments involving the collision of alpha particles with nitrogen atoms that resulted in a proton being knocked out of the nitrogen nucleus. Neutrons were discovered as the radiation product of collisions of alpha particles with beryllium atoms. The resulting radiation was discovered to consist of particles having a mass approximately equal to that of a proton and having no charge (neutral).

2.8   Oxygen consists of three different isotopes, each having eight protons but a different number of neutrons.

2.9   The percentages of the different isotopes in most naturally occurring elements have remained essentially constant over time and in most cases are independent of the origin of the element. Thus, what Dalton actually calculated were average atomic masses (relative masses). He could not weigh individual atoms, but could find the average mass of one atom relative to the average mass of another.

2.10   A mass spectrometer works by allowing a gas, such as neon, to pass through an inlet tube into a chamber where the atoms collide with electrons from an electron beam (cathode rays). The force of a collision can knock an electron from an atom. The positive neon atoms produced this way are drawn toward a negative grid, and some of them pass through slits to form a beam of positive electricity. This beam then travels through a magnetic field (from a magnet whose poles are above and below the positive beam). The magnetic field deflects the positively charged atoms in the positive beam according to their mass-to-charge ratios, and they then travel to a detector at the end of the tube. The beam of neon is split into three beams corresponding to the three different isotopes of neon (Figure 2.11). You obtain the mass of each neon isotope from the magnitude of deflection of its positively charged atoms. You can also obtain the relative number of atoms for the various masses (fractional abundances). The mass spectrum gives us all the information needed to calculate the atomic weight.

2.11   The atomic weight of an element is the average atomic mass for the naturally occurring element expressed in atomic mass units. The atomic weight would be different elsewhere in the universe if the percentages of isotopes in the element were different from those on earth.

2.12   The element in Group IVA and Period 5 is tin (atomic number 50).

2.13   A metal is a substance or mixture that has characteristic luster, or shine, and is generally a good conductor of heat and electricity.

2.14   The formula for ethane is C_2H_6.

2.15   A molecular formula gives the exact number of different atoms of an element in a molecule. A structural formula is a chemical formula that shows how the atoms are bonded to one another in a molecule.

2.16   Organic molecules contain carbon combined with other elements such as hydrogen, oxygen, and nitrogen. An inorganic molecule is composed of elements other than carbon. Some inorganic molecules that contain carbon are carbon monoxide (CO), carbon dioxide (CO_2), carbonates, and cyanides.

2.17   An ionic binary compound: NaCl; a molecular binary compound: H_2O.

2.18 a. The elements are represented by B, F, and I.

   b. The compounds are represented by A, E, and G.

   c. The mixtures are represented by C, D, and H.

   d. The ionic solid is represented by A.

   e. The gas made up of an element and a compound is represented by C.

   f. The mixtures of elements are represented by D and H.

   g. The solid element is represented by F.

   h. The solids are represented by A and F.

   i. The liquids are represented by E, H, and I.

2.19   In the Stock system, CuCl is called copper(I) chloride, and $CuCl_2$ is called copper(II) chloride.

   Some of the advantages of the Stock system are that more than two different ions of the same metal can be named with this system. In the former (older) system, a new suffix other than -ic and -ous must be established and/or memorized.

2.20   A balanced chemical equation has the number of atoms of each element equal on both sides of the arrow. The coefficients are the smallest possible whole numbers.

■ Solutions to Practice Problems

Note on significant figures: If the final answer to a solution needs to be rounded off, it is given first with one nonsignificant figure, and the last significant figure is underlined. The final answer is then rounded to the correct number of significant figures. In multiple-step problems, intermediate answers are given with at least one nonsignificant figure; however, only the final answer has been rounded off.

2.31 a. Argon   b. Zinc   c. Silver   d. Magnesium

2.33 a. K   b. S   c. Fe   d. Mn

2.35   The mass of the electron is found by multiplying the two values:

$$1.605 \times 10^{-19} \text{ C} \times \frac{5.64 \times 10^{-12} \text{ kg}}{1 \text{C}} = 9.0\underline{5}2 \times 10^{-31} \text{ kg} = 9.05 \times 10^{-31} \text{ kg}$$

2.37   The isotope of atom A is the atom with 18 protons, atom C; the atom that has the same mass number as atom A (37) is atom D.

2.39   Each isotope of chlorine (atomic number 17) has 17 protons. The neutral atoms will also each have 17 electrons. The number of neutrons for Cl-35 is 35 – 17 = 18 neutrons. The number of neutrons for Cl-37 is 37 – 17 = 20 neutrons.

2.41   The element with 14 protons in its nucleus is silicon (Si). The mass number = 14 + 14 = 28. The notation for the nucleus is $^{28}_{14}\text{Si}$.

2.43   Since the atomic ratio of nitrogen to hydrogen is 1:3, divide the mass of N by one-third of the mass of hydrogen to find the relative mass of N.

$$\frac{\text{Atomic mass of N}}{\text{Atomic mass of H}} = \frac{7.933 \text{ g}}{1/3 \times 1.712 \text{ g}} = \frac{13.90\underline{1} \text{g N}}{1 \text{g H}} = \frac{13.90}{1}$$

2.45   Multiply each isotopic mass by its fractional abundance, then sum:

X-63:	62.930 × 0.6909	=	43.4283
X-65:	64.928 × 0.3091	=	20.0692
			63.5475 = 63.55 amu

The element is copper, atomic weight 63.546 amu.

2.47   Multiply each isotopic mass by its fractional abundance, then sum:

$$38.964 \times 0.9326 \quad = \quad 36.3378$$
$$39.964 \times 1.00 \times 10^{-4} = \quad 0.0039964$$
$$40.962 \times 0.0673 \quad = \quad \underline{2.75674}$$
$$= \quad 39.09853 = 39.099 \text{ amu}$$

The atomic weight of this element is 39.099 amu. The element is potassium (K).

2.49   According to the picture, there are twenty atoms, five of which are brown and fifteen of which are green. Using the isotopic masses in the problem, the atomic mass of element X is

$$\frac{5}{20}(23.02 \text{ amu}) + \frac{15}{20}(25.147 \text{ amu}) = 5.75\underline{5} + 18.86\underline{0}2 = 24.6\underline{1}5 = 24.62 \text{ amu}$$

2.51 a. C: Group IVA, Period 2; nonmetal   b. Po: Group VIA, Period 6; metal

   c. Cr: Group VIB, Period 4; metal   d. Mg: Group IIA, Period 3; metal

   e. B: Group IIIA, Period 2; metalloid

2.53 a. Tellurium   b. Aluminum

2.55   Examples are:

   a. O (oxygen)   b. Na (sodium)

   c. Fe (iron)   d. Ce (cerium)

2.57   They are different in that the solid sulfur consists of S_8 molecules whereas the hot vapor consists of S_2 molecules. The S_8 molecules are four times as heavy as the S_2 molecules. Hot sulfur is a mixture of S_8 and S_2 molecules, but at high enough temperatures only S_2 molecules are formed. Both hot sulfur and solid sulfur consist of molecules with only sulfur atoms.

2.59   The number of nitrogen atoms in the 1.50 g sample of N_2O is

$$2.05 \times 10^{22} \, N_2O \text{ molecules} \times \frac{2 \, N \text{ atoms}}{1 \, N_2O \text{ molecule}} = 4.10 \times 10^{22} \, N \text{ atoms}$$

The number of nitrogen atoms in 1.00 g of N_2O is

$$1.00 \, g \times \frac{4.10 \times 10^{22} \, N \text{ atoms}}{1.50 \, g \, N_2O} = 2.7\underline{3}3 \times 10^{22} \, N \text{ atoms} = 2.73 \times 10^{22} \, N \text{ atoms}$$

2.61 3.3×10^{21} H atoms x $\dfrac{1\,NH_3\ \text{molecule}}{3\,H\,\text{atoms}}$ = $1.1 \times 10^{21}\ NH_3$ molecules

2.63 a. N_2H_4 b. H_2O_2 c. C_3H_8O d. PCl_3

2.65 a. PCl_5 b. NO_2 c. $C_3H_6O_2$

2.67 $\dfrac{1\ \text{Fe atom}}{1\ Fe(NO_3)_2\ \text{unit}}$ x $\dfrac{1\ Fe(NO_3)_2\ \text{unit}}{2\ NO_3^-\ \text{ions}}$ x $\dfrac{1\ NO_3^-\ \text{ion}}{3\ O\ \text{atoms}}$ = $\dfrac{1\ \text{Fe atom}}{6\ O\ \text{atoms}}$ = $\dfrac{1}{6}$

Thus, the ratio of iron atoms to oxygen atoms is one Fe atom to six O atoms.

2.69 a. $Fe(CN)_3$ b. K_2SO_4 c. Li_3N d. Ca_3P_2

2.71 a. Na_2SO_4: sodium sulfate (Group IA forms only 1+ cations).

   b. CaS: calcium sulfide (Group IIA forms only 2+ cations).

   c. CuCl: copper(I) chloride (Group IB forms 1+ and 2+ cations).

   d. Cr_2O_3: chromium(III) oxide (Group VIB forms numerous oxidation states).

2.73 a. Lead(II) permanganate: $Pb(MnO_4)_2$ (Permanganate is in Table 2.6).

   b. Barium hydrogen carbonate: $Ba(HCO_3)_2$ (The HCO_3^- ion is in Table 2.6).

   c. Cesium sulfide: Cs_2S (Group 1A ions form 1+ cations).

   d. Iron(II) acetate: $Fe(C_2H_3O_2)_2$ (The acetate ion = 1- [Table 2.6]; for the sum of charges to be zero, two must be used).

2.75 a. Molecular   b. Ionic   c. Molecular   d. Ionic

2.77 a. Dinitrogen monoxide   b. Tetraphosphorus dec(a)oxide

   c. Arsenic trichloride   d. Dichlorine hept(a)oxide

2.79 a. NBr_3 b. XeO_4 c. OF_2 d. Cl_2O_5

2.81 a. Selenium trioxide   b. Disulfur dichloride

   c. Carbon monoxide

2.83 a. Bromic acid: $HBrO_3$ **b.** Hyponitrous acid: $H_2N_2O_2$

    c. Disulfurous acid: $H_2S_2O_5$ **d.** Arsenic acid: H_3AsO_4

2.85 $Na_2SO_4 \cdot 10H_2O$ is sodium sulfate decahydrate.

2.87   Iron(II) sulfate heptahydrate is $FeSO_4 \cdot 7H_2O$.

2.89 $Pb(NO_3)_2 \times \dfrac{6 \text{ O atoms}}{1Pb(NO_3)_2 \text{ unit}} + K_2CO_3 \times \dfrac{3 \text{ O atoms}}{1K_2CO_3 \text{ unit}} = 9 \text{ O atoms}$

2.91 a. Balance: $Sn + NaOH \rightarrow Na_2SnO_2 + H_2$

If Na is balanced first by writing a two in front of NaOH, the entire equation is balanced.

$$Sn + 2NaOH \rightarrow Na_2SnO_2 + H_2$$

b. Balance: $Al + Fe_3O_4 \rightarrow Al_2O_3 + Fe$

First balance O (it appears once on each side) by writing a three in front of Fe_3O_4 and a four in front of Al_2O_3:

$$Al + 3Fe_3O_4 \rightarrow 4Al_2O_3 + Fe$$

Now balance Al against the eight Al's on the right and Fe against the nine Fe's on the left:

$$8Al + 3Fe_3O_4 \rightarrow 4Al_2O_3 + 9Fe$$

c. Balance: $CH_3OH + O_2 \rightarrow CO_2 + H_2O$

First balance H (it appears once on each side) by writing a two in front of H_2O:

$$CH_3OH + O_2 \rightarrow CO_2 + 2H_2O$$

To avoid fractional coefficients for O, multiply the equation by two:

$$2CH_3OH + 2O_2 \rightarrow 2CO_2 + 4H_2O$$

Finally, balance O by changing $2O_2$ to "$3O_2$"; this balances the entire equation:

$$2CH_3OH + 3O_2 \rightarrow 2CO_2 + 4H_2O$$

(continued)

d. Balance: $P_4O_{10} + H_2O \rightarrow H_3PO_4$

First balance P (it appears once on each side) by writing a four in front of H_3PO_4:

$$P_4O_{10} + H_2O \rightarrow 4H_3PO_4$$

Finally, balance H by writing a six in front of H_2O; this balances the entire equation:

$$P_4O_{10} + 6H_2O \rightarrow 4H_3PO_4$$

e. Balance: $PCl_5 + H_2O \rightarrow H_3PO_4 + HCl$

First balance Cl (it appears once on each side) by writing a five in front of HCl:

$$PCl_5 + H_2O \rightarrow H_3PO_4 + 5HCl$$

Finally, balance H by writing a four in front of H_2O; this balances the entire equation:

$$PCl_5 + 4H_2O \rightarrow H_3PO_4 + 5HCl$$

2.93   Balance: $Ca_3(PO_4)_2(s) + H_2SO_4(aq) \rightarrow CaSO_4(s) + H_3PO_4(aq)$

Balance Ca first with a three in front of $CaSO_4$:

$$Ca_3(PO_4)_2(s) + H_2SO_4(aq) \rightarrow 3CaSO_4(s) + H_3PO_4(aq)$$

Next, balance the P with a two in front of H_3PO_4:

$$Ca_3(PO_4)_2(s) + H_2SO_4(aq) \rightarrow 3CaSO_4(s) + 2H_3PO_4(aq)$$

Finally, balance the S with a three in front of H_2SO_4; this balances the equation:

$$Ca_3(PO_4)_2(s) + 3H_2SO_4(aq) \rightarrow 3CaSO_4(s) + 2H_3PO_4(aq)$$

2.95   Balance: $NH_4Cl(aq) + Ba(OH)_2(aq) \rightarrow NH_3(g) + BaCl_2(aq) + H_2O(l)$

Balance O first with a two in front of H_2O:

$$NH_4Cl + Ba(OH)_2 \rightarrow NH_3 + BaCl_2 + 2H_2O$$

Balance H with a two in front of NH_4Cl and a two in front of NH_3; this balances the equation:

$$2NH_4Cl(aq) + Ba(OH)_2(aq) \xrightarrow{\Delta} 2NH_3(g) + BaCl_2(aq) + 2H_2O(l)$$

■ Solutions to General Problems

2.97   Calculate the ratio of oxygen for one g (fixed amount) of nitrogen in both compounds:

$$A: \quad \frac{2.755\,g\,O}{1.206\,g\,N} = \frac{2.2844\,g\,O}{1\,g\,N} \qquad B: \quad \frac{4.714\,g\,O}{1.651\,g\,N} = \frac{2.8552\,g\,O}{1\,g\,N}$$

Next, find the ratio of oxygen per gram of nitrogen for the two compounds.

$$\frac{g\,O\,in\,B/1\,g\,N}{g\,O\,in\,A/1\,g\,N} = \frac{2.8552\,g\,O}{2.2844\,g\,O} = \frac{1.2498\,g\,O}{1\,g\,O}$$

B contains 1.25 times as many O atoms as A does (there are five O's in B for every four O's in A).

2.99   The smallest difference is between -1.12×10^{-18} C and 9.60×10^{-19} C and is equal to -1.6×10^{-19} C. If this charge is equivalent to one electron, the number of excess electrons on a drop may be found by dividing the negative charge by the charge of one electron.

Drop 1: $\quad \dfrac{-3.20 \times 10^{-19}\,C}{-1.6 \times 10^{-19}\,C} = 2.0 \cong 2$ electrons

Drop 2: $\quad \dfrac{-6.40 \times 10^{-19}\,C}{-1.6 \times 10^{-19}\,C} = 4.0 \cong 4$ electrons

Drop 3: $\quad \dfrac{-9.60 \times 10^{-19}\,C}{-1.6 \times 10^{-19}\,C} = 6.0 \cong 6$ electrons

Drop 4: $\quad \dfrac{-1.12 \times 10^{-18}\,C}{-1.6 \times 10^{-19}\,C} = 7.0 \cong 7$ electrons

2.101   For the Eu atom to be neutral, the number of electrons must equal the number of protons, so a neutral europium atom has 63 electrons. The +3 charge on the Eu^{3+} indicates there are three more protons than electrons, so the number of electrons = 63 - 3 = 60.

2.103   The number of protons = mass number - number of neutrons = 81 - 46 = 35. The element with Z = 35 is bromine (Br).

The ionic charge = number of protons - number of electrons = 35 - 36 = -1.

Symbol: $^{81}_{35}Br^-$.

2.105 The sum of the fractional abundances must equal one. Let y equal the fractional abundance of ^{63}Cu. Then the fractional abundance of ^{65}Cu equals (1 - y). We write one equation in one unknown:

$$\text{Atomic weight} = 63.546 = 62.9298y + 64.9278(1 - y)$$

$$63.546 = 64.9278 - 1.9980y$$

$$y = \frac{64.9278 - 63.546}{1.9980} = 0.69159$$

The fractional abundance of ^{63}Cu = 0.69159 = 0.6916.

The fractional abundance of ^{65}Cu = 1 - 0.69159 = 0.30841 = 0.3084.

2.107 a. Bromine, Br   b. Hydrogen, H   c. Niobium, Nb   d. Fluorine, F

2.109 a. Chromium(III) ion   b. Lead(IV) ion   c. Copper(I) ion   d. Copper(II) ion

2.111 All possible ionic compounds: Na_2SO_4, NaCl, $NiSO_4$, and $NiCl_2$.

2.113 a. Tin(II) phosphate   b. Ammonium nitrite

   c. Magnesium hydroxide   d. Chromium(II) sulfate

2.115 a. Hg_2S [Mercury(I) exists as the polyatomic Hg_2^{2+} ion (Table 2.6).]

   b. $Co_2(SO_3)_3$ c. $(NH_4)_2Cr_2O_7$ d. AlN

2.117 a. Arsenic tribromide   b. hydrogen selenide (dihydrogen selenide)

   c. Diphosphorus pent(a)oxide   d. Silicon dioxide

2.119 a. Balance the C and H first:

   $$C_2H_6 + O_2 \rightarrow 2CO_2 + 3H_2O$$

   Avoid a fractional coefficient for O on the left by doubling all coefficients except O_2's, and then balance the O's:

   $$2C_2H_6 + 7O_2 \rightarrow 4CO_2 + 6H_2O$$

   (continued)

b. Balance the P first:

$$P_4O_6 + H_2O \rightarrow 4H_3PO_3$$

Then balance the O (or H), which also gives the H (or O) balance:

$$P_4O_6 + 6H_2O \rightarrow 4H_3PO_4$$

c. Balancing the O first is the simplest approach. (Starting with K and Cl and then O will cause the initial coefficient for $KClO_3$ to be changed in balancing O last.)

$$4KClO_3 \rightarrow KCl + 3KClO_4$$

d. Balance the N first:

$$(NH_4)_2SO_4 + NaOH \rightarrow 2NH_3 + H_2O + Na_2SO_4$$

Then balance the Na, followed by O; this also balances the H:

$$(NH_4)_2SO_4 + 2NaOH \rightarrow 2NH_3 + 2H_2O + Na_2SO_4$$

e. Balance the N first:

$$2NBr_3 + NaOH \rightarrow N_2 + NaBr + HOBr$$

Note that NaOH and HOBr each have one O and that NaOH and NaBr each have one Na; thus the coefficients of all three must be equal; from $2NBr_3$ this coefficient must = 6Br/2 = 3:

$$2NBr_3 + 3NaOH \rightarrow N_2 + 3NaBr + 3HOBr$$

2.121   Let: x = number of protons. Then $1.21x$ is the number of neutrons. Since the mass number is 62, you get

$$62 = x + 1.21x = 2.21x$$

Thus, $x = 28.054$, or 28. The element is nickel (Ni). Since the ion has a +2 charge, there are 26 electrons.

2.123   The average atomic mass would be

Natural carbon:	12.011 x 1/2	=	6.005500
Carbon-13:	13.00335 x 1/2	=	6.501675
	Average	=	12.507175

The average atomic mass of the sample is 12.507 amu.

■ Solutions to Cumulative-Skills Problems

2.125   The spheres occupy a diameter of 2 x 1.86 Å = 3.72 Å. The line of sodium atoms would stretch a length of

$$\text{length} = \frac{3.72\ \text{Å}}{1\,\text{Na atom}} \times 2.619 \times 10^{22}\ \text{Na atoms} = 9.7\underline{4}2 \times 10^{22}\ \text{Å}$$

Now, convert this to miles.

$$9.7\underline{4}2 \times 10^{22}\ \text{Å} \times \frac{10^{-10}\ \text{m}}{1\,\text{Å}} \times \frac{1\,\text{mile}}{1.609 \times 10^{3}\ \text{m}}$$

$$= 6.0\underline{5}5 \times 10^{9}\ \text{miles} = 6.06 \times 10^{9}\ \text{miles}$$

2.127 $NiSO_4 \cdot 7H_2O(s) \rightarrow NiSO_4 \cdot 6H_2O(s) + H_2O(g)$

       [8.753]   =   [8.192 g   +   (8.753 - 8.192 = 0.561 g)]

The 8.192 g of $NiSO_4 \cdot 6H_2O$ must contain 6 x 0.561 = 3.366 g H_2O.

Mass of anhydrous $NiSO_4$ = 8.192 g $NiSO_4 \cdot 6H_2O$ - 3.366 g $6H_2O$ = 4.826 g

2.129   Mass of O = 0.6015 L x $\dfrac{1.330\ \text{g O}}{1\text{L}}$ = 0.799995 g

$$15.9994\ \text{amu O} \times \frac{3.177\ \text{g X}}{0.799\underline{9}95\ \text{g O}} = 63.5\underline{3}8\ \text{amu X} = 63.54\ \text{amu}$$

The atomic weight of X is 63.54 amu; X is copper.

3. CALCULATIONS WITH CHEMICAL FORMULAS AND EQUATIONS

■ Solutions to Exercises

Note on significant figures: If the final answer to a solution needs to be rounded off, it is given first with one nonsignificant figure, and the last significant figure is underlined. The final answer is then rounded to the correct number of significant figures. In multiple-step problems, intermediate answers are given with at least one nonsignificant figure; however, only the final answer has been rounded off.

3.1 a. NO_2

1 x AW of N			=	14.0067 amu
2 x AW of O	=	2 x 15.9994	=	31.9988 amu
		MW of NO_2	=	46.0055 = 46.0 amu (3 s.f.)

b. $C_6H_{12}O_6$

6 x AW of C	=	6 x 12.011	=	72.066 amu
12 x AW of H	=	12 x 1.0079	=	12.0948 amu
6 x AW of O	=	6 x 15.9994	=	95.9964 amu
		MW of $C_6H_{12}O_6$	=	180.1572 amu = 180. amu (3 s.f.)

c. NaOH

1 x AW of Na		=	22.98977 amu
1 x AW of O		=	15.9994 amu
1 x AW of H		=	1.0079 amu
	MW of NaOH	=	39.9971 amu = 40.0 amu (3 s.f.)

d. $Mg(OH)_2$

1 x AW of Mg			=	24.305 amu
2 x AW of O	=	2 x 15.9994	=	31.9988 amu
2 x AW of H	=	2 x 1.0079	=	2.0158 amu
		MW of $Mg(OH)_2$	=	58.3196 amu = 58.3 amu (3 s.f.)

3.2 a. The molecular model represents a molecule made up of one S and three O. The chemical formula is SO_3. Using the same approach as Example 3.1 in the text, calculating the formula weight yields 80.07 amu.

b. The molecular model represents one S, four O, and two H. The chemical formula is then H_2SO_4. The formula weight is 98.09 amu.

3.3 a. The atomic weight of Ca = 40.08 amu; thus, the molar mass = 40.08 g/mol, and one mol of Ca = 6.022×10^{23} Ca atoms.

$$\text{Mass of one Ca} = \frac{40.08 \text{ g}}{1 \text{ mol Ca}} = \frac{1 \text{ mol}}{6.022 \times 10^{23} \text{ atoms}} = 6.65\underline{5}6 \times 10^{-23}$$

$$= 6.656 \times 10^{-23} \text{ g/atom}$$

b. The molecular weight of C_2H_5OH, or C_2H_6O, = $(2 \times 12.01) + (6 \times 1.008) + 16.00$ = 46.0\underline{6}8. Its molar mass = 46.07 g/mol, and one mol = 6.022×10^{23} molecules of C_2H_6O.

$$\text{Mass of one } C_2H_6O = \frac{46.07 \text{ g}}{1 \text{ mol } C_2H_6O} = \frac{1 \text{ mol}}{6.022 \times 10^{23} \text{ atoms}}$$

$$= 7.65\underline{0}3 \times 10^{-23} = 7.650 \times 10^{-23} \text{ g/molecule}$$

3.4 The molar mass of H_2O_2 is 34.02 g/mol. Therefore,

$$0.909 \text{ mol } H_2O_2 \times \frac{34.02 \text{ g } H_2O_2}{1 \text{ mol } H_2O_2} = 30.9\underline{2} = 30.9 \text{ g } H_2O_2$$

3.5 The molar mass of HNO_3 is 63.01 g/mol. Therefore,

$$28.5 \text{ g } HNO_3 \times \frac{1 \text{ mol } HNO_3}{63.01 \text{ g } HNO_3} = 0.452\underline{3} = 0.452 \text{ mol } HNO_3$$

3.6 Convert the mass of HCN from milligrams to grams. Then, convert grams of HCN to moles of HCN. Finally, convert moles of HCN to the number of HCN molecules.

$$56 \text{ mg HCN} \times \frac{1 \text{ g}}{1000 \text{ mg}} \times \frac{1 \text{ mol HCN}}{27.02 \text{ g HCN}} \times \frac{6.022 \times 10^{23} \text{ HCN molecules}}{1 \text{ mol HCN}}$$

$$= 1.\underline{2}48 \times 10^{21} = 1.2 \times 10^{21} \text{ HCN molecules}$$

3.7   The molecular weight of NH_4NO_3 = 80.05; thus, its molar mass = 80.05 g/mol. Hence

$$\text{Percent N} = \frac{28.02 \text{ g}}{80.05 \text{ g}} \times 100\% = 35.\underline{0}0 = 35.0\%$$

$$\text{Percent H} = \frac{4.032 \text{ g}}{80.05 \text{ g}} \times 100\% = 5.0\underline{3}6 = 5.04\%$$

$$\text{Percent O} = \frac{48.00 \text{ g}}{80.05 \text{ g}} \times 100\% = 59.\underline{9}6 = 60.0\%$$

3.8   From the previous exercise, NH_4NO_3 is 35.0 percent N (fraction N = 0.350), so the mass of N in 48.5 g of NH_4NO_3 is

$$48.5 \text{ g } NH_4NO_3 \times (0.350 \text{ g N/1 g } NH_4NO_3) = 16.\underline{9}75 = 17.0 \text{ g N}$$

3.9   First, convert the mass of CO_2 to moles of CO_2. Next, convert this to moles of C (one mol of CO_2 is equivalent to one mol C). Finally, convert to mass of carbon, changing mg to g first:

$$5.80 \times 10^{-3} \text{g } CO_2 \times \frac{1 \text{mol } CO_2}{44.01 \text{g } CO_2} \times \frac{1 \text{mol C}}{1 \text{mol } CO_2} \times \frac{12.01 \text{g C}}{1 \text{mol C}}$$

$$= 1\,5\underline{8}3 \times 10^{-3} \text{g C}$$

Do the same series of calculations for water, noting that one mol H_2O contains two mol H.

$$1.58 \times 10^{-3} \text{g } H_2O \times \frac{1 \text{mol } H_2O}{18.02 \text{ g } H_2O} \times \frac{2 \text{ mol H}}{1 \text{mol } H_2O} \times \frac{1.008 \text{ g H}}{1 \text{mol H}}$$

$$= 1.7\underline{6}7 \times 10^{-4} \text{g H}$$

The mass percentages of C and H can be calculated using the masses from the previous calculations:

$$\text{Percent C} = \frac{1.583 \text{ mg}}{3.87 \text{ mg}} \times 100\% = 40.\underline{9}0 = 40.9\% \text{ C}$$

$$\text{Percent H} = \frac{0.1767 \text{ mg}}{3.87 \text{ mg}} \times 100\% = 4.5\underline{6}58 = 4.57\% \text{ H}$$

The mass percentage of O can be determined by subtracting the sum of the above percentages from 100 percent:

$$\text{Percent O} = 100.000\% - (40.90 + 4.5658) = 54.\underline{5}342 = 54.5\% \text{ O}$$

3.10   Convert the masses to moles that are proportional to the subscripts in the empirical formula:

$$33.4 \text{ g S} \times \frac{1 \text{ mol S}}{32.07 \text{ g S}} = 1.0\underline{4}14 \text{ mol S}$$

$$(83.5 - 33.4) \text{ g O} \times \frac{1 \text{ mol O}}{16.00 \text{ g O}} = 3.1\underline{3}2 \text{ mol O}$$

Next, obtain the smallest integers from the moles by dividing each by the smallest number of moles:

For O: $\dfrac{3.1312}{1.0414} = 3.01$ For S: $\dfrac{1.0414}{1.0414} = 1.00$

The empirical formula is SO_3.

3.11   For a 100.0-g sample of benzoic acid, 68.8 g are C, 5.0 g are H, and 26.2 g are O. Using the molar masses, convert these masses to moles:

$$68.8 \text{ g C} \times \frac{1 \text{ mol C}}{12.01 \text{ g C}} = 5.7\underline{2}9 \text{ mol C}$$

$$5.0 \text{ g H} \times \frac{1 \text{ mol H}}{1.008 \text{ g H}} = 4.\underline{9}6 \text{ mol H}$$

$$26.2 \text{ g O} \times \frac{1 \text{ mol O}}{16.00 \text{ g O}} = 1.6\underline{3}8 \text{ mol O}$$

These numbers are in the same ratio as the subscripts in the empirical formula. They must be changed to integers. First, divide each one by the smallest mol number:

For C : $\dfrac{5.729}{1.638} = 3.497$ For H : $\dfrac{4.96}{1.638} = 3.03$

For O : $\dfrac{1.638}{1.638} = 1.000$

Rounding off, we obtain $C_{3.5}H_{3.0}O_{1.0}$. Multiplying the numbers by two gives whole numbers for an empirical formula of $C_7H_6O_2$.

3.12   For a 100.0-g sample of acetaldehyde, 54.5 g are C, 9.2 g are H, and 36.3 g are O. Using the molar masses, convert these masses to moles:

$$54.5 \text{ g C} \times \frac{1 \text{ mol C}}{12.01 \text{ g C}} = 4.5\underline{3}7 \text{ mol C}$$

$$9.2 \text{ g H} \times \frac{1 \text{ mol H}}{1.008 \text{ g H}} = 9.\underline{1}2 \text{ mol H}$$

$$36.3 \text{ g O} \times \frac{1 \text{ mol O}}{16.00 \text{ g O}} = 2.2\underline{6}8 \text{ mol O}$$

These numbers are in the same ratio as the subscripts in the empirical formula. They must be changed to integers. First, divide each one by the smallest mol number:

$$\text{For C}: \frac{4.537}{2.268} = 2.000 \qquad \text{For H}: \frac{9.12}{2.268} = 4.02$$

$$\text{For O}: \frac{2.268}{2.268} = 1.000$$

Rounding off, we obtain C_2H_4O, the empirical formula, which is also the molecular formula.

3.13 H_2 + Cl_2 →   2HCl
       1 molec. (mol) H_2 +   1 molec. (mol) Cl_2 →   2 molec. (mol) HCl (molec., mole interp.)
       2.016 g H_2 +   70.9 g Cl_2 →   2 x 36.5 g HCl (mass interp.)

3.14   Equation: $Na + H_2O \rightarrow 1/2H_2 + NaOH$, or $2Na + 2H_2O \rightarrow H_2 + 2NaOH$. From this equation, one mol of Na corresponds to one-half mol of H_2, or two mol of Na corresponds to one mol of H_2. Therefore,

$$7.81 \text{ g } H_2 \times \frac{1 \text{ mol } H_2}{2.016 \text{ g } H_2} \times \frac{2 \text{ mol Na}}{1 \text{ mol } H_2} \times \frac{22.99 \text{ g Na}}{1 \text{ mol Na}} = 17\underline{8}.1 = 178 \text{ g Na}$$

3.15   Balanced equation: $2ZnS + 3O_2 \rightarrow 2ZnO + 2SO_2$

Convert grams of ZnS to moles of ZnS. Then, determine the relationship between ZnS and O_2 (2ZnS is equivalent to $3O_2$). Finally, convert to mass O_2.

$$5.00 \times 10^3 \text{ g ZnS} \times \frac{1 \text{ mol ZnS}}{97.46 \text{ g ZnS}} \times \frac{3 \text{ mol } O_2}{2 \text{ mol ZnS}} \times \frac{32.00 \text{ g } O_2}{1 \text{ mol } O_2} \times \frac{1 \text{ kg}}{1000 \text{ g}}$$

$$= 2.4\underline{6}3 = 2.46 \text{ kg } O_2$$

3.16   Balanced equation: $2HgO \rightarrow 2Hg + O_2$

Convert the mass of O_2 to mol of O_2. Using the fact that one mol of O_2 is equivalent to two mol of Hg, determine the number of mol of Hg, and convert to mass of Hg.

$$6.47 \text{ g } O_2 \times \frac{1 \text{ mol } O_2}{32.00 \text{ g } O_2} \times \frac{2 \text{ mol Hg}}{1 \text{ mol } O_2} \times \frac{200.59 \text{ g Hg}}{1 \text{ mol Hg}} = 81.\underline{1}1$$

$$= 81.1 \text{ g Hg}$$

3.17   First, determine the limiting reactant by calculating the moles of $AlCl_3$ that would be obtained if Al and HCl were totally consumed:

$$0.15 \text{ mol Al} \times \frac{2 \text{ mol } AlCl_3}{2 \text{ mol Al}} = 0.1\underline{5}0 \text{ mol } AlCl_3$$

$$0.35 \text{ mol HCl} \times \frac{2 \text{ mol } AlCl_3}{6 \text{ mol HCl}} = 0.1\underline{1}66 \text{ mol } AlCl_3$$

Because the HCl produces the smaller amount of $AlCl_3$, the reaction will stop when HCl is totally consumed but before the Al is consumed. The limiting reactant is therefore HCl. The amount of $AlCl_3$ produced must be 0.1$\underline{1}$66, or 0.12 mol.

3.18   First, determine the limiting reactant by calculating the moles of ZnS produced by totally consuming Zn and S_8:

$$7.36 \text{ g Zn} \times \frac{1 \text{ mol Zn}}{65.39 \text{ g Zn}} \times \frac{8 \text{ mol ZnS}}{8 \text{ mol Zn}} = 0.11\underline{2}56 \text{ mol ZnS}$$

$$6.45 \text{ g } S_8 \times \frac{1 \text{ mol } S_8}{256.56 \text{ g } S_8} \times \frac{8 \text{ mol ZnS}}{1 \text{ mol } S_8} = 0.20\underline{1}1 \text{ mol ZnS}$$

The reaction will stop when Zn is totally consumed; S_8 is in excess, and not all of it is converted to ZnS. The limiting reactant is therefore Zn. Now convert the moles of ZnS obtained from the Zn to grams of ZnS:

$$0.11256 \text{ mol ZnS} \times \frac{97.46 \text{ g ZnS}}{1 \text{ mol ZnS}} = 10.\underline{9}7 = 11.0 \text{ g ZnS}$$

3.19   First, write the balanced equation:

$$CH_3OH + CO \rightarrow HC_2H_3O_2$$

Convert grams of each reactant to moles of acetic acid:

$$15.0 \text{ g } CH_3OH \times \frac{1 \text{ mol } CH_3OH}{32.04 \text{ g } CH_3OH} \times \frac{1 \text{ mol } HC_2H_3O_2}{1 \text{ mol } CH_3OH} = 0.4681 \text{ mol } HC_2H_3O_2$$

$$10.0 \text{ g } CO \times \frac{1 \text{ mol } CO}{28.01 \text{ g } CO} \times \frac{1 \text{ mol } HC_2H_3O_2}{1 \text{ mol } CH_3OH} = 0.3570 \text{ mol } HC_2H_3O_2$$

Thus, CO is the limiting reactant, and 0.03570 mol $HC_2H_3O_2$ is obtained. The mass of product is

$$0.3570 \text{ mol } HC_2H_3O_2 \times \frac{60.05 \text{ g } HC_2H_3O_2}{1 \text{ mol } HC_2H_3O_2} = 21.44 \text{ g } HC_2H_3O_2$$

The percentage yield is:

$$\frac{19.1 \text{ g actual yield}}{21.44 \text{ g theoretical yield}} \times 100\% = 89.08 = 89.1\%$$

$$\text{Mass percentage} = \frac{0.25359 \text{ g } HC_2H_3O_2}{5.00 \text{ g vinegar}} \times 100\% = 5.071 = 5.07\%$$

■ Answers to Review Questions

3.1   The molecular weight is the sum of the atomic weights of all the atoms in a molecule of the substance whereas the formula weight is the sum of the atomic weights of all the atoms in one formula unit of the compound, whether the compound is molecular or not. A given substance could have both a molecular weight and a formula weight if it existed as discrete molecules.

3.2   To obtain the formula weight of a substance, sum up the atomic weights of all atoms in the formula.

3.3   A mole of N_2 contains Avogadro's number (6.02×10^{23}) of N_2 molecules and $2 \times 6.02 \times 10^{23}$ of N atoms. One mole of $Fe_2(SO_4)_3$ contains three moles of SO_4^{2-} ions; and it contains twelve moles of O atoms.

3.4   A sample of the compound of known mass is burned, and CO_2 and H_2O are obtained as products. Next, you relate the masses of CO_2 and H_2O to the masses of carbon and hydrogen. Then, you calculate the mass percentages of C and H. You find the mass percentage of O by subtracting the mass percentages of C and H from 100.

3.5   The empirical formula is obtained from the percentage composition by assuming for the purposes of the calculation a sample of 100 g of the substance. Then, the mass of each element in the sample equals the numerical value of the percentage. Convert the masses of the elements to moles of the elements using the atomic mass of each element. Divide the moles of each by the smallest number to obtain the smallest ratio of each atom. If necessary, find a whole-number factor to multiply these results by to obtain integers for the subscripts in the empirical formula.

3.6   The empirical formula is the formula of a substance written with the smallest integer (whole number) subscripts. Each of the subscripts in the formula $C_6H_{12}O_2$ can be divided by two, so the empirical formula of the compound is C_3H_6O.

3.7   The number of empirical formula units in a compound, n, equals the molecular weight divided by the empirical formula weight.

$$n = \frac{34.0 \text{ amu}}{17.0 \text{ amu}} = 2.00$$

The molecular formula of hydrogen peroxide is therefore $(HO)_2$, or H_2O_2.

3.8   The coefficients in a chemical equation can be interpreted directly in terms of molecules or moles. For the mass interpretation, you will need the molar masses of CH_4, O_2, CO_2, and H_2O, which are 16.0, 32.0, 44.0, and 18.0 g/mol, respectively. A summary of the three interpretations is given below the balanced equation:

CH_4	+	$2O_2$	→	CO_2	+	$2H_2O$
1 molecule	+	2 molecules	→	1 molecule	+	2 molecules
1 mole	+	2 moles	→	1 mole	+	2 moles
16.0 g	+	2 x 32.0 g	→	44.0 g	+	2 x 18.0 g

3.9   A chemical equation yields the mole ratio of a reactant to a second reactant or product. Once the mass of a reactant is converted to moles, this can be multiplied by the appropriate mole ratio to give the moles of a second reactant or product. Multiplying this number of moles by the appropriate molar mass gives mass. Thus, the masses of two different substances are related by a chemical equation.

3.10   The limiting reactant is the reactant that is entirely consumed when the reaction is complete. Because the reaction stops when the limiting reactant is used up, the moles of product are always determined by the starting number of moles of the limiting reactant.

3.11 Two examples are given in the book. The first involves making cheese sandwiches. Each sandwich requires two slices of bread and one slice of cheese. The limiting reactant is the cheese because some bread is left unused. The second example is assembling automobiles. Each auto requires one steering wheel, four tires, and other components. The limiting reactant is the tires, since they will run out first.

3.12 Since the theoretical yield represents the maximum amount of product that can be obtained by a reaction from given amounts of reactants under any conditions, in an actual experiment you can never obtain more than this amount.

■ Solutions to Practice Problems

Note on significant figures: If the final answer to a solution needs to be rounded off, it is given first with one nonsignificant figure, and the last significant figure is underlined. The final answer is then rounded to the correct number of significant figures. In multiple-step problems, intermediate answers are given with at least one nonsignificant figure; however, only the final answer has been rounded off.

3.21 a. Formula weight of CH_3OH = AW of C + 4(AW of H) + AW of O. Using the values of atomic weights in the periodic table (inside front cover) rounded to four significant figures and rounding the answer to three significant figures, we have

FW = 12.01 amu + (4 x 1.008 amu) + 16.00 amu = 32.0$\underline{4}$2 = 32.0 amu

b. FW of NO_3 = AW of N + 3(AW of O)

= 14.01 amu + (3 x 16.00 amu) = 62.01 = 62.0 amu

c. FW of K_2CO_3 = 2(AW of K) + AW of C + 3(AW of O)

= (2 x 39.10 amu) + 12.01 amu + (3 x 16.00 amu)

= 138.2$\underline{1}$0 = 138 amu

d. FW of $Ni_3(PO_4)_2$ = 3(AW of Ni) + 2(AW of P) + 8(AW of O)

= (3 x 58.70 amu) + (2 x 30.97 amu) + (8 x 16.00 amu)

= 366.0$\underline{4}$0 = 366 amu

3.23 a. SO_2
1 x AW of S			=	32.07 amu	
2 x AW of O	=	2 x 16.00	=	32.00 amu	
		MW of NO_2	=	64.0$\underline{7}$ amu	= 64.1 amu (3 s.f.)

b. PCl_3
1 x AW of P			=	30.97 amu	
3 x AW of Cl	=	3 x 35.45	=	106.35 amu	
		MW of PCl_3	=	137.32 amu	= 137 amu (3 s.f.)

3.25   First, find the formula weight of NH_4NO_3 by adding the respective atomic weights. Then convert it to the molar mass:

$$FW \text{ of } NH_4NO_3 = 2(AW \text{ of } N) + 4(AW \text{ of } H) + 3(AW \text{ of } O)$$

$$= (2 \times 14.01 \text{ amu}) + (4 \times 1.008 \text{ amu}) + (3 \times 16.00 \text{ amu})$$

$$= 80.052 \text{ amu}$$

The molar mass of $NH_4NO_3 = 80.05$ g/mol.

3.27 a. The atomic weight of Na equals 22.99 amu; thus, the molar mass equals 22.99 g/mol. Because one mol of Na atoms equals 6.022×10^{23} Na atoms, we calculate

$$\text{Mass of one Na atom} = \frac{22.99 \text{ g/mol}}{6.022 \times 10^{23} \text{ atom/mol}} = 3.8177 \times 10^{-23} \text{ g/atom}$$

  b. The atomic weight of N equals 14.01 amu; thus, the molar mass equals 14.01 g/mol. Because one mol of N atoms equals 6.022×10^{23} N atoms, we calculate

$$\text{Mass of one N atom} = \frac{14.01 \text{ g/mol}}{6.022 \times 10^{23} \text{ atom/mol}} = 2.3264 \times 10^{-23} \text{ g/atom}$$

  c. The formula weight of $CH_3Cl = [12.01 + (3 \times 1.008) + 35.45] = 50.48$ amu; thus, the molar mass equals 50.48 g/mol. Because one mol of CH_3Cl molecules equals 6.022×10^{23} CH_3Cl molecules, we calculate

$$\text{Mass of one } CH_3Cl \text{ molecule} = \frac{50.48 \text{ g/mol}}{6.022 \times 10^{23} \text{ molecules/mol}}$$

$$= 8.3826 \times 10^{-23} \text{ g/molecule}$$

  d. The formula weight of $Hg(NO_3)_2 = 200.59 + (2 \times 14.01) + (6 \times 16.00)] = 324.61$ amu; thus, the molar mass equals 324.61 g/mol. Because one formula weight of $Hg(NO_3)_2$ equals 6.022×10^{23} $Hg(NO_3)_2$ formula units, we calculate

$$\text{Mass of one } Hg(NO_3)_2 = \frac{324.61 \text{ g/mol}}{6.022 \times 10^{23} \text{ units/mol}} = 5.3904 \times 10^{-22} \text{ g/unit}$$

3.29   First, find the formula weight (in amu) using the periodic table (inside front cover):

$$\text{FW of } (CH_3CH_2)_2O = (4 \times 12.01 \text{ amu}) + (10 \times 1.008 \text{ amu}) + 16.00 \text{ amu}$$

$$= 74.12 \text{ amu}$$

$$\text{Mass of one } (CH_3CH_2)_2O \text{ molecule} = \frac{74.12 \text{ g/mol}}{6.022 \times 10^{23} \text{ molecules/mol}}$$

$$= 1.23\underline{0}8 \times 10^{-22} \text{ g/molecule}$$

3.31   From the table of atomic weights, we obtain the following molar masses for parts a through d: Na = 22.99 g/mol; S = 32.07 g/mol; C = 12.01 g/mol; H = 1.008 g/mol; Cl = 35.45 g/mol; and N = 14.01 g/mol.

a. 0.15 mol Na x $\dfrac{22.99 \text{ g}}{1 \text{ mol Na}}$ = 3.$\underline{4}$48 = 3.4 g Na

b. 0.594 mol S x $\dfrac{32.07 \text{ g S}}{1 \text{ mol S}}$ = 19.$\underline{0}$4 = 19.0 g S

c. Using molar mass = 84.93 g/mol for CH_2Cl_2, we obtain

$$2.78 \text{ mol } CH_2Cl_2 \times \frac{84.93 \text{ g } CH_2Cl_2}{1 \text{ mol } CH_2Cl_2} = 236.\underline{1} = 236 \text{ g } CH_2Cl_2$$

d. Using molar mass = 68.14 g/mol for $(NH_4)_2S$, we obtain

$$38 \text{ mol } (NH_4)_2S \times \frac{68.14 \text{ g } (NH_4)_2S}{1 \text{ mol } (NH_4)_2S} = 2.\underline{5}8 \times 10^3 = 2.6 \times 10^3 \text{ g } (NH_4)_2S$$

3.33   First, find the molar mass of H_3BO_3: (3 x 1.008 amu) + 10.81 amu + (3 x 16.00 amu) = 61.83. Therefore, the molar mass of H_3BO_3 = 61.83 g/mol. The mass of H_3BO_3 is calculated as follows:

$$0.543 \text{ mol } H_3BO_3 \times \frac{61.83 \text{ g } H_3BO_3}{1 \text{ mol } H_3BO_3} = 33.\underline{5}7 = 33.6 \text{ g } H_3BO_3$$

3.35   From the table of atomic weights, we obtain the following rounded molar masses for parts a through d: C = 12.01 g/mol; Cl = 35.45 g/mol; H = 1.008 g/mol; Al = 26.98 g/mol; and O = 16.00 g/mol.

a. $2.86 \text{ g C} \times \dfrac{1 \text{mol C}}{12.01 \text{g C}} = 0.238\underline{1} = 0.238 \text{ mol C}$

b. $7.05 \text{ g Cl}_2 \times \dfrac{1 \text{mol Cl}_2}{70.90 \text{ g Cl}_2} = 0.099\underline{4}3 = 0.0994 \text{ mol Cl}_2$

c. The molar mass of C_4H_{10} = (4 x 12.01) + (10 x 1.008) = 58.12 g C_4H_{10}/mol C_4H_{10}. The mass of C_4H_{10} is calculated as follows:

$$76 \text{ g C}_4\text{H}_{10} \times \dfrac{1 \text{mol C}_4\text{H}_{10}}{58.12 \text{ g C}_4\text{H}_{10}} = 1.\underline{3}07 = 1.3 \text{ mol C}_4\text{H}_{10}$$

d. The molar mass of $Al_2(CO_3)_3$ = (2 x 26.98) + (3 x 12.01) + (9 x 16.00 g) = 233.99 g/mol $Al_2(CO_3)_3$. The mass of $Al_2(CO_3)_3$ is calculated as follows:

$$26.2 \text{ g Al}_2(\text{CO}_3)_3 \times \dfrac{1 \text{ mol Al}_2(\text{CO}_3)_3}{233.99 \text{ g Al}_2(\text{CO}_3)_3} = 0.11\underline{1}9 = 0.112 \text{ mol Al}_2(\text{CO}_3)_3$$

3.37   Calculate the formula weight of calcium sulfate: 40.08 amu + 32.07 amu + (4 x 16.00 amu) = 136.15 amu. Therefore, the molar mass of $CaSO_4$ is 136.15 g/mol. Use this to convert the mass of $CaSO_4$ to moles:

$$0.791 \text{g CaSO}_4 \times \dfrac{1 \text{mol CaSO}_4}{136.15 \text{ g CaSO}_4} = 5.8\underline{1}1 \times 10^{-3} = 5.81 \times 10^{-3} \text{ mol CaSO}_4$$

Calculate the molecular weight of water: (2 x 1.008 amu) + 16.00 amu = 18.02 amu. Therefore, the molar mass of H_2O equals 18.02 g/mol. Use this to convert the rest of the sample to moles of water:

$$0.209 \text{ g H}_2\text{O} \times \dfrac{1 \text{mol H}_2\text{O}}{18.02 \text{ g H}_2\text{O}} = 1.1\underline{5}9 \times 10^{-2} = 1.16 \times 10^{-2} \text{ mol H}_2\text{O}$$

Because 0.01159 mol is about twice 0.005811 mol, both numbers of moles are consistent with the formula, $CaSO_4 \cdot 2H_2O$.

3.39 The following rounded atomic weights are used: Li = 6.94 g/mol; Br = 79.90 g/mol; N = 14.01 g/mol; H = 1.008 g/mol; Pb = 207.2 g/mol; Cr = 52.00 g/mol; O = 16.00 g/mol; and S = 32.07 g/mol. Also, Avogadro's number is 6.022×10^{23} atoms, so

a. No. Li atoms = $8.21\, g\, Li \times \dfrac{6.022 \times 10^{23}\ atoms}{6.941\, g\, Li}$ = $7.1\underline{2}2 \times 10^{23}$ atoms

b. No. Br atoms = $32.0\, g\, Br_2 \times \dfrac{2 \times 6.022 \times 10^{23}\ atoms}{(2 \times 79.90)\, g\, Br_2}$ = $2.4\underline{1}2 \times 10^{23}$ atoms

c. No. NH_3 molecules = $45\, g\, NH_3 \times \dfrac{6.022 \times 10^{23}\ molecules}{17.03\, g\, NH_3}$ = $1.\underline{5}9 \times 10^{24}$ molecules

d. No. $PbCrO_4$ units = $201\, g\, PbCrO_4 \times \dfrac{6.022 \times 10^{23}\ units}{323.2\, g\, PbCrO_4}$ = $3.7\underline{4}5 \times 10^{23}$ units

e. No. SO_4^{2-} ions = $14.3\, g\, Cr_2(SO_4)_3 \times \dfrac{3 \times 6.022 \times 10^{23}\ ions}{392.21\, g\, Cr_2(SO_4)_3}$ = $6.5\underline{8}7 \times 10^{22}$ ions

3.41 Calculate the molecular weight of CCl_4: 12.01 amu + (4 x 35.45 amu) = 153.81 amu. Use this and Avogadro's number to express it as 153.81 g/N_A to calculate the number of molecules:

$$7.58\ mg\ CCl_4 \times \frac{1\, g}{1000\ mg} \times \frac{6.022 \times 10^{23}\ molecules}{153.81\, g\, CCl_4} = 2.9\underline{6}8 \times 10^{19}$$

$$= 2.97 \times 10^{19}\ molecules$$

3.43 Mass percentage carbon = $\dfrac{mass\ of\ C\ in\ sample}{mass\ of\ sample} \times 100\%$

Percent carbon = $\dfrac{1.584\, g}{1.836\, g} \times 100\%$ = $86.2\underline{7}4$ = 86.27%

3.45 Mass percentage phosphorus oxychloride = $\dfrac{mass\ of\ POCl_3\ in\ sample}{mass\ of\ sample} \times 100\%$

Percent $POCl_3$ = $\dfrac{1.72\ mg}{8.53\ mg} \times 100\%$ = $20.\underline{1}6$ = 20.2%

3.47 Start with the definition for percentage nitrogen, and rearrange this equation to find the mass of N in the fertilizer.

$$\text{Mass percentage nitrogen} = \frac{\text{mass of N in fertilizer}}{\text{mass of fertilizer}} \times 100\%$$

$$\text{Mass N} = \frac{\text{mass \% N}}{100\%} \times \text{mass of fertilizer} = \frac{14.0\%}{100\%} \times 4.15 \text{ kg} = 0.58\underline{1}0$$

$$= 0.581 \text{ kg N}$$

3.49 Convert moles to mass using the molar masses from the respective atomic weights. Then, calculate the mass percentages from the respective masses.

$$0.0898 \text{ mol Al} \times \frac{26.98 \text{ g Al}}{1 \text{ mol Al}} = 2.4\underline{2}2 \text{ g Al}$$

$$0.0381 \text{ mol Mg} \times \frac{24.31 \text{ g Mg}}{1 \text{ mol Mg}} = 0.92\underline{6}2 \text{ g Mg}$$

$$\text{Percent Al} = \frac{\text{mass of Al}}{\text{mass of alloy}} = \frac{2.422 \text{ g Al}}{3.349 \text{ g alloy}} \times 100\% = 72.\underline{3}4 = 72.3\% \text{ Al}$$

$$\text{Percent Mg} = \frac{\text{mass of Mg}}{\text{mass of alloy}} = \frac{0.9262 \text{ g Mg}}{3.349 \text{ g alloy}} \times 100\% = 27.\underline{6}55 = 27.7\% \text{ Mg}$$

3.51 In each part, the numerator consists of the mass of the element in one mole of the compound; the denominator is the mass of one mole of the compound. Use the atomic weights of C = 12.01 g/mol; O = 16.00 g/mol; Na = 22.99 g/mol; H = 1.008 g/mol; P = 30.97 g/mol; Co = 58.93 g/mol; and N = 14.01 g/mol.

a. $\text{Percent C} = \dfrac{\text{mass of C}}{\text{mass of CO}} = \dfrac{12.01 \text{ g C}}{28.01 \text{ g CO}} \times 100\% = 42.878 = 42.9\%$

$\text{Percent O} = 100.000\% - 42.8\underline{7}8\% \text{C} = 57.122 = 57.1\%$

b. $\text{Percent C} = \dfrac{\text{mass of C}}{\text{mass of CO}_2} = \dfrac{12.01 \text{ g C}}{44.01 \text{ g CO}_2} \times 100\% = 27.\underline{2}89 = 27.3\%$

$\text{Percent O} = 100.000\% - 27.2\underline{8}9\% \text{ C} = 72.711 = 72.7\%$

(continued)

c. Percent Na = $\dfrac{\text{mass of Na}}{\text{mass of NaH}_2\text{PO}_4}$ = $\dfrac{22.99 \text{ g Na}}{119.98 \text{ g NaH}_2\text{PO}_4}$ x 100%

$= 19.161 = 19.2\%$

Percent H = $\dfrac{\text{mass of H}}{\text{mass of NaH}_2\text{PO}_4}$ = $\dfrac{2.016 \text{ g H}}{119.98 \text{ g NaH}_2\text{PO}_4}$ x 100% = 1.6802 = 1.68%

Percent P = $\dfrac{\text{mass of P}}{\text{mass of NaH}_2\text{PO}_4}$ = $\dfrac{30.97 \text{ g P}}{119.98 \text{ g NaH}_2\text{PO}_4}$ x 100% = 25.812 = 25.8%

Percent O = 100.000% - (19.161 + 1.6802 + 25.812) = 53.346 = 53.3%

d. Percent Co = $\dfrac{\text{mass of Co}}{\text{mass of Co(NO}_3)_2}$ = $\dfrac{58.93 \text{ g Co}}{182.95 \text{ g Co(NO}_3)_2}$ x 100%

$= 32.211 = 32.2\%$

Percent N = $\dfrac{\text{mass of N}}{\text{mass of Co(NO}_3)_2}$ = $\dfrac{2 \times 14.01 \text{ g N}}{182.95 \text{ g Co(NO}_3)_2}$ x 100%

$= 15.316 = 15.3\%$

Percent O = 100.000% - (32.211 + 15.316) = 52.473 = 52.5%

3.53 The molecular model of toluene contains seven carbon atoms and eight hydrogen atoms, so the molecular formula of toluene is C_7H_8. The molar mass of toluene is 92.134 g/mol. The mass percentages are

Percent C = $\dfrac{\text{mass of C}}{\text{mass of C}_7\text{H}_8}$ = $\dfrac{7 \times 12.01 \text{ g}}{92.134 \text{ g}}$ x 100% = 91.247 = 91.2%

Percent H = 100% - 91.247 = 8.753 = 8.75%

3.55 Find the moles of C in each amount in one-step operations. Calculate the moles of each compound using the molar mass; then multiply by the number of moles of C per mole of compound:

Mol C (glucose) = 6.01 g x $\dfrac{1 \text{ mol}}{180.2 \text{ g}}$ x $\dfrac{6 \text{ mol C}}{1 \text{ mol glucose}}$ = 0.200 mol

Mol C (ethanol) = 5.85 g x $\dfrac{1 \text{ mol}}{46.07 \text{ g}}$ x $\dfrac{2 \text{ mol C}}{1 \text{ mol ethanol}}$ = 0.254 mol (more C)

3.57 First, calculate the mass of C in the glycol by multiplying the mass of CO_2 by the molar mass of C and the reciprocal of the molar mass of CO_2. Then, calculate the mass of H in the glycol by multiplying the mass of H_2O by the molar mass of 2H and the reciprocal of the molar mass of H_2O. Then, use the masses to calculate the mass percentages. Calculate O by difference.

$$9.06 \text{ mg } CO_2 \times \frac{1 \text{ mol } CO_2}{44.01 \text{ g } CO_2} \times \frac{12.01 \text{ g C}}{1 \text{ mol C}} = 2.472 \text{ mg C}$$

$$5.58 \text{ mg } H_2O \times \frac{1 \text{ mol } H_2O}{18.02 \text{ g } H_2O} \times \frac{2 \text{ H}}{1 H_2O} \times \frac{1.008 \text{ g H}}{1 \text{ mol H}} = 0.6243 \text{ mg H}$$

Mass O = 6.38 mg - (2.472 + 0.6243) = 3.284 mg O

Percent C = (2.472 mg C/6.38 mg glycol) × 100% = 38.74 = 38.7%

Percent H = (0.6243 mg H/6.38 mg glycol) × 100% = 9.785 = 9.79%

Percent O = (3.284 mg O/6.38 mg glycol) × 100% = 51.47 = 51.5%

3.59 Start by calculating the moles of Os and O; then divide each by the smaller number of moles to obtain integers for the empirical formula.

$$\text{Mol Os} = 2.16 \text{ g Os} \times \frac{1 \text{ mol Os}}{190.2 \text{ g Os}} = 0.01136 \text{ mol (smaller number)}$$

$$\text{Mol O} = (2.89 - 2.16) \text{ g O} \times \frac{1 \text{ mol O}}{16.00 \text{ g O}} = 0.0456 \text{ mol}$$

Integer for Os = 0.01136 ÷ 0.01136 = 1.000

Integer for O = 0.0456 ÷ 0.01136 = 4.01

Within experimental error, the empirical formula is OsO_4.

3.61 Assume a sample of 100.0 g of potassium manganate. By multiplying this by the percentage composition, we obtain 39.6 g of K, 27.9 g of Mn, and 32.5 g of O. Convert each of these masses to moles by dividing by molar mass.

$$\text{Mol K} = 39.6 \text{ g K} \times \frac{1 \text{ mol K}}{39.10 \text{ g K}} = 1.013 \text{ mol}$$

(continued)

$$\text{Mol Mn} = 29.7 \text{ g Mn} \times \frac{1 \text{ mol Mn}}{54.94 \text{ g Mn}} = 0.50\underline{7}8 \text{ mol (smallest number)}$$

$$\text{Mol O} = 32.5 \text{ g O} \times \frac{1 \text{ mol O}}{16.00 \text{ g O}} = 2.0\underline{3}1 \text{ mol}$$

Now, divide each number of moles by the smallest number to obtain the smallest set of integers for the empirical formula.

Integer for K = 1.013 ÷ 0.5078 = 1.998, or 2

Integer for Mn = 0.5078 ÷ 0.5078 = 1.000, or 1

Integer for O = 2.031 ÷ 0.5078 = 3.999, or 4

The empirical formula is thus K_2MnO_4.

3.63   Assume a sample of 100.0 g of acrylic acid. By multiplying this by the percentage composition, we obtain 50.0 g C, 5.6 g H, and 44.4 g O. Convert each of these masses to moles by dividing by the molar mass.

$$\text{Mol C} = 50.0 \text{ g C} \times \frac{1 \text{ mol C}}{12.01 \text{ g C}} = 4.1\underline{6}3 \text{ mol}$$

$$\text{Mol H} = 5.6 \text{ g H} \times \frac{1 \text{ mol H}}{1.008 \text{ g H}} = 5.\underline{5}6 \text{ mol}$$

$$\text{Mol O} = 44.0 \text{ g O} \times \frac{1 \text{ mol O}}{16.00 \text{ g O}} = 2.7\underline{7}5 \text{ mol (smallest number)}$$

Now, divide each number of moles by the smallest number to obtain the smallest number of moles and the tentative integers for the empirical formula.

Tentative integer for C = 4.163 ÷ 2.775 = 1.50, or 1.5

Tentative integer for H = 5.56 ÷ 2.775 = 2.00, or 2

Tentative integer for O = 2.775 ÷ 2.775 = 1.00, or 1

(continued)

Because 1.5 is not a whole number, multiply each tentative integer by two to obtain the final integer for the empirical formula:

C: 2 x 1.5 = 3

H: 2 x 2 = 4

O: 2 x 1 = 2

The empirical formula is thus $C_3H_4O_2$.

3.65 a. Assume for the calculation that you have 100.0 g; of this quantity, 92.25 g is C and 7.75 g is H. Now, convert these masses to moles:

$$92.25\,g\,C \times \frac{1\,mol\,C}{12.01\,g\,C} = 7.68\underline{1}09\,mol\,C$$

$$7.75\,g\,H \times \frac{1\,mol\,H}{1.008\,g\,H} = 7.6\underline{8}8\,mol\,H$$

Usually, you divide all the mole numbers by the smaller one, but in this case both are equal, so the ratio of the number of C atoms to the number of H atoms is 1:1. Thus, the empirical formula for both compounds is CH.

b. Obtain n, the number of empirical formula units in the molecule, by dividing the molecular weight of 52.03 amu and 78.05 amu by the empirical formula weight of 13.018 amu:

$$\text{For } 52.03 : n = \frac{52.03\,amu}{13.018\,amu} = 3.99\underline{6}8, \text{ or } 4$$

$$\text{For } 78.05 : n = \frac{78.05\,amu}{13.018\,amu} = 5.99\underline{5}5, \text{ or } 6$$

The molecular formulas are: for 52.03, $(CH)_4$ or C_4H_4; and for 78.05, $(CH)_6$ or C_6H_6.

3.67 The formula weight corresponding to the empirical formula C_2H_6N may be found by adding the respective atomic weights.

Formula weight = (2 x 12.01 amu) + (6 x 1.008 amu) + 14.01 amu

= 44.08 amu

(continued)

Dividing the molecular weight by the formula weight gives the number of times the C_2H_6N unit occurs in the molecule. Because the molecular weight is an average of 88.5 ([90 + 87] ÷ 2), this quotient is

$$88.5 \text{ amu} \div 44.1 \text{ amu} = 2.\underline{0}06, \text{ or } 2$$

Therefore, the molecular formula is $(C_2H_6N)_2$, or $C_4H_{12}N_2$.

3.69   Assume a sample of 100.0 g of oxalic acid. By multiplying this by the percentage composition, we obtain 26.7 g C, 2.2 g H, and 71.1 g O. Convert each of these masses to moles by dividing by the molar mass.

$$\text{Mol C} = 26.7 \text{ g C} \times \frac{1 \text{ mol C}}{12.01 \text{ g C}} = 2.2\underline{2}3 \text{ mol}$$

$$\text{Mol H} = 2.2 \text{ g H} \times \frac{1 \text{ mol H}}{1.008 \text{ g H}} = 2.\underline{1}8 \text{ mol (smallest number)}$$

$$\text{Mol O} = 71.1 \text{ g O} \times \frac{1 \text{ mol O}}{16.00 \text{ g O}} = 4.4\underline{4}3 \text{ mol}$$

Now, divide each number of moles by the smallest number to obtain the smallest set of integers for the empirical formula.

Integer for C = 2.223 ÷ 2.18 = 1.02, or 1

Integer for H = 2.18 ÷ 2.18 = 1.00, or 1

Integer for O = 4.443 ÷ 2.18 = 2.0\underline{3}8, or 2

The empirical formula is thus CHO_2. The formula weight corresponding to this formula may be found by adding the respective atomic weights:

Formula weight = 12.01 amu + 1.008 amu + (2 x 16.00 amu) = 45.02 amu

Dividing the molecular weight by the formula weight gives the number of times the CHO_2 unit occurs in the molecule. Because the molecular weight is 90 amu, this quotient is

$$90 \text{ amu} \div 45.02 \text{ amu} = 2.\underline{0}0, \text{ or } 2$$

The molecular formula is thus $(CHO_2)_2$, or $C_2H_2O_4$.

3.71 C_2H_4 $+$ $3O_2$ $\rightarrow$ $2CO_2$ $+$ $2H_2O$

1 molecule C_2H_4 $+$ 3 molecules O_2 $\rightarrow$ 2 molecules CO_2 $+$ 2 molecules H_2O

1 mole C_2H_4 $+$ 3 moles O_2 $\rightarrow$ 2 moles CO_2 $+$ 2 moles H_2O

28.052 g C_2H_4 $+$ 3 x 32.00 g O_2 $\rightarrow$ 2 x 44.01 g CO_2 $+$ 2 x 18.016 g H_2O

3.73 By inspecting the balanced equation, obtain a conversion factor of eight mol CO_2 to two mol C_4H_{10}. Multiply the given amount of 0.30 moles of C_4H_{10} by the conversion factor to obtain the moles of H_2O.

$$0.30 \text{ mol } C_4H_{10} \times \frac{8 \text{ mol } CO_2}{2 \text{ mol } C_4H_{10}} = 1.\underline{2}0 = 1.2 \text{ mol } CO_2$$

3.75 By inspecting the balanced equation, obtain a conversion factor of three mol O_2 to two mol Fe_2O_3. Multiply the given amount of 3.91 mol Fe_2O_3 by the conversion factor to obtain moles of O_2.

$$3.91 \text{ mol } Fe_2O_3 \times \frac{3 \text{ mol } O_2}{2 \text{ mol } Fe_2O_3} = 5.8\underline{6}5 = 5.87 \text{ mol } O_2$$

3.77 $3 NO_2 + H_2O \rightarrow 2 HNO_3 + NO$

three mol of NO_2 are equivalent to two mol of HNO_3 (from equation).

one mol of NO_2 is equivalent to 46.01 g NO_2 (from molecular weight of NO_2).

one mol of HNO_3 is equivalent to 63.02 g HNO_3 (from molecular weight of HNO_3).

$$7.50 \text{ g } HNO_3 \times \frac{1 \text{ mol } HNO_3}{63.02 \text{ g } HNO_3} \times \frac{3 \text{ mol } NO_2}{2 \text{ mol } HNO_3} \times \frac{46.01 \text{ g } NO_2}{1 \text{ mol } NO_2} = 8.2\underline{1}3$$

$$= 8.21 \text{ g } NO_2$$

3.79 $WO_3 + 3H_2 \rightarrow W + 3H_2O$

one mol of W is equivalent to three moles of H_2 (from equation).

one mol of H_2 is equivalent to 2.016 g H_2 (from molecular weight of H_2).

one mol of W is equivalent to 183.8 g W (from atomic weight of W).

(continued)

4.81 kg of H_2 is equivalent to 4.81×10^3 g of H_2.

$$4.81 \times 10^3 \, g \, H_2 \times \frac{1 \, mol \, H_2}{2.016 \, g \, H_2} \times \frac{1 \, mol \, W}{3 \, mol \, H_2} \times \frac{183.85 \, g \, W}{1 \, mol \, W} = 1.4\underline{6}2 \times 10^5$$

$$= 1.46 \times 10^5 \, g \, W$$

3.81   Write the equation, and set up the calculation below the equation (after calculating the two molecular weights):

$$CS_2 + 3Cl_2 \rightarrow CCl_4 + S_2Cl_2$$

$$62.7 \, g \, Cl_2 \times \frac{1 \, mol \, Cl_2}{70.90 \, g \, Cl_2} \times \frac{1 \, mol \, CS_2}{3 \, mol \, Cl_2} \times \frac{76.15 \, g \, CS_2}{1 \, mol \, CS_2} = 22.\underline{4}48$$

$$= 22.4 \, g \, CS_2$$

3.83   Write the equation, and set up the calculation below the equation (after calculating the two molecular weights):

$$2N_2O_5 \rightarrow 4NO_2 + O_2$$

$$1.315 \, g \, O_2 \times \frac{1 \, mol \, O_2}{32.00 \, g \, O_2} \times \frac{4 \, mol \, NO_2}{1 \, mol \, O_2} \times \frac{46.01 \, g \, NO_2}{1 \, mol \, NO_2}$$

$$= 7.56\underline{2}8 = 7.563 \, g \, NO_2$$

3.85   First determine whether KO_2 or H_2O is the limiting reactant by calculating the moles of O_2 that each would form if it were the limiting reactant. Identify the limiting reactant by the smaller number of moles of O_2 formed.

$$0.15 \, mol \, H_2O \times \frac{3 \, mol \, O_2}{2 \, mol \, H_2O} = 0.2\underline{2}5 \, mol \, O_2$$

$$0.25 \, mol \, KO_2 \times \frac{3 \, mol \, O_2}{4 \, mol \, KO_2} = 0.1\underline{8}7 \, mol \, O_2 \text{ (}KO_2 \text{ is the limiting reactant)}$$

The moles of O_2 produced = 0.19 mol.

3.87   First determine whether CO or H_2 is the limiting reactant by calculating the moles of CH_3OH that each would form if it were the limiting reactant. Identify the limiting reactant by the smaller number of moles of CH_3OH formed. Use the molar mass of CH_3OH to calculate the mass of CH_3OH formed. Then, calculate the mass of the unconsumed reactant.

$$CO + 2H_2 \rightarrow CH_3OH$$

$$10.2 \text{ g } H_2 \times \frac{1 \text{ mol } H_2}{2.016 \text{ g } H_2} \times \frac{1 \text{ mol } CH_3OH}{2 \text{ mol } H_2} = 2.5\underline{2}9 \text{ mol } CH_3OH$$

$$35.4 \text{ g CO} \times \frac{1 \text{ mol CO}}{28.01 \text{ g CO}} \times \frac{1 \text{ mol } CH_3OH}{1 \text{ mol CO}} = 1.2\underline{6}3 \text{ mol } CH_3OH$$

CO is the limiting reactant.

$$\text{Mass } CH_3OH \text{ formed} = 1.2\underline{6}3 \text{ mol } CH_3OH \times \frac{32.042 \text{ g } CH_3OH}{1 \text{ mol } CH_3OH}$$

$$= 40.\underline{4}7 = 40.5g \ CH_3OH$$

Hydrogen is left unconsumed at the end of the reaction. The mass of H_2 that reacts can be calculated from the moles of product obtained:

$$1.2\underline{6}3 \text{ mol } CH_3OH \times \frac{2 \text{ mol } H_2}{1 \text{ mol } CH_3OH} \times \frac{2.016 \text{ g } H_2}{1 \text{ mol } H_2} = 5.0\underline{9}2 \text{ g } H_2$$

The unreacted H_2 = 10.2 g total H_2 - 5.092 g reacted H_2 = 5.$\underline{1}$08 = 5.1 g H_2.

3.89   First, determine which of the three reactants is the limiting reactant by calculating the moles of $TiCl_4$ that each would form if it were the limiting reactant. Identify the limiting reactant by the smallest number of moles of $TiCl_4$ formed. Use the molar mass of $TiCl_4$ to calculate the mass of $TiCl_4$ formed.

$$3TiO_2 + 4C + 6Cl_2 \rightarrow 3TiCl_4 + 2CO_2 + 2CO$$

$$4.15 \text{ g } TiO_2 \times \frac{1 \text{ mol } TiO_2}{79.88 \text{ g } TiO_2} \times \frac{3 \text{ mol } TiCl_4}{3 \text{ mol } TiO_2} = 0.051\underline{9}5 \text{ mol } TiCl_4$$

$$5.67 \text{ g C} \times \frac{1 \text{ mol C}}{12.01 \text{ g C}} \times \frac{3 \text{ mol } TiCl_4}{4 \text{ mol C}} = 0.35\underline{4}07 \text{ mol } TiCl_4$$

$$6.78 \text{ g } Cl_2 \times \frac{1 \text{ mol } Cl_2}{70.90 \text{ g } Cl_2} \times \frac{3 \text{ mol } TiCl_4}{6 \text{ mol } Cl_2} = 0.047\underline{8}1 \text{ mol } TiCl_4$$

(continued)

Cl_2 is the limiting reactant.

$$\text{Mass TiCl}_4 \text{ formed} = 0.047\underline{8}1 \text{ mol TiCl}_4 \times \frac{189.68 \text{ g TiCl}_4}{1 \text{ mol TiCl}_4} = 9.0\underline{6}8$$

$$= 9.07 \text{ g TiCl}_4$$

3.91 First, determine which of the two reactants is the limiting reactant by calculating the moles of aspirin that each would form if it were the limiting reactant. Identify the limiting reactant by the smallest number of moles of aspirin formed. Use the molar mass of aspirin to calculate the theoretical yield in grams of aspirin. Then calculate the percentage yield.

$$C_7H_6O_3 + C_4H_6O_3 \rightarrow C_9H_8O_4 + C_2H_4O_2$$

$$4.00 \text{ g } C_4H_6O_3 \times \frac{1 \text{ mol } C_4H_6O_3}{102.09 \text{ g } C_4H_6O_3} \times \frac{1 \text{ mol } C_9H_8O_4}{1 \text{ mol } C_4H_6O_3} = 0.039\underline{1}8 \text{ mol } C_9H_8O_4$$

$$2.00 \text{ g } C_7H_6O_3 \times \frac{1 \text{ mol } C_7H_6O_3}{138.12 \text{ g } C_7H_6O_3} \times \frac{1 \text{ mol } C_9H_8O_4}{1 \text{ mol } C_7H_6O_3} = 0.014\underline{4}8 \text{ mol } C_9H_8O_4$$

Thus, $C_7H_6O_3$ is the limiting reactant. The theoretical yield of $C_9H_8O_4$ is

$$0.014\underline{4}8 \text{ mol } C_9H_8O_4 \times \frac{180.15 \text{ g } C_9H_8O_4}{1 \text{ mol } C_9H_8O_4} = 2.6\underline{0}9 \text{ g } C_9H_8O_4$$

The percentage yield is

$$\text{Percent yield} = \frac{\text{actual yield}}{\text{theoretical yield}} \times 100\% = \frac{1.86 \text{ g}}{2.609 \text{ g}} \times 100\% = 71.\underline{2}9 = 71.3\%$$

■ Solutions to General Problems

3.93 For 1 mol of caffeine, there are eight mol of C, ten mol of H, four mol of N, and two mol of O. Convert these amounts to masses by multiplying them by their respective molar masses:

$$
\begin{aligned}
8 \text{ mol C} \times 12.01 \text{ g C/1 mol C} &= 96.08 \text{ g C} \\
10 \text{ mol H} \times 1.008 \text{ g H/1 mol H} &= 10.08 \text{ g H} \\
4 \text{ mol N} \times 14.01 \text{ g N/1 mol N} &= 56.04 \text{ g N} \\
2 \text{ mol O} \times 16.00 \text{ g O/1 mol O} &= \underline{32.00 \text{ g O}} \\
1 \text{ mol of caffeine (total)} &= 194.20 \text{ g (molar mass)}
\end{aligned}
$$

Each mass percentage is calculated by dividing the mass of the element by the molar mass of caffeine and multiplying by 100 percent: Mass percentage = (mass element ÷ mass caffeine) x 100%.

Mass percentage C = (96.08 g ÷ 194.20 g) x 100% = 49.5% (3 s.f.)

Mass percentage H = (10.08 g ÷ 194.20 g) x 100% = 5.19% (3 s.f.)

Mass percentage N = (56.04 g ÷ 194.20 g) x 100% = 28.9% (3 s.f.)

Mass percentage O = (32.00 g ÷ 194.20 g) x 100% = 16.5% (3 s.f.)

3.95 Assume a sample of 100.0 g of dichlorobenzene. By multiplying this by the percentage composition, we obtain 49.1 g C, 2.7 g of H, and 48.2 g of Cl. Convert each mass to moles by dividing by the molar mass:

$$
49.1 \text{ g C} \times \frac{1 \text{ mol C}}{12.01 \text{ g C}} = 4.0\underline{8}8 \text{ mol C}
$$

$$
2.7 \text{ g H} \times \frac{1 \text{ mol H}}{1.008 \text{ g H}} = 2.\underline{6}8 \text{ mol H}
$$

$$
48.2 \text{ g Cl} \times \frac{1 \text{ mol Cl}}{35.45 \text{ g Cl}} = 1.3\underline{6}0 \text{ mol Cl}
$$

Divide each number of moles by the smallest number to obtain the smallest set of integers for the empirical formula.

Integer for C = 4.088 mol ÷ 1.360 mol = 3.00, or 3

Integer for H = 2.68 mol ÷ 1.360 mol = 1.97, or 2

Integer for Cl = 1.360 mol ÷ 1.360 mol = 1.00, or 1

(continued)

The empirical formula is thus C_3H_2Cl. Find the formula weight by adding the atomic weights:

Formula weight = $(3 \times 12.01$ amu$) + (2 \times 1.008$ amu$) + 35.45$ amu

= 73.4$\underline{9}$6 = 73.50 amu

Divide the molecular weight by the formula weight to find the number of times the C_3H_2Cl unit occurs in the molecule. Because the molecular weight is 147 amu, this quotient is

147 amu ÷ 73.50 amu = 2.00, or 2

The molecular formula is $(C_3H_2Cl)_2$, or $C_6H_4Cl_2$.

3.97   Find the percent composition of C and S from the analysis:

$$0.01665 \text{ g CO}_2 \times \frac{1 \text{ mol CO}_2}{44.01 \text{ g CO}_2} \times \frac{1 \text{ mol C}}{1 \text{ mol CO}_2} \times \frac{12.01 \text{ g C}}{1 \text{ mol C}} = 0.00454\underline{4} \text{ g C}$$

Percent C = (0.004544 g C ÷ 0.00796 g comp.) × 100% = 57.$\underline{0}$9%

$$0.01196 \text{ g BaSO}_4 \times \frac{1 \text{ mol BaSO}_4}{233.39 \text{ g BaSO}_4} \times \frac{1 \text{ mol S}}{1 \text{ mol BaSO}_4} \times \frac{32.07 \text{ g S}}{1 \text{ mol S}}$$

= 0.00164$\underline{3}$ g S

Percent S = (0.001643 g S ÷ 0.00431 g comp.) × 100% = 38.$\underline{1}$2%

Percent H = 100.00% - (57.09 + 38.12)% = 4.$\underline{7}$9%

We now obtain the empirical formula by calculating moles from the grams corresponding to each mass percentage of element:

$$57.09 \text{ g C} \times \frac{1 \text{ mol C}}{12.01 \text{ g C}} = 4.75\underline{4} \text{ mol C}$$

$$38.12 \text{ g S} \times \frac{1 \text{ mol S}}{32.07 \text{ g C}} = 1.18\underline{9} \text{ mol S}$$

$$4.79 \text{ g H} \times \frac{1 \text{ mol H}}{1.008 \text{ g H}} = 4.7\underline{5}2 \text{ mol H}$$

(continued)

Dividing the moles of the elements by the smallest number (1.189), we obtain for C: 3.997, or 4; for S: 1.000, or 1; and for H: 3.996, or 4. Thus, the empirical formula is C_4H_4S (formula weight = 84). Because the formula weight was given as 84 amu, the molecular formula is also C_4H_4S.

3.99 For g $CaCO_3$, use this equation: $CaCO_3 + H_2C_2O_4 \rightarrow CaC_2O_4 + H_2O + CO_2$.

$$0.472 \text{ g } CaC_2O_4 \times \frac{1 \text{ mol } CaC_2O_4}{128.10 \text{ g } CaC_2O_4} \times \frac{1 \text{ mol } CaCO_3}{1 \text{ mol } CaC_2O_4} \times \frac{100.09 \text{ g } CaCO_3}{1 \text{ mol } CaCO_3}$$

$$= 0.3688 \text{ g } CaCO_3$$

$$\text{Mass percentage } CaCO_3 = \frac{\text{mass } CaCO_3}{\text{mass limestone}} \times 100\% = \frac{0.3688 \text{ g}}{0.438 \text{ g}} \times 100\%$$

$$= 84.19 = 84.2\%$$

3.101 Calculate the theoretical yield using this equation: $2C_2H_4 + O_2 \rightarrow 2C_2H_4O$.

$$10.6 \text{ g } C_2H_4 \times \frac{1 \text{ mol } C_2H_4}{28.05 \text{ g } C_2H_4} \times \frac{2 \text{ mol } C_2H_4O}{2 \text{ mol } C_2H_4} \times \frac{44.05 \text{ g } C_2H_4O}{1 \text{ mol } C_2H_4O}$$

$$= 16.65 \text{ g } C_2H_4O$$

$$\text{Percent yield} = \frac{\text{actual yield}}{\text{theoretical yield}} \times 100\% = \frac{9.91 \text{ g}}{16.65 \text{ g}} \times 100\% = 59.53 = 59.5\%$$

3.103 To find Zn, use these equations:

$$2C + O_2 \rightarrow 2CO \text{ and } ZnO + CO \rightarrow Zn + CO_2$$

Two mol C produces two mol CO; because one mol ZnO reacts with one mol CO, two mol ZnO will react with two mol CO. Thus, two mol C is equivalent to two mol ZnO, or one mol C is equivalent to one mol ZnO.

Using this to calculate mass of C from mass of ZnO, we have

$$75.0 \text{ g } ZnO \times \frac{1 \text{ mol } ZnO}{81.39 \text{ g } ZnO} \times \frac{1 \text{ mol } C}{1 \text{ mol } ZnO} \times \frac{12.01 \text{ g } C}{1 \text{ mol } C} = 11.07 \text{ g } C$$

(continued)

Thus, all of the ZnO is used up in reacting with just 11.07 g of C, making ZnO the limiting reactant. Use the mass of ZnO to calculate the mass of Zn formed:

$$75.0 \text{ g ZnO} \times \frac{1 \text{ mol ZnO}}{81.39 \text{ g ZnO}} \times \frac{1 \text{ mol Zn}}{1 \text{ mol ZnO}} \times \frac{65.39 \text{ g Zn}}{1 \text{ mol Zn}} = 60.2\underline{5}6 = 60.3 \text{ g Zn}$$

3.105 For $CaO + 3C \rightarrow CaC_2 + CO$, find the limiting reactant in terms of moles of CaC_2 obtainable:

$$\text{Mol CaC}_2 = 2.60 \times 10^3 \text{ g C} \times \frac{1 \text{ mol C}}{12.01 \text{ g C}} \times \frac{1 \text{ mol CaC}_2}{3 \text{ mol C}} = 72.\underline{1}6 \text{ mol}$$

$$\text{Mol CaC}_2 = 2.60 \times 10^3 \text{ g CaO} \times \frac{1 \text{ mol CaO}}{56.08 \text{ g CaO}} \times \frac{1 \text{ mol CaC}_2}{1 \text{ mol CaO}} \quad 46.\underline{3}62 \text{ mol}$$

Because CaO is the limiting reactant, calculate the mass of CaC_2 from it:

$$\text{Mass CaC}_2 = 46.\underline{3}62 \text{ mol CaC}_2 \times \frac{64.10 \text{ g CaC}_2}{1 \text{ mol CaC}_2} = 2.9\underline{7}1 \times 10^3$$

$$= 2.97 \times 10^3 \text{ g CaC}_2$$

3.107 From the equation $2Na + H_2O \rightarrow 2NaOH + H_2$, convert the mass of H_2 to mass of Na, and then use the mass to calculate the percentage:

$$0.108 \text{ g H}_2 \times \frac{1 \text{ mol H}_2}{2.016 \text{ g H}_2} \times \frac{2 \text{ mol Na}}{1 \text{ mol H}_2} \times \frac{22.99 \text{ g Na}}{1 \text{ mol Na}} = 2.4\underline{6}3 \text{ g Na}$$

$$\text{Percent Na} = \frac{\text{mass Na}}{\text{mass amalgam}} \times 100\% = \frac{2.463 \text{ g}}{15.23 \text{ g}} \times 100\% = 16.\underline{1}7 = 16.2\%$$

■ Solutions to Cumulative-Skills Problems

3.109 Let y equal the mass of CuO in the mixture. Then 0.500 g - y equals the mass of Cu_2O in the mixture. Multiplying the appropriate conversion factors for Cu times the mass of each oxide will give one equation in one unknown for the mass of 0.425 g Cu:

$$0.425 = y \left[\frac{63.55 \text{ g Cu}}{79.55 \text{ g CuO}} \right] + (0.500 - y) \left[\frac{127.10 \text{ g Cu}}{143.10 \text{ g Cu}_2\text{O}} \right]$$

Simplifying the equation by dividing the conversion factors and combining terms gives:

$0.425 = 0.79887 \, y + 0.888190 \, (0.500 - y)$

$0.08932 \, y = 0.019095$

$y = 0.2139 = 0.21 \text{ g} = \text{mass of CuO}$

3.111 If one heme molecule contains one iron atom, then the number of moles of heme in 35.2 mg heme must be the same as the number of moles of iron in 3.19 mg of iron. Start by calculating the moles of Fe (equals moles heme):

$$3.19 \times 10^{-3} \text{ g Fe} \times \frac{1 \text{ mol Fe}}{55.85 \text{ g Fe}} = 5.712 \times 10^{-5} \text{ mol Fe or heme}$$

$$\text{Molar mass of heme} = \frac{35.2 \times 10^{-3} \text{ g}}{5.712 \times 10^{-5} \text{ mol}} = 616.2 = 616 \text{ g/mol}$$

The molecular weight of heme is 616 amu.

3.113 Use the data to find the molar mass of the metal and anion. Start with X_2.

Mass X_2 in MX = 4.52 g - 3.41 g = 1.11 g

Molar mass X_2 = 1.11 g ÷ 0.0158 mol = 70.25 g/mol

Molar mass X = 70.25 ÷ 2 = 35.14 = 35.1 g/mol

Thus X is Cl, chlorine.

Moles of M in 4.52 g MX = 0.0158 x 2 = 0.0316 mol

Molar mass of M = 3.41 g ÷ 0.0316 mol = 107.9 = 108 g/mol

Thus M is Ag, silver.

3.115   After finding the volume of the alloy, convert it to mass Fe using density and percent Fe. Then use Avogadro's number and the atomic weight for the number of atoms.

$$\text{Vol.} = 10.0 \text{ cm} \times 20.0 \text{ cm} \times 15.0 \text{ cm} = 3.00 \times 10^3 \text{ cm}^3$$

$$\text{Mass Fe} = 3.00 \times 10^3 \text{ cm}^3 \times \frac{8.17 \text{ g alloy}}{1 \text{ cm}^3} \times \frac{54.7 \text{ g Fe}}{100.0 \text{ g alloy}} = 1.3\underline{4}07 \times 10^4 \text{ g}$$

$$\text{No. of Fe atoms} = 1.3\underline{4}07 \times 10^{-4} \text{ g Fe} \times \frac{1 \text{ mol Fe}}{55.85 \text{ g Fe}} \times \frac{6.022 \times 10^{23} \text{ Fe atoms}}{1 \text{ mol Fe}}$$

$$= 1.4\underline{4}56 \times 10^{26} = 1.45 \times 10^{26} \text{ Fe atoms}$$

4. CHEMICAL REACTIONS

■ Solutions to Exercises

Note on significant figures: If the final answer to a solution needs to be rounded off, it is given first with one nonsignificant figure, and the last significant figure is underlined. The final answer is then rounded to the correct number of significant figures. In multiple-step problems, intermediate answers are given with at least one nonsignificant figure; however, only the final answer has been rounded off.

4.1   a. According to Table 4.1, all compounds that contain sodium, Na^+, are soluble. Thus, NaBr is soluble in water.

     b. According to Table 4.1, most compounds that contain hydroxides, OH^-, are insoluble in water. However, $Ba(OH)_2$ is listed as one of the exceptions to this rule, so it is soluble in water.

     c. Calcium carbonate is $CaCO_3$. According to Table 4.1, most compounds that contain carbonate, CO_3^{2-}, are insoluble. $CaCO_3$ is not one of the exceptions, so it is insoluble in water.

4.2   a. The problem states that HNO_3 is a strong electrolyte, but $Mg(OH)_2$ is a solid, so retain its formula. On the product side, $Mg(NO_3)_2$ is a soluble ionic compound, but water is a nonelectrolyte, so retain its formula. The resulting complete ionic equation is

$$2H^+(aq) + 2NO_3^-(aq) + Mg(OH)_2(s) \rightarrow 2H_2O(l) + Mg^{2+}(aq) + 2NO_3^-(aq)$$

The corresponding net ionic equation is

$$2H^+(aq) + Mg(OH)_2(s) \rightarrow 2H_2O(l) + Mg^{2+}(aq)$$

(continued)

b. Both reactants are soluble ionic compounds, and on the product side, $NaNO_3$ is also a soluble ionic compound. $PbSO_4$ is a solid, so retain its formula. The resulting complete ionic equation is

$$Pb^{2+}(aq) + 2NO_3^-(aq) + 2Na^+(aq) + SO_4^{2-}(aq) \rightarrow$$
$$PbSO_4(s) + 2Na^+(aq) + 2NO_3^-(aq)$$

The corresponding net ionic equation is

$$Pb^{2+}(aq) + SO_4^{2-}(aq) \rightarrow PbSO_4(s)$$

4.3   The formulas of the compounds are NaI and $Pb(C_2H_3O_2)_2$. Exchanging anions, you get sodium acetate, $NaC_2H_3O_2$, and lead(II) iodide, PbI_2. The equation for the exchange reaction is

$$NaI + Pb(C_2H_3O_2)_2 \rightarrow NaC_2H_3O_2 + PbI_2$$

From Table 4.1, you see that NaI is soluble, $Pb(C_2H_3O_2)_2$ is soluble, $NaC_2H_3O_2$ is soluble, and PbI_2 is insoluble. Thus, lead(II) iodide precipitates. The balanced molecular equation with phase labels is

$$Pb(C_2H_3O_2)_2(aq) + 2NaI(aq) \rightarrow PbI_2(s) + 2NaC_2H_3O_2(aq)$$

To get the net ionic equation, you write the soluble ionic compounds as ions, and cancel the spectator ions, ($C_2H_3O_2^-$ and Na^+). The final result is

$$Pb^{2+}(aq) + 2I^-(aq) \rightarrow PbI_2(s)$$

4.4   a. H_3PO_4 is not listed as a strong acid in Table 4.3, so it is a weak acid.

b. Hypochlorous acid, HClO, is not one of the strong acids listed in Table 4.3, therefore we assume that HClO is a weak acid.

c. As noted in Table 4.3, $HClO_4$ is a strong acid.

d. As noted in Table 4.3, $Sr(OH)_2$ is a strong base.

4.5   The salt consists of the cation from the base (Li^+) and the anion from the acid (CN^-); its formula is LiCN. You will need to add H_2O as a product to complete and balance the molecular equation:

$$HCN(aq) + LiOH(aq) \rightarrow LiCN(aq) + H_2O(l)$$

Note that LiOH (a strong base) and LiCN (a soluble ionic substance) are strong electrolytes; HCN is a weak electrolyte (it is not one of the strong acids in Table 4.3). After eliminating the spectator ions (Li^+ and CN^-), the net ionic equation is

$$HCN(aq) + OH^-(aq) \rightarrow H_2O(l) + CN^-(aq)$$

4.6   The first step in the neutralization is described by the following molecular equation:

$$H_2SO_4(aq) + KOH(aq) \rightarrow KHSO_4(aq) + H_2O(l)$$

The corresponding net ionic equation is

$$H^+(aq) + OH^-(aq) \rightarrow H_2O(l)$$

The reaction of the acid salt $KHSO_4$ is given by the following molecular equation:

$$KHSO_4(aq) + KOH(aq) \rightarrow K_2SO_4(aq) + H_2O(l)$$

The corresponding net ionic equation is

$$HSO_4^-(aq) + OH^-(aq) \rightarrow H_2O(l) + SO_4^{2-}(aq)$$

4.7   First, write the molecular equation for the exchange reaction noting that the products of the reaction would be soluble $Ca(NO_3)_2$ and H_2CO_3. The carbonic acid decomposes to water and carbon dioxide gas. The molecular equation for the process is

$$CaCO_3(s) + 2HNO_3(aq) \rightarrow Ca(NO_3)_2(aq) + H_2O(l) + CO_2(g)$$

The corresponding net ionic equation is

$$CaCO_3(s) + 2H^+(aq) \rightarrow Ca^{2+}(aq) + H_2O(l) + CO_2(g)$$

4.8 a. For potassium dichromate, $K_2Cr_2O_7$,

$$2 \times (\text{oxidation number of K}) + 2 \times (\text{oxidation number of Cr})$$
$$+ 7 \times (\text{oxidation number of O}) = 0$$

For oxygen,

$$2 \times (+1) + 2 \times (\text{oxidation number of Cr}) + 7 \times (-2) = 0$$

Therefore,

$$2 \times \text{oxidation number of Cr} = -2 \times (+1) - 7 \times (-2) = +12$$

or, oxidation number of Cr = +6.

b. For the permanganate ion, MnO_4^-,

$$(\text{Oxidation number of Mn}) + 4 \times (\text{oxidation number of O}) = -1$$

For oxygen,

$$(\text{oxidation number of Mn}) + 4 \times (-2) = -1$$

Therefore,

$$\text{Oxidation number of Mn} = -1 - [4 \times (-2)] = +7$$

4.9 Identify the oxidation states of the elements.

$$\overset{0}{Ca} + \overset{0}{Cl_2} \rightarrow \overset{+2\ -1}{CaCl_2}$$

Break the reaction into two half reactions making sure that both mass and charge are balanced.

$$Ca \rightarrow Ca^{2+} + 2e^-$$

$$Cl_2 + 2e^- \rightarrow 2Cl^-$$

Since each half-reaction has two electrons, it is not necessary to multiply the reactions by any factors to cancel them out. Adding the two half-reactions together and canceling out the electrons, you get

$$Ca(s) + Cl_2(g) \rightarrow CaCl_2(s)$$

4.10 Convert mass of NaCl (molar mass, 58.44 g) to moles of NaCl. Then divide moles of solute by liters of solution. Note that 25.0 mL = 0.0250 L.

$$0.0678 \text{ g NaCl} \times \frac{1 \text{ mol NaCl}}{58.44 \text{ g NaCl}} = 1.1\underline{6}0 \times 10^{-3} \text{ mol NaCl}$$

$$\text{Molarity} = \frac{1.1\underline{6}0 \times 10^{-3} \text{ mol NaCl}}{0.0250 \text{ L soln}} = 0.0464\underline{1} = 0.0464 M$$

4.11 Convert grams of NaCl (molar mass, 58.44 g) to moles NaCl and then to volume of NaCl solution.

$$0.0958 \text{ g NaCl} \times \frac{1 \text{ mol NaCl}}{58.44 \text{ g NaCl}} \times \frac{1 \text{ L soln}}{0.163 \text{ mol NaCl}} \times \frac{1000 \text{ mL}}{1 \text{ L}}$$

$$= 10.\underline{0}6 = 10.1 \text{ mL NaCl}$$

4.12 One (1) liter of solution is equivalent to 0.15 mol NaCl. The amount of NaCl in 50.0 mL of solution is

$$50.0 \text{ mL} \times \frac{1 \text{ L}}{1000 \text{ mL}} \times \frac{0.15 \text{ mol NaCl}}{1 \text{ L soln}} = 0.007\underline{5}0 \text{ mol NaCl}$$

Convert to grams using the molar mass of NaCl (58.44 g/mol).

$$0.00750 \text{ mol NaCl} \times \frac{58.4 \text{ g NaCl}}{1 \text{ mol NaCl}} = 0.4\underline{3}8 = 0.44 \text{ g NaCl}$$

4.13 Use the rearranged version of the dilution formula from the text to calculate the initial volume of 1.5 M sulfuric acid required:

$$V_i = \frac{M_f V_f}{M_i} = \frac{0.18 \text{ M} \times 100.0 \text{ mL}}{1.5 \text{ M}} = 1\underline{2}.0 = 12 \text{ mL}$$

4.14   There are two different reactions taking place in forming the CaC_2O_4 (molar mass 128.10 g/mol) precipitate. These are

$$CaCO_3(s) + 2HCl(aq) \rightarrow CaCl_2(aq) + CO_2(g) + H_2O(l)$$

$$CaCl_2(aq) + Na_2C_2O_4(aq) \rightarrow CaC_2O_4(s) + 2NaCl(aq)$$

The overall stoichiometry of the reactions is one mol $CaCO_3$/one mol CaC_2O_4. Also note that each $CaCO_3$ contains one Ca atom, so this gives an overall conversion factor of one mol Ca/one mol CaC_2O_4. The mass of Ca can now be calculated.

$$0.1402 \text{ g } CaC_2O_4 \times \frac{1 \text{ mol } CaC_2O_4}{128.10 \text{ g } CaC_2O_4} \times \frac{1 \text{ mol Ca}}{1 \text{ mol } CaC_2O_4} \times \frac{40.08 \text{ g Ca}}{1 \text{ mol Ca}}$$

$$= 0.043866 \text{ g Ca}$$

Now, calculate the percentage of calcium in the 128.3 mg (0.1283 g) limestone:

$$\frac{0.043866 \text{ g Ca}}{0.1283 \text{ g limestone}} \times 100\% = 34.190 = 34.19\%$$

4.15   Convert the volume of Na_3PO_4 to moles using the molarity of Na_3PO_4. Note that 45.7 ml = 0.0457 L.

$$0.0457 \text{ L } Na_3PO_4 \times \frac{0.265 \text{ mol } Na_3PO_4}{1 \text{ L}} = 0.01211 \text{ mol } Na_3PO_4$$

Finally, calculate the amount of $NiSO_4$ required to react with this amount of Na_3PO_4:

$$0.1211 \text{ mol } Na_3PO_4 \times \frac{3 \text{ mol } NiSO_4}{2 \text{ mol } Na_3PO_4} \times \frac{1 \text{ L } NiSO_4}{0.375 \text{ M } NiSO_4} = 0.04844 \text{ L } (48.4 \text{ mL})$$

4.16   Convert the volume of NaOH solution (0.0391 L) to moles NaOH (from the molarity of NaOH). Then, convert moles NaOH to moles $HC_2H_3O_2$ (from the chemical equation). Finally, convert moles of $HC_2H_3O_2$ (molar mass 60.05 g/mol) to grams $HC_2H_3O_2$.

$$0.0391 \text{ L NaOH} \times \frac{0.108 \text{ mol NaOH}}{1 \text{ L}} \times \frac{1 \text{ mol } HC_2H_3O_2}{1 \text{ mol NaOH}} \times \frac{60.05 \text{ g } HC_2H_3O_2}{1 \text{ mol } HC_2H_3O_2}$$

$$= 0.25359 \text{ g}$$

The mass percentage of acetic acid in the vinegar can now be calculated.

$$\text{Percentage Mass} = \frac{0.25359 \text{ g } HC_2H_3O_2}{5.00 \text{ g vinegar}} \times 100\% = 5.071 = 5.07\%$$

■ Answers to Review Questions

4.1   Some electrolyte solutions are strongly conducting because they are almost completely ionized and others are weakly conducting because they are weakly ionized. The former solutions will have many more ions to conduct electricity than will the latter solutions if both are present at the same concentrations.

4.2   A strong electrolyte is an electrolyte that exists in solution almost entirely as ions. An example is NaCl. When NaCl dissolves in water, it dissolves almost completely to give Na^+ and Cl^- ions. A weak electrolyte is an electrolyte that dissolves in water to give a relatively small percentage of ions. An example is NH_3. When NH_3 dissolves in water, it reacts very little with the water, so the level of NH_3 is relatively high, and the level of the NH_4^+ and OH^- ions is relatively low.

4.3   Soluble means the ability of a substance to dissolve in water. A compound is insoluble if it does not dissolve appreciably in water. An example of a soluble ionic compound is sodium chloride, NaCl, and for an insoluble ionic compound, an example is calcium carbonate, $CaCO_3$.

4.4   The advantage of using a molecular equation to represent an ionic equation is that it states explicitly what chemical species have been added and what chemical species are obtained as products. It also makes stoichiometric calculations easy to perform. The disadvantages are (1) the molecular equation does not represent the fact that the reaction actually involves ions, and (2) the molecular equation does not indicate which species exist as ions and which exist as molecular solids or molecular gases.

4.5   A spectator ion is an ion that does not take part in the reaction. In the following ionic reaction, the Na^+ and Cl^- are spectator ions:

$$Na^+(aq) + OH^-(aq) + H^+(aq) + Cl^-(aq) \rightarrow Na^+(aq) + Cl^-(aq) + H_2O(l)$$

4.6   A net ionic equation is an ionic equation from which spectator ions have been canceled. The value of such an equation is that it shows the reaction that actually occurs at the ionic level. An example is the ionic equation representing the reaction of calcium chloride ($CaCl_2$) with potassium carbonate(K_2CO_3).

$$CaCl_2(aq) + K_2CO_3(aq) \rightarrow CaCO_3(s) + 2\ KCl(aq):$$

$$Ca^{2+}(aq) + CO_3^{2-}(aq) \rightarrow CaCO_3(s) \qquad \text{(net ionic equation)}$$

4.7   The three major types of chemical reactions are precipitation reactions, acid-base reactions, and oxidation-reduction reactions. Oxidation-reduction reactions can be further classified as combination reactions, decomposition reactions, displacement reactions, and combustion reactions. Brief descriptions and examples of each are given below.

A precipitation reaction is a reaction that appears to involve the exchange of parts of the reactants. An example is: $2KCl(aq) + Pb(NO_3)_2(aq) \rightarrow 2KNO_3(aq) + PbI_2(s)$.

An acid-base reaction, or neutralization reaction, results in an ionic compound and possibly water. An example is: $HCl(aq) + NaOH(aq) \rightarrow NaCl(aq) + H_2O(l)$.

A combination reaction is a reaction in which two substances combine to form a third substance. An example is: $2Na(s) + Cl_2(g) \rightarrow 2NaCl(s)$.

A decomposition reaction is a reaction in which a single compound reacts to give two or more substances. An example is: $2HgO(s) \xrightarrow{\Delta} 2Hg(l) + O_2(g)$.

A displacement reaction, or single replacement reaction, is a reaction in which an element reacts with a compound displacing an element from it. An example is: $Cu(s) + 2AgNO_3(aq) \rightarrow 2Ag(s) + Cu(NO_3)_2(aq)$.

A combustion reaction is a reaction of a substance with oxygen, usually with rapid release of heat to produce a flame. The products include one or more oxides. An example is: $CH_4(g) + 2O_2(g) \rightarrow CO_2(g) + 2H_2O(l)$.

4.8   To prepare crystalline AgCl and NaNO$_3$, first make solutions of AgNO$_3$ and NaCl by weighing equivalent molar amounts of both solid compounds. Then mix the two solutions together, forming a precipitate of silver chloride and a solution of soluble sodium nitrate. Filter off the silver chloride, and wash it with water to remove the sodium nitrate solution. Then allow it to dry to obtain pure crystalline silver chloride. Finally, take the filtrate containing the sodium nitrate, and evaporate it, leaving pure crystalline sodium nitrate.

4.9   An example of a neutralization reaction is

$$HBr + KOH \rightarrow KBr + H_2O(l)$$
$\quad$acid$\quad$base$\quad\quad$salt

4.10   An example of a polyprotic acid is carbonic acid, H_2CO_3. The successive neutralization is given by the following molecular equations:

$$H_2CO_3(aq) + NaOH(aq) \rightarrow NaHCO_3(aq) + H_2O(l)$$

$$NaHCO_3(aq) + NaOH(aq) \rightarrow Na_2CO_3(aq) + H_2O(l)$$

4.11   Since an oxidation-reduction reaction is an electron transfer reaction, one substance must lose the electrons and be oxidized while another substance must gain electrons and be reduced.

4.12   A displacement reaction is an oxidation-reduction reaction in which a free element reacts with a compound, displacing an element from it.

$$Cu(s) + 2AgNO_3(aq) \rightarrow 2Ag(s) + Cu(NO_3)_2(aq)$$

Ag^+ is the oxidizing agent, and Cu is the reducing agent.

4.13   The number of moles present does not change when the solution is diluted.

4.14   The reaction is

$$HCl + NaOH \rightarrow NaCl + H_2O$$

After titration, the volume of hydrochloric acid is converted to moles of HCl using the molarity. Since the stoichiometry of the reaction is one mole HCl to one mole NaOH, these quantities are equal.

moles HCl = moles NaOH = molarity x volume

You could then multiply by the molar mass of NaOH to obtain the amount in the mixture.

■ Solutions to Practice Problems

Note on significant figures: If the final answer to a solution needs to be rounded off, it is given first with one nonsignificant figure, and the last significant figure is underlined. The final answer is then rounded to the correct number of significant figures. In multiple-step problems, intermediate answers are given with at least one nonsignificant figure; however, only the final answer has been rounded off.

4.23 a. Insoluble   b. Soluble   c. Soluble   d. Soluble

4.25 a. Insoluble   b. Soluble; The ions present would be NH_4^+ and SO_4^{2-}.

   c. Insoluble   d. Soluble; The ions present would be Na^+ and CO_3^{2-}.

4.27 a. $H^+(aq) + OH^-(aq) \rightarrow H_2O(l)$

   b. $Ag^+(aq) + Br^-(aq) \rightarrow AgBr(s)$

(continued)

c. $S^{2-}(aq) + 2H^+(aq) \rightarrow H_2S(g)$

d. $OH^-(aq) + NH_4^+(aq) \rightarrow NH_3(g) + H_2O(l)$

4.29   Molecular equation: $Pb(NO_3)_2(aq) + Na_2SO_4(aq) \rightarrow PbSO_4(s) + 2NaNO_3(aq)$

Net ionic equation: $Pb^{2+}(aq) + SO_4^{2-}(aq) \rightarrow PbSO_4(s)$

4.31 a. $FeSO_4(aq) + NaCl(aq) \rightarrow NR$

b. $Na_2CO_3(aq) + MgBr_2(aq) \rightarrow MgCO_3(s) + 2NaBr(aq)$

$CO_3^{2-}(aq) + Mg^{2+}(aq) \rightarrow MgCO_3(s)$

c. $MgSO_4(aq) + 2NaOH(aq) \rightarrow Mg(OH)_2(s) + Na_2SO_4(aq)$

$Mg^{2+}(aq) + 2OH^-(aq) \rightarrow Mg(OH)_2(s)$

d. $NiCl_2(aq) + NaBr(aq) \rightarrow NR$

4.33 a. $Ba(NO_3)_2(aq) + Li_2SO_4(aq) \rightarrow BaSO_4(s) + 2LiNO_3(aq)$

$Ba^{2+}(aq) + SO_4^{2-}(aq) \rightarrow BaSO_4(s)$

b. $Ca(NO_3)_2(aq) + NaBr(aq) \rightarrow NR$

c. $Al_2(SO_4)_3(aq) + 6NaOH(aq) \rightarrow 2Al(OH)_3(s) + 3Na_2SO_4(aq)$

$Al^{3+}(aq) + 3OH^-(aq) \rightarrow Al(OH)_3(s)$

d. $3CaBr_2(aq) + 2Na_3PO_4(aq) \rightarrow Ca_3(PO_4)_2(s) + 6NaBr(aq)$

$3Ca^{2+}(aq) + 2PO_4^{3-}(aq) \rightarrow Ca_3(PO_4)_2(s)$

4.35 a. Weak acid   b. Strong base   c. Strong acid   d. Weak acid

4.37 a. $NaOH(aq) + HNO_3(aq) \rightarrow H_2O(l) + NaNO_3(aq)$

$H^+(aq) + OH^-(aq) \rightarrow H_2O(l)$

b. $2HCl(aq) + Ba(OH)_2(aq) \rightarrow 2H_2O(l) + BaCl_2(aq)$

$H^+(aq) + OH^-(aq) \rightarrow H_2O(l)$

(continued)

c. $2HC_2H_3O_2(aq) + Ca(OH)_2(aq) \rightarrow 2H_2O(l) + Ca(C_2H_3O_2)_2(aq)$

   $HC_2H_3O_2(aq) + OH^-(aq) \rightarrow H_2O(l) + C_2H_3O_2^-(aq)$

d. $NH_3(aq) + HNO_3(aq) \rightarrow NH_4NO_3(aq)$

   $NH_3(aq) + H^+(aq) \rightarrow NH_4^+(aq)$

4.39 a. $2HBr(aq) + Ca(OH)_2(aq) \rightarrow 2H_2O(l) + CaBr_2(aq)$

   $H^+(aq) + OH^-(aq) \rightarrow H_2O(l)$

b. $3HNO_3(aq) + Al(OH)_3(s) \rightarrow 3H_2O(l) + Al(NO_3)_3(aq)$

   $3H^+(aq) + Al(OH)_3(s) \rightarrow 3H_2O(l) + Al^{3+}(aq)$

c. $2HCN(aq) + Ca(OH)_2(aq) \rightarrow 2H_2O(l) + Ca(CN)_2(aq)$

   $HCN(aq) + OH^-(aq) \rightarrow H_2O(l) + CN^-(aq)$

d. $HCN(aq) + LiOH(aq) \rightarrow H_2O(l) + LiCN(aq)$

   $HCN(aq) + OH^-(aq) \rightarrow H_2O(l) + CN^-(aq)$

4.41 a. $2KOH(aq) + H_3PO_4(aq) \rightarrow K_2HPO_4(aq) + 2H_2O(l)$

   $2OH^-(aq) + H_3PO_4(aq) \rightarrow HPO_4^{2-}(aq) + 2H_2O(l)$

b. $3H_2SO_4(aq) + 2Al(OH)_3(s) \rightarrow 6H_2O(l) + Al_2(SO_4)_3(aq)$

   $3H^+(aq) + Al(OH)_3(s) \rightarrow 3H_2O(l) + Al^{3+}(aq)$

c. $2HC_2H_3O_2(aq) + Ca(OH)_2(aq) \rightarrow 2H_2O(l) + Ca(C_2H_3O_2)_2(aq)$

   $HC_2H_3O_2(aq) + OH^-(aq) \rightarrow H_2O(l) + C_2H_3O_2^-(aq)$

d. $H_2SO_3(aq) + NaOH(aq) \rightarrow H_2O(l) + NaHSO_3(aq)$

   $H_2SO_3(aq) + OH^-(aq) \rightarrow HSO_3^-(aq) + H_2O(l)$

4.43   Molecular equations: $2H_2SO_3(aq) + Ca(OH)_2(aq) \rightarrow 2H_2O(l) + Ca(HSO_3)_2(aq)$

$Ca(HSO_3)_2(aq) + Ca(OH)_2(aq) \rightarrow 2H_2O(l) + 2CaSO_3(s)$

Ionic equations: $H_2SO_3(aq) + OH^-(aq) \rightarrow H_2O(l) + HSO_3^-(aq)$

$Ca^{2+}(aq) + HSO_3^-(aq) + OH^-(aq) \rightarrow CaSO_3(s) + H_2O(l)$

4.45 a. Molecular equation: $CaS(aq) + 2HBr(aq) \rightarrow CaBr_2(aq) + H_2S(g)$

Ionic equation: $S^{2-}(aq) + 2H^+(aq) \rightarrow H_2S(g)$

b. Molecular equation: $MgCO_3(s) + 2HNO_3(aq) \rightarrow$
$Mg(NO_3)_2(aq) + CO_2(g) + H_2O(l)$

Ionic equation: $MgCO_3(s) + 2H^+(aq) \rightarrow Mg^{2+}(aq) + CO_2(g) + H_2O(l)$

c. Molecular equation: $K_2SO_3(aq) + H_2SO_4(aq) \rightarrow K_2SO_4(aq) + SO_2(g) + H_2O(l)$

Ionic equation: $SO_3^{2-}(aq) + 2H^+(aq) \rightarrow SO_2(g) + H_2O(l)$

4.47   Molecular equation: $FeS(s) + 2HCl(aq) \rightarrow H_2S(g) + FeCl_2(aq)$

Ionic equation: $FeS(s) + 2H^+(aq) \rightarrow H_2S(g) + Fe^{2+}(aq)$

4.49 a. Because all three O's = a total of -6, both Ga's = +6; thus, the oxidation number of Ga = +3.

b. Because both O's = a total of -4, the oxidation number of Nb = +4.

c. Because the four O's = a total of -8 and K = +1, the oxidation number of Br = +7.

d. Because the four O's = a total of -8 and the 2 K's = +2, the oxidation number of Mn = +6.

4.51 a. Because the charge of -1 = [x_N + 2 (from 2 H's)], x_N must equal -3.

b. Because the charge of -1 = [x_I - 6 (from 3 O's)], x_I must equal +5.

c. Because the charge of -1 = [x_{AI} - 8 (4 O's) + 4 (4 H's)], x_{AI} must equal +3.

d. Because the charge of 0 = [x_{CI} - 8 (4 O's) + 1 (1 H's)], x_{CI} must equal +7.

4.53 a. From the list of common polyatomic anions in Table 2.6, the formula of the ClO_3 anion must be ClO_3^-. Thus, the oxidation state of Mn is Mn^{2+} (see also Table 4.5). Since the oxidation state of O is -2 and the net ionic charge is -1, the oxidation state of chlorine is determined by $x_{Cl} - 6 = -1$, so x_{Cl} must equal +5.

b. From the list of common polyatomic anions in Table 2.6, the formula of the CrO_4 anion must be CrO_4^{2-}. Thus, the oxidation state of Fe is Fe^{3+}. Since the oxidation state of O is -2 and the net ionic charge is -2, the oxidation state of Cr is determined by $x_{Cr} - 8 = -2$, so x_{Cr} must equal +6.

c. From the list of common polyatomic anions in Table 2.6, the formula of the Cr_2O_7 anion must be $Cr_2O_7^{2-}$. Thus, the oxidation state of Hg is Hg^{2+}. Since the oxidation state of O is -2 and the net ionic charge is -2, the oxidation state of Cr determined by $2x_{Cr} - 14 = -2$, so x_{Cr} must equal +6.

d. From the list of common polyatomic anions in Table 2.6, the formula of the PO_4 anion must be PO_4^{3-}. Thus, the oxidation state of Co is Co^{2+}. Since the oxidation state of O is -2 and the net ionic charge is -3, the oxidation state of P is determined by $x_P - 8 = -3$, so x_P must equal +5.

4.55 a. Phosphorus changes from an oxidation number of zero in P_4 to +5 in P_4O_{10}, losing electrons and acting as a reducing agent. Oxygen changes from an oxidation number of zero in O_2 to -2 in P_4O_{10}, gaining electrons and acting as an oxidizing agent.

b. Cobalt changes from an oxidation number of zero in Co(s) to +2 in $CoCl_2$, losing electrons and acting as a reducing agent. Chlorine changes from an oxidation number of zero in Cl_2 to -1 in $CoCl_2$, gaining electrons and acting as an oxidizing agent.

4.57 a. Al changes from oxidation number zero to +3; Al is the reducing agent.
   F changes from oxidation number zero to -1; F_2 is the oxidizing agent.

b. Hg changes from oxidation state +2 to 0; Hg^{2+} is the oxidizing agent.
   N changes from oxidation state +3 to +5; NO_2^- is the reducing agent.

4.59 a. First, identify the species being oxidized and reduced, and assign the appropriate oxidation states. Since $CuCl_2$ and $AlCl_3$ are both soluble ionic compounds, Cl^- is a spectator ion and can be removed from the equation. The resulting net ionic equation is

$$\overset{+2}{Cu^{2+}}(aq) + \overset{0}{Al}(s) \rightarrow \overset{+3}{Al^{3+}}(aq) + \overset{0}{Cu}(s)$$

Next, write the half-reactions in an unbalanced form.

$$Al \rightarrow Al^{3+} \qquad \text{(oxidation)}$$
$$Cu^{2+} \rightarrow Cu \qquad \text{(reduction)}$$

(continued)

Next, balance the charge in each equation by adding electrons to the more positive side to create balanced half-reactions.

$$Al \rightarrow Al^{3+} + 3e^- \qquad \text{(oxidation half-reaction)}$$

$$Cu^{2+} + 2e^- \rightarrow Cu \qquad \text{(reduction half-reaction)}$$

Multiply each half-reaction by a factor that will cancel out the electrons.

$$2 \times (Al \rightarrow Al^{3+} + 3e^-)$$

$$\underline{3 \times (Cu^{2+} + 2e^- \rightarrow Cu)}$$

$$3Cu^{2+} + 2Al + 6e^- \rightarrow 2Al^{3+} + 3Cu + 6e^-$$

Therefore, the balanced oxidation-reduction reaction is

$$3Cu^{2+} + 2Al \rightarrow 2Al^{3+} + 3Cu$$

Finally, add six Cl⁻ ions to each side, and add phase labels. The resulting balanced equation is

$$3CuCl_2(aq) + 2Al(s) \rightarrow 2AlCl_3(aq) + 3Cu(s)$$

b. First, identify the species being oxidized and reduced, and assign the appropriate oxidation states.

$$\overset{+3}{Cr^{3+}}(aq) + \overset{0}{Zn}(s) \rightarrow \overset{0}{Cr}(s) + \overset{+2}{Zn^{2+}}(aq)$$

Next, write the half-reactions in an unbalanced form.

$$Zn \rightarrow Zn^{2+} \qquad \text{(oxidation)}$$

$$Cr^{3+} \rightarrow Cr \qquad \text{(reduction)}$$

Next, balance the charge in each equation by adding electrons to the more positive side to create balanced half-reactions.

$$Zn \rightarrow Zn^{2+} + 2e^- \qquad \text{(oxidation half-reaction)}$$

$$Cr^{3+} + 3e^- \rightarrow Cr \qquad \text{(reduction half-reaction)}$$

Multiply each half-reaction by a factor that will cancel out the electrons.

$$3 \times (Zn \rightarrow Zn^{2+} + 2e^-)$$

$$\underline{2 \times (Cr^{3+} + 3e^- \rightarrow Cr)}$$

$$2Cr^{3+} + 3Zn + 6e^- \rightarrow 2Cr + 3Zn^{2+} + 6e^-$$

Therefore, the balanced oxidation-reduction reaction, including phase labels, is

$$2Cr^{3+}(aq) + 3Zn(s) \rightarrow 2Cr(s) + 3Zn^{2+}(aq)$$

4.61   Molarity = $\dfrac{\text{moles solute}}{\text{liters solution}}$ = $\dfrac{0.0512 \text{ mol}}{0.0250 \text{ L}}$ = 2.0$\underline{4}$8 = 2.05 M

4.63   Find the number of moles of solute ($KMnO_4$) using the molar mass of 158.03 g $KMnO_4$ per one mol $KMnO_4$:

$$0.798 \text{ g } KMnO_4 \times \dfrac{1 \text{ mol } KMnO_4}{158.03 \text{ g } KMnO_4} = 5.0\underline{4}97 \times 10^{-3} \text{ mol } KMnO_4$$

$$\text{Molarity} = \dfrac{\text{moles solute}}{\text{liters of solution}} = \dfrac{5.0497 \times 10^{-3} \text{ mol}}{0.0500 \text{ L}} = 0.10\underline{0}99 = 0.101 \text{ M}$$

4.65 $0.150 \text{ mol } CuSO_4 \times \dfrac{1 \text{ L solution}}{0.120 \text{ mol } CuSO_4} = 1.2\underline{5}0 = 1.25 \text{ L solution}$

4.67 $0.0353 \text{ g KOH} \times \dfrac{1 \text{ mol KOH}}{56.10 \text{ g KOH}} \times \dfrac{1 \text{ L solution}}{0.0176 \text{ mol KOH}} = 0.0357\underline{5}1 \text{ L} = 35.8 \text{ mL}$

4.69   From the molarity, one L of heme solution is equivalent to 0.0019 mol of heme solute. Before starting the calculation, note that 150 mL of solution is equivalent to 150×10^{-3} L of solution:

$$150 \times 10^{-3} \text{ L soln} \times \dfrac{0.0019 \text{ mol heme}}{1 \text{ L solution}} = 2.\underline{8}50 \times 10^{-4} = 2.9 \times 10^{-4} \text{ mol heme}$$

4.71   Multiply the volume of solution by molarity to convert it to moles; then convert to mass of solute by multiplying by the molar mass:

$$100.0 \times 10^{-3} \text{ L soln} \times \dfrac{0.025 \text{ mol } Na_2Cr_2O_7}{1 \text{ L solution}} \times \dfrac{262.0 \text{ g } Na_2Cr_2O_7}{1 \text{ mol } Na_2Cr_2O_7}$$

$$= 0.6\underline{5}50 = 0.66 \text{ g } Na_2Cr_2O_7$$

4.73   Use the rearranged version of the dilution formula to calculate the initial volume of 15.8 M HNO_3 required:

$$V_i = \dfrac{M_f V_f}{M_i} = \dfrac{0.12 \text{ M} \times 1000 \text{ mL}}{15.8 \text{ M}} = 7.\underline{5}9 = 7.6 \text{ mL}$$

4.75   The initial concentration of KCl (molar mass, 74.55 g/mol) is

$$3.50 \text{ g KCl} \times \frac{1 \text{ mol KCl}}{74.55 \text{ g KCl}} \times \frac{1}{0.0100 \text{ L}} = 4.6\underline{9}4 \text{ M}$$

Using the dilution factor, $M_1V_1 = M_2V_2$, with $V_2 = 10.0$ mL + 60.0 mL = 70.0 mL, after the solutions are mixed, the concentration of KCl is

$$M_2 = \frac{M_1V_1}{V_2} = \frac{4.694 \text{ M} \times 10.0 \text{ mL}}{70.0 \text{ mL}} = 0.670\underline{6}9 \text{ M KCl}$$

For $CaCl_2$, the concentration is

$$M_2 = \frac{M_1V_1}{V_2} = \frac{0.500 \text{ M} \times 10.0 \text{ mL}}{70.0 \text{ mL}} = 0.428\underline{5}7 \text{ M CaCl}_2$$

Therefore, the concentrations of the ions are 0.671 M K^+ and 0.429 M Ca^{2+}. For Cl^-, it is 0.670\underline{6}9 M + 2 x 0.428\underline{5}7 M = 1.52\underline{7}8 M = 1.528 M

4.77   Use the appropriate conversion factors to convert the mass of $BaSO_4$ to the mass of Ba^{2+} ions:

$$0.513 \text{ g BaSO}_4 \times \frac{1 \text{ mol BaSO}_4}{233.40 \text{ g BaSO}_4} \times \frac{1 \text{ mol Ba}^{2+}}{1 \text{ mol BaSO}_4} \times \frac{137.33 \text{ g Ba}^{2+}}{1 \text{ mol Ba}^{2+}}$$

$$= 0.301\underline{8}4 \text{ g Ba}^{2+}$$

Then calculate the percentage of barium in the 458 mg (0.458 g) compound:

$$\frac{0.30184 \text{ g Ba}^{2+}}{0.458 \text{ g}} \times 100\% = 65.\underline{9}039 = 65.9\% \text{ Ba}^{2+}$$

4.79 a. The mass of chloride ion in the AgCl from the copper chloride compound is:

$$86.00 \text{ mg AgCl} \times \frac{35.45 \text{ mg Cl}^-}{143.32 \text{ mg AgCl}} = 21.2\underline{7}1 \text{ mg Cl}^-$$

The percentage of chlorine in the 59.40 mg sample is:

$$\frac{21.2\underline{7}1 \text{ mg Cl}^-}{59.40 \text{ mg sample}} \times 100\% = 35.8\underline{0}9 = 35.81\% \text{ Cl}^-$$

(continued)

b. Of the various approaches, it is as easy to calculate the theoretical percentage of Cl^- in both CuCl and $CuCl_2$ as it is to use another approach:

$$CuCl: \frac{35.45 \text{ mg } Cl^-}{99.00 \text{ mg } CuCl} \times 100\% = 35.8\underline{0}8\%$$

$$CuCl_2: \frac{70.90 \text{ mg } Cl^-}{134.45 \text{ mg } CuCl_2} \times 100\% = 52.7\underline{3}3\%$$

The compound is obviously CuCl.

4.81   First, calculate the moles of chlorine in the compound:

$$0.3048 \text{ g AgCl} \times \frac{1 \text{ mol AgCl}}{143.32 \text{ g AgCl}} \times \frac{1 \text{ mol } Cl^-}{1 \text{ mol AgCl}} = 0.002126\underline{7} \text{ mol } Cl^-$$

Then, calculate the g Fe^{x+} from the g Cl^-:

$$g \, Fe^{x+} = 0.1348 \text{ g comp} - \left(0.0021267 \text{ mol } Cl^- \times \frac{35.45 \text{ g } Cl^-}{1 \text{ mol } Cl^-} \right)$$

$$= 0.0594\underline{0}8 \text{ g } Fe^{x+}$$

Now, calculate the moles of Fe^{x+} using the molar mass:

$$0.0594\underline{0}8 \text{ g } Fe^{x+} \times \frac{1 \text{ mol } Fe^{x+}}{55.85 \text{ g } Fe^{x+}} = 0.001063\underline{7} \text{ mol } Fe^{x+}$$

Finally, divide the mole numbers by the smallest mole number:

$$\text{For Cl: } \frac{0.002127 \text{ mol } Cl^-}{0.0010637 \text{ mol}} = 2.00; \text{ for } Fe^{x+}: \frac{0.0010637 \text{ mol } Fe^{x+}}{0.0010637 \text{ mol}} = 1.00$$

Thus, the formula is $FeCl_2$.

4.83 Using molarity, convert the volume of Na_2CO_3 to moles of Na_2CO_3; then use the equation to convert to moles of HNO_3, and finally to volume:

$$2HNO_3 + Na_2CO_3 \rightarrow 2NaNO_3 + H_2O + CO_2$$

$$44.8 \times 10^{-3} \text{ L Na}_2\text{CO}_3 \times \frac{0.150 \text{ mol Na}_2\text{CO}_3}{1 \text{ L soln}} \times \frac{2 \text{ mol HNO}_3}{1 \text{ mol Na}_2\text{CO}_3}$$

$$\times \frac{1 \text{ L HNO}_3}{0.250 \text{ mol HNO}_3} = 0.05376 \text{ L} = 0.0538 \text{ L} = 53.8 \text{ mL}$$

4.85 The reaction is $H_2SO_4 + 2NaHCO_3 \rightarrow Na_2SO_4 + 2H_2O + CO_2$.

$$8.20 \text{ g NaHCO}_3 \times \frac{1 \text{ mol NaHCO}_3}{84.01 \text{ g NaHCO}_3} \times \frac{1 \text{ mol H}_2\text{SO}_4}{2 \text{ mol NaHCO}_3} \times \frac{1 \text{ L soln}}{0.150 \text{ mol H}_2\text{SO}_4}$$

$$= 0.3253 \text{ L (325 mL) soln}$$

4.87 First, find the mass of H_2O_2 required to react with $KMnO_4$.

$$5H_2O_2 + 2KMnO_4 + 3H_2SO_4 \rightarrow 5O_2 + 2MnSO_4 + K_2SO_4 + 8H_2O$$

$$51.7 \times 10^{-3} \text{ L KMnO}_4 \times \frac{0.145 \text{ mol KMnO}_4}{1 \text{ L soln}} \times \frac{5 \text{ mol H}_2\text{O}_2}{2 \text{ mol KMnO}_4} \times \frac{34.02 \text{ g H}_2\text{O}_2}{1 \text{ mol H}_2\text{O}_2}$$

$$= 0.6375 \text{ g H}_2\text{O}_2$$

Percent H_2O_2 = (mass H_2O_2 ÷ mass sample) × 100%

$$= (0.6375 \text{ g} \div 20.0 \text{ g}) \times 100\% = 3.187 = 3.19\%$$

■ Solutions to General Problems

4.89 For the reaction of magnesium metal and hydrobromic acid, the equations are as follows.

Molecular equation: $Mg(s) + 2HBr(aq) \rightarrow H_2(g) + MgBr_2(aq)$

Ionic equation: $Mg(s) + 2H^+(aq) \rightarrow H_2(g) + Mg^{2+}(aq)$

4.91   For the reaction of nickel(II) sulfate and lithium hydroxide, the equations are as follows.

    Molecular equation: $NiSO_4(aq) + 2LiOH(aq) \rightarrow Ni(OH)_2(s) + Li_2SO_4(aq)$

    Ionic equation: $Ni^{2+}(aq) + 2OH^-(aq) \rightarrow Ni(OH)_2(s)$

4.93 a. Molecular equation: $LiOH(aq) + HCN(aq) \rightarrow LiCN(aq) + H_2O(l)$

       Ionic equation: $OH^-(aq) + HCN(aq) \rightarrow CN^-(aq) + H_2O(l)$

    b. Molecular equation: $Li_2CO_3(aq) + 2HNO_3(aq) \rightarrow 2LiNO_3(aq) + CO_2(g) + H_2O(l)$

       Ionic equation: $CO_3^{2-}(aq) + 2H^+(aq) \rightarrow CO_2(g) + H_2O(l)$

    c. Molecular equation: $LiCl(aq) + AgNO_3(aq) \rightarrow LiNO_3(aq) + AgCl(s)$

       Ionic equation: $Cl^-(aq) + Ag^+(aq) \rightarrow AgCl(s)$

    d. Molecular equation: $MgSO_4(aq) + LiCl(aq) \rightarrow$ NR

                         (Li_2SO_4 and $MgCl_2$ are soluble.)

4.95 a. Molecular equation: $Sr(OH)_2(aq) + 2HC_2H_3O_2(aq) \rightarrow$

                                   $Sr(C_2H_3O_2)_2(aq) + 2H_2O(l)$

      Ionic equation: $HC_2H_3O_2(aq) + OH^-(aq) \rightarrow C_2H_3O_2^-(aq) + H_2O(l)$

    b. Molecular equation: $NH_4I(aq) + CsCl(aq) \rightarrow$ NR

                    (NH_4Cl and CsI are soluble.)

    c. Molecular equation: $NaNO_3(aq) + CsCl(aq) \rightarrow$ NR

                  ($NaCl$ and $CsNO_3$ are soluble.)

    d. Molecular equation: $NH_4I(aq) + AgNO_3(aq) \rightarrow NH_4NO_3(aq) + AgI(s)$

       Ionic equation: $I^-(aq) + Ag^+(aq) \rightarrow AgI(s)$

4.97 For each preparation, the compound to be prepared is given first, followed by the compound from which it is to be prepared. Then the method of preparation is given, followed by the molecular equation for the preparation reaction. Steps such as evaporation, etc., are not given in the molecular equation.

a. To prepare $CuCl_2$ from $CuSO_4$, add a solution of $BaCl_2$ to a solution of the $CuSO_4$, precipitating $BaSO_4$. The $BaSO_4$ can be filtered off, leaving aqueous $CuCl_2$, which can be obtained in solid form by evaporation. Molecular equation:

$$CuSO_4(aq) + BaCl_2(aq) \rightarrow BaSO_4(s) + CuCl_2(aq)$$

b. To prepare $Ca(C_2H_3O_2)_2$ from $CaCO_3$, add a solution of acetic acid, $HC_2H_3O_2$, to the solid $CaCO_3$, forming CO_2, H_2O, and aqueous $Ca(C_2H_3O_2)_2$. The aqueous $Ca(C_2H_3O_2)_2$ can be converted to the solid form by evaporation, which also removes the CO_2 and H_2O products. Molecular equation:

$$CaCO_3(s) + 2HC_2H_3O_2(aq) \rightarrow Ca(C_2H_3O_2)_2(aq) + CO_2(g) + H_2O(l)$$

c. To prepare $NaNO_3$ from Na_2SO_3, add a solution of nitric acid, HNO_3, to the solid Na_2SO_3, forming SO_2, H_2O, and aqueous $NaNO_3$. The aqueous $NaNO_3$ can be converted to the solid by evaporation, which also removes the SO_2 and H_2O products. Molecular equation:

$$Na_2SO_3(s) + 2HNO_3(aq) \rightarrow 2NaNO_3(aq) + SO_2(g) + H_2O(l)$$

d. To prepare $MgCl_2$ from $Mg(OH)_2$, add a solution of hydrochloric acid (HCl) to the solid $Mg(OH)_2$, forming H_2O and aqueous $MgCl_2$. The aqueous $MgCl_2$ can be converted to the solid form by evaporation. Molecular equation:

$$Mg(OH)_2(s) + 2HCl(aq) \rightarrow MgCl_2(aq) + 2H_2O(l)$$

4.99   a. Decomposition   b. Decomposition   c. Combination   d. Displacement

4.101 a. $Pb(NO_3)_2 + H_2SO_4$ $[\rightarrow PbSO_4(s) + HNO_3(aq)\ ]$

     $Pb(NO_3)_2 + MgSO_4$ $[\rightarrow PbSO_4(s) + Mg(NO_3)_2(aq)\ ]$

     $Pb(NO_3)_2 + Ba(OH)_2$ $[\rightarrow Pb(OH)_2(s) + Ba(NO_3)_2(aq)\ ]$

     b. $Ba(OH)_2 + MgSO_4$ $[\rightarrow BaSO_4(s) + Mg(OH)_2(s)\ ]$

     c. $Ba(OH)_2 + H_2SO_4$ $[\rightarrow BaSO_4(s) + H_2O(l)\ ]$

4.103   Divide the mass of $CaCl_2$ by its molar mass and volume to find molarity:

$$4.50 \text{ g CaCl}_2 \times \frac{1 \text{ mol CaCl}_2}{110.98 \text{ g CaCl}_2} \times \frac{1}{1.000 \text{ L soln}} = 0.04054 = 0.0405 \text{ M CaCl}_2$$

The $CaCl_2$ dissolves to form Ca^{2+} and $2Cl^-$ ions. Therefore, the molarities of the ions are 0.0405 M Ca^{2+} and 2 x 0.04054, or 0.0811, M Cl^- ions.

4.105   Divide the mass of $K_2Cr_2O_7$ by its molar mass and volume to find molarity. Then calculate the volume needed to prepare 1.00 L of a 0.100 M solution.

$$89.3 \text{ g K}_2\text{Cr}_2\text{O}_7 \times \frac{1 \text{ mol K}_2\text{Cr}_2\text{O}_7}{294.20 \text{ g K}_2\text{Cr}_2\text{O}_7} = 0.3035 \text{ mol K}_2\text{Cr}_2\text{O}_7$$

$$\text{Molarity} = \frac{0.3035 \text{ mol K}_2\text{Cr}_2\text{O}_7}{1.00 \text{ L}} = 0.3035 \text{ M}$$

$$V_i = \frac{M_f \times V_f}{M_i} = \frac{0.100 \text{ M} \times 1.00 \text{ L}}{0.3035 \text{ M}} = 0.3294 \text{ L (329 mL)}$$

4.107   Assume a volume of 1.000 L (1000 cm³) for the 6.00 percent NaBr solution, and convert to moles and then to molarity.

$$1000 \text{ cm}^3 \times \frac{1.046 \text{ g soln}}{1 \text{ cm}^3} \times \frac{6.00 \text{ g NaBr}}{100 \text{ g soln}} \times \frac{1 \text{ mol NaBr}}{102.89 \text{ g NaBr}} = 0.6099 \text{ mol}$$

$$\text{Molarity NaBr} = \frac{0.6099 \text{ mol NaBr}}{1.000 \text{ L}} = 0.6099 = 0.610 \text{ M}$$

4.109   First, calculate the moles of $BaCl_2$:

$$1.128 \text{ g BaSO}_4 \times \frac{1 \text{ mol BaSO}_4}{233.40 \text{ g BaSO}_4} \times \frac{1 \text{ mol BaCl}_2}{1 \text{ mol BaSO}_4} = 0.0048329 \text{ mol BaCl}_2$$

Then, calculate the molarity from the moles and volume (0.0500 L):

$$\text{Molarity} = \frac{0.0048329 \text{ mol BaCl}_2}{0.0500 \text{ L}} = 0.096658 = 0.0967 \text{ M}$$

4.111   First, calculate the g of thallium(I) sulfate:

$$0.2122 \text{ g TlI} \times \frac{1 \text{ mol TlI}}{331.28 \text{ g TlI}} \times \frac{1 \text{ mol Tl}_2\text{SO}_4}{2 \text{ mol TlI}} \times \frac{504.83 \text{ g Tl}_2\text{SO}_4}{1 \text{ mol Tl}_2\text{SO}_4}$$

$$= 0.161\underline{6}8 \text{ g Tl}_2\text{SO}_4$$

Then, calculate the percent Tl_2SO_4 in the rat poison:

$$\text{Percent Ti}_2\text{SO}_4 = \frac{0.16168 \text{ g}}{0.7590 \text{ g}} \times 100\% = 21.3\underline{0}1 = 21.30\%$$

4.113   The mass of copper(II) ion and mass of sulfate ion in the 98.77 mg sample is:

$$0.09877 \text{ g} \times 0.3250 = 0.03210\underline{0} \text{ g Cu}^{2+} \text{ ion}$$

$$0.11666 \text{ g BaSO}_4 \times \frac{96.07 \text{ g SO}_4^{2-}}{233.40 \text{ g BaSO}_4} = 0.04801\underline{9} = 0.04802 \text{ g SO}_4^{2-}$$

The mass of water left = $0.09877 \text{ g} - (0.03210 \text{ g Cu}^{2+} + 0.04802 \text{ g SO}_4^{2-})$

$$= 0.01865 \text{ g H}_2\text{O}$$

The moles of water left = $0.01865 \text{ g} \div 18.02 \text{ g/mol} = 1.035 \times 10^{-3} \text{ mol}$

The moles of Cu^{2+} = $0.03210 \text{ g} \div 63.55 \text{ g/mol} = 5.05\underline{1}1 \times 10^{-4} \text{ mol}$

The ratio of water to Cu^{2+}, or $CuSO_4$ = $1.035 \times 10^{-3} \div 5.0511 \times 10^{-4} = 2.05$ or 2

The formula is thus $CuSO_4 \cdot 2H_2O$

4.115   For these calculations, the relative numbers of moles of gold and chlorine must be determined. These can be found from the masses of the two elements in the sample:

Total mass = mass of Au + mass of Cl = 328 mg

The mass of chlorine in the precipitated AgCl = the mass of chlorine in the compound of gold and chlorine. The mass of Cl in the 0.464 g of AgCl is

$$0.464 \text{ g AgCl} \times \frac{1 \text{ mol AgCl}}{143.32 \text{ g AgCl}} \times \frac{1 \text{ mol Cl}}{1 \text{ mol AgCl}} \times \frac{35.45 \text{ g Cl}}{1 \text{ mol Cl}}$$

$$= 0.114\underline{7}6 \text{ g Cl } (114.76 \text{ mg Cl})$$

(continued)

$$\text{Mass Percent Cl} = \frac{\text{mass Cl}}{\text{mass comp}} \times 100\% = \frac{114.76 \text{ mg}}{328 \text{ mg}} \times 100\% = 35.0\% \text{ Cl}$$

To find the empirical formula, convert each mass to moles:

Mass Au = 328 mg - 114.76 mg Cl = 213.23 mg Au

$$0.11476 \text{ g Cl} \times \frac{1 \text{ mol Cl}}{35.45 \text{ g Cl}} = 0.003238 \text{ mol Cl}$$

$$0.21323 \text{ g Au} \times \frac{1 \text{ mol Au}}{196.97 \text{ g Au}} = 0.001082 \text{ mol Au}$$

Divide both numbers of moles by the smaller number (0.001082) to find the integers:

Integer for Cl: 0.003238 mol ÷ 0.001082 mol = 2.99, or 3

Integer for Au: 0.001082 mol ÷ 0.001082 mol = 1.00, or 1

The empirical formula is $AuCl_3$.

4.117 From the equations $NH_3 + HCl \rightarrow NH_4Cl$ and $NaOH + HCl \rightarrow NaCl + H_2O$, we write

Mol NH_3 = mol $HCl(NH_3)$

Mol NaOH = mol HCl(NaOH)

We can calculate the mol NaOH and the sum [mol $HCl(NH_3)$ + mol HCl(NaOH)] from the titration data. Because the sum equals mol NH_3 plus mol NaOH, we can calculate the unknown mol of NH_3 from the difference: Mol NH_3 = sum - mol NaOH.

$$\text{Mol HCl (NaOH)} + \text{mol HCl (NH}_3) = 0.0463 \text{ L} \times \frac{0.0213 \text{ mol HCl}}{1.000 \text{ L}}$$

$$= 0.009862 \text{ mol HCl}$$

$$\text{Mol NaOH} = 0.0443 \text{ L} \times \frac{0.128 \text{ mol NaOH}}{1.000 \text{ L}} = 0.005670 \text{ mol NaOH}$$

Mol $HCl(NH_3)$ = 0.009862 mol - 0.005670 mol = 0.004192 mol

Mol NH_3 = mol $HCl(NH_3)$ = 0.004192 mol NH_3

(continued)

Because all the N in the $(NH_4)_2SO_4$ was liberated as and titrated as NH_3, the amount of N in the fertilizer is equal to the amount of N in the NH_3. Thus, the moles of NH_3 can be used to calculate the mass percentage of N in the fertilizer:

$$0.004192 \text{ mol } NH_3 \times \frac{1 \text{ mol N}}{1 \text{ mol } NH_3} \times \frac{14.01 \text{ g N}}{1 \text{ mol N}} = 0.05873 \text{ g N}$$

$$\text{Percent N} = \frac{\text{mass N}}{\text{mass fert.}} \times 100\% = \frac{0.05873 \text{ g N}}{0.608 \text{ g}} \times 100\% = 9.659 = 9.66\%$$

■ Solutions to Cumulative-Skills Problems

4.119 For this reaction, the formulas are listed first, followed by the molecular and net ionic equations, the names of the products, and the molecular equation for another reaction giving the same precipitate.

Lead(II) nitrate is $Pb(NO_3)_2$, and cesium sulfate is Cs_2SO_4.

Molecular equation: $Pb(NO_3)_2(aq) + Cs_2SO_4(aq) \rightarrow PbSO_4(s) + 2CsNO_3(aq)$

Net ionic equation: $Pb^{2+}(aq) + SO_4^{2-}(aq) \rightarrow PbSO_4(s)$

$PbSO_4$ is lead(II) sulfate, and $CsNO_3$ is cesium nitrate.

Molecular equation: $Pb(NO_3)_2(aq) + Na_2SO_4(aq) \rightarrow PbSO_4(s) + 2NaNO_3(aq)$

4.121 Net ionic equation: $2Br^-(aq) + Cl_2(g) \rightarrow 2Cl^-(aq) + Br_2(l)$

Molecular equation: $CaBr_2(aq) + Cl_2(g) \rightarrow CaCl_2(aq) + Br_2(l)$

Masses: 40.0 g   14.2 g   22.2 g   (40.0 + 14.2 - 22.2) g

Combining the three known masses gives the unknown mass of Br_2. Now, use a ratio of the known masses of $CaBr_2$ to Br_2 to convert pounds of Br_2 to grams of $CaBr_2$:

$$10.0 \text{ lb } Br_2 \times \frac{40.0 \text{ g } CaBr_2}{(40.0 + 14.2 - 22.2) \text{ g } Br_2} \times \frac{453.6 \text{ g}}{1 \text{ lb}}$$

$$= 5670 = 5.67 \times 10^3 \text{ g } CaBr_2$$

4.123 Molecular equation: $Hg(NO_3)_2 + H_2S \rightarrow HgS(s) + 2HNO_3(aq)$

Net ionic equation: $Hg^{2+} + H_2S \rightarrow HgS(s) + 2H^+(aq)$

The acid formed is nitric acid, a strong acid. The other product is mercury(II) sulfide.

Mass HNO_3 = (81.15 g + 8.52 g) - 58.16 g = 31.5$\underline{1}$ g

Mass of solution = 550.$\underline{0}$ g H_2O + 31.5$\underline{1}$ g HNO_3 = 581.$\underline{5}$1 = 581.5 g

4.125 Let the number of Fe^{3+} ions equals y; then the number of Fe^{2+} ions equals (7 - y). Since Fe_7S_8 is neutral, the number of positive charges must equal the number of negative charges. If the signs are omitted, then:

Total charge on both Fe^{2+} and Fe^{3+} = total charge on all eight sulfide ions

$3y + 2(7 - y) = 8 \times 2$

$y + 14 = 16$

$y = 2$

So the ratio of Fe^{2+} to Fe^{3+} = 5/2

4.127 Use the density, formula weight, and percentage to convert to molarity. Then, combine the 0.200 mol with mol/L to obtain the volume in liters.

$$\frac{0.807 \text{ g soln}}{1 \text{ mL}} \times \frac{0.940 \text{ g ethanol}}{1.00 \text{ g soln}} \times \frac{1 \text{ mol ethanol}}{46.07 \text{ g ethanol}} \times \frac{1000 \text{ mL}}{1 \text{ L}}$$

$$= \frac{16.\underline{4}65 \text{ mol ethanol}}{1 \text{ L ethanol}}$$

L ethanol = 0.200 mol ethanol $\times \dfrac{1 \text{ L ethanol}}{16.465 \text{ mol ethanol}}$ = 0.012$\underline{1}$46 = 0.0121 L

4.129 Convert the 2.183 g of Ag to mol AgI, which is chemically equivalent to moles of KI. Use that to calculate the molarity of the KI.

2.183 g AgI $\times \dfrac{1 \text{ mol AgI}}{234.77 \text{ g AgI}}$ = 9.29$\underline{8}$4 x 10^{-3} mol AgI (eq to mol KI)

Molarity = $\dfrac{9.2984 \times 10^{-3} \text{ mol KI}}{0.0100 \text{ L}}$ = 0.92$\underline{9}$8 = 0.930 M

4.131 Convert the 6.026 g of $BaSO_4$ to mol $BaSO_4$; then, from the equation, deduce that three mol $BaSO_4$ is equivalent to one mol $M_2(SO_4)_3$ and is equivalent to two mol of M. Use that with 1.200 g of the metal M to calculate the atomic weight of M.

$$6.026 \text{ g BaSO}_4 \times \frac{1 \text{ mol BaSO}_4}{233.40 \text{ g BaSO}_4} \times \frac{2 \text{ mol M}}{3 \text{ mol BaSO}_4} = 0.017212 \text{ mol M}$$

$$\text{Atomic wt of M in g/mol} = \frac{1.200 \text{ g M}}{0.017212 \text{ mol M}} = 69.719 \text{ g/mol (gallium)}$$

4.133 Use the density, formula weight, percentage, and volume to convert to mol H_3PO_4. Then, from the equation $P_4O_{10} + 6H_2O \rightarrow 4H_3PO_4$, deduce that four mol H_3PO_4 is equivalent to one mol of P_4O_{10}, and use that to convert to mol P_4O_{10}.

$$1500 \text{ mL} \times \frac{1.025 \text{ g soln}}{1 \text{ mL}} \times \frac{0.0500 \text{ g H}_3PO_4}{1 \text{ g soln}} \times \frac{1 \text{ mol H}_3PO_4}{98.00 \text{ g H}_3PO_4}$$

$$= 0.7844 \text{ mol H}_3PO_4$$

$$0.7844 \text{ mol H}_3PO_4 \times \frac{1 \text{ mol P}_4O_{10}}{4 \text{ mol H}_3PO_4} = 0.1961 \text{ mol P}_4O_{10}$$

$$\text{Mass P}_4O_{10} = 0.1961 \text{ mol P}_4O_{10} \times \frac{283.92 \text{ g P}_4O_{10}}{1 \text{ mol P}_4O_{10}}$$

$$= 55.677 = 55.7 \text{ g P}_4O_{10}$$

4.135 Convert the 0.1068 g of hydrogen to mol H_2; then, deduce from the equation that three mol H_2 is equivalent to two mol Al. Use the moles of Al to calculate mass Al and the percentage Al.

$$0.1068 \text{ g H}_2 \times \frac{1 \text{ mol H}_2}{2.016 \text{ g H}_2} \times \frac{2 \text{ mol Al}}{3 \text{ mol H}_2} = 0.0353174 \text{ mol Al}$$

$$\text{Percent Al} = \frac{0.0353174 \text{ mol Al} \times \frac{26.98 \text{ g Al}}{1 \text{ mol Al}}}{1.118 \text{ g alloy}} \times 100\% = 85.229 = 85.23\%$$

4.137 Use the formula weight of $Al_2(SO_4)_3$ to convert to mol $Al_2(SO_4)_3$. Then deduce from the equation that one mol $Al_2(SO_4)_3$ is equivalent to three mol H_2SO_4, and calculate the moles of H_2SO_4 needed. Combine density, percentage, and formula weight to obtain molarity of H_2SO_4. Then, combine molarity and moles to obtain volume.

$$37.4 \text{ g } Al_2(SO_4)_3 \times \frac{1 \text{ mol } Al_2(SO_4)_3}{342.17 \text{ g } Al_2(SO_4)_3} \times \frac{3 \text{ mol } H_2SO_4}{1 \text{ mol } Al_2(SO_4)_3}$$

$$= 0.32\underline{7}9 \text{ mol } H_2SO_4$$

$$\frac{1.104 \text{ g soln}}{1 \text{ mL}} \times \frac{15.0 \text{ g } H_2SO_4}{100 \text{ g soln}} \times \frac{1 \text{ mol } H_2SO_4}{98.09 \text{ g } H_2SO_4} \times \frac{1000 \text{ mL}}{1 \text{ L}}$$

$$= 1.6\underline{8}8 \text{ mol } H_2SO_4 \text{ /L}$$

$$0.3279 \text{ mol } H_2SO_4 \times \frac{1 \text{ L } H_2SO_4}{1.688 \text{ mol } H_2SO_4} = 0.19\underline{4}2 = 0.194 \text{ L } (194 \text{ mL})$$

4.139 The equations for the neutralization are $2HCl + Mg(OH)_2 \rightarrow MgCl_2 + 2H_2O$ and $3HCl + Al(OH)_3 \rightarrow AlCl_3 + 3H_2O$. Calculate the moles of HCl, and set equal to the total moles of hydroxide ion, OH^-.

$$0.0485 \text{ L HCl} \times \frac{0.187 \text{ mol HCl}}{1 \text{ L HCl}} = 0.00906\underline{9}5 \text{ mol HCl}$$

$$0.0090695 \text{ mol HCl} = 2 \text{ [mol } Mg(OH)_2] + 3 \text{ [mol } Al(OH)_3]$$

Rearrange the last equation, and solve for the moles of $Al(OH)_3$.

(Eq 1) mol $Al(OH)_3$ = 0.0030231 mol HCl - 2/3 [mol $Mg(OH)_2$]

The total mass of chloride salts is equal to the sum of the masses of $MgCl_2$ (molar mass = 95.21 g/mol) and $AlCl_3$ (molar mass = 133.33 g/mol).

$$[95.21 \text{ g/mol} \times \text{mol } MgCl_2] + [133.33 \text{ g/mol} \times \text{mol } AlCl_3] = 0.4200 \text{ g}$$

Since the mol $Mg(OH)_2$ equals mol $MgCl_2$, and mol $Al(OH)_3$ equals mol $AlCl_3$, you can substitute these quantities into the last equation and get

$$[95.21 \text{ g/mol} \times \text{mol } Mg(OH)_2] + [133.33 \text{ g/mol} \times \text{mol } Al(OH)_3] = 0.4200 \text{ g}$$

(continued)

Substitute Eq 1 into this equation for the mol of $Al(OH)_3$:

[95.21 g/mol x mol $Mg(OH)_2$]

+ [133.33 g/mol x (0.0030231 mol - 2/3 mol $Mg(OH)_2$)] = 0.4200 g

Rearrange the equation, and solve for the moles of $Mg(OH)_2$.

6.323 mol $Mg(OH)_2$ + 0.403070 = 0.4200

mol $Mg(OH)_2$ = 0.01693 ÷ 6.323 = 0.0026728 mol

Calculate the mass of $Mg(OH)_2$ in the antacid tablet (molar mass = 58.33 g/mol).

$$0.0026728 \text{ mol } Mg(OH)_2 \text{ x } \frac{58.33 \text{ g } Mg(OH)_2}{1 \text{ mol } Mg(OH)_2} = 0.15590 \text{ g } Mg(OH)_2$$

Use Eq 1 to find the moles and mass of $Al(OH)_3$ (molar mass 78.00 g/mol) in a similar fashion.

Mol $Al(OH)_3$ = 0.0030231 - 2/3[0.0026728 mol $Mg(OH)_2$] = 0.0012412 mol

$$0.0012412 \text{ mol } Al(OH)_3 \text{ x } \frac{78.00 \text{ g } Al(OH)_3}{1 \text{ mol } Al(OH)_3} = 0.096681 \text{ g } Al(OH)_3$$

The percent $Mg(OH)_2$ in the antacid is

Percent $Mg(OH)_2$ = [0.15590 g ÷ (0.15590 + 0.096681) g] x 100%

= 61.72 = 61.7%

5. THE GASEOUS STATE

■ Solutions to Exercises

Note on significant figures: If the final answer to a solution needs to be rounded off, it is given first with one nonsignificant figure, and the last significant figure is underlined. The final answer is then rounded to the correct number of significant figures. In multiple-step problems, intermediate answers are given with at least one nonsignificant figure; however, only the final answer has been rounded off.

5.1   First, the conversion to atm ($57 \text{ kPa} = 57 \times 10^3 \text{ Pa}$).

$$57 \times 10^3 \text{ Pa} \times \frac{1 \text{ atm}}{1.01325 \times 10^5 \text{ Pa}} = 0.5\underline{6}2 = 0.56 \text{ atm}$$

Next, convert to mmHg.

$$57 \times 10^3 \text{ Pa} \times \frac{760 \text{ mmHg}}{1.01325 \times 10^5 \text{ Pa}} = 4\underline{2}7.5 = 4.3 \times 10^2 \text{ mmHg}$$

5.2   Application of Boyle's law gives

$$V_f = V_i \times \frac{P_i}{P_f} = 20.0 \text{ L} \times \frac{1.00 \text{ atm}}{0.830 \text{ atm}} = 24.\underline{0}96 = 24.1 \text{ L}$$

5.3   First, convert the temperatures to the Kelvin scale.

$$T_i = (19 + 273) = 292 \text{ K}$$
$$T_f = (25 + 273) = 298 \text{ K}$$

Following is the data table

$V_i = 4.38 \text{ dm}^3$	$P_i = 101 \text{ kPa}$	$T_i = 292 \text{ K}$
$V_f = ?$	$P_f = 101 \text{ kPa}$	$T_f = 298 \text{ K}$

Apply Charles's law to obtain

$$V_f = V_i \times \frac{T_i}{T_f} = 4.38 \text{ dm}^3 \times \frac{298 \text{ K}}{292 \text{ K}} = 4.4\underline{7}0 = 4.47 \text{ dm}^3$$

5.4   First, convert the temperatures to kelvins.

$$T_i = (24 + 273) = 297 \text{ K}$$
$$T_f = (35 + 273) = 308 \text{ K}$$

Following is the data table

$V_i = 5.41 \text{ dm}^3$	$P_i = 101.5 \text{ kPa}$	$T_i = 297 \text{ K}$
$V_f = ?$	$P_f = 102.8 \text{ kPa}$	$T_f = 308 \text{ K}$

Apply both Boyle's law and Charles's law combined to get

$$V_f = V_i \times \frac{P_i}{P_f} \times \frac{T_f}{T_i} = 5.41 \text{dm}^3 \times \frac{101.5 \text{ kPa}}{102.8 \text{ kPa}} \times \frac{308 \text{ K}}{297 \text{ K}} = 5.5\underline{3}9 = 5.54 \text{ dm}^3$$

5.5   Use the ideal gas law, PV = nRT, and solve for n:

$$n = \frac{PV}{RT} = \left(\frac{V}{RT}\right)P$$

Note that everything in parentheses is constant. Therefore, you can write

$$n = \text{constant} \times P$$

Or, expressing this as a proportion, you get

$$n \propto P$$

5.6 First, convert the mass of O_2 to moles O_2 (molar mass 32.00 g/mol) and convert temperature to kelvins.

$$T = 23 + 273 = 296 \text{ K}$$

$$3.03 \text{ kg } O_2 \times \frac{1000 \text{ g}}{1 \text{ kg}} \times \frac{1 \text{ mol } O_2}{32.00 \text{ g } O_2} = 94.\underline{6}88 \text{ mol } O_2$$

Summarize the data in a table.

Variable	Value
P	?
V	50.0 L
T	296 K
n	94.$\underline{6}$88 mol

Solve the ideal gas equation for P, and substitute the data to get

$$P = \frac{nRT}{V} = \frac{(94.\underline{6}88)(0.08206 \text{ L} \cdot \text{atm/K} \cdot \text{mol})(296 \text{ K})}{50.0 \text{ L}} = 46.\underline{0}0 = 46.0 \text{ atm}$$

5.7 The data given are

Variable	Value
P	752 mmHg x $\frac{1 \text{ atm}}{760 \text{ mmHg}}$ = 0.98$\underline{9}$47 atm
V	1 L (exact number)
T	(21 + 273) = 294 K
n	?

Using the ideal gas law, solve for n, the moles of helium.

$$n = \frac{PV}{RT} = \frac{(0.98\underline{9}47 \text{ atm})(1 \text{ L})}{(0.08206 \text{ L} \cdot \text{atm/K} \cdot \text{mol})(294 \text{ K})} = 0.041\underline{0}1 \text{ mol}$$

Now convert mol He to grams

$$0.041\underline{0}1 \text{ mol He} \times \frac{4.00 \text{ g He}}{1 \text{ mol He}} = 0.16\underline{4}04 \text{ g He}$$

Therefore, the density of He at 21°C and 752 mmHg is 0.164 g/L.

The difference in mass between one liter of air and one liter of helium is

Mass air - mass He = 1.188 g - 0.16$\underline{4}$04 g = 1.02$\underline{3}$96 = 1.024 g difference

5.8 Tabulate the values of the variables:

Variable	Value
P	0.862 atm
V	1 L (exact number)
T	(25 + 273) = 298 K
n	?

From the ideal gas law, PV = nRT, you obtain

$$n = \frac{PV}{RT} = \frac{(0.862 \text{ atm})(1 \text{ L})}{(0.08206 \text{ L} \cdot \text{atm/K} \cdot \text{mol})(298 \text{ K})} = 0.03525 \text{ mol}$$

Dividing the mass of the vapor by moles gives you the mass per mole (the molar mass).

$$\text{Molar mass} = \frac{\text{grams vapor}}{\text{moles vapor}} = \frac{2.26 \text{ g}}{0.03525 \text{ mol}} = 64.114 \text{ g/mol}$$

Therefore, the molecular weight is 64.1 amu.

5.9 First, determine the number of moles of Cl_2 from the mass of HCl (molar mass 36.46 g/mol) and from the stoichiometry of the chemical equation:

$$9.41 \text{ g HCl} \times \frac{1 \text{ mol HCl}}{36.46 \text{ g HCl}} \times \frac{5 \text{ mol Cl}_2}{16 \text{ mol HCl}} = 0.080653 \text{ mol Cl}_2$$

Tabulate the values of the variables:

Variable	Value
P	787 mmHg $\times \dfrac{1 \text{ atm}}{760 \text{ mmHg}}$ = 1.0355 atm
T	(40 + 273) K = 313 K
n	0.080653 mol
V	?

Rearrange the ideal gas law to obtain V:

$$V = \frac{nRT}{P} = \frac{(0.080653 \text{ mol})(0.08206 \text{ L} \cdot \text{atm/K} \cdot \text{mol})(313 \text{ K})}{1.0355 \text{ atm}} = 2.001 = 2.00 \text{ L}$$

5.10   Each gas obeys the ideal gas law. In each case, convert grams to moles and substitute into the ideal gas law to determine the partial pressure of each.

$$1.031 \text{ g } O_2 \times \frac{1 \text{ mol } O_2}{32.00 \text{ g } O_2} = 0.0322188 \text{ mol } O_2$$

$$P = \frac{nRT}{V} = \frac{(0.0322188)(0.08206 \text{ L} \cdot \text{atm/K} \cdot \text{mol})(291 \text{ K})}{10.0 \text{ L}} = 0.076936 \text{ atm}$$

$$0.572 \text{ g } CO_2 \times \frac{1 \text{ mol } CO_2}{44.01 \text{ g } CO_2} = 0.012997 \text{ mol } CO_2$$

$$P = \frac{nRT}{V} = \frac{(0.012997)(0.08206 \text{ L} \cdot \text{atm/K} \cdot \text{mol})(291 \text{ K})}{10.0 \text{ L}} = 0.031036 \text{ atm}$$

The total pressure is equal to the sum of the partial pressures:

$$P = P_{O_2} + P_{CO_2} = 0.076936 + 0.031036 = 0.10797 = 0.1080 \text{ atm}$$

The mole fraction of oxygen in the mixture is

$$\text{Mole fraction } O_2 = \frac{P_{N_2}}{P} = \frac{0.07694 \text{ atm}}{0.10802 \text{ atm}} = 0.7122 = 0.712$$

5.11   Determine the number of moles of O_2 from the mass of $KClO_3$ and from the stoichiometry of the chemical reaction.

$$1.300 \text{ g } KClO_3 \times \frac{1 \text{ mol } KClO_3}{122.5 \text{ g } KClO_3} \times \frac{3 \text{ mol } O_2}{2 \text{ mol } KClO_3} = 0.0159184 \text{ mol } O_2$$

The vapor pressure of water at 23°C is 21.1 mmHg (Table 5.6). Find the partial pressure of O_2 using Dalton's law:

$$P = P_{O_2} + P_{H_2O}$$

$$P_{O_2} = P - P_{H_2O} = (745 - 21.1) \text{ mmHg} = 723.9 \text{ mmHg}$$

Solve for the volume using the ideal gas law.

(continued)

Variable	Value
P	723.9 mmHg x $\dfrac{1 \text{ atm}}{760 \text{ mmHg}}$ = 0.9525 atm
V	?
T	(23 + 273) = 296 K
n	0.0159184

From the ideal gas law, PV = nRT, you have

$$V = \frac{nRT}{P} = \frac{(0.0159184 \text{ mol})(0.08206 \text{ L} \cdot \text{atm/K} \cdot \text{mol})(296 \text{ K})}{0.9525 \text{ atm}}$$

$$= 0.4059 = 0.406 \text{ L}$$

5.12   The absolute temperature is (22 + 273) = 293 K. In SI units, the molar mass of carbon tetrachloride, CCl_4, is 153.8 x 10^{-3} kg/mol. Therefore,

$$u = \sqrt{\frac{3RT}{M}} = \sqrt{\frac{3 \times 8.31 \text{ kg} \cdot \text{m}^2/(\text{s}^2 \cdot \text{K} \cdot \text{mol}) \times 295 \text{ K}}{153.8 \times 10^{-3} \text{ kg/mol}}}$$

$$= 218.7 = 219 \text{ m/s}$$

5.13   Determine the rms molecular speed for N_2 at 455°C (728 K):

$$u = \sqrt{\frac{3RT}{M}} = \sqrt{\frac{3 \times 8.31 \text{kg} \cdot \text{m}^2/(\text{s}^2 \cdot \text{K} \cdot \text{mol}) \times 728 \text{ K}}{28.02 \times 10^{-3} \text{ kg/mol}}} = 804.81 \text{m/s}$$

After writing this equation with the same speed for H_2, square both sides and solve for T. The molar mass of H_2 in SI units is 2.016 x 10^{-3} kg/mol. Therefore,

$$T = \frac{u^2 M}{3R} = \frac{(804.81)^2 (2.016 \times 10^{-3} \text{ kg/mol})}{(3)(8.31 \text{kg} \cdot \text{m}^2/\text{s}^2 \cdot \text{K} \cdot \text{mol})} = 52.379 = 52.4 \text{ K}$$

Because the average kinetic energy of a molecule is proportional to only T, the temperature at which an H_2 molecule has the same average kinetic energy as an N_2 molecule at 455°C is exactly the same temperature, 455°C.

5.14　The two rates of effusion are inversely proportional to the square roots of their molar masses, so you can write

$$\frac{\text{Rate of effusion of } O_2}{\text{Rate of effusion of He}} = \sqrt{\frac{M_m(\text{He})}{M_m(O_2)}}$$

where $M_m(\text{He})$ is the molar mass of He (4.00 g/mol) and $M_m(O_2)$ is the molar mass of O_2 (32.00 g/mol). Substituting these values into the formula gives

$$\frac{\text{Rate of effusion of } O_2}{\text{Rate of effusion of He}} = \sqrt{\frac{4.00 \text{ g/mol}}{32.00 \text{ g/mol}}} = 0.35\underline{3}55$$

Rearranging gives

rate of effusion of O_2 = 0.35$\underline{3}$55 x rate of effusion of He.

Now, the problem states that the rate of effusion can be given in terms of volume of gas effused per second, so

$$\frac{\text{Volume of } O_2}{\text{time for } O_2} = 0.35\underline{3}55 \text{ x } \frac{\text{Volume of He}}{\text{time for He}}$$

Substituting in the values gives

$$\frac{10.0 \text{ mL}}{\text{time for } O_2} = 0.35\underline{3}55 \text{ x } \frac{10.0 \text{ mL}}{3.52 \text{ s}}$$

Rearranging gives

$$\text{time for } O_2 = \frac{3.52 \text{ s}}{0.35\underline{3}55} = 9.9\underline{5}6 = 9.96 \text{ s}$$

5.15　The problem states the rate of effusion is inversely proportional to the time it takes for a gas to effuse, so you can write

$$\frac{\text{Rate of effusion of } H_2}{\text{Rate of effusion of gas}} = \frac{\text{time for gas}}{\text{time for } H_2} = \sqrt{\frac{M_m(\text{gas})}{M_m(H_2)}} = 4.67$$

Rearranging and solving for $M_m(H_2)$ gives

$$M_m(\text{gas}) = (4.67)^2 \text{ x } M_m(H_2) = (4.67)^2 \text{ x } 2.016 \text{ g/mol} = 43.\underline{9}6 = 44.0 \text{ g/mol}$$

Thus, the molecular weight of the gas is 44.0 amu.

5.16   From Table 5.7, a = 5.570 L^2•atm/mol^2 and b = 0.06499 L/mol. Substitute these values into the van der Waals equation along with R = 0.08206 L•atm/K•mol, T = 273.2 K, n = 1.000 mol, and V = 22.41 L.

$$P = \frac{nRT}{(V - nb)} - \frac{n^2a}{V^2}$$

$$P = \frac{(1.00 \text{ mol})(0.08206 \text{ L} \cdot \text{atm/ K} \cdot \text{mol})(273.2 \text{ K})}{22.41 \text{L} - (1.00 \text{ mol})(0.06499 \text{ L/mol})} -$$

$$\frac{(1.00 \text{ mol})^2(5.570 \text{ L}^2 \cdot \text{atm/ mol}^2)}{(22.41 \text{L})^2}$$

$$= 1.00\underline{33} - 0.0110\underline{91} = 0.99\underline{221} = 0.992 \text{ atm}$$

Using the ideal gas law, P = 1.00\underline{04} atm (larger).

■ Answers to Review Questions

5.1   Pressure is the force exerted per unit area of surface. Force is further defined as mass multiplied by acceleration. The SI unit of mass is kg, of acceleration is m/s^2, and of area is m^2. Therefore, the SI unit of pressure (Pascal) is given by

$$\text{Pressure} = \frac{\text{force}}{\text{area}} = \frac{\text{mass x acceleration}}{\text{area}} = \frac{\text{kg x m/s}^2}{m^2} = \frac{\text{kg}}{m \cdot s^2} = \text{Pa}$$

5.2   A manometer is a device that measures the pressure of a gas in a vessel. The gas pressure in the flask is proportional to the difference in heights between the liquid levels in the manometer (Figure 5.4).

5.3   The general relationship between the pressure (P) and the height (h) of the liquid in a manometer is P = dgh. Therefore, the variables that determine the height of the liquid in a manometer are the density (d) of the liquid and the pressure of the gas being measured. The acceleration of gravity (g) is a constant, 9.81 m/s^2.

5.4   From Boyle's law, PV = constant. Because this is true for conditions P_i and V_i as well as conditions P_f and V_f, we can write

$$P_fV_f = P_iV_i = \text{constant}$$

(continued)

Dividing both sides of this equation by P_f gives

$$V_f = V_i \times \frac{P_i}{P_f}$$

5.5 A linear relationship between variables such as x and y is given by the mathematical relation

$$y = a + bx$$

The variable y is directly proportional to x only if a = 0.

5.6 First, find the equivalent of absolute zero on the Fahrenheit scale. Converting -273.15°C to degrees Fahrenheit, you obtain -459.67°F. Since the volume of a gas varies linearly with the temperature, you get the following linear relationship.

$$V = a + bt_F$$

where t_F is the temperature on the Fahrenheit scale. Since the volume of a gas is zero at absolute zero, you get

$$0 = a + b(-459.67), \text{ or } a = 459.67b$$

The equation can now be rewritten as

$$V = 459.67b + bt = b(459.67 + t_F) = bT_F$$

where T_F is the temperature in the new absolute scale based on the Fahrenheit scale. The relationship is

$$T_F = t_F + 459.67$$

5.7 From Charles's law, V = constant x T. Because this is true for conditions T_i and V_i as well as conditions T_f and V_f, we can write

$$\frac{V_f}{T_f} = \frac{V_i}{T_i} = \text{constant}$$

Multiplying both sides of the equation by T_f gives

$$V_f = V_i \times \frac{T_f}{T_i}$$

5.8   Avogadro's law states that equal volumes of any two gases at the same temperature and pressure contain the same number of molecules. The law of combining volumes states that the volumes of reactant gases at a given pressure and temperature are in ratios of small whole numbers. The combining-volume law may be explained from Avogadro's law using the reaction $N_2 + 3H_2 \rightarrow 2NH_3$ as follows: In Avogadro's terms, this equation says that Avogadro's number of N_2 molecules reacts with three times Avogadro's number of H_2 molecules to form two times Avogadro's number of NH_3 molecules. From Avogadro's law, it follows that one volume of N_2 reacts with three volumes of H_2 to form two volumes of NH_3. This result is true for all gas reactions.

5.9   The standard conditions are 0°C and one atm pressure (STP).

5.10   The molar gas volume, V_m, is the volume of one mole of gas at any given temperature and pressure. At standard conditions (STP), the molar gas volume equals 22.4 L.

5.11   Boyle's law ($V \propto 1/P$) and Charles's law ($V \propto T$) can be combined and expressed in a single statement: The volume occupied by a given amount of gas is proportional to the absolute temperature divided by the pressure. In equation form, this is

$$V = \text{constant} \times \frac{T}{P}$$

The constant is independent of temperature and pressure, but does depend on the amount of gas. For one mole, the constant will have a specific value, denoted as R. The molar volume, V_m, is

$$V_m = R \times \frac{T}{P}$$

Because V_m has the same value for all gases, we can write this equation for n moles of gas if we multiply both sides by n. This yields the equation

$$nV_m = \frac{nRT}{P}$$

Because V_m is the volume per mole, nV_m is the total volume V. Substituting gives

$$V = \frac{nRT}{P} \qquad \text{or} \qquad PV = nRT$$

5.12   The variables in the ideal gas law are P, V, n, and T. The SI units of these variables are pascals (P), cubic meters (V), moles (n), and kelvins (T).

5.13   Use the value of R from Table 5.5, and the conversion factor 1 atm = 760 mmHg. This gives

$$0.082058 \frac{L \cdot atm}{K \cdot mol} \times \frac{760 \, mmHg}{1 \, atm} = 62.3640 = 62.364 \frac{L \cdot mmHg}{K \cdot mol}$$

5.14   Six empirical gas laws can be obtained. They can be stated as follows:

P x V = constant   (T and n constant)

$\dfrac{P}{T}$ = constant   (V and n constant)

$\dfrac{P}{n}$ = constant   (T and V constant)

$\dfrac{V}{T}$ = constant   (P and n constant)

$\dfrac{V}{n}$ = constant   (P and T constant)

n x T = constant   (P and V constant)

5.15   The postulates and supporting evidence are the following:

(1) Gases are composed of molecules whose sizes are negligible compared with the distance between them.

(2) Molecules move randomly in straight lines in all directions and at various speeds.

(3) The forces of attraction or repulsion between two molecules (intermolecular forces) in a gas are very weak or negligible, except when they collide.

(4) When molecules collide with one another, the collisions are elastic.

(5) The average kinetic energy of a molecule is proportional to the absolute temperature.

One example of evidence that supports the kinetic theory of gases is Boyle's law. Constant temperature means the average molecular force from collision remains constant. If you increase the volume, you decrease the number of collisions per unit wall area, thus lowering the pressure in accordance with Boyle's law. Another example is Charles's law. If you raise the temperature, you increase the average molecular force from a collision with the wall, thus increasing the pressure. For the pressure to remain constant, it is necessary for the volume to increase so the frequency of collisions with the wall decreases. Thus, when you raise the temperature of a gas while keeping the pressure constant, the volume increases in accordance with Charles's law.

5.16   Boyle's law requires the temperature be constant. Postulate 5 of the kinetic theory holds that the average kinetic energy of a molecule is constant at constant temperature. Therefore, the average molecular force from collisions is constant. If we increase the volume of a gas, this decreases the number of molecules per unit volume and so decreases the frequency of collisions per unit wall area, causing the pressure to decrease in accordance with Boyle's law.

5.17   According to kinetic theory, the pressure of a gas results from the bombardment of container walls by molecules.

5.18   The rms speed of a molecule equals $(3RT/M_m)^{1/2}$, where M_m is the molar mass of the gas. The rms speed does not depend on the molar volume.

5.19   A gas appears to diffuse more slowly because it never travels very far in one direction before it collides with another molecule and moves in another direction. Thus, it must travel a very long, crooked path as the result of collisions.

5.20   Effusion is the process in which a gas flows through a small hole in a container. It results from the gas molecules encountering the hole by chance, rather than by colliding with the walls of the container. The faster the molecules move, the more likely they are to encounter the hole. Thus, the rate of effusion depends on the average molecular speed, which depends inversely on molecular mass.

5.21   The behavior of a gas begins to deviate significantly from that predicted by the ideal gas law at high pressures and relatively low temperatures.

5.22   The constant "a" is the proportionality constant in the van der Waals equation related to intermolecular forces The term "nb" represents the volume occupied by n moles of molecules.

■ Solutions to Practice Problems

Note on significant figures: If the final answer to a solution needs to be rounded off, it is given first with one nonsignificant figure, and the last significant figure is underlined. The final answer is then rounded to the correct number of significant figures. In multiple-step problems, intermediate answers are given with at least one nonsignificant figure; however, only the final answer has been rounded off.

5.31   Use the conversion factor 1 atm = 760 mmHg.

$$0.047 \text{ atm} \times \frac{760 \text{ mmHg}}{1 \text{ atm}} = 3\underline{5}.7 = 36 \text{ mmHg}$$

5.33 Using Boyle's law, solve for V_f of the neon gas at 1.292 atm pressure.

$$V_f = V_i \times \frac{P_i}{P_f} = 3.15\text{ L} \times \frac{0.951\text{ atm}}{1.292\text{ atm}} = 2.3\underline{1}8 = 2.32\text{ L}$$

5.35 Using Boyle's law, let V_f = volume at 0.974 atm (P_f), V_i = 50.0 L, and P_i = 19.8 atm.

$$V_f = V_i \times \frac{P_i}{P_f} = 50.0\text{ L} \times \frac{19.8\text{ atm}}{0.974\text{ atm}} = 10\underline{1}6.4 = 1.02 \times 10^3\text{ L}$$

5.37 Using Boyle's law, let P_i = pressure of 315 cm^3 of gas, and solve for it:

$$P_i = P_f \times \frac{V_f}{V_i} = 2.51\text{ kPa} \times \frac{0.0457\text{ cm}^3}{315\text{ cm}^3} = 3.6\underline{4}1 \times 10^{-4} = 3.64 \times 10^{-4}\text{ kPa}$$

5.39 Use Charles's law: T_i = 18°C + 273 = 291 K, and T_f = 0°C + 273 = 273 K.

$$V_f = V_i \times \frac{T_f}{T_i} = 3.92\text{ mL} \times \frac{273\text{ K}}{291\text{K}} = 3.6\underline{7}7 = 3.68\text{ mL}$$

5.41 Use Charles's law: T_i = 22°C + 273 = 295 K, and T_f = -197°C + 273 = 76 K.

$$V_f = V_i \times \frac{T_f}{T_i} = 2.54\text{ L} \times \frac{76\text{ K}}{295\text{ K}} = 0.6\underline{5}4 = 0.65\text{ L}$$

5.43 Use Charles's law: T_i = 25°C + 273 = 298 K_i, and V_f is the difference between the vessel's volume of 39.5 cm^3 and the 7.7 cm^3 of ethanol that is forced into the vessel.

$$T_f = T_i \times \frac{V_f}{V_i} = \frac{(39.5 - 7.7)\text{ cm}^3}{39.5\text{ cm}^3} = 23\underline{9}.9\text{ K }(-3\underline{3}.2\text{ or } -33°C)$$

5.45 Use the combined law: T_i = 31°C + 273 = 304 K, and T_f = 0°C + 273 = 273 K.

$$V_f = V_i = \times \frac{P_i}{P_f} \times \frac{T_f}{T_i} = 35.5\text{ mL} \times \frac{753\text{ mmHg}}{760\text{ mmHg}} \times \frac{273\text{ K}}{304\text{ K}} = 31.\underline{5}8 = 31.6\text{ mL}$$

5.47   The balanced equation is

$$4NH_3 + 5O_2 \rightarrow 4NO + 6H_2O$$

The ratio of moles of NH_3 to moles of NO = 4 to 4, or 1 to 1, so one volume of NH_3 will produce one volume of NO at the same temperature and pressure.

5.49   Solve the ideal gas law for V:

$$V = \frac{nRT}{P} = nRT\left(\frac{1}{P}\right)$$

If the temperature and number of moles are held constant, then the product nRT is constant, and volume is inversely proportional to pressure:

$$V = \text{constant} \times \frac{1}{P}$$

5.51   Calculate the moles of oxygen, and then solve the ideal gas law for P:

$$n = 91.3\,g \times \frac{1\text{ mol }O_2}{32.00\text{ g }O_2} = 2.8\underline{5}3\text{ mol }O_2$$

$$P = \frac{nRT}{V} = \frac{(2.8\underline{5}3\text{ mol})(0.08206\text{ L}\cdot\text{atm}/\text{K}\cdot\text{mol})(294\text{ K})}{8.58\text{ L}} = 8.0\underline{2}2 = 8.02\text{ atm}$$

5.53   Using the moles of chlorine, solve the ideal gas law for V:

$$V = \frac{nRT}{P} = \frac{(3.50\text{ mol})(0.08206\text{ L}\cdot\text{atm}/\text{K}\cdot\text{mol})(307\text{ K})}{4.00\text{ atm}} = 22.0\underline{4} = 22.0\text{ L}$$

5.55   Solve the ideal gas law for temperature in K, and convert to °C:

$$T = \frac{PV}{nR} = \frac{(3.50\text{ atm})(4.00\text{ L})}{(0.410\text{ mol})(0.08206\text{ L}\cdot\text{atm}/\text{K}\cdot\text{mol})} = 41\underline{6}.1 = 416\text{ K}$$

$$°C = 416 - 273 = 143°C$$

5.57   Because density equals mass per unit volume, calculating the mass of 1 L (exact number) of gas will give the density of the gas. Start from the ideal gas law, and calculate n; then convert the moles of gas to grams using the molar mass.

$$n = \frac{PV}{RT} = \frac{(751/760 \text{ atm}) (1 \text{ L})}{(0.08206 \text{ L} \cdot \text{atm/K} \cdot \text{mol}) (304 \text{ K})} = 0.03961 \text{ mol}$$

$$0.03961 \text{ mol} \times \frac{17.03 \text{ g}}{1 \text{ mol}} = 0.67458 = 0.675 \text{ g}$$

Therefore, the density of NH_3 at 31°C is 0.675 g/L.

5.59   Calculate the mass of 1 L (exact number) using the ideal gas law; convert to mass.

$$n = \frac{PV}{RT} = \frac{(0.897 \text{ atm}) (1 \text{ L})}{(0.08206 \text{ L} \cdot \text{atm/K} \cdot \text{mol}) (297 \text{ K})} = 0.036804 \text{ mol}$$

$$0.036804 \text{ mol} \times \frac{58.12 \text{ g}}{1 \text{ mol}} = 2.139 = 2.14 \text{ g}$$

Therefore, the density of C_4H_{10} is 2.14 g/L.

5.61   The ideal gas law gives n moles, which then are divided into the mass of 1.585 g for molar mass.

$$n = \frac{PV}{RT} = \frac{(753/760 \text{ atm}) (1 \text{ L})}{(0.08206 \text{ L} \cdot \text{atm/K} \cdot \text{mol}) (363 \text{ K})} = 0.033262 \text{ mol}$$

$$\text{Molar mass} = \frac{1.585 \text{ g}}{0.033262 \text{ mol}} = 47.651 \text{ g/mol}$$

The molecular weight is 47.7 amu.

5.63   The moles in 250 mL (0.250 L) of the compound are obtained from the ideal gas law. The 2.56 g mass of the gas then is divided by the moles to obtain the molar mass and molecular weight.

$$n = \frac{PV}{RT} = \frac{(786/760 \text{ atm}) (0.250 \text{ L})}{(0.08206 \text{ L} \cdot \text{atm/K} \cdot \text{mol}) (394 \text{ K})} = 0.007996 \text{ mol}$$

$$\text{Molar mass} = \frac{2.56 \text{ g}}{0.007996 \text{ mol}} = 3.201 \times 10^2 = 3.20 \times 10^2 \text{ g/mol}$$

The molecular weight is 3.20×10^2 amu.

5.65   For a gas at a given temperature and pressure, the density depends on molecular weight (or for a mixture, the average molecular weight). Thus, at the same temperature and pressure, the density of NH_4Cl gas would be greater than that of a mixture of NH_3 and HCl because the average molecular weight of NH_3 and HCl would be lower than that of NH_4Cl.

5.67   The 0.050 mol CaC_2 will form 0.025 mol C_2H_2. The volume is found from the ideal gas law:

$$Vol = \frac{(0.050 \text{ mol})(0.08206 \text{ L} \cdot \text{atm/K} \cdot \text{mol})(299 \text{ K})}{684/760 \text{ atm}} = 1.3\underline{6} = 1.4 \text{ L}$$

5.69   Use the equation to obtain the moles of CO_2 and then the ideal gas law to obtain the volume.

$$2LiOH(s) + CO_2(g) \rightarrow Li_2CO_3(aq) + H_2O(l)$$

$$327 \text{ g LiOH} \times \frac{1 \text{ mol LiOH}}{23.95 \text{ g LiOH}} \times \frac{1 \text{ mol CO}_2}{2 \text{ mol LiOH}} = 6.8\underline{2}67 \text{ mol CO}_2$$

$$V = \frac{(6.8\underline{2}67 \text{ mol})(0.08206 \text{ L} \cdot \text{atm/K} \cdot \text{mol})(294 \text{ K})}{781/760 \text{ atm}} = 16\underline{0}.2 = 160. \text{ L}$$

5.71   Use the equation to obtain the moles of ammonia and then the ideal gas law to obtain the volume.

$$2NH_3(g) + CO_2(g) \rightarrow NH_2CONH_2(aq) + H_2O(l)$$

$$908 \text{ g urea} \times \frac{1 \text{ mol urea}}{60.06 \text{ g urea}} \times \frac{2 \text{ mol NH}_3}{1 \text{ mol urea}} = 30.\underline{2}3 \text{ mol NH}_3$$

$$V = \frac{(30.23 \text{ mol})(0.08206 \text{ L} \cdot \text{atm/K} \cdot \text{mol})(298 \text{ K})}{3.00 \text{ atm}} = 24\underline{6}.4 = 246 \text{ L}$$

5.73 Use the equation to obtain the moles of ammonia and then the ideal gas law to obtain the volume.

$$2NH_3(g) + H_2SO_4(aq) \rightarrow (NH_4)_2SO_4(aq)$$

$$150.0 \text{ g } (NH_4)_2SO_4 \times \frac{1 \text{ mol } (NH_4)_2SO_4}{132.1 \text{ g } (NH_4)_2SO_4} \times \frac{2 \text{ mol } NH_3}{1 \text{ mol } (NH_4)_2SO_4}$$

$$= 2.2710 \text{ mol } NH_3$$

$$V = \frac{nRT}{P} = \frac{(2.2710 \text{ mol})(0.08206 \text{ L} \cdot \text{atm/K} \cdot \text{mol})(288 \text{ K})}{1.15 \text{ atm}} = 46.67 = 46.7 \text{ L}$$

5.75 Calculate the partial pressure of each gas; then add the pressures since the total pressure is equal to the sum of the partial pressures:

$$P(He) = \frac{(0.0200 \text{ mol})(0.08206 \text{ L} \cdot \text{atm/K} \cdot \text{mol})(283 \text{ K})}{2.50 \text{ L}} = 0.18578 \text{ atm}$$

$$P(He) = \frac{(0.0100 \text{ mol})(0.08206 \text{ L} \cdot \text{atm/K} \cdot \text{mol})(283 \text{ K})}{2.50 \text{ L}} = 0.09289 \text{ atm}$$

The total pressure = 0.18587 = 0.09289 = 0.27867 = 0.279

5.77 Convert mass of O_2 and mass of He to moles. Use the ideal gas law to calculate the partial pressures, and then add to obtain the total pressures.

$$0.00103 \text{ g } O_2 \times \frac{1 \text{ mol } O_2}{32.00 \text{ g } O_2} = 3.219 \times 10^{-5} \text{ mol } O_2$$

$$0.00056 \text{ g He} \times \frac{1 \text{ mol He}}{4.00 \text{ g He}} = 1.40 \times 10^{-4} \text{ mol}$$

$$P = \frac{nRT}{V} = \frac{(3.219 \times 10^{-5} \text{ mol})(0.08206 \text{ L} \cdot \text{atm/K} \cdot \text{mol})(288 \text{ K})}{0.2000 \text{ L}}$$

$$= 0.003804 \text{ atm } O_2$$

$$P = \frac{nRT}{V} = \frac{(1.40 \times 10^{-4} \text{ mol})(0.08206 \text{ L} \cdot \text{atm/K} \cdot \text{mol})(288 \text{ K})}{0.2000 \text{ L}} = 0.01654 \text{ atm He}$$

$$P = P_{O_2} + P_{He} = 0.003804 \text{ atm} + 0.01654 \text{ atm} = 0.0203 = 0.020 \text{ atm}$$

5.79   For each gas, P(gas) = P x (mole fraction of gas).

$P(H_2)$ = 760 mmHg x 0.250   = 190.0   = 190 mmHg

$P(CO_2)$ = 760 mmHg x 0.650   = 494.0   = 494 mmHg

$P(HCl)$ = 760 mmHg x 0.054   = 41.04   = 41 mmHg

$P(HF)$ = 760 mmHg x 0.028   = 21.28   = 21 mmHg

$P(SO_2)$ = 760 mmHg x 0.017   = 12.92   = 13 mmHg

$P(H_2S)$ = 760 mmHg x 0.001   = 0.76   = 0.8 mmHg

5.81   The total pressure is the sum of the partial pressures of CO and H_2O, so

$$P_{CO} = P - P_{water} = 689 \text{ mmHg} - 23.8 \text{ mmHg} = 665.2 \text{ mmHg}$$

$$n_{CO} = \frac{P_{CO}V}{RT} = \frac{(665/760 \text{ atm})(3.85 \text{ L})}{(0.08206 \text{ L} \cdot \text{atm}/\text{K} \cdot \text{mol})(298 \text{ K})} = 0.1378 \text{ mol CO}$$

$$0.1378 \text{ g CO} \times \frac{1 \text{ mol HCOOH}}{1 \text{ mol CO}} \times \frac{46.03 \text{ g HCOOH}}{1 \text{ mol HCOOH}}$$

$$= 6.342 = 6.34 \text{ g HCOOH}$$

5.83   Substitute 298 K (25°C) and 398 K (125°C) into Maxwell's distribution:

$$u_{25} = \sqrt{\frac{3RT}{M_m}} = \sqrt{\frac{3 \times 8.31 \text{ kg} \cdot \text{m}^2/(\text{s}^2 \cdot \text{K} \cdot \text{mol}) \times 298 \text{ K}}{28.02 \times 10^{-3} \text{ kg/mol}}}$$

$$= 5.149 \times 10^2 = 5.15 \times 10^2 \text{ m/s}$$

$$u_{125} = \sqrt{\frac{3RT}{M_m}} = \sqrt{\frac{3 \times 8.31 \text{ kg} \cdot \text{m}^2/(\text{s}^2 \cdot \text{K} \cdot \text{mol}) \times 398 \text{ K}}{28.02 \times 10^{-3} \text{ kg/mol}}}$$

$$= 5.9507 \times 10^2 = 5.95 \times 10^2 \text{ m/s}$$

Graph as in Figure 5.23.

5.85   Substitute 330 K (57°C) into Maxwell's distribution:

$$u_{330\ K} = \sqrt{\frac{3RT}{M_m}} = \sqrt{\frac{3 \times 8.31\ kg \cdot m^2/(s^2 \cdot K \cdot mol) \times 330\ K}{352 \times 10^{-3}\ kg/mol}}$$

$$= 1.5\underline{2}8 \times 10^2 = 1.53 \times 10^2\ m/s$$

5.87   Because $u(CO_2) = u(H_2)$, we can equate the two right-hand sides of the Maxwell distributions:

$$\sqrt{\frac{3RT(CO_2)}{M_m(CO_2)}} = \sqrt{\frac{3RT(H_2)}{M_m(H_2)}}$$

Squaring both sides, rearranging to solve for $T(CO_2)$, and substituting numerical values, we have

$$T(CO_2) = T(H_2) \times \frac{M_m(CO_2)}{M_m(H_2)} = 298\ K \times \frac{44.01\ g/mol}{2.016\ g/mol} = 65\underline{0}5.4\ K$$

Thus the temperature is

$$T = 65\underline{0}5.4 - 273 = 62\underline{3}2 = 6.23 \times 10^3\ ^\circ C$$

5.89   Because the ratio is the same at any temperature, $T(N_2) = T(O_2)$. Write a ratio of two Maxwell distributions after omitting $T(O_2)$ and $T(H_2)$ in each distribution:

$$\frac{u_{N_2}}{u_{O_2}} = \frac{\sqrt{\dfrac{3RT(N_2)}{M_m}}}{\sqrt{\dfrac{3RT(O_2)}{M_m}}} = \frac{\sqrt{\dfrac{3 \times 8.31\ kg \cdot m^2/(s^2 \cdot K \cdot mol)}{28.02 \times 10^{-3}\ kg/mol}}}{\sqrt{\dfrac{3 \times 8.31\ kg \cdot m^2/(s^2 \cdot K \cdot mol)}{32.00 \times 10^{-3}\ kg/mol}}}$$

$$\frac{u_{N_2}}{u_{O_2}} = \sqrt{\frac{32.00}{28.02}} = \frac{1.06\underline{8}6}{1} = \frac{1.069}{1}$$

5.91   Because the ratio is the same at any temperature, $T(H_2) = T(I_2)$. A ratio of two Maxwell distributions can be written as in the previous two problems, but this also can be simplified by canceling the 3 x 8.31 terms and rearranging the denominators to give

$$\frac{u_{H_2}}{u_{I_2}} = \frac{\sqrt{M_m(I_2)}}{\sqrt{M_m(H_2)}} = \sqrt{\frac{253.8}{2.016}} = \frac{11.2205}{1} = \frac{11.22}{1}$$

Because hydrogen diffuses 11.22 times as fast as iodine, the time it would take would be 1/11.22 of the time required for iodine:

$$t(H_2) = 39 \text{ s} \times (1/11.22) = 3.\underline{4}7 = 3.5 \text{ s}$$

5.93   Because the diffusion occurs at the same temperature, $T(gas) = T(Ar)$. A ratio of two Maxwell distributions can be written, but it can be simplified by canceling the 3 x 8.31 terms and rearranging the denominators. To simplify the definition of the rates, we assume the time is one second, and define the rate(Ar) as 9.23 mL/1 s and the rate(gas) as 4.83 mL/1 s. Then we write a ratio of two Maxwell distributions:

$$\frac{u_{gas}}{u_{Ar}} = \frac{4.83 \text{ mL} / 1 \text{ s}}{9.23 \text{ mL} / 1 \text{ s}} = \frac{\sqrt{M_m(Ar)}}{\sqrt{M_m(gas)}} = \frac{\sqrt{39.95}}{\sqrt{M_m(gas)}}$$

Squaring both sides gives

$$0.27\underline{3}83 = \frac{39.95}{M_m(gas)}$$

Solving for the molar mass gives

$$M_m(gas) = \frac{39.95}{0.27383} = 14\underline{5}.8 \text{ g/mol; molecular weight} = 146 \text{ amu}$$

5.95   Solving the van der Waals equation for n = 1 and T = 355.2 K for P gives

$$P = \frac{RT}{(V - b)} - \frac{a}{V^2} = \frac{(0.08206 \text{ L} \bullet \text{atm/K} \bullet \text{mol})(355.2 \text{ K})}{(30.00 \text{ L} - 0.08710 \text{ L})} - \frac{12.56 \text{ L}^2 \bullet \text{atm}}{(30.00 \text{ L})^2}$$

$$P = 0.9744\underline{1}9 - 0.01395\underline{5}5 = 0.960\underline{4}6 = 0.9605 \text{ atm}$$

$$P(\text{ideal gas law}) = 0.9715\underline{9} \text{ atm}$$

5.97   To calculate a/V^2 in the van der Waals equation, we obtain V from the ideal gas law at 1.00 atm:

$$V = \frac{RT}{P} = \frac{(0.08206 \text{ L} \cdot \text{atm/K} \cdot \text{mol})(273 \text{ K})}{1.00 \text{ atm}} = 22.40 \text{ L}$$

$$\frac{a}{V^2} = \frac{5.570}{(22.4 \text{ L})^2} = 1.1\underline{1}0 \times 10^{-2}$$

At 1.00 atm, V = 22.4 L, and $a/V^2 = 1.110 \times 10^{-2}$. Substituting into the van der Waals equation:

$$V = \frac{RT}{P + \dfrac{a}{V^2}} + b = \frac{(0.08206 \text{ L} \cdot \text{atm/K} \cdot \text{mol})(273 \text{ K})}{(1.00 + 1.110 \times 10^{-2}) \text{atm}} + 0.06499$$

$$= 22.\underline{22} = 22.2 \text{ L}$$

At 10.0 atm, the van der Waals equation gives 2.08 L. The ideal gas law gives 22.4 L for 1.00 atm and 2.24 L for 10.0 atm.

■ Solutions to General Problems

5.99   Calculate the mass of one cm^2 of the 20.5 m of water above the air in the glass. The volume is the product of the area of one cm^2 and the height of 20.5×10^2 cm (20.5 m) of water. The density of 1.00 g/cm^3 must be used to convert volume to mass:

$$m = d \times V$$

$$m = 1.00 \text{ g/cm}^3 \times (1.00 \text{ cm}^2 \times 20.5 \times 10^2 \text{ cm}) = 2.05 \times 10^3 \text{ g, or } 2.05 \text{ kg}$$

The pressure exerted on an object at the bottom of the column of water is

$$P = \frac{\text{force}}{\text{area}} = \frac{(m)(g)}{\text{area}} = \frac{(2.05 \text{ kg})(9.807 \text{ m/s}^2)}{(1.00 \text{ cm}^2)\left(\dfrac{10^{-2} \text{ m}}{1 \text{cm}}\right)^2} = 2.01 \times 10^5 \text{ kg/ms}^2$$

$$= 2.01 \times 10^5 \text{ Pa}$$

(continued)

The total pressure on the air in the tumbler equals the barometric pressure and the water pressure:

$$P = 1.00 \times 10^2 \text{ kPa} + 2.01 \times 10^2 \text{ kPa} = 3.01 \times 10^2 \text{ kPa}$$

Multiply the initial volume by a factor accounting for the change in pressure to find V_f:

$$V_f = V_i \times \frac{P_i}{P_f} = 243 \text{ cm}^3 \times \left(\frac{1.00 \times 10^2 \text{ kPa}}{3.01 \times 10^2 \text{ kPa}}\right) = 80.\underline{7}3 = 80.7 \text{ cm}^3$$

5.101   Use the combined gas law and solve for V_f:

$$V_f = V_i \times \frac{P_i}{P_f} \times \frac{T_f}{T_i} = 201 \text{ mL} \times \frac{738 \text{ mmHg}}{760 \text{ mmHg}} \times \frac{273 \text{ K}}{294 \text{ K}}$$

$$= 18\underline{1}.24 = 181 \text{ mL}$$

5.103   Use the combined gas law and solve for V_f:

$$V_f = V_i \times \frac{P_i}{P_f} \times \frac{T_f}{T_i} = 5.0 \text{ dm}^3 \times \frac{100.0 \text{ kPa}}{79.0 \text{ kPa}} \times \frac{293 \text{ K}}{287 \text{ K}} = 6.\underline{4}6 = 6.5 \text{ dm}^3$$

5.105   Use the ideal gas law to calculate the moles of helium, and combine this with Avogadro's number to obtain the number of helium atoms:

$$n = \frac{PV}{RT} = \frac{(765/760 \text{ atm})(0.01205 \text{ L})}{(0.08206 \text{ L} \bullet \text{atm/ K} \bullet \text{mol})(296 \text{ K})} = 4.9\underline{9}3 \times 10^{-4} \text{ mol}$$

$$4.993 \times 10^{-4} \text{ mol He} \times \frac{6.022 \times 10^{23} \text{ He}^{2+} \text{ ions}}{1 \text{ mol He}} \times \frac{1 \text{ atom}}{1 \text{ He}^{2+} \text{ ion}} = 3.0\underline{0}67 \times 10^{20}$$

$$= 3.01 \times 10^{20} \text{ atoms}$$

5.107   Calculate the molar mass, M_m, by dividing the mass of one L of air by the moles of the gas from the ideal gas equation:

$$M_m = \frac{\text{mass}}{n} = 1.2929 \text{ g air} \times \frac{(0.082058 \text{ L} \bullet \text{atm /K} \bullet \text{mol}) (273.15 \text{ K})}{(1 \text{atm})(1 \text{L})}$$

$$= 28.97\underline{9}2 \text{ g/mol} = 28.979 \text{ g/ mol (amu)}$$

5.109 Use the ideal gas law to calculate the moles of CO_2. Then convert to mass of LiOH.

$$n = \frac{PV}{RT} = \frac{(1.00 \text{ atm})(5.8 \times 10^2 \text{ L})}{(0.08206 \text{ L} \cdot \text{atm/K} \cdot \text{mol})(273 \text{ K})} = 25.89 \text{ mol } CO_2$$

$$25.89 \text{ mol } CO_2 \times \frac{2 \text{ mol LiOH}}{1 \text{ mol } CO_2} \times \frac{23.95 \text{ g LiOH}}{1 \text{ mol LiOH}} = 1240 = 1.2 \times 10^3 \text{ g LiOH}$$

5.111 Convert mass to moles of $KClO_3$, and then use the equation below to convert to moles of O_2. Use the ideal gas law to convert moles of O_2 to pressure at 25°C (298 K).

$$2KClO_3(s) \rightarrow 2KCl(s) + 3O_2(g)$$

$$170.0 \text{ g } KClO_3 \times \frac{1 \text{ mol } KClO_3}{122.55 \text{ g } KClO_3} \times \frac{3 \text{ mol } O_2}{2 \text{ mol } KClO_3} = 2.080 \text{ mol } O_2$$

$$P = \frac{nRT}{V} = \frac{(2.080 \text{ mol})(0.08206 \text{ L} \cdot \text{atm/K} \cdot \text{mol})(298 \text{ K})}{2.50 \text{ L}}$$

$$= 20.353 = 20.4 \text{ atm } O_2$$

5.113 Find the number of moles of CO_2 first. Then convert this to moles HCl and molarity of HCl.

$$n = \frac{(727 / 760 \text{ atm}) \times 0.141 \text{ L}}{0.08206 \text{ L} \cdot \text{atm/K} \cdot \text{mol} \times 300 \text{ K}} = 0.0054788 \text{ mol}$$

mol HCl = 0.0054788 mol CO_2 × (2 mol HCl / 1 mol CO_2) = 0.010957 mol

M HCl = 0.010957 mol HCl ÷ 0.0249 L HCl = 0.4400 = 0.440 mol/L HCl

5.115 The number of moles of carbon dioxide is:

$$n = \frac{646 / 760 \text{ atm} \times 0.1500 \text{ L}}{0.08206 \text{ L} \cdot \text{atm/K} \cdot \text{mol} \times 300 \text{ K}} = 0.005179 \text{ mol}$$

The number of moles of molecular acid used is:

0.1250 mol/L × 0.04141 L = 0.005176 mol acid

Thus, the acid is H_2SO_4 since one mole of H_2SO_4 reacts to form one mole of CO_2

5.117 Use Maxwell's distribution to calculate the temperature in kelvins; then convert to °C.

$$T = \frac{u^2 M_m}{3R} = \frac{(0.605 \times 10^3 \text{ m/s})^2 \, (17.03 \times 10^{-3} \text{ kg/mol})}{3 \, (8.31 \text{ kg} \cdot \text{m}^2/\text{s}^2 \cdot \text{K} \cdot \text{mol})}$$

$$= 25\underline{0}.0 = 200. \text{ K } (-23°C)$$

5.119 Calculate the ratio of the root-mean-square molecular speeds, which is the same as the ratio of the rates of effusion:

$$\frac{u \text{ of } U(235)F_6}{u \text{ of } U(238)F_6} = \frac{\sqrt{M_m[U(238)F_6]}}{\sqrt{M_m[U(235)F_6]}} = \sqrt{\frac{352.04 \text{ g/mol}}{349.03 \text{ g/mol}}}$$

$$= \frac{1.004\,3\underline{0}2}{1} = \frac{1.0043}{1}$$

5.121 First, calculate the apparent molar masses at each pressure using the ideal gas law. Only the calculation of the apparent molar mass for 0.2500 atm will be shown; the other values will be summarized in a table.

$$n = \frac{PV}{RT} = \frac{(0.2500 \text{ atm})(3.1908 \text{ L})}{(0.082057 \text{ L} \cdot \text{atm/K} \cdot \text{mol})(273.15 \text{ K})} = 3.55\underline{8}96 \times 10^{-2} \text{ mol}$$

$$\text{Apparent molar mass} = \frac{1.000 \text{ g}}{3.55896 \times 10^{-2} \text{ mol}} = 28.0\underline{9}8 = 28.10 \text{ g/mol}$$

The following table summarizes the apparent molar masses calculated as above for all P's; these data are plotted in the graph to the right of the table.

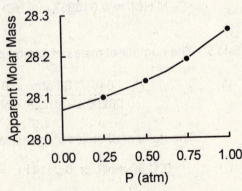

P (atm)	App. Molar Mass (g/mol)
0.2500	28.10
0.5000	28.14
0.7500	28.19
1.0000	28.26

Extrapolation back to P = 0 gives 28.07 g/mol for the molar mass of the unknown gas (CO).

5.123 Use $CO + 1/2O_2 \rightarrow CO_2$, instead of $2CO$. First, find the moles of CO and O_2 by using the ideal gas law.

$$n_{CO} = \frac{PV}{RT} = \frac{(0.500 \text{ atm}) (2.00 \text{ L})}{(0.08206 \text{ L} \cdot \text{atm/K} \cdot \text{mol}) (300 \text{ K})} = 0.040\underline{6}2 \text{ mol}$$

$$n_{O_2} = \frac{PV}{RT} = \frac{(1.00 \text{ atm}) (1.00 \text{ L})}{(0.08206 \text{ L} \cdot \text{atm/K} \cdot \text{mol}) (300 \text{ K})} = 0.04\underline{0}62 \text{ mol}$$

There are equal amounts of CO and O_2, but (from the equation) only half as many moles of O_2 as CO are required for the reaction. Therefore, when 0.04062 moles of CO have been consumed, only 0.04062/2 moles of O_2 will have been used up. Then 0.04062/2 mol O_2 will remain, and 0.04062 mol CO_2 will have been produced. At the end,

$$n_{CO} = 0 \text{ mol}; \quad n_{O_2} = 0.0202 \text{ mol}; \quad \text{and } n_{CO_2} = 0.04062 \text{ mol}$$

However, the total volume with the valve open is 3.00 L, so the partial pressures of O_2 and CO_2 must be calculated from the ideal gas law for each:

$$\frac{nRT}{V} = \frac{(0.0203 \text{ mol } O_2)(0.08206 \text{ L} \cdot \text{atm/K} \cdot \text{mol})(300 \text{ K})}{3.00 \text{ L}}$$

$$= 0.16\underline{6}58 = 0.167 \text{ atm } O_2$$

$$\frac{nRT}{V} = \frac{(0.04062 \text{ mol } CO_2)(0.08206 \text{ L} \cdot \text{atm/K} \cdot \text{mol})(300 \text{ K})}{3.00 \text{ L}}$$

$$= 0.33\underline{3}32 = 0.333 \text{ atm } CO_2$$

■ Solutions to Cumulative-Skills Problems

5.125 Assume a 100.0-g sample, giving 85.2 g CH_4 and 14.8 g C_2H_6. Convert each to moles:

$$85.2 \text{ g } CH_4 \times \frac{1 \text{ mol } CH_4}{16.04 \text{ g } CH_4} = 5.3\underline{1}1 \text{ mol } CH_4$$

$$14.8 \text{ g } C_2H_6 \times \frac{1 \text{ mol } CH_4}{16.04 \text{ g } CH_4} = 0.49\underline{2}1 \text{ mol } C_2H_6$$

(continued)

$$V_{CH_4} = \frac{(5.311 \text{ mol})(0.08206 \text{ L} \cdot \text{atm/K} \cdot \text{mol})(291 \text{ K})}{(771/760) \text{ atm}} = 12\underline{5}.03 \text{ L}$$

$$V_{C_2H_6} = \frac{(0.4921 \text{ mol})(0.08206 \text{ L} \cdot \text{atm/K} \cdot \text{mol})(291 \text{ K})}{(771/760) \text{ atm}} = 11.\underline{5}8 \text{ L}$$

The density is calculated as follows:

$$d = \frac{85.2 \text{ g CH}_4 + 14.8 \text{ g C}_2\text{H}_6}{(125.03 + 11.5\underline{8}) \text{ L}} = 0.731\underline{9} = 0.732 \text{ g/L}$$

5.127   First, subtract the height of mercury equivalent to the 25.00 cm (250 mm) of water inside the tube from 771 mmHg to get P_{gas}. Then, subtract the vapor pressure of water, 18.7 mmHg, from P_{gas} to get P_{O_2}

$$h_{Hg} = \frac{(h_W)(d_W)}{d_{Hg}} = \frac{250 \text{ mm} \times 0.99987 \text{ g/cm}^3}{13.596 \text{ g/cm}^3} = 18.\underline{3}8 \text{ mmHg}$$

$$P_{gas} = P - P_{25 \text{ cm water}} = 771 \text{ mmHg} - 18.38 \text{ mmHg} = 75\underline{2}.62 \text{ mmHg}$$

$$P_{O_2} = 752.62 \text{ mmHg} - 18.7 \text{ mmHg} = 73\underline{3}.92 \text{ mmHg}$$

$$n = \frac{PV}{RT} = \frac{(733.92/760 \text{ atm})(0.0310 \text{ L})}{(0.08206 \text{ L} \cdot \text{atm/K} \cdot \text{mol})(294 \text{ K})} = 0.001240\underline{8} \text{ mol O}_2$$

$$\text{Mass} = (2 \times 0.0012408) \text{ mol Na}_2\text{O}_2 \times \frac{77.98 \text{ g Na}_2\text{O}_2}{1 \text{ mol Na}_2\text{O}_2}$$

$$= 0.193\underline{5} = 0.194 \text{ g Na}_2\text{O}_2$$

5.129   First find the moles of CO_2:

$$n = \frac{PV}{RT} = \frac{(785/760 \text{ atm})(1.94 \text{ L})}{(0.08206 \text{ L} \cdot \text{atm/K} \cdot \text{mol})(298 \text{ K})} = 0.081\underline{9}4 \text{ mol C}_2\text{O}$$

Set up one equation in one unknown: x = mol $CaCO_3$; (0.08194 - x) = mol $MgCO_3$.

$$7.85 \text{ g} = (100.1 \text{ g/mol})x + (84.32 \text{ g/mol})(0.08194 - x)$$

$$x = \frac{(7.85 - 6.9092)}{(100.1 - 84.32)} = 0.05\underline{9}62 \text{ mol CaCO}_3$$

(continued)

$(0.08194 - x) = 0.02232$ mol $MgCO_3$

$$CaCO_3 = \frac{0.05962 \text{ mol } CaCO_3 \times 100.1 \text{ g/mol}}{7.85 \text{ g}} \times 100\% = 76.02 = 76\%$$

Percent $MgCO_3$ = 100.00% - 76.02% = 23.98 = 24%

5.131 Write a mass-balance equation to solve for the moles of each gas using 28.01 g/mol for the molar mass of N_2 and 20.18 g/mol for the molar mass of Ne. Let y equal mol of each gas:

$28.01\, y + 20.18\, y = 10.0$ g

$y = (10.0 \div 48.19) = 0.20751$ mol

Total moles (n) = 0.41502 mol (use below)

Use the ideal gas law to calculate the volume, which can then be used to calculate density.

$$\frac{(0.41502 \text{ mol})(0.08206 \text{ L} \cdot \text{atm/K} \cdot \text{mol})(500 \text{ K})}{15.00 \text{ atm}} = 1.1352$$

d = 10.0 g ÷ 1.1352 L = 8.808 = 8.81 g/L

5.133 Rearrange the equation $PM_m = dRT$ to find the quantity RT/M_m. Then, plug into the equation for the root-mean-square speed.

$$\frac{RT}{M_m} = \frac{P}{d}$$

$$u = \sqrt{\frac{3RT}{M_m}} = \sqrt{\frac{3P}{d}} = \sqrt{\frac{3\left(\frac{675}{760}\right) \text{ atm} \times (8.31 \text{ kg} \cdot \text{m}^2/\text{s}^2 \cdot \text{K} \cdot \text{mol})}{(3.00 \times 10^{-3} \text{ kg/L})(0.08206 \text{ L} \cdot \text{atm/K} \cdot \text{mol})}}$$

$$= 2.9990 \times 10^2 = 3.00 \times 10^2 \text{ m/s}$$

6. THERMOCHEMISTRY

■ Solutions to Exercises

Note on significant figures: If the final answer to a solution needs to be rounded off, it is given first with one nonsignificant figure, and the last significant figure is underlined. The final answer is then rounded to the correct number of significant figures. In multiple-step problems, intermediate answers are given with at least one nonsignificant figure; however, only the final answer has been rounded off.

.6.1   Substitute into the formula $E_k = 1/2\ mv^2$ using SI units:

$$E_k = 1/2 \times 9.11 \times 10^{-31}\ kg \times (5.0 \times 10^6\ m/s)^2 = 1.1\underline{3} \times 10^{-17} = 1.1 \times 10^{-17}\ J$$

$$1.13 \times 10^{-17}\ J \times \frac{1\ cal}{4.184\ J} = 2.7\underline{2} \times 10^{-18} = 2.7 \times 10^{-18}\ cal$$

6.2   Heat is evolved; therefore, the reaction is exothermic. The value of q is -1170 kJ.

6.3   The thermochemical equation is

$$2N_2H_4(l) + N_2O_4(l) \rightarrow 3N_2(g) + 4H_2O(g);\ \Delta H = -1049\ kJ$$

ΔH equals -1049 kJ for 1 mol N_2H_4 and for 2 mol N_2H_4.

6.4   a. $N_2H_4(l) + 1/2N_2O_4(l) \rightarrow 3/2N_2(g) + 2H_2O(g);\ \Delta H = -524.5\ kJ$

b. $4H_2O(g) + 3N_2(g) \rightarrow 2N_2H_4(l) + N_2O_4(l);\ \Delta H = 1049\ kJ$

6.5   The reaction is

$$2N_2H_4(l) + N_2O_4(l) \rightarrow 3N_2(g) + 4H_2O(g); \; \Delta H = -1049 \text{ kJ}$$

$$10.0 \text{ g } N_2H_4 \times \frac{1 \text{ mol } N_2H_4}{32.02 \text{ g}} \times \frac{1 \text{ mol } N_2O_4}{2 \text{ mol } N_2H_4} \times \frac{-1049 \text{ kJ}}{1 \text{ mol } N_2H_4} = -163.80 = -164 \text{ kJ}$$

6.6   Substitute into the equation $q = s \times m \times \Delta t$ to obtain the heat transferred. The temperature change is

$$\Delta t = t_f - t_i = 100.0°C - 20.0°C = 80.0°C$$

Therefore,

$$q = s \times m \; \Delta t = 0.449 \text{ J/(g•°C)} \times 5.00 \text{ g} \times 80.0°C = 1.796 \times 10^2 = 1.80 \times 10^2 \text{ J}$$

6.7   The total mass of the solution is obtained by adding the volumes together and by using the density of water (1.000 g/mL). This gives 33 + 42 = 75 mL, or 75 g. The heat absorbed by the solution is

$$q = s \times m \times \Delta t = 4.184 \text{ J/(g•°C)} \times 75 \text{ g} \times (31.8°C - 25.0°C) = 2133.8 \text{ J}$$

The heat released by the reaction, q_{rxn}, is equal to the negative of this value, or -2133.8 J. To obtain the enthalpy change for the reaction, you need to calculate the moles of HCl that reacted. This is

Mol HCl = 1.20 mol/L × 0.033 L = 0.0396 mol

The enthalpy change for the reaction can now be calculated.

$$\Delta H = \frac{-2133.8 \text{ J}}{0.0396 \text{ mol}} = -53884 \text{ J/mol} = -54 \text{ kJ/mol}$$

Expressing this result as a thermochemical equation,

$$HCl(aq) + NaOH(aq) \rightarrow NaCl(aq) + H_2O(l); \; \Delta H = -54 \text{ kJ}$$

6.8   Use Hess's law to find ΔH for $4Al(s) + 3MnO_2(s) \rightarrow 2Al_2O_3 + 3Mn(s)$ from the following data for reactions 1 and 2:

$$2Al(s) + 3/2O_2(g) \rightarrow Al_2O_3(s); \quad \Delta H = -1676 \text{ kJ} \quad \text{(reaction 1)}$$

$$Mn(s) + O_2(g) \rightarrow MnO_2(s); \quad \Delta H = -520 \text{ kJ} \quad \text{(reaction 2)}$$

If you take reaction 1 and multiply it by 2, you obtain

$$4Al(s) + 3O_2(g) \rightarrow 2Al_2O_3(s); \quad \Delta H = 2 \times (-1676 \text{ kJ}) = -3352 \text{ kJ}$$

Since the desired reaction has three MnO_2 on the left side, reverse reaction 2 and multiply it by three. The result is

$$3MnO_2(s) \rightarrow 3Mn(s) + 3O_2(g) \quad \Delta H = -3 \times (-520 \text{ kJ}) = 1560 \text{ kJ}$$

If you add the two reactions and corresponding enthalpy changes, you obtain the enthalpy change of the desired reaction

$4Al(s) + 3O_2(g)$	$\rightarrow \quad 2Al_2O_3(s)$	$\Delta H = -3352 \text{ kJ}$
$3MnO_2(s)$	$\rightarrow \quad 3Mn(s) + 3O_2(g)$	$\Delta H = 1560 \text{ kJ}$
$4Al(s) + 3MnO_2(s)$	$\rightarrow \quad 2Al_2O_3(s) + 3Mn(s)$	$\Delta H = -1792 \text{ kJ}$

6.9   The vaporization process, with the ΔH°_f values given below the substances, is

$$H_2O(l) \rightarrow H_2O(g)$$
$$-285.8 \quad\quad -241.8 \quad\quad \text{(kJ)}$$

The calculation is

$$\Delta H^\circ_{vap} = \Sigma \, n \, \Delta H^\circ_f \text{ (products)} - \Sigma \, m \, \Delta H^\circ_f \text{ (reactants)}$$

$$= \Delta H^\circ_f [H_2O(g)] - \Delta H^\circ_f [H_2O(l)]$$

$$= (-241.8 \text{ kJ}) - (-285.8 \text{ kJ}) = 44.0 \text{ kJ}$$

6.10   The reaction, with the ΔH°_f values given below the substances, is

$$3NO_2(g) \; + \; H_2O(l) \; \rightarrow \; 2HNO_3(aq) \; + \; NO(g)$$

    33.10   -285.8   -207.4   90.29   (kJ)

The calculation is

$$\Delta H^\circ_{rxn} \; = \; \Sigma \, n \, \Delta H^\circ_f \, (\text{products}) \; - \; \Sigma \, m \, \Delta H^\circ_f \, (\text{reactants})$$

$$= \; [2 \; \Delta H^\circ_f (HNO_3) \; + \; \Delta H^\circ_f (NO)] \; - \; [3 \; \Delta H^\circ_f (NO_2) \; + \; \Delta H^\circ_f (H_2O)]$$

$$= \; [2(-207.4) + (90.29)] \, kJ \; - \; [3(33.10) + (-285.8)] \, kJ \; = \; -138.\underline{01} \; = \; -138.0 \, kJ$$

6.11   The net chemical reaction, with the ΔH°_f values given below the substances, is

$$2 \, NH_4^+(aq) \; + \; 2OH^-(aq) \; \rightarrow \; 2NH_3(g) \; + 2H_2O(l)$$

    2(-132.5)   2(-230.0)   2(-45.90)   2(-285.8)   (kJ)

The calculation is

$$\Delta H^\circ_{rxn} \; = \; [2 \; \Delta H^\circ_f (NH_3) \; + \; 2 \; \Delta H^\circ_f (H_2O)] \; - \; [2 \; \Delta H^\circ_f (NH_4^+) \; + \; 2 \; \Delta H^\circ_f (OH^-)]$$

$$= \; [2(-45.90) \; + \; 2(-285.8)] \; - \; [2(-132.5) \; + \; 2(-230.0)] \; = \; 61.\underline{60} \; = \; 61.6 \, kJ$$

■ Answers to Review Questions

6.1   Energy is the potential or capacity to move matter. Kinetic energy is the energy associated with an object by virtue of its motion. Potential energy is the energy an object has by virtue of its position in a field of force. Internal energy is the sum of the kinetic and potential energies of the particles making up a substance.

6.2   In terms of SI base units, a joule is $kg \cdot m^2/s^2$. This is equivalent to the units for the kinetic energy ($\frac{1}{2}mv^2$) of an object of mass m (in kg) moving with speed v (in m/s).

6.3   Originally, a calorie was defined as the amount of energy required to raise the temperature of one gram of water by one degree Celsius. At present, the calorie is defined as 4.184 J.

6.4   At either of the two highest points above the earth in a pendulum's cycle, the energy of the pendulum is all potential energy and is equal to the product mgh (m = mass of pendulum, g = constant acceleration of gravity, and h = height of pendulum). As the pendulum moves downward, its potential energy decreases from mgh to near zero, depending on how close it comes to the earth's surface. During the downward motion, its potential energy is converted to kinetic energy. When it reaches the lowest point (middle) of its cycle, the pendulum has its maximum kinetic energy and minimum potential energy. As it rises above the lowest point, its kinetic energy begins to be converted to potential energy. When it reaches the other high point in its cycle, the energy of the pendulum is again all potential energy. By the law of conservation of energy, this energy cannot be lost, only converted to other forms. At rest, the energy of the pendulum has been transferred to the surroundings in the form of heat.

6.5   As the heat flows into the gas, the gas molecules gain energy and move at a faster average speed. The internal energy of the gas increases.

6.6   An exothermic reaction is a chemical reaction or a physical change in which heat is evolved (q is negative). For example, burning one mol of methane, $CH_4(g)$, yields carbon dioxide, water, and 890.3 kJ of heat. An endothermic reaction is a chemical reaction or physical change in which heat is absorbed (q is positive). For example, the reaction of one mol of barium hydroxide with ammonium nitrate absorbs 170.4 kJ of heat in order to form ammonia, water, and barium nitrate.

6.7   Changes in internal energy depend only on the initial and final states of the system, which are determined by variables such as temperature and pressure. Such changes do not depend on any previous history of the system.

6.8   The enthalpy change equals the heat of reaction at constant pressure.

6.9   At constant pressure, the enthalpy change is positive (the enthalpy increases) for an endothermic reaction.

6.10   It is important to give the states when writing an equation for ΔH because ΔH depends on the states of all reactants and products. If any state changes, ΔH changes.

6.11   When the equation for the reaction is doubled, the enthalpy is also doubled. When the equation is reversed, the sign of ΔH is also reversed.

6.12   First, convert the 10.0 g of water to moles of water, using its molar mass (18.02 g/mol). Next, using the equation, multiply the moles of water by the appropriate mole ratio (1 mol CH_4 / 2 mol H_2O). Finally, multiply the moles of CH_4 by the heat of the reaction (-890.3 kJ/mol CH_4).

6.13   The heat capacity (C) of a substance is the quantity of heat needed to raise the temperature of the sample of substance one degree Celsius (or one kelvin). The specific heat of a substance is the quantity of heat required to raise the temperature of one gram of a substance by one degree Celsius (or one kelvin) at constant pressure.

6.14   A simple calorimeter consists of an insulated container (for example, a pair of styrene coffee cups as in Figure 6.12) with a thermometer. The heat of the reaction is obtained by conducting the reaction in the calorimeter. The temperature of the mixture is measured before and after the reaction. The heat capacity of the calorimeter and its contents also must be measured.

6.15   Hess's law states that, for a chemical equation that can be written as the sum of two or more steps, the enthalpy change for the overall equation is the sum of the enthalpy changes for the individual steps. In other words, no matter how you go from reactants to products, the enthalpy change for the overall chemical change is the same. This is because enthalpy is a state function.

6.16   No, you can still obtain the enthalpy for the desired reaction. You will need to come up with a system of reactions that can be combined to give the desired reaction and know the enthalpy changes for each of the steps. Then, using Hess's law, you can obtain the enthalpy change for the reaction under study.

6.17   The thermodynamic standard state refers to the standard thermodynamic conditions chosen for substances when listing or comparing thermodynamic data: one atm pressure and the specified temperature (usually 25°C).

6.18   The reference form of an element is the most stable form (physical state and allotrope) of the element under standard thermodynamic conditions. The standard enthalpy of formation of an element in its reference form is zero.

6.19   The standard enthalpy of formation of a substance, ΔH°_f, is the enthalpy change for the formation of one mole of the substance in its standard state from its elements in their reference form and in their standard states.

6.20   The equation for the formation of $H_2S(g)$ is

$$H_2(g) \ + \ 1/8 S_8(\text{rhombic}) \ \rightarrow \ H_2S(g)$$

6.21   The reaction of $C(g) + 4H(g) \rightarrow CH_4(g)$ is not an appropriate equation for calculating the ΔH°_f of methane because the most stable form of each element is not used. Both $H_2(g)$ and $C(\text{graphite})$ should be used instead of $H(g)$ and $C(g)$, respectively.

6.22   A fuel is any substance that is burned or similarly reacted to provide heat and other forms of energy. The fossil fuels are petroleum (oil), gas, and coal. They were formed millions of years ago when aquatic plants and animals were buried and compressed by layers of sediment at the bottoms of swamps and seas. Over time this organic matter was converted by bacterial decay and pressure to fossil fuels.

6.23 One of the ways of converting coal to methane involves the water-gas reaction.

$$C(s) + H_2O(g) \rightarrow CO(g) + H_2(g)$$

In this reaction, steam is passed over hot coal. This mixture is then reacted over a catalyst to give methane.

$$CO(g) + 3H_2(g) \rightarrow CH_4(g) + H_2O(g)$$

6.24 Some possible rocket fuel/oxidizer combinations are H_2/O_2 and hydrazine/dinitrogen tetroxide. The chemical equations for their reactions are

$$H_2(g) + 1/2O_2(g) \rightarrow H_2O(g); \quad \Delta H° = -242 \text{ kJ}$$

$$2N_2H_4(l) + N_2O_4(l) \rightarrow 3N_2(g) + 4H_2O(g); \quad \Delta H° = -1049 \text{ kJ}$$

■ Solutions to Practice Problems

Note on significant figures: If the final answer to a solution needs to be rounded off, it is given first with one nonsignificant figure, and the last significant figure is underlined. The final answer is then rounded to the correct number of significant figures. In multiple-step problems, intermediate answers are given with at least one nonsignificant figure; however, only the final answer has been rounded off.

6.35 The heat released, in kcal, is

$$-445.1 \text{ kJ} \times \frac{1000 \text{ J}}{1 \text{ kJ}} \times \frac{1 \text{ cal}}{4.184 \text{ J}} \times \frac{1 \text{ kcal}}{1000 \text{ cal}} = -106.3\underline{8} = -106.4 \text{ kcal}$$

6.37 The kinetic energy, in J, is

$$E_k = 1/2 \times 4.85 \times 10^3 \text{ lb} \times \frac{0.4536 \text{ kg}}{1 \text{ lb}} \times \left[\frac{57 \text{ mi}}{1 \text{ h}}\right]^2 \times \left[\frac{1609 \text{ m}}{1 \text{ mi}}\right]^2 \times \left[\frac{1 \text{ h}}{3600 \text{ s}}\right]^2$$

$$= 7.\underline{1}3 \times 10^5 = 7.1 \times 10^5 \text{ J}$$

The kinetic energy, in calories, is

$$E_k = 7.\underline{1}3 \times 10^5 \text{ J} \times \frac{1 \text{ cal}}{4.184 \text{ J}} = 1.\underline{7}0 \times 10^5 = 1.7 \times 10^5 \text{ cal}$$

6.39   To insert the mass of one molecule of ClO_2 in the formula, multiply the molar mass by the reciprocal of Avogadro's number. The kinetic energy in J is

$$E_k = 1/2 \times \frac{67.45 \text{ g}}{1 \text{ mol}} \times \frac{1 \text{ mol}}{6.022 \times 10^{23} \text{ molec.}} \times \frac{1 \text{ kg}}{1000 \text{ g}} \times \left[\frac{306 \text{ m}}{1 \text{ s}}\right]^2$$

$$= 5.2\underline{4}3 \times 10^{-21} = 5.24 \times 10^{-21} \text{ J/molec.}$$

6.41   Endothermic reactions absorb heat, so the sign of q will be positive because energy must be gained by the system from the surroundings. The flask will feel cold to the touch.

6.43   The gain of 66.2 kJ of heat per two mol NO_2 means the reaction is endothermic. Because energy is gained by the system from the surroundings, q is positive and is +66.2 kJ for two mol of NO_2 reacting.

6.45   The reaction of Fe(s) with HCl must yield H_2 and $FeCl_3$. To balance the hydrogen, 2HCl must be written first as a reactant. However, to balance the chlorine, 3 x 2HCl must be written finally as a reactant, and $2FeCl_3$ must be written as a product:

$$2Fe(s) + 6HCl(aq) \rightarrow 2FeCl_3(aq) + 3H_2(g)$$

To write a thermochemical equation, the sign of ΔH must be negative because heat is evolved:

$$2Fe(s) + 6HCl(aq) \rightarrow 2FeCl_3(aq) + 3H_2(g); \quad \Delta H = 2 \times -89.1 \text{ kJ} = -178.2 \text{ kJ}$$

6.47   The first equation is

$$P_4(s) + 5O_2(g) \rightarrow P_4O_{10}(s); \quad \Delta H = -3010 \text{ kJ}$$

The second equation is

$$P_4O_{10}(s) \rightarrow P_4(s) + 5O_2(g); \quad \Delta H = ?$$

The second equation has been obtained by reversing the first equation. Therefore, to obtain ΔH for the second equation, ΔH for the first equation must be reversed in sign: -(-3010) = +3010 kJ.

6.49   The first equation is

$$\frac{1}{4}P_4O_{10}(s) + \frac{3}{2}H_2O(l) \rightarrow H_3PO_4(aq); \quad \Delta H = -96.2 \text{ kJ}$$

The second equation is

$$P_4O_{10}(s) + 6H_2O(l) \rightarrow 4H_3PO_4(aq); \quad \Delta H = ?$$

The second equation has been obtained from the first by multiplying each coefficient by four. Therefore, to obtain the ΔH for the second equation, the ΔH for the first equation must be multiplied by four: -96.2 x 4 = -384.8 kJ

6.51   Because nitric oxide is written as NO in the equation, the molar mass of NO equals 30.01 g per mol NO. From the equation, two mol NO evolve 114 kJ heat. Divide the 114 kJ by the 30.01 g/mol NO to obtain the amount of heat evolved per gram of NO:

$$\frac{-114 \text{ kJ}}{2 \text{ mol NO}} \times \frac{1 \text{ mol NO}}{30.01 \text{ g NO}} = 1.8\underline{9}9 = \frac{-1.90 \text{ kJ}}{\text{g NO}}$$

6.53   The molar mass of ammonia is 17.03 g/mol. From the equation, four moles of NH_3 evolve 1267 kJ of heat. Divide 35.8 g NH_3 by its molar mass and the four moles of NH_3 in the equation to obtain the amount of heat evolved:

$$35.8 \text{ g NH}_3 \times \frac{1 \text{ mol NH}_3}{17.03 \text{ g NH}_3} \times \frac{-1267 \text{ kJ}}{4 \text{ mol NH}_3} = -6.6\underline{5}8 \times 10^2 = -6.66 \times 10^2 \text{ kJ}$$

6.55   The molar mass of C_3H_8 is 44.06 g/mol. From the equation, one mol C_3H_8 evolves 2043 kJ heat. This gives

$$-369 \text{ kJ} \times \frac{1 \text{ mol C}_3\text{H}_8}{-2043 \text{ kJ}} \times \frac{44.06 \text{ g C}_3\text{H}_8}{1 \text{ mol C}_3\text{H}_8} = 7.9\underline{5}7 = 7.96 \text{ g C}_3\text{H}_8$$

6.57   Multiply the 180 g (0.180 kg) of water by the specific heat of 4.18 J/(g•°C) and by Δt to obtain heat in joules:

$$180 \text{ g} \times (96°C - 19°C) \times \frac{4.18 \text{ J}}{1 \text{ g}•°C} = 5\underline{7}934 = 5.8 \times 10^4 \text{ J}$$

6.59   Use the 2.26×10^3 J/g (2.26 kJ/g) heat of vaporization to calculate the heat of condensation. Then, use it to calculate Δt, the temperature change.

$$\text{Heat of condensation} = \frac{2.26 \times 10^3 \text{ J}}{1 \text{ g}} \times 168 \text{ g} = 3.7\underline{9}68 \times 10^5 \text{ J}$$

$$\text{Temperature change} = \Delta t = \frac{3.7\underline{9}68 \times 10^5 \text{ J}}{6.44 \times 10^4 \text{ g}} \times \frac{1 \text{g} \cdot {}^\circ\text{C}}{1.015 \text{ J}} = 5.8\underline{0}85 = 5.81{}^\circ\text{C}$$

6.61   The enthalpy change for the reaction is equal in magnitude and opposite in sign to the heat-energy change occurring from the cooling of the solution and calorimeter.

$$q_{calorimeter} = (1071 \text{ J/}^\circ\text{C})(21.56{}^\circ\text{C} - 25.00{}^\circ\text{C}) = -368\underline{4}.2 \text{ J}$$

Thus, 15.3 g $NaNO_3$ is equivalent to -3684.2 J heat energy. The amount of heat absorbed by 1.000 mol $NaNO_3$ is calculated from +3684.2 J (opposite sign):

$$1.000 \text{ mol NaNO}_3 \times \frac{85.00 \text{ g NaNO}_3}{1 \text{ mol NaNO}_3} \times \frac{3684.2 \text{ J}}{15.3 \text{ g NaNO}_3} = 2.0\underline{4}68 \times 10^4 \text{ J}$$

Thus, the enthalpy change, ΔH, for the reaction is 2.05×10^4 J, or 20.5 kJ, per mol $NaNO_3$.

6.63   The energy change for the reaction is equal in magnitude and opposite in sign to the heat energy produced from the warming of the solution and the calorimeter.

$$q_{calorimeter} = (9.63 \text{ kJ/}^\circ\text{C})(33.73 - 25.00{}^\circ\text{C}) = +84.\underline{0}69 \text{ kJ}$$

Thus, 2.84 g C_2H_5OH is equivalent to 84.069 kJ heat energy. The amount of heat released by 1.000 mol C_2H_5OH is calculated from -84.069 kJ (opposite sign):

$$1.00 \text{ mol C}_2\text{H}_5\text{OH} \times \frac{46.07 \text{ g C}_2\text{H}_5\text{OH}}{1 \text{ mol C}_2\text{H}_5\text{OH}} \times \frac{-84.069 \text{ kJ}}{2.84 \text{ g C}_2\text{H}_5\text{OH}} = -136\underline{3}.75 \text{ kJ}$$

Thus, the enthalpy change, ΔH, for the reaction is -1.36×10^3 kJ/mol ethanol.

6.65   Using the equations in the data, reverse the direction of the first reaction, and reverse the sign of its ΔH. Then multiply the second equation by two, multiply its ΔH by two, and add. Setup:

$$N_2(g) + 2H_2O(l) \rightarrow N_2H_4(l) + O_2(g); \qquad \Delta H = (-622.2 \text{ kJ}) \times (-1)$$

$$2H_2(g) + O_2(g) \rightarrow 2H_2O(l); \qquad \Delta H = (-285.8 \text{ kJ}) \times (2)$$

_____

$$N_2(g) + 2H_2(g) \rightarrow N_2H_4(l); \qquad \Delta H = 50.6 \text{ kJ}$$

6.67 Using the equations in the data, multiply the second equation by two, and reverse its direction; do the same to its ΔH. Then multiply the first equation by two and its ΔH by two. Finally, multiply the third equation by three and its ΔH by three. Then add. Setup:

$$4NH_3(g) \rightarrow 2N_2(g) + 6H_2(g); \quad \Delta H = (-91.8 \text{ kJ}) \times (-2)$$

$$2N_2(g) + 2O_2(g) \rightarrow 4NO(g); \quad \Delta H = (180.6 \text{ kJ}) \times (2)$$

$$6H_2(g) + 3O_2(g) \rightarrow 6H_2O(g); \quad \Delta H = (-483.7 \text{ kJ}) \times (3)$$

$$4NH_3(g) + 5O_2(g) \rightarrow 4NO(g) + 6H_2O(g); \quad \Delta H = -906.3 \text{ kJ}$$

6.69 After reversing the first equation in the data, add all the equations. Setup:

$$C_2H_4(g) + 3O_2(g) \rightarrow 2CO_2(g) + 2H_2O(l); \quad \Delta H = (-1411 \text{ kJ})$$

$$2CO_2(g) + 3H_2O(l) \rightarrow C_2H_6(g) + 7/2O_2(g); \quad \Delta H = (-1560) \times (-1)$$

$$H_2(g) + 1/2O_2(g) \rightarrow H_2O(l); \quad \Delta H = (-286 \text{ kJ})$$

$$C_2H_4(g) + H_2(g) \rightarrow C_2H_6(g); \quad \Delta H = -137 \text{ kJ}$$

6.71 Write the ΔH° values (Appendix C) underneath each compound in the balanced equation:

$$C_2H_5OH(l) \rightarrow C_2H_5OH(g)$$
$$-277.7 \qquad\qquad -235.4 \qquad (kJ)$$

$$\Delta H^\circ_{vap} = [\,\Delta H^\circ_f(C_2H_5OH)\,] - [\,\Delta H^\circ_f(C_2H_5OH)\,] = [-235.4] - [-277.7] = +42.3 \text{ kJ}$$

6.73 Write the ΔH° values (Table 6.2) underneath each compound in the balanced equation:

$$2H_2S(g) + 3O_2(g) \rightarrow 2H_2O(l) + 2SO_2(g)$$
$$2(-20.50) \quad 3(0) \qquad 2(-285.8) \quad 2(-296.8) \qquad (kJ)$$

$$\Delta H^\circ = \Sigma\, n\Delta H^\circ(\text{products}) - \Sigma\, m\Delta H^\circ(\text{reactants})$$

$$= [2(-285.8) + 2(-296.8)] - [2(-20.50) + 3(0)]\, kJ = -112\underline{4}.2 = -1124 \text{ kJ}$$

6.75   Write the $\Delta H°$ values (Appendix C) underneath each compound in the balanced equation:

$$Fe_2O_3(s) \;+\; 3CO(g) \;\rightarrow\; 2Fe(s) \;+\; 3CO_2(g)$$

-825.5   3(-110.5)   2(0)   3(-393.5)   (kJ)

$\Delta H° \;=\; \Sigma \, n\Delta H°(products) \;-\; \Sigma \, m\Delta H°(reactants)$

$= [2(0) \;+\; 3(-393.5)] \;-\; [(-825.5) \;+\; 3(-110.5)] \;=\; -23.5 \text{ kJ}$

6.77   Write the $\Delta H°$ values (Table 6.2) underneath each compound in the balanced equation:

$$HCl(g) \;\rightarrow\; H^+(aq) \;+\; Cl^-(aq)$$

-92.31   0   -167.2 (kJ)

$\Delta H° \;=\; \Sigma \, n\Delta H°(products) \;-\; \Sigma m\Delta H°(reactant)$

$= [(0) \;+\; (-167.2)] \;-\; [-92.31] \;=\; -74.\underline{8}9 \;=\; -74.9 \text{ kJ}$

6.79   Calculate the molar heat of formation from the equation with the $\Delta H°_f$ values below each substance; then convert to the heat for 10.0 g of $MgCO_3$ using the molar mass of 84.3.

$$MgCO_3(s) \;\rightarrow\; MgO(s) \;+\; CO_2(g)$$

-1111.7   -601.2   -393.5   kJ

$\Delta H° = \Sigma \, n\Delta H°(products) \;-\; \Sigma \, m\Delta H°(reactants)$

$\Delta H° = -601.2 \text{ kJ} \;+\; (-393.5 \text{ kJ}) \;-\; (-1111.7 \text{ kJ}) \;=\; 11\underline{7}.0 \text{ kJ}$

$\text{Heat} = 10.0 \text{ g} \times \dfrac{1 \text{ mol}}{84.3 \text{ g}} \times \dfrac{117.0 \text{ kJ}}{\text{mol}} = 13.\underline{8}7 = 13.9 \text{ kJ}$

■ Solutions to General Problems

6.81   The SI units of force must be kg•m/s² (= newton, N) to be consistent with the joule, the SI unit of energy:

$$\dfrac{\text{kg} \cdot \text{m}}{\text{s}^2} \times \text{m} \;=\; \dfrac{\text{kg} \cdot \text{m}^2}{\text{s}^2} \;=\; \text{joule, J}$$

6.83   Using Table 1.5 and 4.184 J/cal, convert the 686 Btu/lb to J/g:

$$\frac{686\ Btu}{1\ lb} \times \frac{252\ cal}{1\ Btu} \times \frac{4.184\ J}{1\ cal} \times \frac{1\ lb}{0.4536\ kg} \times \frac{1\ kg}{10^3\ g}$$

$$= 1.5\underline{9}4 \times 10^3 = 1.59 \times 10^3\ J/g$$

6.85   Substitute into the equation $E_p = mgh$, and convert to SI units.

$$E_p = 1.00\ lb \times \frac{9.807\ m}{s^2} \times 167\ ft \times \frac{0.4536\ kg}{1\ lb} \times \frac{0.9144\ m}{3\ ft}$$

$$= \frac{226.4\underline{3}3\ kg \bullet m^2}{s^2} = 226\ J$$

At the bottom, all the potential energy is converted to kinetic energy, so $E_k = 225.8$ $kg \bullet m^2/s^2$. Because $E_k = 1/2mv^2$, solve for v, the speed (velocity):

$$Speed = \sqrt{\frac{E_k}{1/2 \times m}} = \sqrt{\frac{226.433\ kg \bullet m^2/s^2}{1/2 \times 1.00\ lb \times 0.4536\ kg/lb}} = 31.\underline{6}0 = 31.6\ m/s$$

6.87   The equation is

$$CaCO_3(s) \rightarrow CaO(s) + CO_2(g); \quad \Delta H = 177.9\ kJ$$

Use the molar mass of 100.08 g/mol to convert the heat per mol to heat per 27.3 g.

$$21.3\ g\ CaCO_3 \times \frac{1\ mol\ CaCO_3}{100.08\ g\ CaCO_3} \times \frac{177.9\ kJ}{1\ mol\ CaCO_3} = 37.\underline{8}6 = 37.9\ kJ$$

6.89   The equation is

$$2HCHO_2(l) + O_2(g) \rightarrow 2CO_2(g) + 2H_2O(l)$$

Use the molar mass of 46.03 g/mol to convert -30.3 kJ/5.48 g to ΔH per mol of acid.

$$\frac{-30.3\ kJ}{5.48\ g\ HCHO_2} \times \frac{46.03\ g\ HCHO_2}{1\ mol\ HCHO_2} = -25\underline{4}.508 = -255\ kJ/mol$$

6.91   The heat gained by the water at the lower temperature equals the heat lost by the water at the higher temperature. Each heat term is mass x sp ht x ΔT. This gives

$$(54.9 \text{ g})(4.184 \text{ J/g°C})(T_f - 31.5°C) = (21.0 \text{ g})(4.184 \text{ J/g°C})(52.7°C - T_f)$$

The specific heat and the units can be canceled from both sides to give

$$(54.9)(T_f - 31.5°C) = (21.0)(52.7°C - T_f)$$

After rearranging, you get $75.9 \, T_f = 2836.05$. This gives $T_f = 37.\underline{3}6 = 37.4°C$.

6.93   Divide the 235 J heat by the mass of lead and the Δt to obtain the specific heat.

$$\text{Specific heat} = \frac{235 \text{ J}}{121.6 \text{ g} (35.5°C - 20.4°C)} = 0.127\underline{9}8 = \frac{0.128 \text{ J}}{g \bullet °C}$$

6.95   The energy used to heat the Zn comes from cooling the water. Calculate q for water:

$$q_{wat} = \text{specific heat x mass x } \Delta t$$

$$q_{wat} = \frac{4.18 \text{ J}}{g \bullet °C} \text{ x } 50.0 \text{ g x } (96.68°C - 100.00°C) = -69\underline{3}.88 \text{ J}$$

The sign of q for the Zn is the reverse of the sign of q for water because the Zn is absorbing heat:

$$q_{met} = -(q_{wat}) = -(-693.88) = 69\underline{3}.88 \text{ J}$$

$$\text{Specific heat} = \frac{693.88 \text{ J}}{25.3 \text{ g} (96.68°C - 25.00°C)} = 0.38\underline{2}6 = \frac{0.383 \text{ J}}{g \bullet °C}$$

6.97   First, multiply the molarities by the volumes, in liters, to get the moles of NaOH and the moles of HCl, and thus determine the limiting reactant.

Mol NaOH = M x V = 0.996 M x 0.0141 L = 0.014\underline{0}4 mol

Mol HCl = M x V = 0.905 M x 0.0323 L = 0.029\underline{2}3 mol

Therefore, NaOH is the limiting reactant. The total volume of the system is 14.1 mL + 32.3 mL = 46.4 mL. Since the density of water is 1.00 g/mL, the total mass of the system is 46.4 g. Next, set the heat released by the reaction (mol x enthalpy of reaction) equal to the heat absorbed by the water (m x sp ht. x ΔT). This gives

$$(0.01404 \text{ mol})(55.8 \times 10^3 \text{ J/mol}) = (46.4 \text{ g})(4.184 \text{ J/g} \bullet °C)(T_f - 21.6°C)$$

After dividing, this gives $T_f - 21.6°C = 4.035°C$, or $T_f = 25.\underline{6}3 = 25.6°C$

6.99 Use Δt and the heat capacity of 547 J/°C to calculate q:

$q = C\Delta t = (547 \text{ J/°C})(36.66 - 25.00)°C = 6.3\underline{7}8 \times 10^3 \text{ J } (6.3\underline{7}8 \text{ kJ})$

Energy is released in the solution process in raising the temperature, so ΔH is negative:

$$\Delta H = \frac{-6.378 \text{ kJ}}{6.48 \text{ g LiOH}} \times \frac{23.95 \text{ g LiOH}}{1 \text{ mol LiOH}} = -23.\underline{5}7 = -23.6 \text{ kJ/mol}$$

6.101 Use Δt and the heat capacity of 13.43 kJ/°C to calculate q:

$q = C\Delta t = (13.43 \text{ kJ/°C})(35.84 - 25.00)°C = 145.\underline{5}81 \text{ kJ}$

As in the previous two problems, the sign of ΔH must be reversed, making the heat negative:

$$\Delta H = \frac{-145.581 \text{ kJ}}{10.00 \text{ g HC}_2\text{H}_3\text{O}_2} \times \frac{60.05 \text{ g HC}_2\text{H}_3\text{O}_2}{1 \text{ mol HC}_2\text{H}_3\text{O}_2} = -874.\underline{2}1 = -874.2 \text{ kJ/mol}$$

6.103 Using the equations in the data, reverse the direction of the first reaction, and reverse the sign of its ΔH. Then add the second and third equations and their ΔH's.

$H_2O(g)$ +	$SO_2(g)$	→	$H_2S(g)$ +	$3/2O_2(g)$;	$\Delta H = (-518 \text{ kJ}) \times (-1)$
$H_2(g)$ +	$1/2O_2(g)$	→	$H_2O(g)$;		$\Delta H = (-242 \text{ kJ})$
$1/8S_8(\text{rh.})$ +	$O_2(g)$	→	$SO_2(g)$;		$\Delta H = (-297 \text{ kJ})$
$H_2(g)$ +	$1/8S_8(\text{rh.})$	→	$H_2S(g)$;		$\Delta H = -21 \text{ kJ}$

6.105 Write the $\Delta H°$ values (Table 6.2) underneath each compound in the balanced equation.

$CH_4(g)$ + $H_2O(g)$ → $CO(g)$ + $3H_2(g)$
-74.9   -241.8   -110.5   3(0)   (kJ)

$\Delta H° = [-110.5 + 3(0)] - [(-74.87) + (-241.8)] = 206.\underline{1}7 = 206.2 \text{ kJ}$

6.107   Write the $\Delta H°$ values (Table 6.2) underneath each compound in the balanced equation. The $\Delta H_f°$ of -635 kJ/mol CaO is given in the problem.

$$CaCO_3(s) \rightarrow \quad CaO(s) + CO_2(g)$$

-1206.9   -635   -393.5   (kJ)

$$\Delta H° = [(-635.1) + (-393.5)] - [-1206.9] = 17\underline{8}.3 = 178 \text{ kJ}$$

6.109   Calculate the molar heat of reaction from the equation with the $\Delta H_f°$ values below each substance; then convert to the heat for the reaction at 25°C.

$$2H_2(g) + O_2(g) \rightarrow 2H_2O(l)$$

0.0 kJ   0.0 kJ   -285.8 kJ

$$\Delta H° = 2 \times (-285.8 \text{ kJ}) - 0 - 0 = -571.6 \text{ kJ}$$

The moles of oxygen in 2.000 L with a density of 1.11 g/L is:

$$2.000 \text{ L } O_2 \times \frac{1.11 \text{ g } O_2}{1 \text{ L } O_2} \times \frac{1 \text{ mol } O_2}{32.0 \text{ g } O_2} = 0.069\underline{3}7 \text{ mol } O_2$$

The heat of reaction from 0.06937 mol O_2 is:

$$0.69\underline{3}7 \text{ mol } O_2 \times \frac{-571.6 \text{ kJ}}{\text{mol } O_2} = -39.\underline{6}54 = -39.7 \text{ kJ}$$

6.111   Let $\Delta H_f°$ = the unknown standard enthalpy of formation for sucrose. Then write this symbol and the other $\Delta H_f°$ values underneath each compound in the balanced equation, and solve for $\Delta H_f°$ using the $\Delta H°$ of -5641 kJ for the reaction.

$$C_{12}H_{22}O_{11}(s) + \quad 12O_2(g) \rightarrow 12CO_2(g) + 11H_2O(l)$$

$\Delta H_f°$ 12(0)   12(-393.5)   11(-285.8)   (kJ)

$$\Delta H° = -5641 = [12(-393.5) + 11(-285.8)] - [\Delta H_f° + 12(0)]$$

$$\Delta H_f° = -2225 \text{ kJ/mol sucrose}$$

6.113   First, calculate the heat evolved from the molar amounts represented by the balanced equation. From Appendix C, obtain the individual heats of formation, and write those below the reactants and products in the balanced equation. Then, multiply the molar heats of formation by the number of moles in the balanced equation, and write those products below the molar heats of formation.

$$2Al(s) + 3NH_4NO_3(s) \rightarrow 3N_2(g) + 6H_2O(g) + Al_2O_3(s)$$

0.0 kJ	-365.6 kJ	0.0 kJ	-241.826	-1675.7 kJ (kJ/mol)
0.0 kJ	-1096.8 kJ	0.0 kJ	-1450.96	-1675.7 kJ (kJ/eqn.)

The total heat of reaction of two moles of Al and three moles of NH_4NO_3 is

$$\Delta H = -1450.8 - 1675.7 - (-1096.8 \text{ kJ}) = -202\underline{9}.7 \text{ kJ}$$

Now, 245 kJ represents the following fraction of the total heat of reaction:

$$\frac{245 \text{ kJ}}{2029.7 \text{ kJ}} = 0.12\underline{0}74$$

Thus, 245 kJ requires the fraction 0.12074 of the moles of each reactant: 0.12074 x 2, or 0.24141, mol of Al and 0.12068 x 3, or 0.36209, mol of NH_4NO_3. The mass of each reactant and the mass of the mixture are as follows:

0.24141 mol Al x 26.98 g/mol Al = 6.5$\underline{1}$32 g of Al

0.36209 mol NH_4NO_3 x 80.04 g/mol NH_4NO_3 = 28.$\underline{9}$816 g NH_4NO_3

6.5132 g Al + 28.9816 g NH_4NO_3 = 35.$\underline{4}$9 = 35.5 g of the mixture

■ Solutions to Cumulative-Skills Problems

6.115   The heat lost, q, by the water(s) with a temperature higher than the final temperature must be equal to the heat gained, q, by the water(s) with a temperature lower than the final temperature.

q(lost by water at higher temp) = q(gained by water at lower temp)

$$\Sigma (s \times m \times \Delta t) = \Sigma (s \times m \times \Delta t)$$

Divide both sides of the equation by the specific heat, s, to eliminate this term, and substitute the other values. Since the mass of the water at 50.0°C is greater than the sum of the masses of the other two waters, assume the final temperature will be greater than 37°C and greater than 15°C. Use this to set up the three Δt expressions, and write one equation in one unknown, letting t equal the final temperature.

(continued)

Simplify by omitting the "grams" from 45.0 g, 25.0 g, and 15.0 g.

$$\Sigma (m \times \Delta t) = \Sigma (m \times \Delta t)$$

$$45.0 \times (50.0°C - t) = 25.0 \times (t - 15.0°C) + 15.0 \times (t - 37°C)$$

$$2250°C - 45.0\,t = 25.0\,t - 375°C + 15.0\,t - 555.0°C$$

$$-85.0\,t = -3180°C$$

$$t = 37.\underline{4}1 = 37.4°C \quad \text{(the final temperature)}$$

6.117 First, calculate the mole fraction of each gas in the product, assuming 100 g product:

Mol CO = 33 g CO x 1 mol CO/28.01 g CO = 1.$\underline{1}$78 mol CO

Mol CO_2 = 67 g CO_2 x 1 mol CO_2/44.01 g CO_2 = 1.$\underline{5}$22 mol CO_2

$$\text{Mol frac. CO} = \frac{1.178 \text{ mol CO}}{(1.178 + 1.522) \text{ mol}} = 0.4\underline{3}63$$

$$\text{Mol frac. } CO_2 = \frac{1.522 \text{ mol } CO_2}{(1.178 + 1.522) \text{ mol}} = 0.5\underline{6}37$$

Now, calculate the starting moles of C, which equal the total moles of CO and CO_2:

Starting mol C(s) = 1.00 g C x 1 mol C/12.01 g C

$$= 0.083\underline{2}6 \text{ mol C} = \text{mol CO} + CO_2$$

Use the mol fractions to convert mol CO + CO_2 to mol CO and mol CO_2:

$$\text{Mol CO} = \frac{0.4363 \text{ mol CO}}{1 \text{ mol total}} \times 0.08326 \text{ mol total} = 0.036\underline{3}3 \text{ mol CO}$$

$$\text{Mol } CO_2 = \frac{0.5637 \text{ mol } CO_2}{1 \text{ mol total}} \times 0.08326 \text{ mol total} = 0.046\underline{9}3 \text{ mol } CO_2$$

Now, use enthalpies of formation (Table 6.2) to calculate the heat of combustion for both:

C(s) +	1/2O_2(g)	→	CO(g)
0.03633	excess		0.03633 (mol)
0	0		-110.5 (kJ/mol)
0	0		-4.014 (kJ/0.03633 mol)

(continued)

$$C(s) \; + \; O_2(g) \; \rightarrow \; CO_2(g)$$

0.04693	excess	0.04693 (mol)
0	0	-393.5 (kJ/mol)
0	0	-18.47 (kJ/0.04693 mol)

Total ΔH = -4.014 + (-18.47) = -22.48 = -22 kJ

Heat released = 22 kJ

6.119   The $\Delta H°$ for the reactions to produce CO and CO_2 in each case is equal to the $\Delta H_f°$ of the respective gases. Thus $\Delta H°$ to produce CO is -110.5 kJ/mol, and $\Delta H°$ to produce CO_2 is -393.5 kJ/mol. Since one mol of graphite is needed to produce one mol of either CO or CO_2, one equation in one unknown can be written for the heat produced using m for the moles of CO and (2.00 - m) for moles of CO_2.

- 481 kJ = m x (-110.5 kJ) + (2.00 - m) x (-393.5 kJ)

- 481 = -110.5 m - 787 + 393.5 m

mol CO = m = 1.081;   mass CO = 1.081 x 28 g/mol = 30.26 = 30.3 g

mol CO_2 = (2.00 - m) = 0.919;   mass CO_2 = 0.919 mol x 44.0 g/mol

= 40.4 = 40. g

6.121   The equation is

$$4NH_3(g) \; + \; 5O_2(g) \; \rightarrow \; 4NO(g) \; + \; 6H_2O(g); \quad \Delta H° = -906 \text{ kJ}$$

First, determine the limiting reactant by calculating the moles of NH_3 and of O_2; then, assuming one of the reactants is totally consumed, calculate the moles of the other reactant needed for the reaction.

$$10.0 \text{ g } NH_3 \times \frac{1 \text{ mol } NH_3}{17.03 \text{ g } NH_3} = 0.5872 \text{ mol } NH_3$$

$$20.0 \text{ g } O_2 \times \frac{1 \text{ mol } O_2}{32.00 \text{ g } O_2} = 0.6250 \text{ mol } O_2$$

$$0.6250 \text{ mol } O_2 \times \frac{1 \text{ mol } O_2}{32.00 \text{ g } O_2} = 0.500 \text{ mol } NH_3 \text{ needed}$$

Because NH_3 is present in excess of what is needed, O_2 must be the limiting reactant. Now calculate the heat released on the basis of the complete reaction of 0.622250 mol O_2:

(continued)

$$\Delta H = \frac{-906 \text{ kJ}}{5 \text{ mol O}_2} \times 0.6250 \text{ mol O}_2 = -113.25 = -113 \text{ kJ}$$

The heat released by the complete reaction of the 20.0 g (0.6250 mol) of $O_2(g)$ is 113 kJ.

6.123   The equation is

$$N_2(g) + 3H_2(g) \rightarrow 2NH_3(g); \quad \Delta H° = -91.8 \text{ kJ}$$

a. To find the heat evolved from the production of 1.00 L of NH_3, convert the 1.00 L to mol NH_3 using the density and molar mass (17.03 g/mol). Then convert the moles to heat (ΔH) using $\Delta H°$:

$$\text{moles NH}_3 = \frac{d \times V}{Mwt} = \frac{0.696 \text{ g/L} \times 1 \text{ L}}{17.03 \text{ g/mol}} = 0.04087 \text{ mol NH}_3$$

$$\Delta H = \frac{-91.8 \text{ kJ}}{2 \text{ mol NH}_3} \times 0.04087 \text{ mol NH}_3 = -1.876 \text{ (1.88 kJ heat evolved)}$$

b. First, find the moles of N_2 using the density and molar mass (28.02 g/mol). Then convert to the heat needed to raise the N_2 from 25°C to 400°C:

$$\text{moles N}_2 = \frac{d \times V}{Mwt} = \frac{1.145 \text{ g/L} \times 0.500 \text{ L}}{28.02 \text{ g/mol}} = 0.02043 \text{ mol N}_2$$

$$0.02043 \text{ mol N}_2 \times \frac{29.12 \text{ J}}{\text{mol} \cdot °C} \times (400 - 25)°C = 223.2 \text{ J } (0.2232 \text{ kJ})$$

$$\text{Percent heat for N}_2 = \frac{0.2232 \text{ kJ}}{1.876 \text{ kJ}} \times 100\% = 11.89 = 11.9\%$$

6.125   The glucose equation is

$$C_6H_{12}O_6 + 6O_2 \rightarrow 6CO_2 + 6H_2O; \quad \Delta H° = -2802.8 \text{ kJ}$$

Convert the 2.50×10^3 kcal to mol of glucose using the $\Delta H°$ of -2802.8 kJ for the reaction and the conversion factor of 4.184 kJ/kcal:

$$2.50 \times 10^3 \text{ kcal} \times \frac{4.184 \text{ kJ}}{1.000 \text{ kcal}} \times \frac{1 \text{ mol glucose}}{2802.8 \text{ kJ}} = 3.7319 \text{ mol glucose}$$

(continued)

Next, convert mol glucose to mol LiOH using the above equation for glucose and the equation for LiOH:

$$2LiOH(s) + CO_2(g) \rightarrow Li_2CO_3(s) + H_2O(l)$$

$$3.7\underline{3}19 \text{ mol glucose} \times \frac{6 \text{ mol } CO_2}{1 \text{ mol glucose}} \times \frac{2 \text{ mol LiOH}}{1 \text{ mol } CO_2} = 44.\underline{7}83 \text{ mol LiOH}$$

Finally, use the molar mass of LiOH to convert moles to mass:

$$44.783 \text{ mol LiOH} \times \frac{23.95 \text{ g LiOH}}{1 \text{ mol LiOH}} = 1.0\underline{7}25 \times 10^3 \text{ g } (1.07 \text{ kg}) \text{ LiOH}$$

7. QUANTUM THEORY OF THE ATOM

■ Solutions to Exercises

Note on significant figures: If the final answer to a solution needs to be rounded off, it is given first with one nonsignificant figure, and the last significant figure is underlined. The final answer is then rounded to the correct number of significant figures. In multiple-step problems, intermediate answers are given with at least one nonsignificant figure; however, only the final answer has been rounded off.

7.1 Rearrange the equation $c = \nu\lambda$, which relates wavelength to frequency and the speed of light (3.00×10^8 m/s):

$$\lambda = \frac{c}{\nu} = \frac{3.00 \times 10^8 \text{ m/s}}{3.91 \times 10^{14} \text{ /s}} = 7.6\underline{7}2 \times 10^{-7} = 7.67 \times 10^{-7} \text{ m, or 767 nm}$$

7.2 Rearrange the equation $c = \nu\lambda$, which relates frequency to wavelength and the speed of light (3.00×10^8 m/s). Recognize that 456 nm = 4.56×10^{-7} m.

$$\nu = \frac{c}{\lambda} = \frac{3.00 \times 10^8 \text{ m/s}}{4.56 \times 10^{-7} \text{ m}} = 6.5\underline{7}8 \times 10^{14} = 6.58 \times 10^{14} \text{ /s}$$

7.3 First, use the wavelengths to calculate the frequencies from $c = \nu\lambda$. Then calculate the energies using $E = h\nu$.

$$\nu = \frac{c}{\lambda} = \frac{3.00 \times 10^8 \text{ m/s}}{1.0 \times 10^{-6} \text{ m}} = 3.\underline{0}0 \times 10^{14} \text{ /s}$$

$$\nu = \frac{c}{\lambda} = \frac{3.00 \times 10^8 \text{ m/s}}{1.0 \times 10^{-8} \text{ m}} = 3.\underline{0}0 \times 10^{16} \text{ /s}$$

$$\nu = \frac{c}{\lambda} = \frac{3.00 \times 10^8 \text{ m/s}}{1.0 \times 10^{-10} \text{ m}} = 3.\underline{0}0 \times 10^{18} \text{ /s}$$

$E = h\nu = 6.63 \times 10^{-34} \text{ J•s} \times 3.00 \times 10^{14} \text{ /s} = 1.\underline{9}89 \times 10^{-19} = 2.0 \times 10^{-19}$ J (IR)

$E = h\nu = 6.63 \times 10^{-34} \text{ J•s} \times 3.00 \times 10^{16} \text{ /s} = 1.\underline{9}89 \times 10^{-17} = 2.0 \times 10^{-17}$ J (UV)

$E = h\nu = 6.63 \times 10^{-34} \text{ J•s} \times 3.00 \times 10^{18} \text{ /s} = 1.\underline{9}89 \times 10^{-15} = 2.0 \times 10^{-15}$ J (x-ray)

The x-ray photon (shortest wavelength) has the greatest amount of energy; the infrared photon (longest wavelength) has the least amount of energy.

7.4 From the formula for the energy levels, $E = -R_H/n^2$, obtain the expressions for both E_i and E_f. Then calculate the energy change for the transition from $n = 3$ to $n = 1$ by subtracting the lower value from the upper value. Set this equal to $h\nu$. The result is

$$\left[\frac{-R_H}{9}\right] - \left[\frac{-R_H}{1}\right] = \frac{8R_H}{9} = h\nu$$

The frequency of the emitted radiation is

$$\nu = \frac{8R_H}{9h} = \frac{8}{9} \times \frac{2.179 \times 10^{-18} \text{ J}}{6.63 \times 10^{-34} \text{ J•s}} = 2.9\underline{2}1 \times 10^{15} = 2.92 \times 10^{15} \text{ /s}$$

Since $\lambda = c/\nu$,

$$\lambda = \frac{3.00 \times 10^8 \text{ m/s}}{2.92 \times 10^{15} \text{ /s}} = 1.0\underline{2}7 \times 10^{-7} = 1.03 \times 10^{-7} \text{ m, or 103 nm}$$

7.5   Calculate the frequency from $c = \nu\lambda$, recognizing that 589 nm is 5.89×10^{-7} m.

$$\nu = \frac{c}{\lambda} = \frac{3.00 \times 10^8 \text{ m/s}}{5.89 \times 10^{-7} \text{ m}} = 5.0\underline{9}3 \times 10^{14} = 5.09 \times 10^{14} \text{ /s}$$

Finally, calculate the energy difference.

$$E = h\nu = 6.63 \times 10^{-34} \text{ J•s} \times 5.093 \times 10^{14} \text{ /s} = 3.3\underline{7}66 \times 10^{-19} = 3.38 \times 10^{-19} \text{ J}$$

7.6   To calculate wavelength, use the mass of an electron, $m = 9.11 \times 10^{-31}$ kg, and Planck's constant ($h = 6.63 \times 10^{-34}$ J•s, or 6.63×10^{-34} kg•m^2/s^2).

$$\lambda = \frac{h}{mv} = \frac{6.63 \times 10^{-34} \text{ kg} • \text{m}^2/\text{s}}{9.11 \times 10^{-31} \text{ kg} \times 2.19 \times 10^6 \text{ m/s}} = 3.3\underline{2}3 \times 10^{-10} \text{ m (332 pm)}$$

7.7   a. The value of n must be a positive whole number greater than zero. Here, it is zero. Also, if n is zero, there would be no allowed values for l and m_l.

        b. The values for l can range only from zero to (n - 1). Here, l has a value greater than n.

        c. The values for m_l range from -l to +l. Here, m_l has a value greater than l.

        d. The values for m_s are either + 1/2 or -1/2. Here, it is zero.

■ Answers to Review Questions

7.1   Light is a wave, which is a form of electromagnetic radiation. In terms of waves, light can be described as a continuously repeating change, or oscillation, in electric and magnetic fields that can travel through space. Two characteristics of light are wavelength (often given in nanometers, nm), and frequency. The eye can see the light of wavelengths from about 400 nm (violet) to less than 800 nm (red).

7.2   The relationship among the different characteristics of light waves is $c = \nu\lambda$, where ν is the frequency, λ is the wavelength, and c is the speed of light.

7.3   Starting with the shortest wavelengths, the electromagnetic spectrum consists of gamma rays, x rays, far ultraviolet (UV), near UV, visible light, near infrared (IR), far IR, microwaves, radar, and TV/FM radio waves (longest wavelengths).

7.4   The term "quantized" means the possible values of the energies of an atom are limited to only certain values. Planck was trying to explain the intensity of light of various frequencies emitted by a hot solid at different temperatures. The formula he arrived at was $E = nh\nu$, where E is energy, n is a whole number (n = 1, 2, 3, ...), h is Planck's constant, and ν is frequency.

7.5   Photoelectric effect is the term applied to the ejection of electrons from the surface of a metal or from other materials when light shines on it. Electrons are only ejected when the frequency (or energy) of light is larger than a certain minimum, or threshold, value that is constant for each metal. If a photon has a frequency equal to or greater than this minimum value, then it will eject one electron from the metal surface.

7.6   The wave-particle picture of light regards the wave and particle depictions of light as complementary views of the same physical entity. Neither view alone is a complete description of all the properties of light. The wave picture characterizes light only by wavelength and frequency. The particle picture characterizes light only as having an energy equal to $h\nu$.

7.7   The equation that relates the particle properties of light is $E = h\nu$. The symbol E is energy, h is Planck's constant, and ν is the frequency of the light.

7.8   According to physical theory at Rutherford's time, an electrically charged particle revolving around a center would continuously lose energy as electromagnetic radiation. As an electron in an atom lost energy, it would spiral into the nucleus (in about 10^{-10} s). Thus, the stability of the atom could not be explained.

7.9   According to Bohr, an electron in an atom can have only specific energy values. An electron in an atom can change energy only by going from one energy level (of allowed energy) to another energy level (of allowed energy). An electron in a higher energy level can go to a lower energy level by emitting a photon of an energy equal to the difference in energy. However, when an electron is in its lowest energy level, no further changes in energy can occur. Thus, the electron does not continuously radiate energy as thought at Rutherford's time. These features solve the difficulty alluded to in Question 7.8.

7.10   Emission of a photon occurs when an electron in a higher energy level undergoes a transition to a lower energy level. The energy lost is emitted as a photon.

7.11   Absorption of a photon occurs when a photon of a certain required energy is absorbed by a certain electron in an atom. The energy of the photon must be equal to the energy necessary to excite the electron of the atom from a lower energy level, usually the lowest, to a higher energy level. The photon's energy is converted into electronic energy.

7.12   The diffraction of an electron beam is evidence of electron waves. A practical example of diffraction is the operation of the electron microscope.

7.13   The square of a wave function equals the probability of finding an electron within a region of space. More complex mathematical manipulations of the wave function yield values for other parameters.

7.14   The uncertainty principle says we can no longer think of the electron as having a precise orbit in an atom similar to the orbit of the planets around the sun. This principle says it is impossible to know with absolute precision both the position and the speed of a particle such as an electron. Application of Heisenberg's uncertainty principle shows that the uncertainties of position and speed are significant when the principle is applied to electrons.

7.15   Quantum mechanics vastly changes Bohr's original picture of the hydrogen atom in that we can no longer think of the electron as having a precise orbit around the nucleus in this atom. Recall that Bohr's theory depended on the hydrogen electron having specific energy values and thus specific positions and speeds around the nucleus. But quantum mechanics and the uncertainty principle say it is impossible to know with absolute precision both the speed and the position of an electron. So Bohr's energy levels are only the most probable paths of the electrons.

7.16 a.   The principal quantum number can have an integer value between one and infinity.

   b.   The angular momentum quantum number can have any integer value between zero and (n - 1).

   c.   The magnetic quantum number can have any integer value between -l and +l.

   d.   The spin quantum number can be either +1/2 or -1/2.

7.17   The notation is *4f*. This subshell contains seven orbitals.

7.18   An *s* orbital has a spherical shape. A *p* orbital has two lobes positioned along a straight line through the nucleus at the center of the line (a dumbbell shape).

■ Solutions to Practice Problems

Note on significant figures: If the final answer to a solution needs to be rounded off, it is given first with one nonsignificant figure, and the last significant figure is underlined. The final answer is then rounded to the correct number of significant figures. In multiple-step problems, intermediate answers are given with at least one nonsignificant figure; however, only the final answer has been rounded off. Starting with Problem 7.29, the value 2.998×10^8 m/s will be used for the speed of light.

7.29   Solve $c = \lambda\nu$ for λ:

$$\lambda = \frac{c}{\nu} = \frac{2.998 \times 10^8 \text{ m/s}}{1.365 \times 10^6 \text{ /s}} = 219.\underline{6}3 = 219.6 \text{ m}$$

7.31   Solve $c = \lambda\nu$ for ν. Recognize that 478 nm = 478 x 10^{-9} m, or 4.78 x 10^{-7} m.

$$\nu = \frac{c}{\lambda} = \frac{2.998 \times 10^8 \text{ m/s}}{4.78 \times 10^{-7} \text{ m}} = 6.2\underline{7}1 \times 10^{14} = 6.27 \times 10^{14} \text{ /s}$$

7.33   Radio waves travel at the speed of light, so divide the distance by c:

$$56 \times 10^9 \text{ m} \times \frac{1 \text{ s}}{2.998 \times 10^8 \text{ m}} = 18\underline{6}.7 = 1.9 \times 10^2 \text{ s}$$

7.35   To do the calculation, divide one meter by the number of wavelengths in one meter to find the wavelength of this transition. Then, use the speed of light (with nine digits for significant figures) to calculate the frequency:

$$\lambda = \frac{1 \text{ m}}{1,650,763.73} = 6.507802\underline{1}06 \times 10^{-7} \text{ m}$$

$$\nu = \frac{c}{\lambda} = \frac{2.99792458 \times 10^8 \text{ m/s}}{6.507802106 \times 10^{-7} \text{ m}} = 4.948865\underline{1}62 \times 10^{14} = 4.94886516 \times 10^{14} \text{ /s}$$

7.37   Solve for E, using $E = h\nu$, and use four significant figures for h:

$$E = h\nu = (6.626 \times 10^{-34} \text{ J} \cdot \text{s}) \times (1.365 \times 10^6 \text{ /s}) = 9.04\underline{4}4 \times 10^{-28}$$

$$= 9.044 \times 10^{-28} \text{ J}$$

7.39   Recognize that 535 nm = 535 x 10^{-9} m = 5.35 x 10^{-7} m. Then, calculate ν and E.

$$\nu = \frac{c}{\lambda} = \frac{2.998 \times 10^8 \text{ m/s}}{5.35 \times 10^{-7} \text{ m}} = 5.6\underline{0}37 \times 10^{14} \text{ /s}$$

$$E = h\nu = (6.626 \times 10^{-34} \text{ J} \cdot \text{s}) \times (5.6037 \times 10^{14} \text{/s}) = 3.7\underline{1}3 \times 10^{-19} = 3.71 \times 10^{-19} \text{ J}$$

7.41   First, calculate the wavelength of this transition from the frequency using the speed of light:

$$\lambda = \frac{c}{\nu} = \frac{2.998 \times 10^8 \text{ m/s}}{3.84 \times 10^{14} \text{ /s}} = 7.8\underline{0}72 \times 10^{-7} = 7.81 \times 10^{-7} \text{ m} \quad (781 \text{ pm})$$

Using Figure 7.5, note that 781 nm is just on the edge of the red end of the spectrum and is barely visible to the eye.

7.43   Solve the equation $E = -R_H/n^2$ for both E_5 and E_3; equate to $h\nu$, and solve for ν.

$$E_5 = \frac{-R_H}{5^2} = \frac{-R_H}{25}; \qquad E_3 = \frac{-R_H}{3^2} = \frac{-R_H}{9}$$

$$\left[\frac{-R_H}{25}\right] - \left[\frac{-R_H}{9}\right] = \frac{16\,R_H}{225} = h\nu$$

The frequency of the emitted radiation is:

$$\nu = \frac{16\,R_H}{225\,h} = \frac{16}{225} \times \frac{2.179 \times 10^{-18}\ J}{6.626 \times 10^{-34}\ J\bullet s} = 2.3\underline{3}8 \times 10^{14} = 2.34 \times 10^{14}\ /s$$

7.45   Solve the equation $E = -R_H/n^2$ for both E_2 and E_1.

$$E_2 = \frac{-R_H}{2^2} = \frac{-R_H}{4}; \qquad E_1 = \frac{-R_H}{1^2} = \frac{-R_H}{1}$$

$$\left[\frac{-R_H}{4}\right] - \left[\frac{-R_H}{1}\right] = \frac{3\,R_H}{4} = h\nu$$

The frequency of the emitted radiation is:

$$\nu = \frac{3\,R_H}{4\,h} = \frac{3}{4} \times \frac{2.179 \times 10^{-18}\ J}{6.626 \times 10^{-34}\ J\bullet s} = 2.4\underline{6}6 \times 10^{15}\ /s$$

The wavelength can now be calculated.

$$\lambda = \frac{c}{\nu} = \frac{2.998 \times 10^8\ m/s}{2.466 \times 10^{15}\ /s} = 1.2\underline{1}55 \times 10^{-7} = 1.22 \times 10^{-7}\ m\ \text{(near UV)}$$

7.47   This is the highest energy transition from the n = 6 level, so the electron must undergo a transition to the n = 1 level. Solve the Balmer equation using Bohr's approach:

$$E_6 = \frac{-R_H}{6^2} = \frac{-R_H}{36}; \qquad E_1 = \frac{-R_H}{1^2} = \frac{-R_H}{1}$$

$$h\nu = \left[\frac{-R_H}{36}\right] - \left[\frac{-R_H}{1}\right] = \frac{35\,R_H}{36}$$

(continued)

The frequency of the emitted radiation is:

$$\nu = \frac{35 R_H}{36 h} = \frac{35}{36} \times \frac{2.179 \times 10^{-18} \text{ J}}{6.626 \times 10^{-34} \text{ J} \cdot \text{s}} = 3.197 \times 10^{15} \text{ /s}$$

The wavelength can now be calculated.

$$\lambda = \frac{c}{\nu} = \frac{2.998 \times 10^8 \text{ m/s}}{3.197 \times 10^{15} \text{ /s}} = 9.3769 \times 10^{-8} = 9.38 \times 10^{-8} \text{ m} \quad (93.8 \text{ nm})$$

7.49   Noting that 422.7 nm = 4.227×10^{-7} m, convert the 422.7 nm to frequency. Then convert the frequency to energy using $E = h\nu$.

$$\nu = \frac{c}{\lambda} = \frac{2.998 \times 10^8 \text{ m/s}}{4.227 \times 10^{-7} \text{ m}} = 7.0925 \times 10^{14} \text{ /s}$$

$$E = h\nu = (6.626 \times 10^{-34} \text{ J} \cdot \text{s}) \times 7.0925 \times 10^{14} \text{ /s}) = 4.6994 \times 10^{-19}$$

$$= 4.699 \times 10^{-19} \text{ J}$$

7.51   The mass of a neutron = 1.67493×10^{-27} kg. Its speed or velocity, v, of 4.15 km/s equals 4.15×10^3 m/s. Substitute these parameters into the de Broglie relation, and solve for λ:

$$\lambda = \frac{h}{mv} = \frac{6.626 \times 10^{-34} \text{ kg} \cdot \text{m}^2/\text{s}}{1.67493 \times 10^{-27} \text{ kg} \times 4.15 \times 10^3 \text{ m/s}}$$

$$= 9.532 \times 10^{-11} = 9.53 \times 10^{-11} \text{ m} \quad \text{or} \quad 95.3 \text{ pm}$$

7.53   The mass of an electron equals 9.10953×10^{-31} kg. The wavelength, λ, given as 10.0 pm, is equivalent to 1.00×10^{-11} m. Substitute these parameters into the de Broglie relation, and solve for the frequency, v:

$$v = \frac{h}{m\lambda} = \frac{6.626 \times 10^{-34} \text{ kg} \cdot \text{m}^2/\text{s}}{9.10953 \times 10^{-31} \text{ kg} \times 1.00 \times 10^{-11} \text{ m}} = 7.273 \times 10^7$$

$$= 7.27 \times 10^7 \text{ m/s}$$

7.55 Substitute the 1.45×10^{-1} kg mass of the baseball and the 30.0 m/s velocity, v, into the de Broglie relation, and solve for wavelength (recall that 1 pm = 10^{-12} m).

$$\lambda = \frac{h}{mv} = \frac{6.626 \times 10^{-34} \text{ kg} \cdot \text{m}^2/\text{s}}{1.45 \times 10^{-1} \text{ kg} \times 30.0 \text{ m/s}} = 1.5\underline{2}3 \times 10^{-34} = 1.52 \times 10^{-34} \text{ m}$$

$$= 1.52 \times 10^{-22} \text{ pm}$$

Because this is much smaller than 100 pm, the wavelength is much smaller than the diameter of one atom.

7.57 The possible values of l range from zero to (n - 1), so l may be 0, 1, 2, or 3. The possible values of m_l range from - l to + l, so m_l may be -3, -2, -1, 0, +1, +2, or +3.

7.59 For the M shell, n = 3; there are three subshells in this shell (l = 0, 1, and 2). An f subshell has l = 3; the number of orbitals in this subshell is 2(3) + 1 = 7 (m_l = -3, -2, -1, 0, 1, 2, and 3).

7.61 a. 6d   **b.** 5g   **c.** 4f   **d.** 6p

7.63 a. Not permissible; m_s may be only + 1/2 or -1/2.

b. Not permissible; l can only be as large as (n - 1).

c. Not permissible; m_l may not exceed +2 in magnitude.

d. Not permissible; n may not be zero.

e. Not permissible; m_s may only be + 1/2 or -1/2.

■ Solutions to General Problems

7.65 Use c = $\nu\lambda$ to calculate frequency; then use E = hν to calculate energy.

$$\nu = \frac{c}{\lambda} = \frac{2.998 \times 10^8 \text{ m/s}}{4.61 \times 10^{-7} \text{ m}} = 6.5\underline{0}3 \times 10^{14} = 6.50 \times 10^{14} \text{ /s}$$

$$E = h\nu = (6.626 \times 10^{-34} \text{ J} \cdot \text{s}) \times (6.503 \times 10^{14}/\text{s}) = 4.3\underline{0}9 \times 10^{-19}$$

$$= 4.31 \times 10^{-19} \text{ J}$$

7.67 Calculate the frequency corresponding to 4.10 x 10⁻¹⁹ J. Then convert that to wavelength.

$$\nu = \frac{E}{h} = \frac{4.10 \times 10^{-19} \text{ J}}{6.626 \times 10^{-34} \text{ J} \cdot \text{s}} = 6.1\underline{8}7 \times 10^{14} \text{ /s}$$

$$\lambda = \frac{c}{\nu} = \frac{2.998 \times 10^8 \text{ m/s}}{6.187 \times 10^{14} \text{ /s}} = 4.8\underline{4}5 \times 10^{-7} = 4.85 \times 10^{-7} \text{ m} = 485 \text{ nm (blue)}$$

7.69 Solve for frequency using $E = h\nu$.

$$\nu = \frac{E}{h} = \frac{4.34 \times 10^{-19} \text{ J}}{6.626 \times 10^{-34} \text{ J} \cdot \text{s}} = 6.5\underline{4}9 \times 10^{14} = 6.55 \times 10^{14} \text{ /s}$$

7.71 First calculate E_p, the energy of the 345-nm photon, noting that it is equivalent to 3.45×10^{-7} m.

$$E_p = \frac{hc}{\lambda} = \frac{(6.626 \times 10^{-34} \text{ J} \cdot \text{s})(2.998 \times 10^8 \text{ m/s})}{3.45 \times 10^{-7} \text{ m}} = 5.7\underline{5}78 \times 10^{-19} \text{ J}$$

Now, subtract the work function of Ca = 4.34 x 10⁻¹⁹ J (Problem 7.69) from E_p:

$$5.7578 \times 10^{-19} \text{ J} - 4.34 \times 10^{-19} \text{ J} = 1.4\underline{1}78 \times 10^{-19} \text{ J}$$

Note that, for this situation, $E = 1/2 mv^2$. Recall that the mass of the electron is 9.1095 x 10⁻³¹ kg. Now calculate speed, v:

$$v = \sqrt{\frac{2E}{m}} = \sqrt{\frac{2 \times 1.4178 \times 10^{-19} \text{ J}}{9.1095 \times 10^{-31} \text{ kg}}} = 5.5\underline{7}9 \times 10^5 = 5.58 \times 10^5 \text{ m/s}$$

7.73 This is a transition from the n = 5 level to the n = 2 level. Solve the Balmer equation using Bohr's approach.

$$E_5 = \frac{-R_H}{5^2} = \frac{-R_H}{25}; \qquad E_2 = \frac{-R_H}{2^2} = \frac{-R_H}{4}$$

$$h\nu = \left[\frac{-R_H}{25}\right] - \left[\frac{-R_H}{4}\right] = \frac{21 R_H}{100}$$

(continued)

$$\nu = \frac{21R_H}{100\,h} = \frac{21}{100} \times \frac{2.179 \times 10^{-18}\ J}{6.626 \times 10^{-34}\ J \cdot s} = 6.9059 \times 10^{14}\ /s$$

$$\lambda = \frac{c}{\nu} = \frac{2.998 \times 10^8\ m/s}{6.9059 \times 10^{14}\ /s} = 4.341 \times 10^{-7} = 4.34 \times 10^{-7}\ m\ \ (434\ nm)$$

7.75   Use 397 nm = 3.97×10^{-7} m, and convert to frequency and then to energy.

$$\nu = \frac{c}{\lambda} = \frac{2.998 \times 10^8\ m/s}{3.97 \times 10^{-7}\ m} = 7.551 \times 10^{14}\ /s$$

$$E = h\nu = (6.626 \times 10^{-34}\ J \cdot s) \times (7.551 \times 10^{14}) = 5.0037 \times 10^{-19}\ J$$

Substitute this energy into the Balmer formula recalling that the Balmer series is an emission spectrum, so ΔE is negative:

$$E_2 = \frac{-R_H}{2^2} = \frac{-R_H}{4}; \qquad E_i = \frac{-R_H}{n_i^2}$$

$$\left[\frac{-R_H}{4}\right] - \left[\frac{-R_H}{n_i^2}\right] = (-R_H)\left[\frac{1}{4} - \frac{1}{n_i^2}\right] = E \ \ (\text{of line})$$

$$\left[\frac{1}{4} - \frac{1}{n_i^2}\right] = \frac{5.0037 \times 10^{-19}\ J}{2.179 \times 10^{-18}\ J} = 0.22963$$

$$\frac{1}{n_i^2} = \frac{1}{4} - 0.22963 = 0.02036$$

$$n_i = \left[\frac{1}{0.02036}\right]^{1/2} = 7.007 = 7.0\ (= n)$$

7.77　Employ the Balmer formula using $Z = 2$ for the He^+ ion.

$$E_3 = (2)^2 \frac{-R_H}{3^2} = (4)\frac{-R_H}{9}; \qquad E_2 = (2)^2 \frac{-R_H}{2^2} = (4)\frac{-R_H}{4}$$

$$4\left[\frac{-R_H}{9}\right] - 4\left[\frac{-R_H}{4}\right] = 4 \times \frac{5R_H}{36} = h\nu$$

The frequency of the radiation is

$$\nu = \frac{4 \times 5 R_H}{36\,h} = \frac{20}{36} \times \frac{2.179 \times 10^{-18}\ J}{6.626 \times 10^{-34}\ J \cdot s} = 1.8269 \times 10^{15}\ /s$$

$$\lambda = \frac{c}{\nu} = \frac{2.998 \times 10^8\ m/s}{1.8269 \times 10^{15}\ /s} = 1.6409 \times 10^{-7} = 1.64 \times 10^{-7}\ m \ \ (164\ nm;\ near\ UV)$$

7.79　First, use the wavelength of 10.0 pm (1.00×10^{-11} m) and the mass of 9.1095×10^{-31} kg to calculate the velocity, v. Then, use the kinetic energy equation to calculate kinetic energy from velocity.

$$v = \frac{h}{m\lambda} = \frac{6.626 \times 10^{-34}\ kg \cdot m^2/s}{9.1095 \times 10^{-31}\ kg \ \times \ 1.00 \times 10^{-11}\ m} = 7.2737 \times 10^7\ m/s$$

$$E = 1/2mv^2 = 1/2 \times (9.1095 \times 10^{-31}\ kg) \times (7.273 \times 10^7\ m/s)^2 = 2.409 \times 10^{-15}\ J$$

$$E_{ev} = 2.409 \times 10^{-15}\ J \times \frac{1eV}{1.602 \times 10^{-19}\ J} = 1.503 \times 10^4 = 1.50 \times 10^4\ eV$$

7.81 a. Five　　　　b. Seven　　　c. Three　　　　d. One

7.83　The possible subshells for the $n = 6$ shell are 6s, 6p, 6d, 6f, 6g, and 6h.

■ Solutions to Cumulative-Skills Problems

7.85   First, use Avogadro's number to calculate the energy for one Cl_2 molecule.

$$\frac{239 \text{ kJ}}{1 \text{ mol}} \times \frac{1000 \text{ J}}{1 \text{ kJ}} \times \frac{1 \text{ mol}}{6.022 \times 10^{23} \text{ molecules}} = 3.9687 \times 10^{-19} \text{ J/molecule}$$

Then, convert energy to frequency and finally to wavelength.

$$\nu = \frac{E}{h} = \frac{3.9687 \times 10^{-19} \text{ J}}{6.626 \times 10^{-34} \text{ J} \cdot \text{s}} = 5.9897 \times 10^{14} \text{ /s}$$

$$\lambda = \frac{c}{\nu} = \frac{2.998 \times 10^{8} \text{ m/s}}{5.9897 \times 10^{14} \text{ /s}} = 5.0052 \times 10^{-7} \text{ m} \quad (501 \text{ nm; visible region})$$

7.87   First, calculate the energy needed to heat the 0.250 L of water from 20.0°C to 100.0°C.

$$0.250 \text{ L} \times \frac{1000 \text{ g}}{1 \text{ L}} \times \frac{4.184 \text{ J}}{(\text{g} \cdot ^{\circ}\text{C})} \times (100.0 \, ^{\circ}\text{C} - 20.0 ^{\circ}\text{C}) = 8.368 \times 10^{4} \text{ J}$$

Then, calculate the frequency, the energy of one photon, and the number of photons.

$$\nu = \frac{c}{\lambda} = \frac{2.998 \times 10^{8} \text{ m/s}}{0.125 \text{ m}} = 2.398 \times 10^{9} \text{ /s}$$

E of one photon = $h\nu$ = $(6.626 \times 10^{-34} \text{ J} \cdot \text{s}) \times (2.398 \times 10^{9} \text{/s}) = 1.589 \times 10^{-24} \text{ J}$

No. photons = $h\nu$ = $8.368 \times 10^{4} \text{ J} \times \frac{1 \text{ photon}}{1.589 \times 10^{-24} \text{ J}} = 5.2656 \times 10^{28}$

$$= 5.27 \times 10^{28} \text{ photons}$$

7.89   First, write the following equality for the energy to remove one electron, $E_{removal}$:

$$E_{removal} = E_{425\ nm} - E_k \text{ of ejected photon}$$

Use $E = h\nu$ to calculate the energy of the photon. Then, recall that E_k, the kinetic energy, is $1/2mv^2$. Use this to calculate E_k.

$$E_{425\ nm} = \frac{hc}{\lambda} = \frac{(6.626 \times 10^{-34} \text{ J} \bullet \text{s})(2.998 \times 10^8 \text{ m/s})}{4.25 \times 10^{-7} \text{ m}} = 4.6\underline{7}40 \times 10^{-19} \text{ J}$$

$$E_k = 1/2mv^2 = 1/2 \times (9.1095 \times 10^{-31} \text{ kg}) \times (4.88 \times 10^5 \text{ m/s})^2 = 1.0\underline{8}46 \times 10^{-19} \text{ J}$$

Subtract to find $E_{removal}$, and convert it to kJ/mol:

$$E_{removal} = 4.6740 \times 10^{-19} \text{ J} - (1.0847 \times 10^{-19} \text{ J}) = 3.5\underline{8}93 \times 10^{-19}$$

$$= 3.59 \times 10^{-19} \text{ J/electron}$$

$$E_{removal} = \frac{3.5893 \times 10^{-19} \text{ J}}{1 \text{ e}^-} \times \frac{6.022 \times 10^{23} \text{ e}^-}{1 \text{ mol}} \times \frac{1 \text{ kJ}}{1000 \text{ J}} = 2.1\underline{6}1 \times 10^2$$

$$= 2.16 \times 10^2 \text{ kJ/mol}$$

7.91   First, calculate the energy, E, in joules using the product of voltage and charge:

$$E = (4.00 \times 10^3 \text{ V}) \times (1.602 \times 10^{-19} \text{ C}) = 6.4\underline{0}8 \times 10^{-16} \text{ J}$$

Now, use the kinetic energy equation, $E_k = 1/2mv^2$, and solve for velocity:

$$v = \sqrt{\frac{2E_k}{m}} = \sqrt{\frac{2 \times 6.408 \times 10^{-16} \text{ J}}{9.1095 \times 10^{-31} \text{ kg}}} = 3.7\underline{5}08 \times 10^7 \text{ m/s}$$

$$\lambda = \frac{h}{mv} = \frac{6.626 \times 10^{-34} \text{ J} \bullet \text{s}}{(9.1095 \times 10^{-31} \text{ kg}) \times (3.7508 \times 10^7 \text{ m/s})} = 1.9\underline{3}9 \times 10^{-11}$$

$$= 1.94 \times 10^{-11} \text{ m} \ (19.4 \text{ pm})$$

8. ELECTRON CONFIGURATIONS AND PERIODICITY

■ Solutions to Exercises

Note on significant figures: If the final answer to a solution needs to be rounded off, it is given first with one nonsignificant figure, and the last significant figure is underlined. The final answer is then rounded to the correct number of significant figures. In multiple-step problems, intermediate answers are given with at least one nonsignificant figure; however, only the final answer has been rounded off.

8.1 a. Possible orbital diagram.

   b. Possible orbital diagram.

   c. Impossible orbital diagram; there are two electrons in a $2p$ orbital with the same spin.

   d. Possible electron configuration.

   e. Impossible electron configuration; only two electrons are allowed in an s subshell.

   f. Impossible electron configuration; only six electrons are allowed in a p subshell.

8.2   Look at the periodic table. Start with hydrogen and go through the periods, writing down the subshells being filled, stopping with manganese ($Z = 25$). You obtain the following order:

Order:	$1s$	$2s2p$	$3s3p$	$4s3d4p$
Period:	first	second	third	fourth

Now fill the subshells with electrons, remembering that you have a total of twenty five electrons to distribute. You obtain

$$1s^2 2s^2 2p^6 3s^2 3p^6 4s^2 3d^5, \quad \text{or} \quad 1s^2 2s^2 2p^6 3s^2 3p^6 3d^5 4s^2$$

8.3   Arsenic is a main group element in Period 4, Group VA, of the periodic table. The five outer electrons should occupy the $4s$ and $4p$ subshells; the five valence electrons have the configuration $4s^2 4p^3$.

8.4   Because the sum of the $6s^2$ and $6p^2$ electrons gives four outer (valence) electrons, lead should be in Group IVA, which it is. Looking at the table, you find lead in Period 6. From its position, it would be classified as a main-group element.

8.5   The electronic configuration of phosphorus is $1s^2 2s^2 2p^6 3s^2 3p^3$. The orbital diagram is:

1s   2s   2p   3s   3p

8.6   The radius tends to decrease across a row of the periodic table from left to right, and it tends to increase from the top of a column to the bottom. Therefore, in order of increasing radius,

Be < Mg < Na

8.7   It is more likely that (a), 1000 kJ/mol, is the ionization energy for iodine because ionization energies tend to decrease with atomic number in a group (I is below Cl in Group VIIA).

8.8   Fluorine should have a more negative electron affinity because (1) carbon has only two electrons in the p subshell, (2) the -1 fluoride ion has a stable noble-gas configuration, and (3) the electron can approach the fluorine nucleus more closely than the carbon nucleus. This follows the general trend, which is toward more negative electron affinities from left to right in any period.

■ Answers to Review Questions

8.1   In the original Stern-Gerlach experiment, a beam of silver atoms is directed into the field of a specially designed magnet. (The same can be done with hydrogen atoms.) The beam of atoms is split into two by the magnetic field; half are bent toward one magnetic pole face and the other half toward the other magnetic pole face. This effect shows that the atoms themselves act as magnets with a positive or a negative component as indicated by the positive or negative spin quantum numbers.

8.2 In effect, the electron acts as though it were a sphere of spinning charge (Figure 8.3). Like any circulating electric charge, it creates a magnetic field with a spin axis that has more than one possible direction relative to a magnetic field. Electron spin is subject to a quantum restriction to one of two directions corresponding to the m_s quantum numbers +1/2 and -1/2.

8.3 The Pauli exclusion principle limits the configurations of an atom by excluding configurations in which two or more electrons have the same four quantum numbers. For example, each electron in the same orbital must have different m_s values. This also implies that only two electrons occupy one orbital.

8.4 According to the principles discussed in Section 7.5, the number of orbitals in the g subshell (l = 4) is given by 2l + 1 and is thus equal to nine. Because each orbital can hold a maximum of two electrons, the g subshell can hold a maximum of eighteen electrons.

8.5 The orbitals, in order of increasing energy up to and including the 3p orbitals (but not including the 3d orbitals), are as follows: 1s, 2s, 2p, 3s, and 3p (Figure 8.7).

8.6 The noble-gas core is an inner-shell configuration corresponding to one of the noble gases. The pseudo-noble-gas core is an inner-shell configuration corresponding to one of the noble gases together with $(n-1)d^{10}$ electrons. Like the noble-gas core electrons, the d^{10} electrons are not involved in chemical reactions. The valence electron is an electron (of an atom) located outside the noble-gas core or pseudo-noble-gas core. It is an electron primarily involved in chemical reactions.

8.7 The orbital diagram for the $1s^2 2s^2 2p^4$ ground state of oxygen is

1s   2s   2p

Another possible oxygen orbital diagram, but not a ground state, is

1s   2s   2p

8.8 A diamagnetic substance is a substance that is not attracted by a magnetic field or is very slightly repelled by such a field. This property generally indicates the substance has only paired electrons. A paramagnetic substance is a substance that is weakly attracted by a magnetic field. This property generally indicates the substance has one or more unpaired electrons. Ground-state oxygen has two unpaired 2p electrons and is therefore paramagnetic.

8.9 In Groups IA and IIA, the outer s subshell is being filled: s^1 for Group IA and s^2 for Group IIA. In Groups IIIA to VIIIA, the outer p subshell is being filled: p^1 for IIIA, p^2 for IVA, p^3 for VA, p^4 for VIA, p^5 for VIIA, and p^6 for VIIIA. In the transition elements, the $(n-1)d$ subshell is being filled from d^1 to d^{10} electrons. In the lanthanides and actinides, the f subshell is being filled from f^1 to f^{14} electrons.

8.10 Mendeleev arranged the elements in increasing order of atomic weight, an arrangement that was later changed to atomic numbers. His periodic table was divided into rows (periods) and columns (groups). In his first attempt, he left spaces for what he believed to be undiscovered elements. In row five, under aluminum and above indium in Group III, he left a blank space. This Group III element he called eka-aluminum, and he predicted its properties from those of aluminum and indium. Later, the French chemist de Boisbaudran discovered this element and named it gallium.

8.11 In a plot of atomic radii versus atomic number (Figure 8.16), the major trends that emerge are the following: (1) Within each period (horizontal row), the atomic radius tends to decrease with increasing atomic number or nuclear charge. The largest atom in a period is thus the Group 1A atom, and the smallest atom in a period is thus the noble-gas atom. (2) Within each group (vertical column), the atomic radius tends to increase with the period number.

In a plot of ionization energy versus atomic number (Figure 8.18), the major trends are (1) the ionization energy within a period increases with atomic number, and (2) the ionization energy within a group tends to decrease going down the group.

8.12 The alkaline earth element with the smallest radius is beryllium (Be).

8.13 Group VIIA (halogens) is the main group with the most negative electron affinities. Configurations with filled subshells (ground states of the noble-gas elements) would form unstable negative ions when adding one electron per atom.

8.14 The Na^+ and Mg^{2+} ions are stable because they are isoelectronic with the noble gas neon. If Na^{2+} and Mg^{3+} ions were to exist, they would be very unstable because they would not be isoelectronic with any noble-gas structure and because of the energy needed to remove an electron from an inner shell.

8.15 The elements tend to increase in metallic character from right to left in any period. They also tend to increase in metallic character down any column (group) of elements.

8.16 A basic oxide is an oxide that reacts with acids. An example is calcium oxide, CaO. An acidic oxide is an oxide that reacts with bases. An example is carbon dioxide, CO_2.

8.17 Rubidium is the alkali metal atom with a $5s^1$ configuration.

8.18 Atomic number equals 117 (protons in last known element plus those needed to reach Group VIIA).

8.19   The following elements are in Groups IIIA to VIA:

Group IIIA	Group IVA	Group VA	Group VIA
B: metalloid	C: nonmetal	N: nonmetal	O: nonmetal
Al: metal	Si: metalloid	P: nonmetal	S: nonmetal
Ga: metal	Ge: metalloid	As: metalloid	Se: nonmetal
In: metal	Sn: metal	Sb: metalloid	Te: metalloid
Tl: metal	Pb: metal	Bi: metal	Po: metal

Yes, each column displays the expected increasing metallic character.

8.20   The oxides of the following elements are listed as either acidic, basic, amphoteric, or t.g.n.i. (text gives no information):

Group IIIA	Group IVA	Group VA	Group VIA
B: acidic	C: acidic	N: acidic	O: amphoteric (H_2O)
Al: amphoteric	Si: acidic	P: acidic	S: acidic
Ga: amphoteric	Ge: t.g.n.i.	As: t.g.n.i.	Se: acidic
In: basic	Sn: amphoteric	Sb: t.g.n.i.	Te: t.g.n.i.
Tl: basic	Pb: amphoteric	Bi: t.g.n.i.	Po: t.g.n.i.

8.21 $2K(s) + 2H_2O(l) \rightarrow 2KOH(aq) + H_2(g)$

8.22   Barium should be a soft, reactive metal. Barium should form the basic oxide, BaO. Barium metal, for example, would be expected to react with water according to the equation

$Ba(s) + 2H_2O(l) \rightarrow Ba(OH)_2(aq) + H_2(g)$

8.23   The two oxides of carbon are carbon monoxide, CO, and carbon dioxide, CO_2.

8.24 a. White phosphorus   b. Sulfur   c. Bromine   d. Sodium

■ Solutions to Practice Problems

Note on significant figures: If the final answer to a solution needs to be rounded off, it is given first with one nonsignificant figure, and the last significant figure is underlined. The final answer is then rounded to the correct number of significant figures. In multiple-step problems, intermediate answers are given with at least one nonsignificant figure; however, only the final answer has been rounded off.

8.35 a. Not allowed; the paired electrons in the 2p orbital should have opposite spins.

  b. Allowed; electron configuration is $1s^2 2s^2 2p^4$.

  c. Not allowed; the electrons in the 1s orbital must have opposite spins.

  d. Not allowed; the 2s orbital can hold only two electrons maximum, with opposite spins.

8.37 a. Impossible state; the 2p orbitals can hold no more than six electrons.

  b. Impossible state; the 3s orbital can hold no more than two electrons.

  c. Possible state.

  d. Possible state; however, the 3p and 4s orbitals should be filled before the 3d orbital.

8.39   The six possible orbital diagrams for $1s^2 2p^1$ are

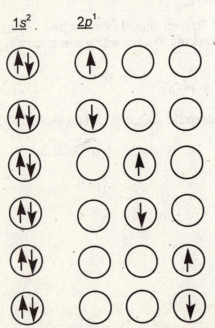

8.41 Iodine (Z = 53): $1s^2 2s^2 2p^6 3s^2 3p^6 3d^{10} 4s^2 4p^6 4d^{10} 5s^2 5p^5$

8.43 Manganese (Z = 25): $1s^2 2s^2 2p^6 3s^2 3p^6 3d^5 4s^2$

8.45 Bromine (Z = 35): $4s^2 4p^5$

8.47 Zirconium (Z = 40): $4d^2 5s^2$

8.49 The highest value of n is six, so thallium (Tl) is in the sixth period. The 5d subshell is filled, and there is a 6p electron, so Tl belongs in an A group. There are three valence electrons, so Tl is in Group IIIA. It is a main-group element.

8.51 Cobalt (Z = 27): [Ar]

3d   4s

8.53 Potassium (Z = 19): [Ar]

4s

All the subshells are filled in the argon core; however, the 4s electron is unpaired, causing the ground state of the potassium atom to be a paramagnetic substance.

8.55 Atomic radius increases going down a column (group), from S to Se, and increases going from right to left in a row, from Se to As. Thus, the order by increasing atomic radius is S, Se, As.

8.57 Ionization energy increases going left to right in a row. Thus, the order by increasing ionization energy is Na, Al, Cl, Ar.

8.59 a. In general, the electron affinity becomes more negative going from left to right within a period. Thus, Br has a more negative electron affinity than As.

  b. In general, a nonmetal has a more negative electron affinity than a metal. Thus, F has a more negative electron affinity than Li.

8.61   Chlorine forms the ClO_3^- ion, so bromine should form the BrO_3^- ion, and potassium
        forms the K^+ ion, so lithium should be Li^+. Thus, the expected formula of lithium
        bromate is $LiBrO_3$.

■ Solutions to General Problems

8.63   Strontium: $1s^2 2s^2 2p^6 3s^2 3p^6 3d^{10} 4s^2 4p^6 5s^2$

8.65   Polonium: $6s^2 6p^4$

8.67   The orbital diagram for arsenic is:

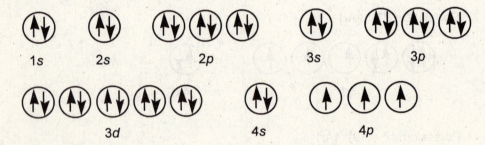

8.69   For eka-lead: $[Rn] 5f^{14} 6d^{10} 7s^2 7p^2$. It is a metal; the oxide is eka-PbO or eka-PbO_2.

8.71   The ionization energy of Fr is ~370 kJ/mol (slightly less than that of Cs).

8.73   Niobium: [Kr]

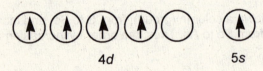

8.75 a. Cl_2 b. Na   c. Sb   d. Ar

8.77   Element with Z = 23: $1s^2 2s^2 2p^6 3s^2 3p^6 3d^3 4s^2$. The element is in Group VB (three of the
        five valence electrons are d electrons) and in Period 4 (largest n is 4). It is a d-block
        transition element.

■ Solutions to Cumulative-Skills Problems

8.79 The equation is:

$$Ba(s) + 2H_2O(l) \rightarrow Ba(OH)_2(aq) + H_2(g)$$

Using the equation, calculate the moles of H_2; then use the ideal gas law to convert to volume.

$$\text{mol } H_2 = 2.50 \text{ g Ba} \times \frac{1 \text{ mol Ba}}{137.33 \text{ g Ba}} \times \frac{1 \text{ mol } H_2}{1 \text{ mol Ba}} = 0.018\underline{2}04 \text{ mol}$$

$$V = \frac{nRT}{P} = \frac{(0.018204 \text{ mol})(0.082057 \text{ L} \cdot \text{atm/K} \cdot \text{mol})(294.2 \text{ K})}{(748/760) \text{ atm}}$$

$$= 0.44\underline{6}51 \text{ L} \quad (447 \text{ mL})$$

8.81 Radium is in Group IIA; hence, the radium cation is Ra^{2+}, and its oxide is RaO. Use the atomic weights to calculate the percentage of Ra in RaO.

$$\text{Percent Ra} = \frac{226 \text{ amu Ra}}{226 \text{ amu Ra} + 16.00 \text{ amu O}} \times 100\% = 93.\underline{38} = 93.4\% \text{ Ra}$$

8.83 Convert 5.00 mg (0.00500 g) Na to moles of Na; then convert to energy using the first ionization energy of 496 kJ/mol Na.

$$\text{mol Na} = 0.00500 \text{ g Na} \times \frac{1 \text{ mol Na}}{22.99 \text{ g Na}} = 2.1\underline{7}4 \times 10^{-4} \text{ mol Na}$$

$$2.1\underline{7}4 \times 10^{-4} \text{ mol Na} \times \frac{496 \text{ kJ}}{1 \text{ mol Na}} = 0.10\underline{7}8 = 0.108 \text{ kJ} = 108 \text{ J}$$

8.85 Use the Bohr formula, where $n_f = \infty$ and $n_i = 1$.

$$\Delta E = -R_H \left[\frac{1}{\infty^2} - \frac{1}{1^2} \right] = -R_H[-1] = R_H = \frac{2.179 \times 10^{-18} \text{ J}}{1 \text{ H atom}}$$

$$\text{I.E.} = \frac{2.179 \times 10^{-18} \text{ J}}{1 \text{ H atom}} \times \frac{6.022 \times 10^{23} \text{ H atoms}}{1 \text{ mol H}} = \frac{1.31\underline{2}19 \times 10^6 \text{ J}}{1 \text{ mol H}}$$

$$= 1.312 \times 10^3 \text{ kJ/mol H}$$

8.87   Add the three equations after reversing the equation for the lattice energy and its ΔH:

$$Na(g) \rightarrow Na^+(g) + e^- \qquad \Delta H = +496 \text{ kJ/mol}$$

$$Cl(g) + e^- \rightarrow Cl^-(g) \qquad \Delta H = -349 \text{ kJ/mol}$$

$$Na^+(g) + Cl^-(g) \rightarrow NaCl(s) \qquad -1(\Delta H = 786 \text{ kJ/mol})$$

$$Na(g) + Cl(g) \rightarrow NaCl(s) \qquad \Delta H = -639 \text{ kJ/mol}$$

9. IONIC AND COVALENT BONDING

■ Solutions to Exercises

9.1   The Lewis symbol for oxygen is $:\overset{\cdot\cdot}{O}\cdot$ and the Lewis symbol for magnesium is $\cdot Mg\cdot$.
The magnesium atom loses two electrons, and the oxygen atom accepts two
electrons. You can represent this electron transfer as follows:

9.2   The electron configuration of the Ca atom is $[Ar]4s^2$. By losing two electrons, the atom
assumes a 2+ charge and the argon configuration, $[Ar]$. The Lewis symbol is Ca^{2+}. The
S atom has the configuration $[Ne]3s^23p^4$. By gaining two electrons, the atom assumes
a 2- charge and the argon configuration $[Ne]3s^23p^6$ and is the same as $[Ar]$. The Lewis
symbol is

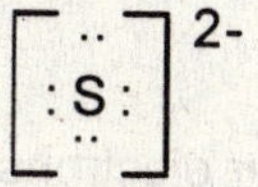

$$\left[:\overset{\cdot\cdot}{\underset{\cdot\cdot}{S}}:\right]^{2-}$$

9.3   The electron configuration of lead (Pb) is $[Xe]4f^{14}5d^{10}6s^26p^2$. The electron
configuration of Pb^{2+} is $[Xe]4f^{14}5d^{10}6s^2$.

9.4   The electron configuration of manganese ($Z = 25$) is $[Ar]3d^54s^2$. To find the ion
configuration, first remove the $4s$ electrons, then the $3d$ electrons. In this case, only
two electrons need to be removed. The electron configuration of Mn^{2+} is $[Ar]3d^5$.

9.5 S^{2-} has a larger radius than S. The anion has more electrons than the atom. The electron-electron repulsion is greater; hence, the valence orbitals expand. The anion radius is larger than the atomic radius.

9.6   The ionic radii increase down any column because of the addition of electron shells. All of these ions are from the Group IIA family; therefore, $Mg^{2+} < Ca^{2+} < Sr^{2+}$.

9.7 Cl^-, Ca^{2+}, and P^{3-} are isoelectronic with an electronic configuration equivalent to [Ar]. In an isoelectronic sequence, the ionic radius decreases with increasing nuclear charge. Therefore, in order of increasing ionic radius, we have Ca^{2+}, Cl^-, and P^{3-}.

9.8   The absolute value of the electronegativity differences are C-O, 1.0; C-S, 0.0; and H-Br, 0.7. Therefore, C-O is the most polar bond.

9.9   First, calculate the total number of valence electrons. C has four, Cl has seven, and F has seven. The total number is $4 + (2 \times 7) + (2 \times 7) = 32$. The expected skeleton consists of a carbon atom surrounded by Cl and F atoms. Distribute the electron pairs to the surrounding atoms to satisfy the octet rule. All 32 electrons (16 pairs) are accounted for.

$$: \overset{\displaystyle ..}{\underset{\displaystyle ..}{Cl}} :$$

$$: \overset{..}{\underset{..}{F}} : \overset{..}{\underset{..}{C}} : \overset{..}{\underset{..}{F}} :$$

$$: \overset{\displaystyle ..}{\underset{\displaystyle ..}{Cl}} :$$

9.10 The total number of electrons in CO_2 is $4 + (2 \times 6) = 16$. Because carbon is more electropositive than oxygen, it is expected to be the central atom. Distribute the electrons to the surrounding atoms to satisfy the octet rule.

$$: \overset{..}{\underset{..}{O}} : C : \overset{..}{\underset{..}{O}} :$$

All sixteen electrons have been used, but notice there are only four electrons on carbon. This is four electrons short of a complete octet, which suggests the existence of double bonds. Move a pair of electrons from each oxygen to the carbon-oxygen bonds.

$$\overset{..}{O} :: C :: \overset{..}{O} \qquad \text{or} \qquad \overset{..}{O} = C = \overset{..}{O}$$

9.11 a. There are (3 x 1) + 6 = 9 valence electrons in H_3O. The H_3O^+ ion has one less electron than is provided by the neutral atoms because the charge on the ion is +1. Hence, there are eight valence electrons in H_3O^+. The electron-dot formula is

$$\left[\begin{array}{c} \text{H} \\ \text{..} \\ \text{H} : \text{O} : \text{H} \\ \text{..} \end{array} \right]^+$$

b. Cl has seven valence electrons, and O has six valence electrons. The total number of valence electrons from the neutral atoms is 7 + (2 x 6) = 19. The charge on the ClO_2^- is –1, which provides one more electron than the neutral atoms. This makes a total of twenty valence electrons. The electron-dot formula for ClO_2^- is

$$\left[: \overset{..}{\underset{..}{O}} : \overset{..}{\underset{..}{Cl}} : \overset{..}{\underset{..}{O}} : \right]^-$$

9.12 The resonance formulas for NO_3^- are

$$\left[\begin{array}{c} : \overset{..}{O} : \\ :: \\ : \overset{..}{O} : \overset{..}{N} : \overset{..}{O} : \end{array} \right]^- \qquad \left[\begin{array}{c} : \overset{..}{O} : \\ \\ : \overset{..}{O} :: N \ : \overset{..}{O} : \end{array} \right]^- \qquad \left[\begin{array}{c} : \overset{..}{O} \\ \\ : \overset{..}{O} : N :: \overset{..}{O} : \end{array} \right]^-$$

9.13 The number of valence electrons in SF_4 is 6 + (4 x 7) = 34. The skeleton structure is a sulfur atom surrounded by fluorine atoms. After the electron pairs are placed on the F atoms to satisfy the octet rule, two electrons remain.

$$\begin{array}{ccc} : \overset{..}{F} & & \overset{..}{F} : \\ \diagdown & & \diagup \\ & S & \\ \diagup & & \diagdown \\ : \underset{..}{F} & & \underset{..}{F} : \end{array}$$

These additional two electrons are put on the sulfur atom because it had *d* orbitals and, therefore, it can expand its octet.

$$\begin{array}{ccc} : \overset{..}{F} & & \overset{..}{F} : \\ \diagdown & & \diagup \\ & : S & \\ \diagup & & \diagdown \\ : \underset{..}{F} & & \underset{..}{F} : \end{array}$$

9.14 Be has two valence electrons, and Cl has seven valence electrons. The total number of valence electrons is 2 + (2 x 7) = 16 in the $BeCl_2$ molecule. Be, a Group IIA element, can have fewer than eight electrons around it. The electron-dot formula of $BeCl_2$ is

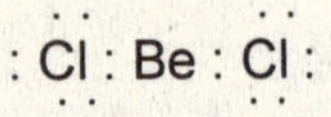

$$: \ddot{Cl} : Be : \ddot{Cl} :$$

9.15 The total number of electrons in H_3PO_4 is 3 + 5 + 24 = 32. Assume a skeleton structure in which the phosphorus atom is surrounded by the more electronegative four oxygen atoms. The hydrogen atoms are then attached to the oxygen atoms. Distribute the electron pairs to the surrounding atoms to satisfy the octet rule. If you assume all single bonds (structure on the left), the formal charge on the phosphorus is +1 and the formal charge on the top oxygen is -1. Using the principle of forming a double bond with a pair of electrons on the atom with the negative formal charge, you obtain the structure on the right. The formal charge on all oxygens in this structure is zero; the formal charge on phosphorus is 5 - 5 = 0. This is the better structure.

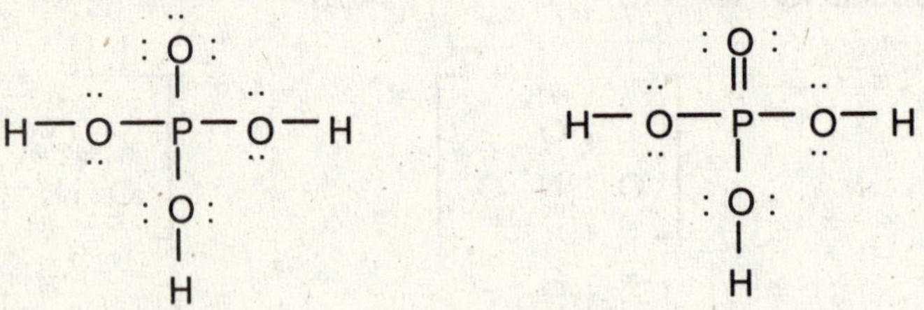

9.16 The bond length can be predicted by adding the covalent radii of the two atoms. For O-H, we have 66 pm + 37 pm = 103 pm.

9.17 As the bond order increases, the bond length decreases. Since the C=O is a double bond, we would expect it to be the shorter one, 123 pm.

9.18

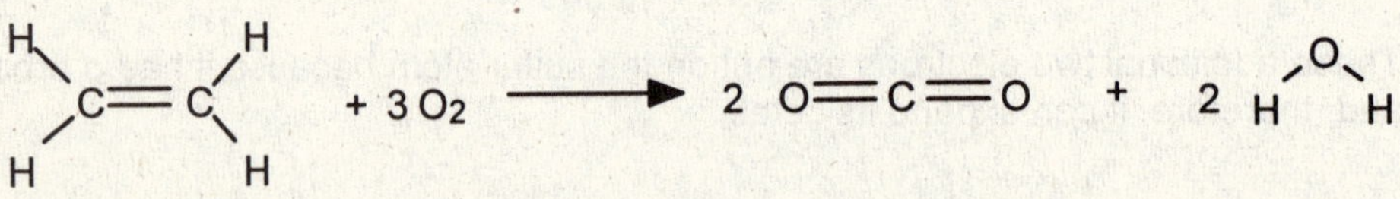

One C=C bond, four C-H bonds, and three O_2 bonds are broken. There are four C=O bonds and four O-H bonds formed.

$$\Delta H = \{[602 + (4 \times 411) + (3 \times 494)] - [(4 \times 799) + (4 \times 459)]\} \text{ kJ} = -1304 \text{ kJ}$$

■ Answers to Review Questions

9.1 As an Na atom approaches a Cl atom, the outer electron of the Na atom is transferred to the Cl atom. The result is an Na^+ and a Cl^- ion. Positively charged ions attract negatively charged ions, so, finally, the NaCl crystal consists of Na^+ ions surrounded by six Cl^- ions surrounded by six Na^+ ions.

9.2 Ions tend to attract as many ions of opposite charge about them as possible. The result is that ions tend to form crystalline solids rather than molecular substances.

9.3 The energy terms involved in the formation of an ionic solid from atoms are the ionization energy of the metal atom, the electron affinity of the nonmetal atom, and the energy of the attraction of the ions forming the ionic solid. The energy of the solid will be low if the ionization energy of the metal is low, the electron affinity of the nonmetal is high, and the energy of the attraction of the ions is large.

9.4 The lattice energy for potassium bromide is the change in energy that occurs when KBr(s) is separated into isolated $K^+(g)$ and $Br^-(g)$ ions in the gas phase.

$$KB(s) \rightarrow K^+(g) + Br^-(g)$$

9.5 A monatomic cation with a charge equal to the group number corresponds to the loss of all valence electrons. This loss of electrons would give a noble-gas configuration, which is especially stable. A monatomic anion with a charge equal to the group number minus eight would have a noble-gas configuration.

9.6 Most of the transition elements have configurations in which the outer s subshell is doubly occupied. These electrons will be lost first, and we might expect each to be lost with almost equal ease, resulting in +2 ions.

9.7 If we assume the ions are spheres that are just touching, the distances between centers of the spheres will be related to the radii of the spheres. For example, in LiI, we assume that the -1 ions are large spheres that are touching. The distance between centers of the I^- ions equals two times the radius of the I^- ion.

9.8 In going across a period, the cations decrease in radius. When we reach the anions, there is an abrupt increase in radius, and then the radii again decrease. Ionic radii increase going down any column of the periodic table.

9.9 As the H atoms approach one another, their $1s$ orbitals begin to overlap. Each electron can then occupy the space around both atoms; that is, the two electrons are shared by the atoms.

9.10

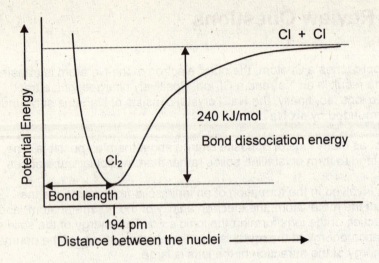

9.11   An example is thionyl chloride, $SOCl_2$:

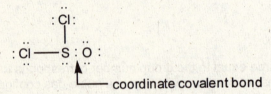

Note that the O atom has eight electrons around it; that is, it has two more electrons than the neutral atom. These two electrons must have come from the S atom. Thus, this bond is a coordinate covalent bond.

9.12   In many atoms of the main-group elements, bonding uses an s orbital and the three p orbitals of the valence shell. These four orbitals are filled with eight electrons, thus accounting for the octet rule.

9.13   Electronegativity increases from left to right (with the exception of the noble gases) and decreases from top to bottom in the periodic table.

9.14   The absolute difference in the electronegativities of the two atoms in a bond gives a rough measure of the polarity of the bond.

9.15   Resonance is used to describe the electron structure of a molecule in which bonding electrons are delocalized. In a resonance description, the molecule is described in terms of two or more Lewis formulas. If we want to retain Lewis formulas, resonance is required because each Lewis formula assumes that a bonding pair of electrons occupies the region between two atoms. We must imagine that the actual electron structure of the molecule is a composite of all resonance formulas.

9.16 Molecules having an odd number of electrons do not obey the octet rule. An example is nitrogen monoxide, NO. The other exceptions fall into two groups. In one group are molecules with an atom having fewer than eight valence electrons around it. An example is borane, BH_3. In the other group are molecules with an atom having more than eight valence electrons around it. An example is sulfur hexafluoride, SF_6.

9.17 As the bond order increases, the bond length decreases. For example, the average carbon-carbon single-bond length is 154 pm, whereas the carbon-carbon double-bond length is 134 pm, and the carbon-carbon triple-bond length is 120 pm.

9.18 Bond energy is the average enthalpy change for the breaking of a bond in a molecule. The enthalpy of a reaction for gaseous reactions can be determined by summing the bond energies of all the bonds that are broken and subtracting the sum of the bond energies of all the bonds that are formed.

■ Solutions to Practice Problems

9.29 a. P has the electron configuration $[Ne]3s^2 3p^3$. It has five electrons in its valence shell. The Lewis formula is

$$\cdot \overset{\cdot \cdot}{\underset{\cdot}{P}} \cdot$$

b. P^{3-} has three more valence electrons than P. It now has eight electrons in its valence shell. The Lewis formula is

$$\left[\; \overset{\cdot \cdot}{\underset{\cdot \cdot}{: P :}} \; \right]^{3-}$$

c. Ga has the electron configuration $[Ar]3d^{10}4s^2 4p^1$. It has three electrons in its valence shell. The Lewis formula is

$$\cdot \overset{\cdot}{Ga} \cdot$$

d. Ga^{3+} has three less valence electrons than Ga. It now has zero electrons in its valence shell. The Lewis formula is

$$Ga^{3+}$$

9.31 a. If the calcium atom loses two electrons, and the bromine atoms gain one electron each, all three atoms will assume noble-gas configurations. This can be represented as follows.

$$: \overset{..}{\underset{..}{Br}} \cdot \; + \; \cdot Ca \cdot \; + \; \cdot \overset{..}{\underset{..}{Br}} : \; \longrightarrow \; \left[: \overset{..}{\underset{..}{Br}} : \right]^{-} \; + \; Ca^{2+} \; + \; \left[: \overset{..}{\underset{..}{Br}} : \right]^{-}$$

b. If the potassium atom loses one electron, and the iodine atom gains one electron, both atoms will assume a noble-gas configuration. This can be represented as follows.

$$K \cdot \; + \; \cdot \overset{..}{\underset{..}{I}} : \; \longrightarrow \; K^{+} \; + \; \left[: \overset{..}{\underset{..}{I}} : \right]^{-}$$

9.33 a. As: $\quad 1s^2 2s^2 2p^6 3s^2 3p^6 3d^{10} 4s^2 4p^3$ $\qquad : \overset{}{\underset{.}{As}} \cdot$

b. As^{3+}: $\quad 1s^2 2s^2 2p^6 3s^2 3p^6 3d^{10} 4s^2$ $\qquad \left[\cdot \overset{}{As} \cdot \right]^{3+}$

c. Se: $\quad 1s^2 2s^2 2p^6 3s^2 3p^6 3d^{10} 4s^2 4p^4$ $\qquad : \overset{}{\underset{.}{Se}} \cdot$

d. Se^{2-}: $\quad 1s^2 2s^2 2p^6 3s^2 3p^6 3d^{10} 4s^2 4p^6$ $\qquad \left[: \overset{..}{\underset{..}{Se}} : \right]^{2-}$

9.35 a. Bi: $\qquad$ [Xe]$4f^{14} 5d^{10} 6s^2 6p^3$

b. Bi^{3+}: $\qquad$ The three $6p$ electrons are lost from the valence shell.

$\qquad$ [Xe]$4f^{14} 5d^{10} 6s^2$

9.37 $\quad$ The +2 ion is formed by the loss of electrons from the $4s$ subshell.

$\qquad$ Ni^{2+}: $\quad$ [Ar]$3d^8$

$\quad$ The +3 ion is formed by the loss of electrons from the $4s$ and $3d$ subshells.

$\qquad$ Ni^{3+}: $\quad$ [Ar]$3d^7$

9.39 a. $Sr^{2+} < Sr$

The cation is smaller than the neutral atom because it has lost all its valence electrons; hence, it has one less shell of electrons. The electron-electron repulsion is reduced, so the orbitals shrink because of the increased attraction of the electrons to the nucleus.

b. $Br < Br^-$

The anion is larger than the neutral atom because it has more electrons. The electron-electron repulsion is greater, so the valence orbitals expand to give a larger radius.

9.41 $S^{2-} < Se^{2-} < Te^{2-}$

All have the same number of electrons in the valence shell. The radius increases with the increasing number of filled shells.

9.43 Smallest Na^+ (Z = 11), F^- (Z = 9), N^{3-} (Z = 7) Largest

These ions are isoelectronic. The atomic radius increases with the decreasing nuclear charge (Z).

9.45

$$2 H \cdot + \cdot \ddot{Se} : \longrightarrow H : \ddot{Se} : \\ \qquad\qquad\qquad\qquad\quad H$$

lone pairs

Bonding electron pairs

9.47 Arsenic is in Group VA on the periodic table and has five valence electrons. Bromine is in Group VIIA and has seven valence electrons. Arsenic forms three covalent bonds, and bromine forms one covalent bond. Therefore, the simplest compound would be $AsBr_3$.

9.49 a. P, N, O

Electronegativity increases from left to right and bottom to top in the periodic table.

b. Na, Mg, Al

Electronegativity increases from left to right within a period.

c. Al, Si, C

Electronegativity increases from left to right and bottom to top in the periodic table.

9.51 $X_O - X_P = 3.5 - 2.1 = 1.4$

$X_{Cl} - X_C = 3.0 - 2.5 = 0.5$

$X_{Br} - X_{As} = 2.8 - 2.0 = 0.8$

The bonds arranged by increasing difference in electronegativity are C-Cl, As-Br, P-O.

9.53 a. P—O
$\delta+$ $\delta-$

b. C—Cl
$\delta+$ $\delta-$

c. As—Br
$\delta+$ $\delta-$

9.55 a. Total number of valence electrons = 7 + 7 = 14. Br—Br is the skeleton. Distribute the remaining twelve electrons.

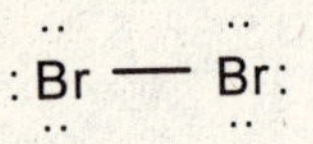

: Br —— Br :

b. Total valence electrons = (2 x 1) + 6 = 8. The skeleton is H—Se—H. Distribute the remaining four electrons.

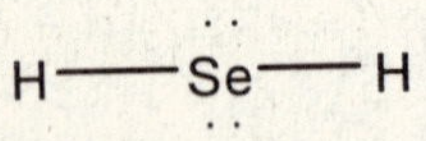

H —— Se —— H

c. Total valence electrons = (3 x 7) + 5 = 26. The skeleton is

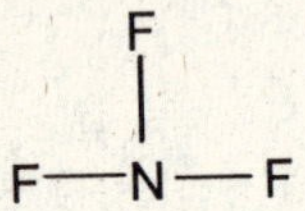

F
|
F—N—F

Distribute the remaining twenty electrons.

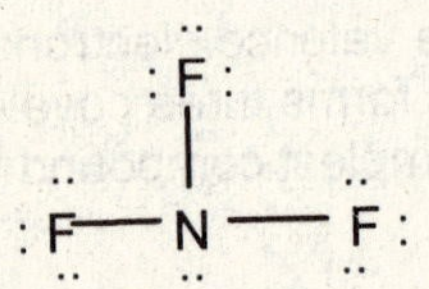

: F :
|
: F—N—F :

9.57 a. Total valence electrons = 2 x 5 = 10. The skeleton is P-P. Distribute the remaining electrons symmetrically:

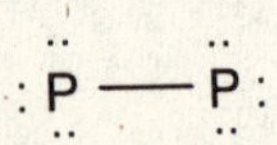

: P —— P :

(continued)

Neither P atom has an octet. There are four fewer electrons than needed. This suggests the presence of a triple bond. Make one lone pair from each P a bonding pair.

$: P \equiv P :$

b. Total valence electrons = 4 + 6 + (2 x 7) = 24. The skeleton is

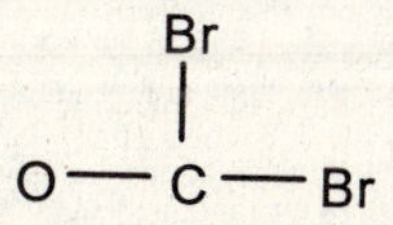

$$O - \underset{\underset{Br}{|}}{C} - Br$$

Distribute the remaining eighteen electrons.

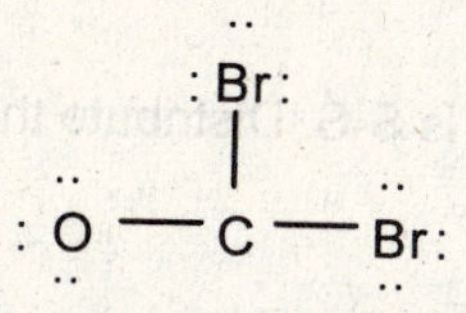

$$: \ddot{O} - \underset{\underset{: \ddot{Br}:}{|}}{C} - \ddot{Br}:$$

Notice that carbon is two electrons short of an octet. This suggests the presence of a double bond. The most likely double bond is between C and O.

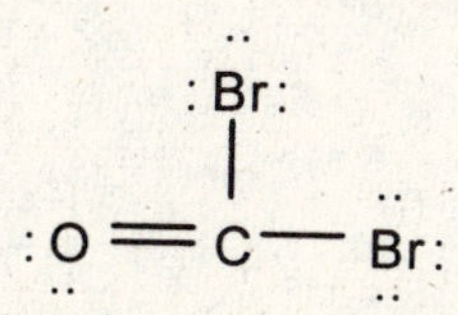

$$: \ddot{O} = \underset{\underset{: \ddot{Br}:}{|}}{C} - \ddot{Br}:$$

c. Total valence electrons = 1 + 5 + (2 x 6) = 18. The skeleton is most likely H-O-N-O. Distribute the remaining twelve electrons:

$$H - \ddot{O} - \ddot{N} - \ddot{O}:$$

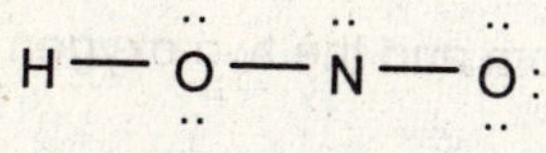

Notice the N atom does not have an octet. It is two electrons short. The most likely double bond is between N and O.

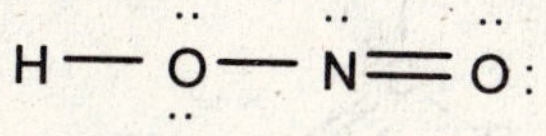

$$H - \ddot{O} - N = \ddot{O}:$$

9.59 a. Total valence electrons = 7 + 6 + 1 = 14. The skeleton is Cl-O. Distribute the remaining twelve electrons.

$$\left[: \ddot{Cl} - \ddot{O}: \right]^-$$

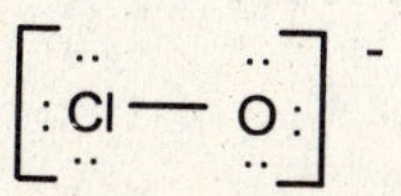

(continued)

b. Total valence electrons = 4 + (3 x 7) + 1 = 26. The skeleton is

$$Cl-Sn-Cl$$
$$Cl$$

Distribute the remaining twenty electrons so that each atom has an octet.

c. Total valence electrons = (2 x 6) + 2 = 14. The skeleton is S-S. Distribute the remaining twelve electrons.

9.61 a. There are two possible resonance structures for HNO_3.

One electron pair is delocalized over the nitrogen atom and the two oxygen atoms.

b. There are three possible resonance structures for SO_3.

One pair of electrons is delocalized over the region of the three sulfur-oxygen bonds.

9.63

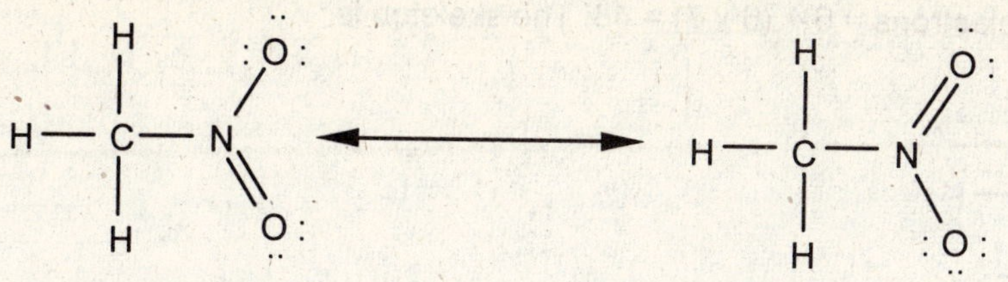

$$H - \overset{\overset{\displaystyle H}{|}}{\underset{\underset{\displaystyle H}{|}}{C}} - N \overset{\displaystyle \ddot{O}:}{\underset{\displaystyle \ddot{O}:}{}} \longleftrightarrow H - \overset{\overset{\displaystyle H}{|}}{\underset{\underset{\displaystyle H}{|}}{C}} - N \overset{\displaystyle \ddot{O}:}{\underset{\displaystyle \ddot{O}:}{}}$$

One pair of electrons is delocalized over the O-N-O bonds.

9.65 a. Total valence electrons = 8 + (2 x 7) = 22. The skeleton is F-Xe-F. Place six electrons around each fluorine atom to satisfy its octet.

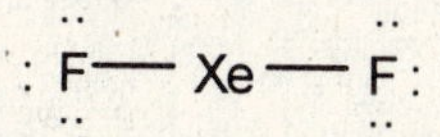

$$: \ddot{F} - Xe - \ddot{F} :$$

There are three electron pairs remaining. Place them on the xenon atom.

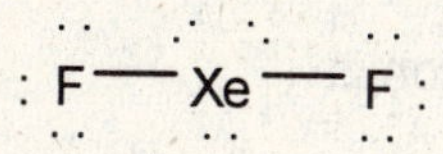

$$: \ddot{F} - Xe - \ddot{F} :$$

b. Total valence electrons = 6 + (4 x 7) = 34. The skeleton is

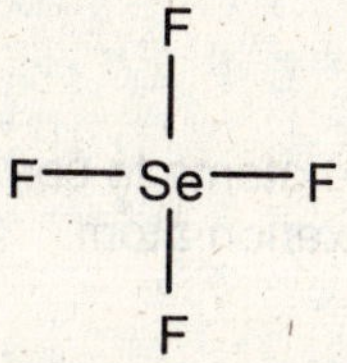

$$\underset{\displaystyle F}{\overset{\displaystyle F}{F - Se - F}}$$

Distribute twenty four of the remaining twenty six electrons on the fluorine atoms. The remaining pair of electrons is placed on the selenium atom.

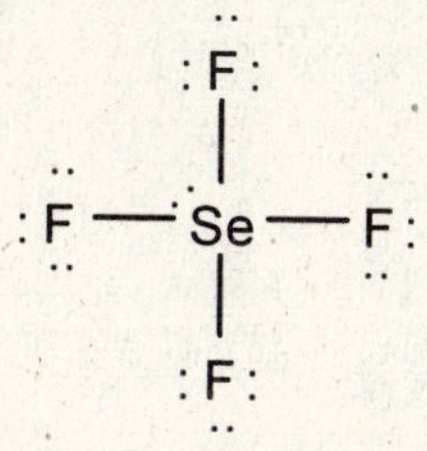

$$\underset{\displaystyle :\ddot{F}:}{\overset{\displaystyle :\ddot{F}:}{:\ddot{F} - \ddot{Se} - \ddot{F}:}}$$

(continued)

c. Total valence electrons = 6 + (6 x 7) = 48. The skeleton is

Distribute the remaining thirty six electrons on the fluorine atoms.

d. Total valence electrons = 8 + (5 x 7) - 1 = 42. The skeleton is

Use thirty of the remaining thirty two electrons on the fluorine atoms to complete their octets. The remaining two electrons form a lone pair on the xenon atom.

9.67 a. Total valence electrons = 3 + (3 x 7) = 24. The skeleton is

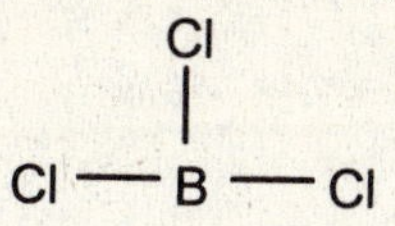

$$
\begin{array}{c}
\text{Cl} \\
| \\
\text{Cl} - \text{B} - \text{Cl}
\end{array}
$$

Distribute the remaining eighteen electrons.

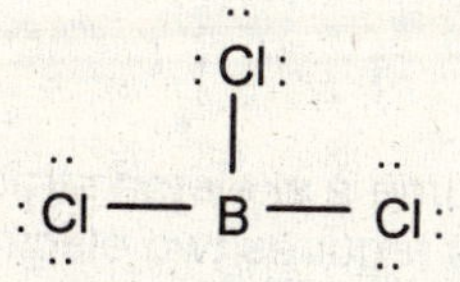

$$
\begin{array}{c}
: \ddot{\text{Cl}} : \\
| \\
: \ddot{\text{Cl}} - \text{B} - \ddot{\text{Cl}} :
\end{array}
$$

Although boron has only six electrons, it has the normal number of covalent bonds.

b. Total valence electrons = 3 + (2 x 7) - 1 = 16. The skeleton is Cl-Tl-Cl. Distribute the remaining twelve electrons.

$$
\left[: \ddot{\text{Cl}} - \text{Tl} - \ddot{\text{Cl}} : \right]^{+}
$$

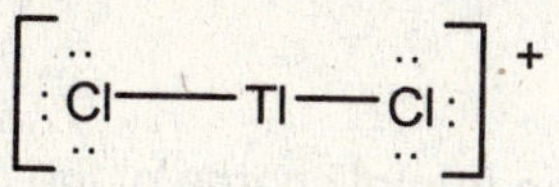

Tl has only four electrons around it.

c. Total valence electrons = 2 + (2 x 7) = 16. The skeleton is Br-Be-Br. Distribute the remaining twelve electrons.

$$
: \ddot{\text{Br}} - \text{Be} - \ddot{\text{Br}} :
$$

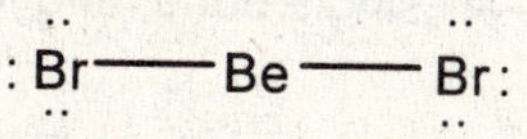

In covalent compounds, beryllium frequently has two bonds, even though it does not have an octet.

9.69 a. The total number of electrons in O_3 is 3 x 6 = 18. Assume a skeleton structure in which one oxygen atom is singly bonded to the other two oxygen atoms. This requires six electrons for the three single bonds, leaving twelve electrons to be used. It is impossible to fill the outer octets of all three oxygen atoms by writing three electron pairs around each, so a double bond must be written between the central oxygen and one of the other oxygen atoms. Then, distribute the electron pairs to the oxygen atoms to satisfy the octet rule. As shown below, there are three bonds and six lone pairs of electrons, or eighteen electrons, in the structure. Thus, all eighteen electrons are accounted for.

(continued)

One of the possible resonance structures is shown below; the other structure would have the double bond written between the left and central oxygen atoms.

$$\ddot{O} - \ddot{O} = \ddot{O}$$

Starting with the left oxygen, the formal charge of this oxygen is 6 - 1 - 6 = -1. The formal charge of just the central oxygen is 6 - 3 - 2 = +1. The formal charge of the right oxygen is 6 - 2 - 4 = 0. The sum of all three is 0.

b. The total number of electrons in CO is 4 + 6 = 10. Assume a skeleton structure in which the oxygen atom is singly bonded to carbon. This requires two electrons for the single bond, and leaves eight electrons to be used. It is impossible to fill the outer octets of the carbon and oxygen by writing four electron pairs around each, so a triple bond must be written between the carbon and oxygen. Then, distribute the electron pairs to both atoms to satisfy the octet rule. As shown below, there are one triple bond and two lone pairs of electrons, or ten electrons, in the structure. Thus, all ten electrons are accounted for. The structure is

$$: C \equiv O :$$

The formal charge of the carbon is 4 - 3 - 2 = -1. The formal charge of the oxygen is 6 - 3 - 2 = +1. The sum of both is 0.

c. The total number of electrons in HNO_3 is 1 + 5 + 18 = 24. Assume a skeleton structure in which the nitrogen atom is singly bonded to two oxygen atoms and doubly bonded to one oxygen. This requires two electrons for the O—H single bond, and leaves eight electrons to be used for the N bonds.

$$H - \ddot{O} - N = \ddot{O} :$$
$$\underset{\displaystyle :\ddot{O}:}{|}$$

The formal charge of the nitrogen is 5 - 4 - 0 = +1. The formal charge of the hydrogen is 1 - 1 - 0 = 0. The formal charge of the oxygen bonded to the hydrogen is 6 - 2 - 4 = 0. The formal charge of the other singly bonded oxygen is 6 - 1 - 6 = -1. The formal charge of the doubly bonded oxygen is 6 - 2 - 4 = 0.

9.71 a. The total number of electrons in SOF_2 is 6 + 6 + 14 = 26. Assume a skeleton structure in which the sulfur atom is singly bonded to the two fluorine atoms. If a S-O single bond is assumed, twenty six electrons are needed. However, there would be formal charges of +1 on the sulfur and -1 on the oxygen. Using a S=O double bond requires only twenty six electrons and results in zero formal charge.

$$:\overset{..}{\underset{..}{F}}-\overset{\overset{\overset{..}{\cdot\cdot}}{}}{\underset{\overset{\parallel}{\underset{..}{:O:}}}{S}}-\overset{..}{\underset{..}{F}}:$$

The formal charge on each of the two fluorine atoms is 7 - 1 - 6 = 0. The formal charge on the oxygen is 6 - 2 - 4 = 0. The formal charge on the sulfur is 6 - 4 - 2 = 0.

b. The total number of electrons in H_2SO_3 is 2 + 6 + 18 = 26. Assume a skeleton structure in which the sulfur atom is singly bonded to the three oxygen atoms. Then, form single bonds from the two hydrogen atoms to each of two oxygen atoms. If a S-O single bond is assumed, twenty six electrons are needed. However, there would be formal charges of +1 on the sulfur and -1 on the oxygen. A S=O double bond also requires twenty six electrons but results in zero formal charge on all atoms.

$$H-\overset{..}{\underset{..}{O}}-\overset{\overset{}{}}{\underset{\overset{\parallel}{\underset{..}{:O:}}}{S}}-\overset{..}{\underset{..}{O}}-H$$

The formal charge of each hydrogen is 1 - 1 = 0. The formal charge of each oxygen bonded to a hydrogen is 6 - 2 - 4 = 0. The formal charge of the oxygen doubly bonded to the sulfur is 6 - 2 - 4 = 0. The formal charge of the sulfur is 6 - 4 - 2 = 0.

c. The total number of electrons in $HClO_2$ is 1 + 7 + 12 = 20. Assume a skeleton structure in which the chlorine atom is singly bonded to the two oxygen atoms, and the hydrogen is singly bonded to one of the oxygen atoms. If all Cl-O single bonds are assumed, twenty electrons are needed, but the chlorine exhibits a formal charge of +1 and one oxygen exhibits a formal charge of -1. Hence, a pair of electrons on the oxygen without the hydrogen is used to form a Cl=O double bond.

$$H-\overset{..}{\underset{..}{O}}-\overset{}{\underset{\overset{\parallel}{\underset{..}{:O:}}}{Cl}}:$$

The formal charge of hydrogen is 1 - 1 = 0. The formal charge of the oxygen bonded to hydrogen is 6 - 2 - 4 = 0. The formal charge of the oxygen that is not bonded to the hydrogen is 6 - 2 - 4 = 0. The formal charge of the chlorine is 7 - 3 - 4 = 0.

9.73 r_F = 64 pm

r_P = 110 pm

$d_{P-F} = r_F + r_P$ = 64 pm + 110 pm = 174 pm

9.75 a. $d_{C-H} = r_C + r_H$ = 77 pm + 37 pm = 114 pm

b. $d_{S-Cl} = r_S + r_{Cl}$ = 104 pm + 99 pm = 203 pm

c. $d_{Br-Cl} = r_{Br} + r_{Cl}$ = 114 pm + 99 pm = 213 pm

d. $d_{Si-O} = r_{Si} + r_O$ = 117 pm + 66 pm = 183 pm

9.77   Methylamine   147pm   Single C-N bond is longer.

Acetonitrile   116pm   Triple C-N bond is shorter.

9.79

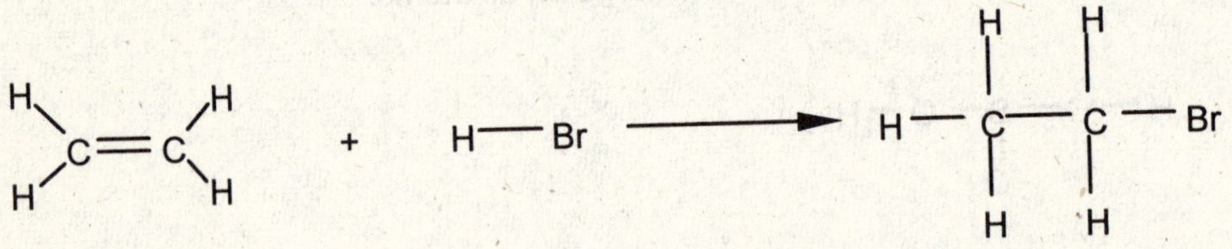

In the reaction, a C=C double bond is converted to a C-C single bond. An H-Br bond is broken, and one C-H bond and one C-Br bond are formed.

$$\Delta H \cong BE(C=C) + BE(H\text{-}Br) - BE(C\text{-}C) - BE(C\text{-}H) - BE(C\text{-}Br)$$
$$= (602 + 362 - 346 - 411 - 285) \text{ kJ} = -78 \text{ kJ}$$

■ Solutions to General Problems

9.81 a. Strontium is a metal, and oxygen is a nonmetal. The binary compound is likely to be ionic. Strontium, in Group IIA, forms Sr^{2+} ions; oxygen, from Group VIA, forms O^{2-} ions. The binary compound has the formula SrO and is named strontium oxide.

b. Carbon and bromine are both nonmetals; hence, the binary compound is likely to be covalent. Carbon usually forms four bonds, and bromine usually forms one bond. The formula for the binary compound is CBr_4. It is called carbon tetrabromide.

(continued)

c. Gallium is a metal, and fluorine is a nonmetal. The binary compound is likely to be ionic. Gallium is in Group IIIA and forms Ga^{3+} ions. Fluorine is in Group VIIA and forms F^- ions. The binary compound is GaF_3 and is named gallium(III) fluoride.

d. Nitrogen and bromine are both nonmetals; hence, the binary compound is likely to be covalent. Nitrogen usually forms three bonds, and bromine usually forms one bond. The formula for the binary compound is NBr_3. It is called nitrogen tribromide.

9.83   Total valence electrons = 5 + (4 x 6) + 3 = 32. The skeleton is

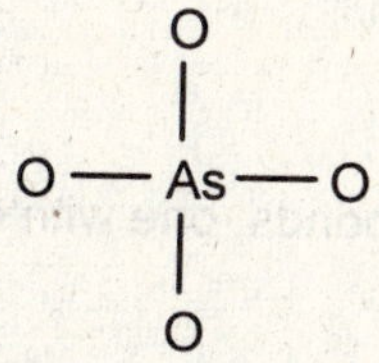

O
|
O — As — O
|
O

Distribute the remaining twenty four electrons to complete the octets around the oxygen atoms.

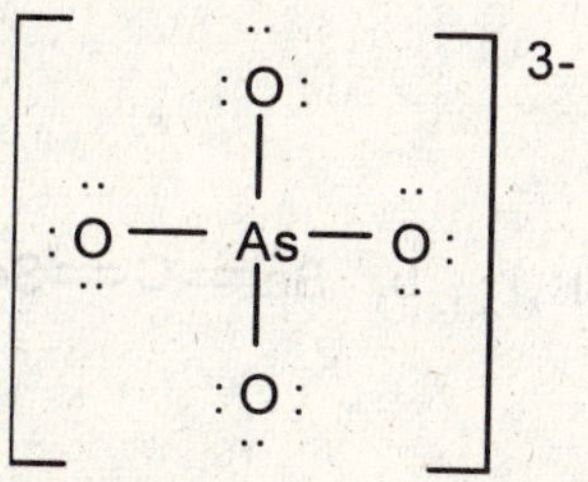

or

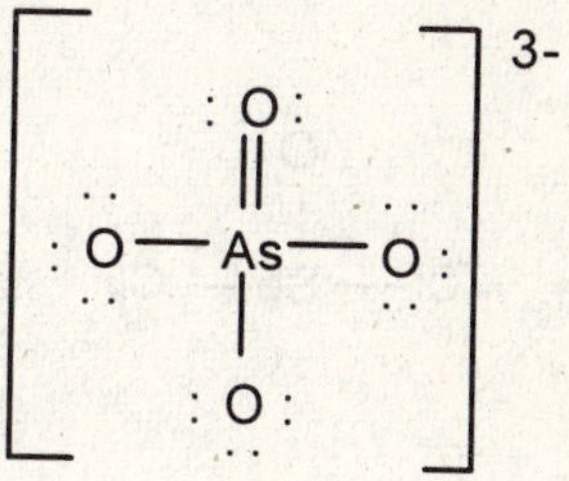

The formula for lead(II) arsenate is $Pb_3(AsO_4)_2$. The structure on the right has no formal charge.

9.85   Total valence electrons = 1 + 7 + (3 x 6) = 26. The skeleton is

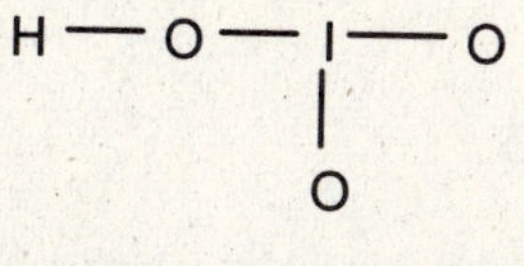

H — O — I — O
|
O

Distribute the remaining eighteen electrons to satisfy the octet rule. The structure on the right has no formal charge.

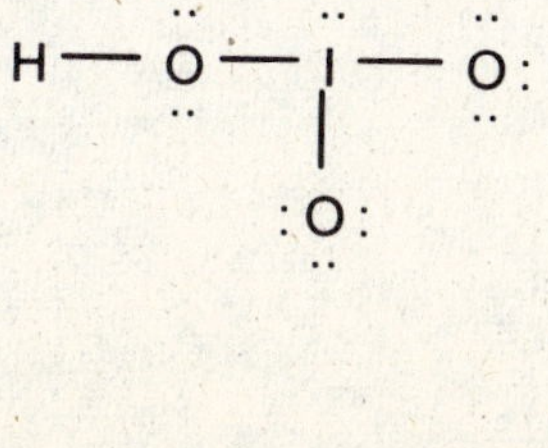

or

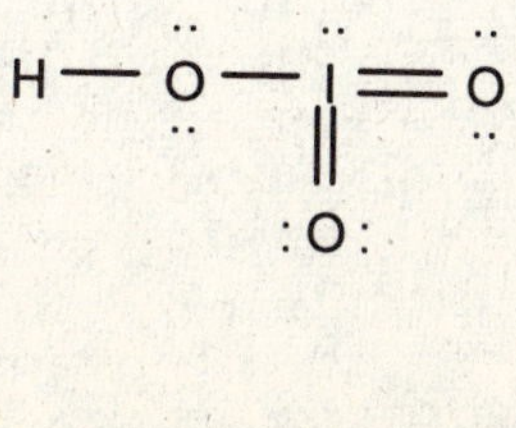

9.87 Total valence electrons = 5 + (2 x 1) + 1 = 8. The skeleton is H-N-H. Distribute the remaining four electrons to complete the octet of the nitrogen atom.

$$\left[H - \ddot{N} - H \right]^-$$

9.89 Total valence electrons = 5 + (2 x 6) - 1 = 16. The skeleton is O-N-O. Distribute the remaining electrons on the oxygen atoms.

$$: \ddot{O} - N - \ddot{O} :$$

The nitrogen atom is short four electrons. Use two double bonds, one with each oxygen.

$$\left[\ddot{O} = N = \ddot{O} : \right]^+$$

9.91 a. $: \ddot{Cl} - \underset{}{\overset{}{Se}} - \ddot{Cl} :$ or $: \ddot{Cl} - \underset{}{\overset{:\ddot{O}:}{Se}} - \ddot{Cl} :$ b. $\ddot{Se} = C = \ddot{Se}$

c. $\left[: \ddot{Cl} - \underset{:\ddot{Cl}:}{\overset{:\ddot{Cl}:}{Ga}} - \ddot{Cl} : \right]^-$

d. $\left[: C \equiv C : \right]^{2-}$

9.93 a. Total valence electrons = 5 + (3 x 7) = 26. The skeleton is

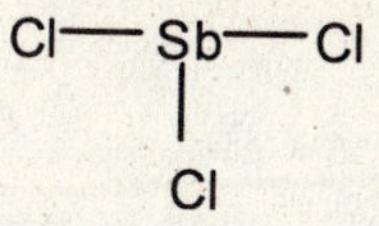

$$Cl-Sb-Cl$$
$$|$$
$$Cl$$

Distribute the remaining twenty electrons to the chlorine atoms and the antimony atom to complete their octets.

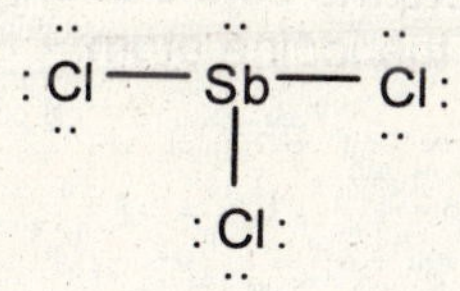

$$:\ddot{C}l-\ddot{S}b-\ddot{C}l:$$
$$|$$
$$:\ddot{C}l:$$

b. Total valence electrons = 7 + 4 + 5 = 16. The skeleton is I-C-N. Distribute the remaining twelve electrons.

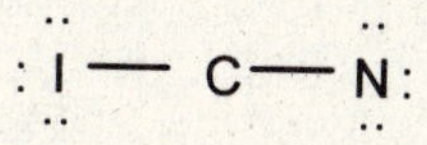

$$:\ddot{I}-C-\ddot{N}:$$

Notice the carbon atom is four electrons short of an octet. Make a triple bond between C and N from the four nonbonding electrons on the nitrogen.

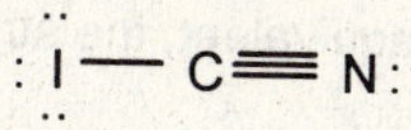

$$:\ddot{I}-C\equiv N:$$

c. Total valence electrons = 7 + (3 x 7) = 28. The skeleton is

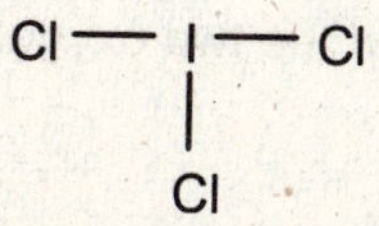

$$Cl-I-Cl$$
$$|$$
$$Cl$$

Distribute eighteen of the remaining twenty two electrons to complete the octets of the chlorine atoms. The four remaining electrons form two sets of lone pairs on the iodine atom.

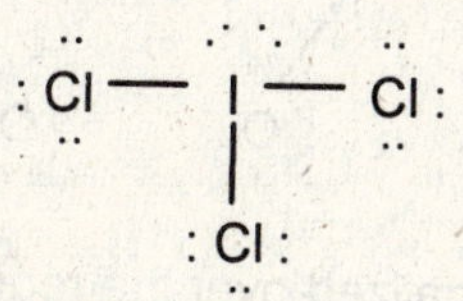

$$:\ddot{C}l-\ddot{I}-\ddot{C}l:$$
$$|$$
$$:\ddot{C}l:$$

(continued)

d. Total valence electrons = 7 + (5 x 7) = 42. The skeleton is

Use thirty of the remaining thirty two electrons to complete the octets of the fluorine atoms. The two electrons remaining form a lone pair on the iodine atom:

9.95 a. One possible electron-dot structure is

Because the selenium-oxygen bonds are expected to be equivalent, the structure must be described in resonance terms.

One electron pair is delocalized over the selenium atom and the two oxygen atoms.

b. The possible electron-dot structures are

At each end of the molecule, a pair of electrons is delocalized over the region of the nitrogen atom and the two oxygen atoms.

9.97 The compound S_2N_2 will have a four-membered ring structure. Calculate the total number of valence electrons of sulfur and nitrogen, which will be 6 + 6 + 5 + 5 = 22. Writing the skeleton of the four-membered ring with single S-N bonds will use up eight electrons, leaving fourteen electrons for electron pairs. After writing an electron pair on each atom, there will be six electrons left for three electron pairs. No matter where these electrons are written, either a nitrogen or a sulfur will be left with less than eight electrons. This suggests one or more double bonds are needed. If only one S=N double bond is written, writing the remaining twelve electrons as six electron pairs will give the sulfur in the S=N double bond a formal charge of +1 (see second line of formulas on next page). Writing two S=N double bonds using the same sulfur for both double bonds will give a formal charge of zero for all four atoms (see first line of formulas). The first two resonance formulas below have a zero formal charge on all atoms. However, one of the sulfur atoms does not obey the octet rule.

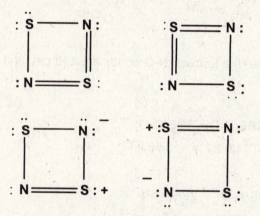

In the two resonance formulas immediately above, all of the atoms obey the octet rule, but there is a positive formal charge on one sulfur atom, and a negative formal charge on one nitrogen atom.

9.99   The possible electron-dot structures are

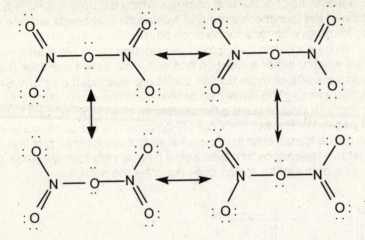

Because double bonds are shorter, the terminal N-O bonds are 118 pm, and the central N-O bonds are 136 pm.

9.101 ΔH = BE(H-H) + BE(O=O) - 2BE(H-O) - BE(O-O)

= (432 + 494 - 2 x 459 - 142) kJ = -134 kJ

9.103 ΔH = BE(N=N) + BE(F-F) - BE(N-N) - 2BE(N-F)

= (418 + 155 - 167 - 2 x 283) kJ = -1.60 x 10^2 kJ

■ Solutions to Cumulative-Skills Problems

Note on significant figures: The final answer to each cumulative-skills problem is given first with one nonsignificant figure (the rightmost significant figure is underlined) and then is rounded to the correct number of figures. Intermediate answers usually also have at least one nonsignificant figure. Atomic weights are rounded to two decimal places except for that of hydrogen.

9.105   The electronegativity differences and bond polarities are:

P- H	0.0	nonpolar
O- H	1.4	polar (acidic)

$H_3PO_3(aq)$ + $2NaOH(aq)$ → $Na_2HPO_3(aq)$ + $2H_2O(l)$

(continued)

$$\frac{mol}{L} = \frac{0.1250 \; mol \; NaOH}{L \; NaOH} \times 0.02250 \; L \; NaOH \times \frac{1}{0.2000 \; L \; H_3PO_3}$$

$$\times \frac{1 \; mol \; H_3PO_3}{2 \; mol \; NaOH} \; 0.0070312 = 0.007031 \; M \; H_3PO_3$$

9.107   After assuming a 100.0 g sample, convert to moles:

$$10.9 \; g \; Mg \times \frac{1 \; mol \; Mg}{24.3 \; g \; Mg} = 0.4485 \; mol \; Mg$$

$$31.8 \; g \; Cl \times \frac{1 \; mol \; Cl}{35.453 \; g \; Cl} = 0.89696 \; mol \; Cl$$

$$57.3 \; g \; O \times \frac{1 \; mol \; O}{16.00 \; g \; O} = 3.581 \; mol \; O$$

Divide by 0.4485:

$$Mg: \; \frac{0.4485}{0.4485} = 1; \; Cl: \; \frac{0.89696}{0.4485} = 2.00; \; O: \; \frac{3.581}{0.4485} = 7.98$$

The simplest formula is $MgCl_2O_8$. However, since $Cl_2O_8{}^{2-}$ is not a well known ion, write the simplest formula as $Mg(ClO_4)_2$, magnesium perchlorate. The Lewis formulas are Mg^{2+} and

9.109 After assuming a 100.0 g sample, convert to moles:

$$25.0 \text{ g C} \times \frac{1 \text{ mol C}}{12.01 \text{ g C}} = 2.0\underline{8}1 \text{ mol C}$$

$$2.1 \text{ g H} \times \frac{1 \text{ mol H}}{1.008 \text{ g H}} = 2.0\underline{8} \text{ mol H}$$

$$39.6 \text{ g F} \times \frac{1 \text{ mol F}}{18.99 \text{ g F}} = 2.0\underline{8}5 \text{ mol F}$$

$$33.3 \text{ g O} \times \frac{1 \text{ mol O}}{16.00 \text{ g O}} = 2.0\underline{8}1 \text{ mol O}$$

The simplest formula is CHOF. Because the molecular mass of 48.0 divided by the formula mass of 48.0 is one, the molecular formula is also CHOF. The Lewis formula is

9.111 First, calculate the number of moles in one liter:

$$n = \frac{PV}{RT} = \frac{1.00 \text{ atm} \times 1.00 \text{ L}}{0.0821 \text{ L} \cdot \text{atm/ (K} \cdot \text{mol)} \times 424 \text{ K}} = 0.028\underline{7}27 \text{ mol}$$

$$MW = \frac{g}{mol} = \frac{7.49 \text{ g}}{0.028727 \text{ mol}} = 26\underline{0}.7 \text{ g/mol}$$

$$MW = 260.7 \text{ amu} = 118.69 \text{ amu Sn} + n \times (35.453 \text{ amu Cl})$$

$$n = \frac{260.7 - 118.69}{35.453} = 4.0\underline{0}55$$

(continued)

The formula is $SnCl_4$, which is molecular because the electronegativity difference is 1.2. The Lewis formula is

$$\begin{array}{c} \ddot{:}\ddot{Cl}: \\ | \\ :\ddot{Cl} - Sn - \ddot{Cl}: \\ | \\ :\ddot{Cl}: \end{array}$$

9.113 $HCN(g)$ $\rightarrow$ $H(g)$ $+$ $C(g)$ $+$ $N(g)$

(135   218   715.0   473) ΔH_f (kJ/mol)

ΔH_{rxn} = 1271 kJ/mol

$BE(C{\equiv}N)$ = ΔH - BE(C-H) = [1271 - 411] kJ/mol

= 860 kJ/mol (Table 9.5 has 887 kJ/mol.)

9.115   Use the O-H bond and its bond energy of 459 kJ/mol to calculate X_o.

$BE(O-H)$ = $1/2[BE(H-) + BE(O-)] + k(X_O - X_H)^2$

459 kJ/mol = 1/2(432 kJ/mol + 142 kJ/mol) + 98.6 kJ $(X_O - X_H)^2$

Collecting the terms gives

$$\frac{459 - 287}{98.6} = (X_O - X_H)^2$$

Taking the square root of both sides gives

1.3$\underline{2}$07 = 1.32 = $(X_O - X_H)$

Because X_H = 2.1, X_O = 2.1 + 1.32 = 3.42 = 3.4.

9.117 $X = \dfrac{I.E. - E.A.}{2} = \dfrac{[1250 - (-349)] \text{ kJ/mol}}{2}$ = 799.$\underline{5}$ kJ/mol

$$\frac{799.5 \text{ kJ/mol}}{230 \text{ kJ/mol}} = 3.47 \text{ (Pauling's } X = 3.0)$$

10. MOLECULAR GEOMETRY AND CHEMICAL BONDING THEORY

■ Solutions to Exercises

10.1 a. A Lewis structure of ClO_3^- is

$$
\left[\; :\!\ddot{O}\!: \atop \substack{ | \\ :\!\ddot{O}\! - Cl - \ddot{O}\!: } \; \right]^{-}
$$

There are four electron pairs in a tetrahedral arrangement about the central atom. Three pairs are bonding, and one pair is nonbonding. The expected geometry is trigonal pyramidal.

b. The Lewis structure of OF_2 is

$$
:\!\ddot{F}\! - \ddot{O}\! - \ddot{F}\!:
$$

There are four electron pairs in a tetrahedral arrangement about the central atom. Two pairs are bonding, and two pairs are nonbonding. The expected geometry is bent.

(continued)

c. The Lewis structure of SiF_4 is

There are four bonding electron pairs in a tetrahedral arrangement around the central atom. The expected geometry is tetrahedral.

10.2   First, distribute the valence electrons to the bonds and the chlorine atoms. Then, distribute the remaining electrons to iodine.

The five electron pairs around iodine should have a trigonal bipyramidal arrangement with two lone pairs occupying equatorial positions. The molecule is T-shaped.

10.3   Both trigonal pyramidal (b) and T-shaped (c) geometries are consistent with a nonzero dipole moment. In trigonal planar geometry, the Br-F contributions to the dipole moment would cancel.

10.4   On the basis of symmetry, SiF_4 (b) would be expected to have a dipole moment of zero. The bonds are all symmetric about the central atom.

10.5   The Lewis structure for ammonia, NH_3, is

There are four pairs of electrons around the nitrogen atom. According to the VSEPR model, these are arranged tetrahedrally around the nitrogen atom, and you should use sp^3 hybrid orbitals. Each N–H bond is formed by the overlap of a $1s$ orbital of a hydrogen atom with one of the singly occupied sp^3 hybrid orbitals of the nitrogen atom.

(continued)

This gives the following bonding description for NH_3:

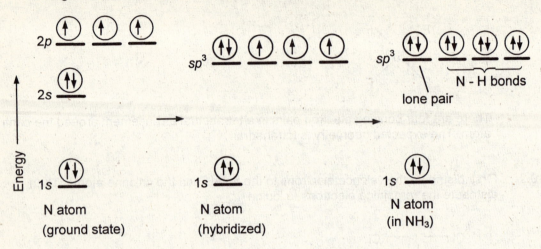

10.6   The Lewis structure for PCl_5 is

The phosphorus atom has five single bonds and no lone pairs around it. This suggests that you use sp^3d hybrid orbitals on phosphorus. Each chlorine atom (valence shell configuration $3s^23p^5$) has one singly occupied $3p$ orbital. The P–Cl bonds are formed by the overlap of a phosphorus sp^3d hybrid orbital with a singly occupied chlorine $3p$ orbital. The hybridization of phosphorus can be represented as follows:

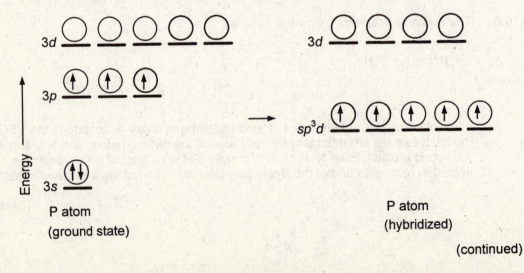

(continued)

The bonding description of phosphorus in PCl_5 is

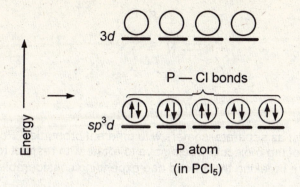

10.7   The Lewis structure of CO_2 is

$$\ddot{O}\!=\!C\!=\!\ddot{O}$$

The double bonds are each a π bond plus a σ bond. Hybrid orbitals are needed to describe the two σ bonds. This suggests sp hybridization. Each sp hybrid orbital on the carbon atom overlaps with a $2p$ orbital on one of the oxygen atoms to form a σ bond. The π bonds are formed by the overlap of a $2p$ orbital on the carbon atom with a $2p$ orbital on one of the oxygen atoms. Hybridization and bonding of the carbon atom are shown as follows:

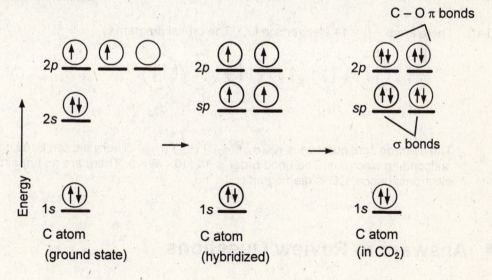

10.8 The structural formulas for the isomers are as follows:

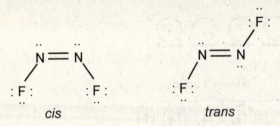

cis   trans

These compounds exist as separate isomers with different properties. For these to interconvert, one end of the molecule would have to rotate with respect to the other end. This would require breaking the π bond and expending considerable energy.

10.9 There are 2 x 6 = 12 electrons in C_2. They occupy the orbitals as shown below.

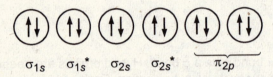

σ_{1s} σ_{1s}^* σ_{2s} σ_{2s}^* π_{2p}

The electron configuration is $KK(\sigma_{2s})^2(\sigma_{2s}^*)^2(\pi_{2p})^4$. There are no unpaired electrons; therefore, C_2 is diamagnetic. There are eight bonding and four antibonding electrons. The bond order is 1/2(8 - 4) = 2.

10.10 There are 6 + 8 = 14 electrons in CO. The orbital diagram is

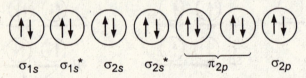

σ_{1s} σ_{1s}^* σ_{2s} σ_{2s}^* π_{2p} σ_{2p}

The electron configuration is $KK(\sigma_{2s})^2(\sigma_{2s}^*)^2(\pi_{2p})^4(\sigma_{2p})^2$. There are ten bonding and four antibonding electrons. The bond order is 1/2(10 - 4) = 3. There are no unpaired electrons; hence, CO is diamagnetic.

■ Answers to Review Questions

10.1 The VSEPR model is used to predict the geometry of molecules. The electron pairs around an atom are assumed to arrange themselves to reduce electron repulsion. The molecular geometry is determined by the positions of the bonding electron pairs.

10.2   The arrangements are linear, trigonal planar, tetrahedral, trigonal bipyramidal, and octahedral.

10.3   A lone pair is "larger" than a bonding pair; therefore, it will occupy an equatorial position where it encounters less repulsion than if it were in an axial position.

10.4   The bonds could be polar, but if they are arranged symmetrically, the molecule will be nonpolar. The bond dipoles will cancel.

10.5   Nitrogen trifluoride has three N–F bonds arranged to form a trigonal pyramid. These bonds are polar and would give a polar molecule with partial negative charges on the fluorine atoms and a partial positive charge on the nitrogen atom. However, there is also a lone pair of electrons on nitrogen that is directed away from the bonds. The result is that the lone pair nearly cancels the polarity of the bonds and gives a molecule with a very small dipole moment.

10.6   Certain orbitals, such as p orbitals and hybrid orbitals, have lobes in given directions. Bonding to these orbitals is directional; that is, the bonding is in preferred directions. This explains why the bonding gives a particular molecular geometry.

10.7   The angle is 109.5°.

10.8   A sigma bond has a cylindrical shape about the bond axis. A pi bond has a distribution of electrons above and below the bond axis.

10.9   In ethylene, C_2H_4, the changes on a given carbon atom may be described as follows:

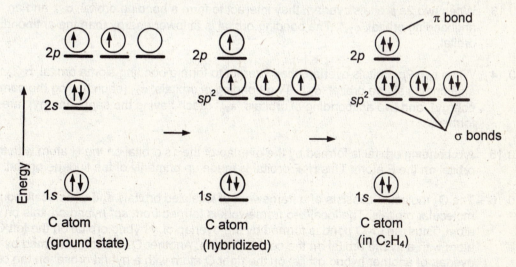

An sp^2 hybrid orbital on one carbon atom overlaps a similar hybrid orbital on the other carbon atom to form a σ bond. The remaining hybrid orbitals on the two carbon atoms overlap 1s orbitals from the hydrogen atoms to form four C-H bonds. The unhybridized $2p$ orbital on one carbon atom overlaps the unhybridized $2p$ orbital on the other carbon atom to form a π bond. The σ and π bonds together constitute a double bond.

10.10   Both of the unhybridized 2p orbitals, one from each carbon atom, are perpendicular to their CH_2 planes. When these orbitals overlap each other, they fix both planes in the same plane. The two ends of the molecule cannot twist around without breaking the π bond, which requires considerable energy. Therefore, it is possible to have stable molecules with the following structures:

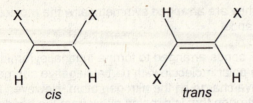

cis   trans

Because these have the same molecular formulas, they are isomers. In this case, they are called *cis-trans* isomers, or geometrical isomers.

10.11   In a bonding orbital, the probability of finding electrons between the two nuclei is high. For this reason, the energy of the bonding orbital is lower than that of the separate atomic orbitals. In an antibonding orbital, the probability of finding electrons between the two nuclei is low. For this reason, the energy of the antibonding orbital is higher than that of the separate atomic orbitals.

10.12   The factors determining the strength of interaction of two atomic orbitals are (1) the energy difference between the interacting orbitals and (2) the magnitude of their overlap.

10.13   When two 2s orbitals overlap, they interact to form a bonding orbital, σ_{2s}, and an antibonding orbital, σ_{2s}^*. The bonding orbital is at lower energy than the antibonding orbital.

10.14   When two 2p orbitals overlap, they interact to form a bonding sigma orbital, σ_{2p}, and an antibonding sigma orbital, σ_{2p}^*. Two bonding *pi* orbitals, π_{2p} (each having the same energy), and two antibonding *pi* orbitals, π_{2p}^* (each having the same energy), are also formed.

10.15   A σ bonding orbital is formed by the overlap of the 1s orbital on the H atom with the 2p orbital on the F atom. This H–F orbital is made up primarily of the fluorine orbital.

10.16   The O_3 molecule consists of a framework of localized orbitals and of delocalized *pi* molecular orbitals. The localized framework is formed from sp^2 hybrid orbitals on each atom. Thus, an O–O bond is formed by the overlap of a hybrid orbital on the left O atom with a hybrid orbital on the center O atom. Another O–O bond is formed by the overlap of another hybrid orbital on the right O atom with a hybrid orbital on the center O atom. The remaining hybrid orbitals are occupied by lone pairs of electrons. Also, there is one unhybridized p orbital on each of the atoms. These p orbitals are perpendicular to the plane of the molecule and overlap sidewise to give three *pi* molecular orbitals that are delocalized. The two orbitals of lowest energy are occupied by pairs of electrons.

■ Solutions to Practice Problems

10.27 The number of electron pairs, lone pairs and geometry are for the central atom.

	Electron-Dot Structure	Number of Electron Pairs	Number of Lone Pairs	Geometry
a.	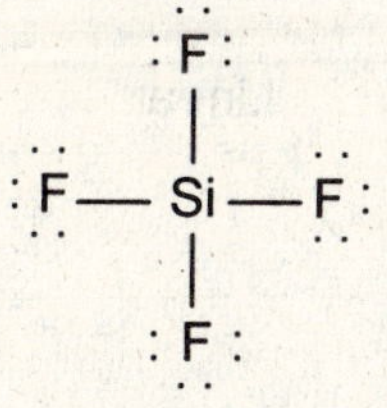	4	0	Tetrahedral
b.	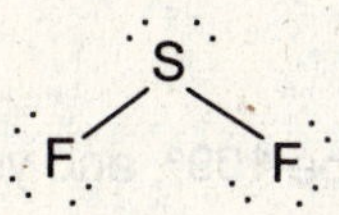	4	2	Bent
c.	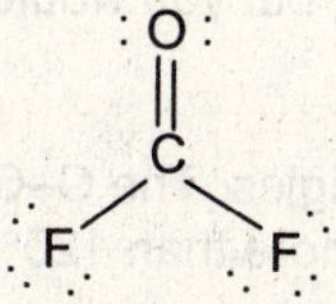	3	0	Trigonal planar
d.	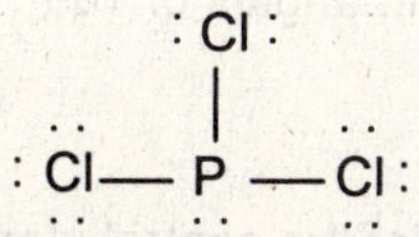	4	1	Trigonal pyramidal

10.29 The number of electron pairs, lone pairs and geometry are for the central atom.

	Electron-Dot Structure	Number of Electron Pairs	Number of Lone Pairs	Geometry
a.	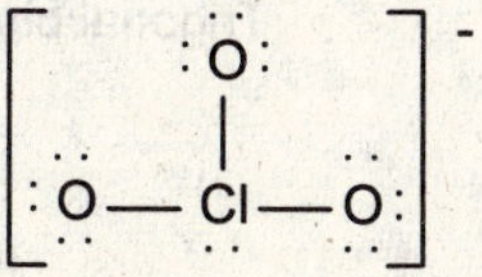	4	1	Trigonal pyramidal

(continued)

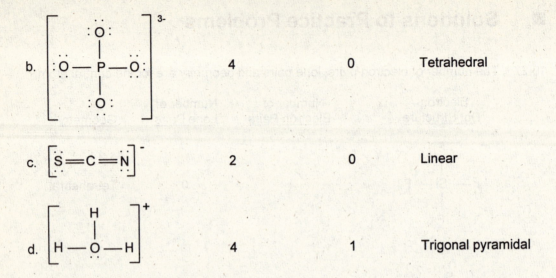

		4	0	Tetrahedral
		2	0	Linear
		4	1	Trigonal pyramidal

10.31 a. CCl$_4$ is tetrahedral, VSEPR predicts that the bond angles will be 109°, and you would expect them to be this angle.

b. SCl$_2$ is bent, and VSEPR theory predicts bond angles of 109°, but you would expect the Cl–S–Cl bond angle to be less.

c. COCl$_2$ is trigonal planar. VSEPR theory predicts 120° bond angles. The C–O bond is a double bond, so you would expect the Cl–C–O bond to be more than 120° and the Cl–C–Cl bond to be less than 120°.

d. AsH$_3$ is trigonal pyramidal, and VSEPR theory predicts bond angles of 109°, but you would expect the H–As–H bond angle to be less.

10.33 The number of electron pairs, lone pairs and geometry are for the central atom.

Electron- Dot Structure	Number of Electron Pairs	Number of Lone Pairs	Geometry
a.	5	0	Trigonal bipyramidal

(continued)

b.	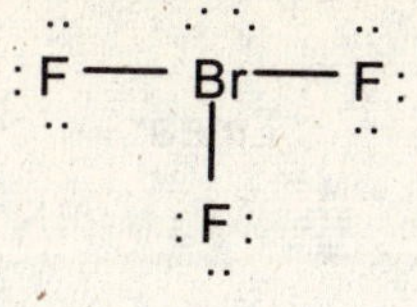	5	2	T-shaped
c.	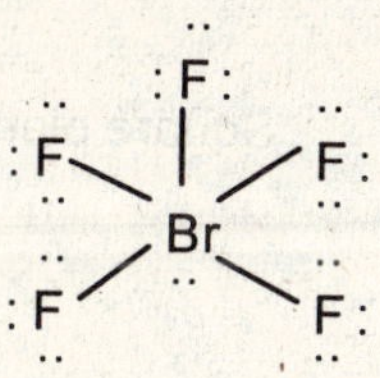	6	1	Square pyramidal
d.	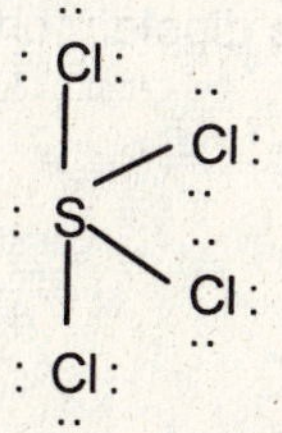	5	1	Seesaw

10.35 The number of electron pairs, lone pairs and geometry are for the central atom.

	Electron Dot Structure	Number of Electron Pairs	Number of Lone Pairs	Geometry
a.	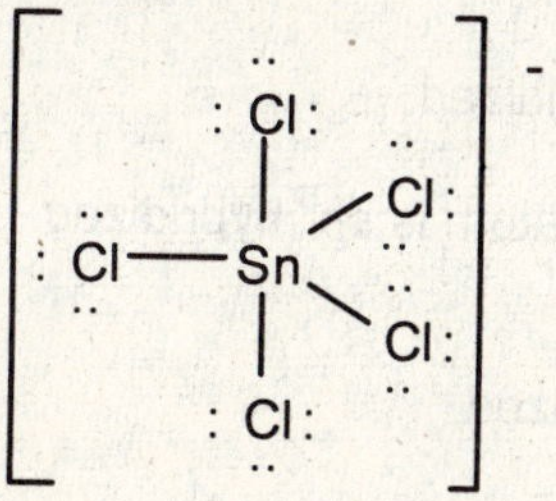	5	0	Trigonal bipyramidal
b.	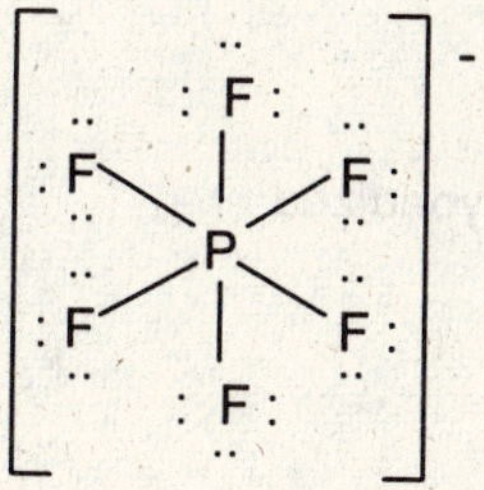	6	0	Octahedral

(continued)

c. $\left[:\!F\!-\!Cl\!-\!F\!: \right]^{-}$ 5   3   Linear

d. $\left[:F, F:, I, :F, F: \right]^{-}$ 6   2   Square planar

10.37 a. Trigonal pyramidal and T-shaped. Trigonal planar would have a dipole moment of zero.

b. Bent. Linear would have a dipole moment of zero.

10.39 a. CS_2 has a linear geometry and has no dipole moment.

b. TeF_2 has a bent geometry and has a dipole moment.

c. $SeCl_4$ has a seesaw geometry and has a dipole moment.

d. XeF_4 has a square planar geometry and has no dipole moment.

10.41 a. SF_2 is an angular model. The sulfur atom is sp^3 hybridized.

b. ClO_3^- is a trigonal pyramidal molecule. The chlorine atom is sp^3 hybridized.

10.43 a. $SeCl_2$ is a bent molecule. The Se atom is sp^3 hybridized.

b. NO_2^- is a bent ion. The N atom is sp^2 hybridized.

c. CO_2 is a linear molecule. The C atom is sp hybridized.

d. COF_2 is a trigonal planar molecule. The C atom is sp^2 hybridized.

10.45 a. The Lewis structure is

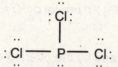

The presence of two single bonds and no lone pairs suggests sp hybridization. Thus, an Hg atom with the configuration $[Xe]4f^{14}5d^{10}6s^2$ is promoted to $[Xe]4f^{14}5d^{10}6s^16p^1$, then hybridized. An Hg–Cl bond is formed by overlapping an Hg hybrid orbital with a $3p$ orbital of Cl.

b. The Lewis structure is

The presence of three single bonds and one lone pair suggests sp^3 hybridization of the P atom. Three hybrid orbitals each overlap a $3p$ orbital of a Cl atom to form a P–Cl bond. The fourth hybrid orbital contains the lone pair.

10.47 a. Xenon has eight valence electrons. Each F atom donates one electron to give a total of ten electrons, or five electron pairs, around the Xe atom. The hybridization is sp^3d.

b. Bromine has seven valence electrons. Each F atom donates one electron to give a total of ten electrons, or five electron pairs, around the Br atom. The hybridization is sp^3d.

c. Phosphorus has five valence electrons. The Cl atoms each donate one electron to give a total of ten electrons, or five electron pairs, around the P atom. The hybridization is sp^3d.

d. Chlorine has seven valence electrons, to which may be added one electron from each F atom minus one electron for the charge on the ion. This gives a total of ten electrons, or five electron pairs, around chlorine. The hybridization is sp^3d.

10.49 The P atom in PCl_6^- has six single bonds around it and no lone pairs. This suggests sp^3d^2 hybridization. Each bond in this ion is a σ bond formed by overlap of an sp^3d^2 hybrid orbital on P with a $3p$ orbital on Cl.

10.51 a. The structural formula of formaldehyde is

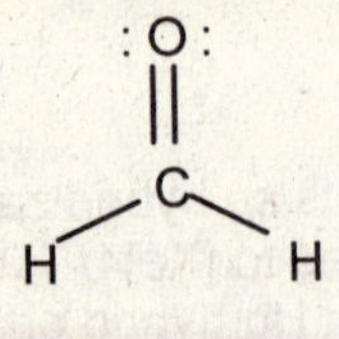

Because the C is bonded to three other atoms, it is assumed to be sp^2 hybridized. One 2p orbital remains unhybridized. The carbon-hydrogen bonds are σ bonds formed by the overlap of an sp^2 hybrid orbital on C with a 1s orbital on H. The remaining sp^2 hybrid orbital on C overlaps with a 2p orbital on O to form a σ bond. The unhybridized 2p orbital on C overlaps with a parallel 2p orbital on O to form a π bond. Together, the σ and π bonds constitute a double bond.

b. The nitrogen atoms are sp hybridized. A σ bond is formed by the overlap of an sp hybrid orbital from each N. The remaining sp hybrid orbitals contain lone pairs of electrons. The two unhybridized 2p orbitals on one N overlap with the parallel unhybridized 2p orbitals on the other N to form two π bonds.

10.53 Each of the N atoms has a lone pair of electrons and is bonded to two atoms. The N atoms are sp^2 hybridized. The two possible arrangements of the O atoms relative to one another are shown below. Because the π bond between the N atoms must be broken to interconvert these two forms, it is to be expected that the hyponitrite ion will exhibit *cis-trans* isomerism.

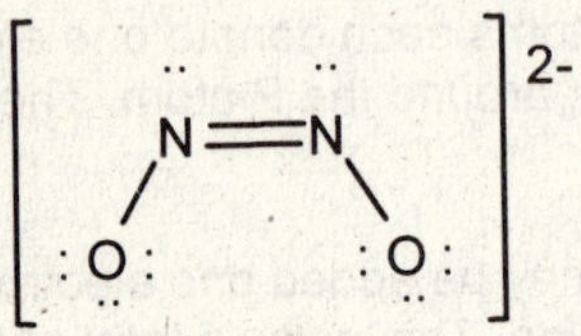

cis 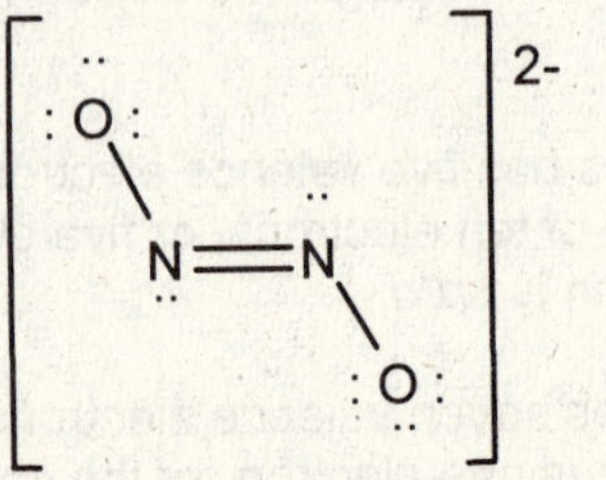   *trans*

10.55 a. Total electrons = 2 x 5 = 10.

The electron configuration is $KK(\sigma_{2s})^2(\sigma_{2s}{}^*)^2(\pi_{2p})^2$.

Bond order = $1/2 (n_b - n_a)$ = 1/2 (6 - 4) = 1

The B_2 molecule is stable. It is paramagnetic because the two electrons in the π_{2p} subshell occupy separate orbitals.

(continued)

b. Total electrons = $(2 \times 5) - 1 = 9$.

The electron configuration is $KK(\sigma_{2s})^2(\sigma_{2s}{}^*)^2(\pi_{2p})^1$.

Bond order = $1/2 (n_b - n_a) = 1/2 (5 - 4) = 1/2$

The $B_2{}^+$ molecule should be stable and is paramagnetic because there is one unpaired electron in the π_{2p} subshell.

c. Total electrons = $(2 \times 8) + 1 = 17$.

The electron configuration is $KK(\sigma_{2s})^2(\sigma_{2s}{}^*)^2(\pi_{2p})^4(\sigma_{2p})^2(\pi_{2p}{}^*)^3$.

Bond order = $1/2 (n_b - n_a) = 1/2 (10 - 7) = 3/2$

The $O_2{}^-$ molecule should be stable and is paramagnetic because there is one unpaired electron in the $\pi_{2p}{}^*$ subshell.

10.57 Total electrons = $6 + 7 + 1 = 14$.

The electron configuration is $KK(\sigma_{2s})^2(\sigma_{2s}{}^*)^2(\pi_{2p})^4(\sigma_{2p})^2$.

Bond order = $1/2 (n_b - n_a) = 1/2 (10 - 4) = 3$

The CN^- ion is diamagnetic.

■ Solutions to General Problems

10.59 The number of electron pairs, lone pairs and geometry are for the central atom.

Electron-Dot Structure	Number of Electron Pairs	Number of Lone Pairs	Geometry
a. $\ddot{:}Cl - Sn - Cl\ddot{:}$	2	1	Bent
b. (structure of C double-bonded to O, single-bonded to two F)	3	0	Trigonal planar

(continued)

c. $\left[\; \ddot{\underset{..}{Cl}} - \ddot{\underset{..}{I}} - \ddot{\underset{..}{Cl}} \; \right]^{-}$ 5   3   Linear

d. $\left[\begin{array}{c} :\ddot{Cl}: \\ :\ddot{Cl} \quad \ddot{Cl}: \\ P \\ :\ddot{Cl} \quad \ddot{Cl}: \\ :\ddot{Cl}: \end{array} \right]^{-}$ 6   0   Octahedral

10.61 a. $:\ddot{\underset{..}{F}} - \ddot{Se} - \ddot{\underset{..}{F}}:$ Bent

b. $:\ddot{\underset{..}{Cl}} - \underset{\underset{H}{|}}{C} - \ddot{\underset{..}{Cl}}:$ Tetrahedral

c. $\begin{array}{c} :\ddot{F}: \\ | \\ :\ddot{Se} \\ | \\ :\ddot{F}: \end{array}$ with $\ddot{F}:$ groups   Seesaw

d. $\left[\begin{array}{c} :\ddot{F}: \\ :\ddot{F} \quad \ddot{F}: \\ Sn \\ :\ddot{F} \quad \ddot{F}: \\ :\ddot{F}: \end{array} \right]^{2-}$ Octahedral

10.63 a.

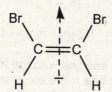

C_a and C_b: Three electron pairs around
They are sp^2 hybridized.

C_c: Four electron pairs around it. It is sp^3
hybridized.

b. : N≡C——C≡N :

Both C atoms are bonded to two other atoms
and have no lone pairs of electrons. They are
sp hybridized.

10.65

Br   Br
 \   /
  C ═══════ C   has a net dipole
 /   \
H   H
     cis

Br   Br
 \ |
  C ═══════ C   has no net dipole. The two C-Br bond dipoles cancel and
 /   \   the two C-H bond dipoles cancel.
H   H

Br   H
 \   /
  C ═══════ C
 /   \
H   Br
     trans

10.67   All four hydrogen atoms of $H_2C=C=CH_2$ cannot lie in the same plane because the
second C=C bond forms perpendicular to the plane of the first C=C bond. By looking at
Figure 10.26, you can see how this is so. The second C=C bond forms in the plane of
the C–H bonds. Thus, the plane of the C–H bonds on the right side will be
perpendicular to the plane of the C–H bonds on the left side. This is shown below
using the π orbitals of the two C=C double bonds.

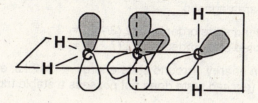

10.69   Total electrons = 2 + 1 - 1 = 2.

The electron configuration is $(\sigma_{1s})^2$.

Bond order = $1/2 \, (n_b - n_a)$ = 1/2 (2) = 1

The HeH⁺ ion is expected to be stable.

10.71 Total electrons $= (2 \times 6) + 2 = 14$.

The electron configuration is $KK(\sigma_{2s})^2(\sigma_{2s}{}^*)^2(\pi_{2p})^4(\sigma_{2p})^+$.

Bond order $= 1/2 \, (n_b - n_a) = 1/2 \, (10 - 4) = 3$

10.73 The molecular orbital configuration of O_2 is $KK(\sigma_{2s})^2(\sigma_{2s}{}^*)^2(\pi_{2p})^4(\sigma_{2p})^2(\pi_{2p}{}^*)^2$. $O_2{}^+$ has one electron less than O_2. The difference is in the number of electrons in the $\pi_{2p}{}^*$ antibonding orbital. This means that the bond order is larger for $O_2{}^+$ than for O_2.

O_2: Bond order $= 1/2 \, (n_b - n_a) = 1/2 \, (10 - 6) = 2$

$O_2{}^+$: Bond order $= 1/2 \, (n_b - n_a) = 1/2 \, (10 - 5) = 5/2$

It is expected that the species with the higher bond order, $O_2{}^+$, has the shorter bond length. In $O_2{}^-$, there is one more electron than in O_2. This additional electron occupies a $\pi_{2p}{}^*$ orbital. Increasing the number of electrons in antibonding orbitals decreases the bond order; hence, $O_2{}^-$ should have a longer bond length than O_2.

10.75 As shown in Figure 10.34, the occupation of molecular orbitals by the N_2 valence electrons is:

$$KK(\sigma_{2s})^2(\sigma_{2s}{}^*)^2(\pi_{2p})^4(\sigma_{2p})^2(\pi_{2p}{}^*)^0$$

To form the first excited state of N_2, a σ_{2p} electron is promoted to the $\pi_{2p}{}^*$ orbital, giving:

$$KK(\sigma_{2s})^2(\sigma_{2s}{}^*)^2(\pi_{2p})^4(\sigma_{2p})^1(\pi_{2p}{}^*)^1$$

The differences in properties are as follows:

Magnetic character: The ground state is diamagnetic (all electrons paired), and the excited state is paramagnetic.

Bond order: ground state order $= 1/2 \, (8 - 2) = 3$
excited state order $= 1/2 \, (7 - 3) = 2$

Bond-dissociation energy: ground state energy $= 942$ kJ; excited state energy is less than 942 kJ (excited state does not possess a stable triple bond like the ground state)

Bond length: ground state length $= 110$ pm; excited state length is more than 110 pm.

■ Solutions to Cumulative-Skills Problems

Note on significant figures: If the final answer to a solution needs to be rounded off, it is given first with one nonsignificant figure, and the last significant figure is underlined. The final answer is then rounded to the correct number of significant figures. In multiple-step problems, intermediate answers are given with at least one nonsignificant figure; however, only the final answer has been rounded off.

10.77   After assuming a 100.0 g sample, convert to moles:

$$60.4 \text{ g Xe} \times \frac{1 \text{ mol Xe}}{131.29 \text{ g Xe}} = 0.46\underline{0}05 \text{ mol Xe}$$

$$22.1 \text{ g O} \times \frac{1 \text{ mol O}}{16.00 \text{ g O}} = 1.3\underline{8}1 \text{ mol O}$$

$$17.5 \text{ g F} \times \frac{1 \text{ mol F}}{18.99 \text{ g F}} = 0.92\underline{1}5 \text{ mol F}$$

Divide by 0.460 :

$$\text{Xe: } \frac{0.460}{0.460} = 1; \quad \text{O: } \frac{1.381}{0.460} = 3.00; \quad \text{F: } \frac{0.9215}{0.460} = 2.00$$

The simplest formula is XeO_3F_2. This is also the molecular formula. The Lewis formula is

Number of electron pairs = 5, number of lone pairs = 0; hence, the geometry is trigonal bipyramidal. Because xenon has five single bonds, it will require five orbitals to describe the bonding. This suggests sp^3d hybridization.

10.79 $U(s) + ClF_n \rightarrow UF_6 + ClF(g)$

$$\text{mol } UF_6 = 3.53 \text{ g } UF_6 \times \frac{1 \text{ mol } UF_6}{352.07 \text{ g } UF_6} = 0.01003 \text{ mol } UF_6$$

$$\text{mol ClF} = n = \frac{PV}{RT} = \frac{2.50 \text{ atm} \times 0.343 \text{ L}}{0.082057 \text{ L} \cdot \text{atm/ (K} \cdot \text{mol)} \times 348K} = 0.3002 \text{ mol ClF}$$

0.010 mol UF_6 = 0.060 mol F, and 0.030 mol ClF = 0.030 mol F; therefore, the total moles of F from ClF_n = 0.090 mol F. Because mol ClF_n must equal mol ClF, mol ClF_n = 0.030 mol and n = 0.090 mol F ÷ 0.030 mol ClF_n = 3. The Lewis formula is

$$:\!F\!-\!Cl\!-\!F\!:$$
$$|$$
$$:\!F\!:$$

Number of electron pairs = 5, number of lone pairs = 2; hence, the geometry is T-shaped. Because chlorine has five electron pairs, it will require five orbitals to describe the bonding. This suggests sp^3d hybridization.

10.81 N_2:   Triple bond; bond length = 110 pm. Geometry is linear; sp-hybrid orbitals are needed for one lone pair and one σ bond.

N_2F_2:   Double bond; bond length = 122 pm. Geometry is trigonal planar; sp^2-hybrid orbitals are needed for one lone pair and two σ bonds.

N_2H_4:   Single bond; bond length = 145 pm. Geometry is tetrahedral; sp^3-hybrid orbitals are needed for one lone pair and three σ bonds.

10.83 HNO_3 resonance formulas:

The geometry around the nitrogen is trigonal planar; therefore, the hybridization is sp^2.

Formation reaction: $H_2(g) + 3O_2(g) + N_2(g) \rightarrow 2HNO_3(g)$

(continued)

$2 \times \Delta H_f° = [BE(H-H) + 3BE(O_2) + BE(N_2)] - [2BE(H-O) + 4BE(N-O) + 2BE(N=O)]$

$= [(432 + 3 \times 494 + 942) - (2 \times 459 + 4 \times 201 + 2 \times 607)$ kJ/2 mol$]$

$= -80$ kJ/2 mol $= -40$ kJ/mol

Resonance energy $= -40$ kJ $- (-135$ kJ$) = 95$ kJ

11. STATES OF MATTER; LIQUIDS AND SOLIDS

■ Solutions to Exercises

Note on significant figures: If the final answer to a solution needs to be rounded off, it is given first with one nonsignificant figure, and the last significant figure is underlined. The final answer is then rounded to the correct number of significant figures. In multiple-step problems, intermediate answers are given with at least one nonsignificant figure; however, only the final answer has been rounded off.

11.1 First, calculate the heat required to vaporize 1.00 kg of ammonia:

$$1.00 \text{ kg NH}_3 \times \frac{1000 \text{ g}}{1 \text{ kg}} \times \frac{1 \text{ mol NH}_3}{17.03 \text{ g NH}_3} \times \frac{23.4 \text{ kJ}}{1 \text{ mol}} = 137\underline{4}.04 \text{ kJ}$$

The amount of water at 0°C that can be frozen to ice at 0°C with this heat is

$$137\underline{4}.04 \text{ kJ} \times \frac{1 \text{ mol H}_2\text{O}}{6.01 \text{ kJ}} \times \frac{18.01 \text{ g H}_2\text{O}}{1 \text{ mol H}_2\text{O}} = 411\underline{7}.54\text{g} = 4.12 \text{ kg H}_2\text{O}$$

11.2 Use the two-point form of the Clausius-Clapeyron equation to calculate P_2:

$$\ln \frac{P_2}{760 \text{ mmHg}} = \frac{26.8 \times 10^3 \text{ J/mol}}{8.31 \text{ J/(K} \cdot \text{mol)}} \left[\frac{1}{319 \text{ K}} - \frac{1}{308 \text{ K}} \right]$$

$$= 3225 \text{ K} \left[\frac{-1.12 \times 10^{-4}}{\text{K}} \right] = -0.36\underline{1}06$$

(continued)

Converting to antilogs gives

$$\frac{P_2}{760 \text{ mmHg}} = \text{antilog}(-0.36106) = e^{-0.36106} = 0.69\underline{6}93$$

$$P_2 = 0.69693 \times 760 \text{ mmHg} = 529.6 = 5.30 \times 10^2 \text{ mmHg}$$

11.3   Use the two-point form of the Clausius-Clapeyron equation to solve for ΔH_{vap}:

$$\ln \frac{757 \text{ mmHg}}{760 \text{ mmHg}} = \frac{\Delta H_{vap}}{8.31 \text{ J/(K} \cdot \text{mol)}} \left[\frac{1}{368 \text{ K}} - \frac{1}{378 \text{ K}} \right]$$

$$0.37\underline{1}69 = \frac{\Delta H_{vap}}{8.31 \text{ J/(K} \cdot \text{mol)}} \left[\frac{7.188 \times 10^{-5}}{\text{K}} \right]$$

$$\Delta H_{vap} = 4.2\underline{9}6 \times 10^4 \text{ J/mol} \quad (43.0 \text{ kJ/mol})$$

11.4 a. Liquefy methyl chloride by a sufficient increase in pressure below 144°C.

   b. Liquefy oxygen by compressing to 50 atm below -119°C.

11.5 a. Propanol has a hydrogen atom bonded to an oxygen atom. Therefore, hydrogen bonding is expected. Because propanol is polar (from the O–H bond), we also expect dipole-dipole forces. Weak London forces exist, too, because such forces exist between all molecules.

   b. Linear carbon dioxide is not polar, so only London forces exist among CO_2 molecules.

   c. Bent sulfur dioxide is polar, so we expect dipole-dipole forces; we also expect the usual London forces.

11.6   The order of increasing vapor pressure is butane (C_4H_{10}), propane (C_3H_8), and ethane (C_2H_6). Because London forces tend to increase with increasing molecular weight, we would expect the molecule with the highest molecular weight to have the lowest vapor pressure.

11.7   Because ethanol has an H atom bonded to an O atom, strong hydrogen bonding exists in ethanol but not in methyl chloride. Hydrogen bonding explains the lower vapor pressure of ethanol compared to methyl chloride.

11.8 a. Zinc, a metal, is a metallic solid.

   b. Sodium iodide, an ionic substance, exists as an ionic solid.

   c. Silicon carbide, a compound in which carbon and silicon might be expected to form covalent bonds to other carbon and silicon atoms, exists as a covalent network solid.

   d. Methane, at room temperature a gaseous molecular compound with covalent bonds, freezes as a molecular solid.

11.9   Only $MgSO_4$ is an ionic solid; C_2H_5OH, CH_4, and CH_3Cl form molecular solids; thus, $MgSO_4$ should have the highest melting point. Of the molecular solids, CH_4 has the lowest molecular weight (16.0 amu) and would be expected to have the lowest melting point. Both C_2H_5OH and CH_3Cl have approximately the same molecular weights (46.0 amu vs. 50.5 amu), but C_2H_5OH exhibits strong hydrogen bonding and, therefore, would be expected to have the higher melting point. The order of increasing melting points is CH_4, CH_3Cl, C_2H_5OH, and $MgSO_4$.

11.10   Each of the four corners of the cell contains one atom, which is shared by a total of four unit cells. Therefore, the corners contribute one whole atom. This is:

$$\frac{Atoms}{Unit\ cell} = 4\ corners \times \frac{1/4\ atom}{1\ corner} = 1\ atom$$

11.11   Use the edge length to calculate the volume of the unit cell. Then, use the density to determine the mass of one atom. Divide the molar mass by the mass of one atom.

$$V = (3.509 \times 10^{-10}\ m)^3 = 4.321 \times 10^{-29}\ m^3$$

$$D = \frac{0.534\ g}{1\ cm^3} \times \left[\frac{100\ cm}{1\ m}\right]^3 = 5.34 \times 10^5\ g/m^3$$

Mass of 1 unit = d x V

= $(5.34 \times 10^5\ g/m^3) \times (4.321 \times 10^{-29}\ m^3) = 2.3074 \times 10^{-23}\ g$

(continued)

There are two atoms in a body-centered cubic unit cell; thus, the mass of one lithium atom is

$$1/2 \times 2.3074 \times 10^{-23} \text{ g} = 1.1537 \times 10^{-23} \text{ g}$$

The known atomic weight of lithium is 6.941 amu, so Avogadro's number is

$$N_A = \frac{6.941 \text{ g/mol}}{1.1537 \times 10^{-23} \text{ g/atom}} = 6.016 \times 10^{23} = 6.02 \times 10^{23} \text{ atoms/mol}$$

11.12   Use Avogadro's number to convert the molar mass of potassium to the mass per one atom.

$$\frac{39.0983 \text{ g K}}{1 \text{ mol K}} \times \frac{1 \text{ mol K}}{6.022 \times 10^{23} \text{ atoms}} = \frac{6.4925 \times 10^{-23} \text{ g K}}{1 \text{ atom}}$$

There are two K atoms per unit cell; therefore, the mass per unit cell is

$$\frac{6.4925 \times 10^{-23} \text{ g K}}{1 \text{ atom}} \times \frac{2 \text{ atoms}}{1 \text{ unit cell}} = \frac{1.2985 \times 10^{-22} \text{ g}}{1 \text{ unit cell}}$$

The density of 0.856 g/cm^3 is equal to the mass of one unit cell divided by its unknown volume, V. After solving for V, determine the edge length from the cube root of the volume.

$$0.856 \text{ g/cm}^3 = \frac{1.2985 \times 10^{-22} \text{ g}}{V}$$

$$V = \frac{1.2985 \times 10^{-22} \text{ g}}{0.856 \text{ g/cm}^3} = 1.517 \times 10^{-22} \text{ cm}^3 \ (1.517 \times 10^{-28} \text{ m}^3)$$

$$\text{Edge length} = \sqrt[3]{1.517 \times 10^{-28} \text{ m}^3} = 5.333 \times 10^{-10} = 5.33 \times 10^{-10} \text{ m} \ (533 \text{ pm})$$

■ Answers to Review Questions

11.1   The six different phase transitions, with examples in parentheses, are melting (snow melting), sublimation (dry ice subliming directly to carbon dioxide gas), freezing (water freezing), vaporization (water evaporating), condensation (dew forming on the ground), and gas-solid condensation or deposition (frost forming on the ground).

11.2   Iodine can be purified by heating it in a beaker covered with a dish containing ice or ice water. Only pure iodine should sublime, crystallizing on the cold bottom surface of the dish above the iodine. The common impurities in iodine do not sublime nor do they vaporize significantly.

11.3   The vapor pressure of a liquid is the partial pressure of the vapor over the liquid, measured at equilibrium. In molecular terms, vapor pressure involves molecules of a liquid vaporizing from the liquid phase, colliding with any surface above the liquid, and exerting pressure on it. The equilibrium is a dynamic one because molecules of the liquid are continually leaving the liquid phase and returning to it from the vapor phase.

11.4   Steam at 100°C will melt more ice than the same weight of water at 100°C because it contains much more energy in the form of its heat of vaporization. It will transfer this energy to the ice and condense in doing so. The condensed steam and the water will both transfer heat to the ice as the temperature then drops.

11.5   The heat of fusion is smaller than the heat of vaporization because melting requires only enough energy for molecules to escape from their sites in the crystal lattice, leaving other molecular attractions intact. In vaporization, sufficient energy must be added to break almost all molecular attractions and also to do the work of pushing back the atmosphere.

11.6   Evaporation leads to cooling of a liquid because the gaseous molecules require heat to evaporate; as they leave the other liquid molecules, they remove the heat energy required to vaporize them. This leaves less energy in the liquid whose temperature then drops.

11.7   As the temperature increases for a liquid and its vapor in the closed vessel, the two, which are separated by a meniscus, gradually become identical. The meniscus first becomes fuzzy and then disappears altogether as the temperature reaches the critical temperature. Above this temperature, only the vapor exists.

11.8   A permanent gas can be liquefied only by lowering the temperature below its critical temperature while compressing the gas.

11.9   The pressure in the cylinder of nitrogen at room temperature (above its critical temperature of -147°C) decreases continuously as gas is released because the number of molecules in the vapor phase, which governs the pressure, decreases continuously. The pressure in the cylinder of propane at room temperature (below its critical temperature) is constant because liquid propane and gaseous propane exist at equilibrium in the cylinder. The pressure will remain constant at the vapor pressure of propane until only gaseous propane remains. At that point, the pressure will decrease until all of the propane is gone.

11.10 The vapor pressure of a liquid depends on the intermolecular forces in the liquid phase since the ease with which a molecule leaves the liquid phase depends on how strongly it is attracted to the other molecules. If such molecules attract each other strongly, the vapor pressure will be relatively low; if they attract each other weakly, the vapor pressure will be relatively high.

11.11   Surface tension makes a liquid act as though it had a skin because, for an object to break through the surface, the surface area must increase. This requires energy, so there is some resistance to the object breaking through the surface.

11.12   London forces, also known as dispersion forces, originate between any two molecules that are weakly attracted to each other by means of small instantaneous dipoles that occur as a result of the varying positions of the electrons during their movement about their nuclei.

11.13   Hydrogen bonding is a weak to moderate attractive force that exists between a hydrogen atom covalently bonded to a very electronegative atom, X (N, O, or F), and a lone pair of electrons on another small, electronegative atom, Y. (X and Y may be the same or different elements.) Hydrogen bonding in water involves a hydrogen atom of one water molecule bonding to a lone pair of electrons on the oxygen atom of another water molecule.

11.14   Molecular substances have relatively low melting points because the forces broken by melting are weak intermolecular attractions in the solid state, not strong bonding attractions.

11.15   A crystalline solid has a well-defined, orderly structure; an amorphous solid has a random arrangement of structural units.

11.16   In a face-centered cubic cell, there are atoms at the center of each face of the unit cell in addition to those at the corners.

11.17   The structure of thallium(I) iodide is a simple cubic lattice for both the metal ions and the anions. Thus, the structure consists of two interpenetrating cubic lattices of cation and anion.

11.18   The coordination number of Cs^+ in CsCl is eight; the coordination number of Na^+ in NaCl is six; and the coordination number of Zn^{2+} in ZnS is four.

11.19   Starting with the edge length of a cubic crystal, we can calculate the volume of a unit cell by cubing the edge length. Then, knowing the density of the crystalline solid, we can calculate the mass of the atoms in the unit cell. Then, the mass of the atoms in the unit cell is divided by the number of atoms in the unit cell, to give the mass of one atom. Dividing the mass of one mole of the crystal by the mass of one atom yields a value for Avogadro's number.

11.20   X rays can strike a crystal and be reflected at various angles; at most angles, the reflected waves will be out of phase and will interfere destructively. At certain angles, however, the reflected waves will be in phase and will interfere constructively, giving rise to a diffraction pattern.

■ Solutions to Practice Problems

Note on significant figures: If the final answer to a solution needs to be rounded off, it is given first with one nonsignificant figure, and the last significant figure is underlined. The final answer is then rounded to the correct number of significant figures. In multiple-step problems, intermediate answers are given with at least one nonsignificant figure; however, only the final answer has been rounded off.

11.31 a. Vaporization

b. Freezing of eggs and sublimation of ice

c. Condensation

d. Gas-solid condensation, deposition

e. Freezing

11.33 Dropping a line from the intersection of a 350-mmHg line with the diethyl ether curve in Figure 11.7 intersects the temperature axis at about 10°C.

11.35 The total amount of energy provided by the heater in 4.54 min is

$$4.54 \text{ min } \times \frac{60 \text{ s}}{1 \text{ min}} \times \frac{3.48 \text{ J}}{\text{s}} = 94\underline{7}.95 \text{ J} \ (0.94\underline{7}95 \text{ kJ})$$

The heat of fusion per mole of I_2 is

$$\frac{0.9479 \text{ kJ}}{15.5 \text{ g } I_2} \times \frac{2 \times 126.9 \text{ g } I_2}{\text{mol } I_2} = 15.\underline{5}2 = \frac{15.5 \text{ kJ}}{\text{mol } I_2}$$

11.37 The heat absorbed per 2.25 g of isopropyl alcohol, C_3H_8O, is

$$2.25 \text{ g } C_3H_8O \times \frac{1 \text{ mol } C_3H_8O}{60.09 \text{ g } C_3H_8O} \times \frac{42.1 \text{ kJ}}{1 \text{ mol } C_3H_8O} = 1.5\underline{7}6 = 1.58 \text{ kJ}$$

11.39 Because all the heat released by freezing the water is used to evaporate the remaining water, you must first calculate the amount of heat released in the freezing:

$$9.31\,g\,H_2O \times \frac{mol\,H_2O}{18.02\,g\,H_2O} \times \frac{6.01\,kJ}{mol\,H_2O} = 3.10505\,kJ$$

Finally, calculate the mass of H_2O that was vaporized by the 3.10505 kJ of heat:

$$3.10505\,kJ \times \frac{18.02\,g\,H_2O}{1\,mol\,H_2O} \times \frac{1\,mol\,H_2O}{44.9\,kJ} = 1.246 = 1.25\,g\,H_2O$$

11.41 Calculate how much heat is released by cooling 64.3 g of H_2O from 55°C to 15°C.

$$Heat\ rel'd\ =\ (64.3\ g)(15\ °C\ -\ 55\ °C)\left(\frac{4.18\ J}{1\ g\bullet°C}\right)$$

$$=\ -1.07509 \times 10^4\ J\ =\ -10.7509\ kJ$$

The heat released is used first to melt the ice and then to warm the liquid from 0 °C to 15 °C. Let the mass of ice equal y grams. Then, for fusion, and for warming, we have

$$Fusion:\ (y\ g\ H_2O) \times \frac{1\ mol\ H_2O}{18.02\ g\ H_2O} \times \frac{6.01\ kJ}{1\ mol\ H_2O} = 0.3335\ y\ kJ$$

$$Warming:\ (y\ g\ H_2O)(15\ °C\ -\ 0\ °C)\left(\frac{4.18\ J}{1\ g\bullet°C}\right) = 62.70\ y\ J\ (0.06270\ y\ kJ)$$

Because the total heat required for melting and warming must equal the heat released by cooling, equate the two, and solve for y.

$$10.7509\ kJ\ =\ 0.3335y\ kJ\ +\ 0.0627y\ kJ\ =\ y(0.3335\ +\ 0.0627)\ kJ$$

$$y\ =\ 10.7509\ kJ \div 0.3962\ kJ\ =\ 27.13\ (grams)$$

Thus, 27 g of ice were added.

11.43 At the normal boiling point, the vapor pressure of a liquid is 760.0 mmHg. Use the Clausius-Clapeyron equation to find P_2 when P_1 = 760.0 mmHg, T_1 = 334.85 K (61.7°C), and T_2 = 309.35 K (36.2°C). Also use ΔH_{vap} = 31.4 x 10^3 J/mol.

$$\ln \frac{P_2}{P_1} = \frac{\Delta H_{vap}}{R}\left(\frac{1}{T_1} - \frac{1}{T_2}\right)$$

$$\ln \frac{P_2}{760 \text{ mmHg}} = \frac{31.4 \times 10^3 \text{ J/mol}}{8.31 \text{ J/K} \cdot \text{mol}}\left(\frac{1}{334.85 \text{ K}} - \frac{1}{309.35 \text{ K}}\right) = -0.93018$$

Taking antilogs of both sides gives

$$\frac{P_2}{760 \text{ mmHg}} = e^{-0.93018} = 0.3944$$

$$P_2 = 0.3944 \times 760 \text{ mmHg} = 299.8 = 3.00 \times 10^2 \text{ mmHg (300. mmHg)}$$

11.45 From the Clausius-Clapeyron equation,

$$\Delta H_{vap} = R\left(\frac{T_2 T_1}{T_2 - T_1}\right)\left[\ln \frac{P_2}{P_1}\right]$$

$$= [8.31 \text{ J/(K} \cdot \text{mol)}]\left[\frac{(553.2 \text{ K})(524.2 \text{ K})}{(553.2 - 524.2) \text{ K}}\right]\left[\ln \frac{760.0 \text{ mmHg}}{400.0 \text{ mmHg}}\right]$$

$$= 5.3336 \times 10^4 \text{ J/ mol} = 53.3 \text{ kJ/ mol}$$

11.47 a. At point A, the substance will be a gas.

  b. The substance will be a gas.

  c. The substance will be a liquid.

  d. no

11.49 The phase diagram for oxygen is shown below. It is plotted from these points: triple point = -219 °C, boiling point = -183 °C, and critical point = -118 °C.

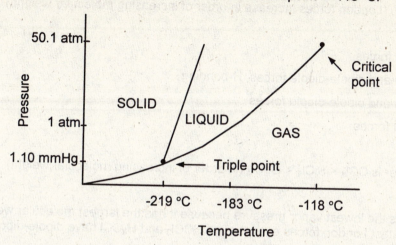

11.51 Liquefied at 25 °C: SO_2 and C_2H_2. To liquefy CH_4, lower its temperature below -82 °C, and then compress it. To liquefy CO, lower its temperature below -140 °C, and then compress it.

11.53 Br_2 phase diagram:

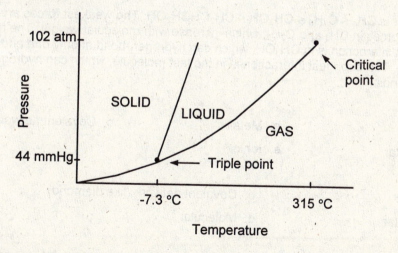

a. Circle "solid." The pressure of 40 mmHg is lower than the pressure at the triple point so the liquid phase cannot exist.

b. Circle "liquid." The pressure of 400 mmHg is above the triple point so the gas will condense to a liquid.

11.55 Yes, the heats of vaporization of 0.9, 5.6, and 20.4 kJ/mol (for H_2, N_2, and Cl_2, respectively) increase in the order of the respective molecular weights of 2.016, 28.02, and 71.0. (London forces increase in order of increasing molecular weight.)

11.57 a. London forces

b. London and dipole-dipole forces, H-bonding

c. London and dipole-dipole forces

d. London forces

11.59 The order is CCl_4 < $SiCl_4$ < $GeCl_4$ (in order of increasing molecular weight).

11.61 CCl_4 has the lowest vapor pressure because it has the largest molecular weight and the greatest London forces even though $HCCl_3$ and H_3CCl have dipole-dipole interactions.

11.63 The order of increasing vapor pressure is $HOCH_2CH_2OH$, FCH_2CH_2OH, FCH_2CH_2F. There is no hydrogen bonding in the third molecule; the second molecule can hydrogen-bond at only one end; and the first molecule can hydrogen-bond at both ends for the strongest interaction.

11.65 The order is CH_4 < C_2H_6 < CH_3OH < CH_2OHCH_2OH. The weakest forces are the London forces in CH_4 and C_2H_6, which increase with molecular weight. The next strongest interaction is in CH_3OH, which can hydrogen-bond at only one end of the molecule. The strongest interaction is in the last molecule, which can hydrogen-bond at both ends.

11.67 a. Metallic   b. Metallic   c. Covalent network

d. Molecular   e. Ionic

11.69 a. Metallic   b. Covalent network (like diamond)

c. Molecular   d. Molecular

11.71 The order is $(C_2H_5)_2O$ < C_4H_9OH < KCl < CaO. Melting points increase in the order of attraction between molecules or ions in the solid state. Hydrogen bonding in C_4H_9OH causes it to melt at a higher temperature than $(C_2H_5)_2O$. Both KCl and CaO are ionic solids with much stronger attraction than the organic molecules. In CaO, the higher charges cause the lattice energy to be higher than in KCl.

11.73 a. Low-melting and brittle   b. High-melting, hard, and brittle
     c. Malleable and electrically conducting   d. Hard and high-melting

11.75 a. LiCl   b. SiC   c. CHI_3 d. Co

11.77 In a simple cubic lattice with one atom at each lattice point, there are atoms only at the corners of unit cells. Each corner is shared by eight unit cells, and there are eight corners per unit cell. Therefore, there is one atom per unit cell.

11.79 Calculate the volume of the unit cell, change density to g/m^3, and then convert volume to mass, using density:

$$\text{Volume} = (2.866 \times 10^{-10} \text{ m})^3 = 2.354 \times 20^{-29} \text{ m}^3$$

$$\frac{7.87 \text{ g}}{1 \text{ cm}^3} \times \left(\frac{100 \text{ cm}}{1 \text{ m}}\right)^3 = 7.87 \times 10^6 \text{ g/m}^3$$

Mass of one cell = $(7.87 \times 10^6 \text{ g/m}^3) \times (2.354 \times 10^{-29} \text{ m}^3) = 1.8\underline{5}26 \times 10^{-22}$ g

Because Fe is a body-centered cubic cell, there are two Fe atoms in the cell, and

Mass of one Fe atom = $(1.8526 \times 10^{-22} \text{ g}) \div 2 = 9.2\underline{6}3 \times 10^{-23}$ g

Using the molar mass to calculate the mass of one Fe atom, you find the agreement is good:

$$\frac{55.85 \text{ g Fe}}{1 \text{ mol Fe}} \times \frac{1 \text{ mol Fe}}{6.022 \times 10^{23} \text{ Fe atoms}} = 9.27\underline{4}3 \times 10^{-23} \text{ g/Fe atom}$$

11.81 There are four Cu atoms in the face-centered cubic structure, so the mass of one cell is

$$4 \text{ Cu atoms} \times \frac{1 \text{ mol Cu}}{6.022 \times 10^{23} \text{ Cu atoms}} \times \frac{63.5 \text{ g Cu}}{1 \text{ mol Cu}} = 4.2\underline{1}8 \times 10^{-22} \text{ g}$$

$$\text{Cell volume} = \frac{4.218 \times 10^{-22} \text{ g}}{8.93 \text{ g/cm}^3} = 4.7\underline{2}3 \times 10^{-23} \text{ cm}^3$$

All edges are the same length in a cubic cell, so the edge length, l, is

$$l = \sqrt[3]{V} = \sqrt[3]{4.723 \times 10^{-23} \text{ cm}^3} = 3.6\underline{1}4 \times 10^{-8}$$

$$= 3.61 \times 10^{-8} \text{ cm} \quad (361 \text{ pm})$$

11.83 Calculate the volume from the edge length of 407.9 pm (4.079 x 10⁻⁸ cm), and then use it to calculate the mass of the unit cell:

$$\text{Cell volume} = (4.079 \times 10^{-8} \text{ cm}^3 = 6.7869 \times 10^{-23} \text{ cm}^3$$

$$\text{Cell mass} = (19.3 \text{ g/cm}^3)(6.7869 \times 20^{-23} \text{ cm}^3) = 1.3\underline{0}98 \times 10^{-21} \text{ g}$$

Calculate the mass of one gold atom:

$$1 \text{ Au atom} \times \frac{1 \text{ mol Au}}{6.022 \times 10^{23} \text{ Au atoms}} \times \frac{196.97 \text{ g Au}}{1 \text{ mol Au}} = 3.27\underline{0}8 \times 10^{-22} \text{ g}$$

$$\frac{1.3098 \times 10^{-21} \text{ g}}{1 \text{ unit cell}} \times \frac{1 \text{ Au atom}}{3.2708 \times 10^{-22} \text{ g Au}} = \frac{4.004 \text{ Au atoms}}{\text{unit cell}}$$

Since there are four atoms per unit cell, it is a face-centered cubic.

11.85 Calculate the volume from the edge (316.5 pm = 3.165 x 10⁻⁸ cm). Use it to calculate the mass:

$$\text{Cell volume} = (3.165 \times 10^{-8} \text{ cm})^3 = 3.17\underline{0}5 \times 10^{-23} \text{ cm}^3$$

For a body-centered cubic lattice, there are two atoms per cell, so their mass is

$$2 \text{ W atoms} \times \frac{1 \text{ mol W}}{6.022 \times 10^{23} \text{ W atoms}} \times \frac{183.8 \text{ g W}}{1 \text{ mol W}} = 6.10\underline{4}3 \times 10^{-22} \text{ g W}$$

$$\text{Density} = \frac{6.1043 \times 10^{-22} \text{ g W}}{3.1705 \times 10^{-23} \text{ cm}^3} = 19.2\underline{5}3 = 19.25 \text{ g/cm}^3$$

11.87 Use Avogadro's number to calculate the number of atoms in 1.74 g (= d x 1.000 cm³):

$$1.74 \text{ g Mg} \times \frac{1 \text{ mol Mg}}{24.305 \text{ g Mg}} \times \frac{6.022 \times 10^{23} \text{ Mg atoms}}{1 \text{ mol Mg}}$$

$$= 4.311 \times 10^{22} \text{ Mg atoms}$$

(continued)

Because the space occupied by the Mg atoms = 0.741 cm^3, each atom's volume is

$$\text{Volume 1 Mg atom} = \frac{0.741 \text{ cm}^3}{4.311 \times 10^{22} \text{ Mg atoms}} = 1.7\underline{1}9 \times 10^{-23} \text{ cm}^3$$

$$\text{Volume} = \frac{4\pi r^3}{3}$$

so

$$r = \sqrt[3]{\frac{3V}{4\pi}}$$

$$r = \sqrt[3]{\frac{3}{4\pi}(1.719 \times 10^{-23} \text{ cm}^3)} = 1.6\underline{0}1 \times 10^{-8} = 1.60 \times 10^{-8} \text{ cm}$$

$$= 1.60 \times 10^2 \text{ pm}$$

■ Solutions to General Problems

11.89   Water vapor deposits directly to solid water (frost) without forming liquid water. After heating, most of the frost melted to liquid water, which then vaporized to water vapor. Some of the frost may have sublimed directly to water vapor.

11.91   From Table 5.5, the vapor pressures are 18.7 mmHg at 21 °C and 12.8 mmHg at 15 °C. If the moisture did not begin to condense until the air had been cooled to 15 °C, then the partial pressure of water in the air at 21 °C must have been 12.8 mmHg. The relative humidity is

$$\text{Percent relative humidity} = \frac{12.8 \text{ mmHg}}{18.7 \text{ mmHg}} \times 100\% = 68.\underline{4}4 = 68.4 \text{ percent}$$

11.93 After labeling the problem data as below, use the Clausius-Clapeyron equation to obtain ΔH_{vap}, which can then be used to calculate the boiling point.

At T_1 = 299.3 K, P_1 = 100.0 mmHg; at T_2 = 333.8 K, P_2 = 400.0 mmHg

$$\ln \frac{400.0 \text{ mmHg}}{100.0 \text{ mmHg}} = \frac{\Delta H_{vap}}{8.31 \text{ J}/(\text{K} \cdot \text{mol})} \left[\frac{333.8 \text{ K} - 299.3 \text{ K}}{333.8 \text{ K} \times 299.3 \text{ K}} \right]$$

$$1.386\underline{2} = \Delta H_{vap} (4.153\underline{5} \times 10^{-5} \text{ mol/J})$$

$$\Delta H_{vap} = 33.37 \times 10^3 \text{ J/mol} (33.4 \text{ kJ/mol})$$

Now, use this value of ΔH_{vap} and the following data to calculate the boiling point:

At T_1 = 299.3 K, P_1 = 100.0 mmHg; at T_2 (boiling pt.), P_2 = 760 mmHg

$$\ln \frac{760.0 \text{ mmHg}}{100.0 \text{ mmHg}} = \frac{33.4 \times 10^3 \text{ J/mol}}{8.314 \text{ J}/(\text{K} \cdot \text{mol})} \left[\frac{1}{299.3 \text{ K}} - \frac{1}{T_2} \right]$$

$$2.0281 = 4.0144 \times 10^3 \text{ K} \left[\frac{1}{299.3 \text{ K}} - \frac{1}{T_2} \right]$$

$$\frac{1}{T_2} = \frac{1}{299.3 \text{ K}} - \frac{2.0281}{4.0144 \times 10^3 \text{ K}} = 2.8\underline{3}59 \times 10^{-3}/\text{K}$$

$$T_2 = 352.\underline{6} = 353 \text{ K} (80 \text{ °C})$$

11.95 a. As this gas is compressed at 20 °C, it will condense into a liquid because 20 °C is above the triple point but below the critical point.

  b. As this gas is compressed at -70 °C, it will condense directly to the solid phase because the temperature of -70 °C is below the triple point.

  c. As this gas is compressed at 40 °C, it will not condense because 40 °C is above the critical point.

11.97 In propanol, hydrogen bonding exists between the hydrogen of the OH group and the lone pair of electrons of oxygen of the OH group of an adjacent propanol molecule. For two adjacent propanol molecules, the hydrogen bond may be represented as follows:

C_3H_7–O–H•••O(H)C_3H_7

11.99 Ethylene glycol molecules are capable of hydrogen bonding to each other, whereas pentane molecules are not. The greater intermolecular forces in ethylene glycol are reflected in greater resistance to flow (viscosity) and high boiling point.

11.101 Aluminum (Group IIIA) forms a metallic solid. Silicon (Group IVA) forms a covalent network solid. Phosphorus (Group VA) forms a molecular solid. Sulfur (Group VIA) forms a molecular (amorphous) solid.

11.103 a. Lower: KCl. The lattice energy should be lower for ions with a lower charge. A lower lattice energy implies a lower melting point.

b. Lower: CCl_4. Both are molecular solids, so the compound with the lower molecular weight should have weaker London forces and, therefore, the lower melting point.

c. Lower: Zn. Melting points for Group IIB metals are lower than for metals near the middle of the transition-metal series.

d. Lower: C_2H_5Cl. Ethyl chloride cannot hydrogen-bond, but acetic acid can. The compound with the weaker intermolecular forces has the lower melting point.

11.105 The face-centered cubic structure means one atom is at each lattice point. All edges are the same length in such a structure, so the volume is

$$\text{Volume} = l^3 = (3.839 \times 10^{-8} \text{ cm})^3 = 5.6579 \times 10^{-23} \text{ cm}^3$$

$$\text{Mass of unit cell} = dV = (22.42 \text{ g/cm}^3)(5.6579 \times 10^{-23} \text{ cm}^3) = 1.2685 \times 10^{-21} \text{ g}$$

There are four atoms in a face-centered cubic cell, so

$$\text{Mass of one Ir atom} = \text{mass of unit cell} \div 4 = (1.2685 \times 10^{-21} \text{ g}) \div 4$$

$$= 3.1712 \times 10^{-22} \text{ g}$$

$$\text{Molar mass of Ir} = (3.1712 \times 10^{-22} \text{ g/Ir atom}) \times (6.022 \times 10^{23} \text{ Ir atoms/mol})$$

$$= 190.96 = 191.0 \text{ g/mol} \quad (\text{The atomic weight} = 191.0 \text{ amu.})$$

11.107 From Problem 11.81, the cell edge length (l) is 361.4 pm. There are four copper atom radii along the diagonal of a unit-cell face. Because the diagonal square $= l^2 + l^2$ (Pythagorean theorem),

$$4r = \sqrt{2 \, l^2} = \sqrt{2} \, l, \text{ or } r = \frac{\sqrt{2}}{4}(361.4 \text{ pm}) = 127.7 = 128 \text{ pm}$$

11.109 The body diagonal (diagonal passing through the center of the cell) is four times the radius, r, of a sphere. Also, from the geometry of a cube and the Pythagorean theorem, the body diagonal equals $\sqrt{3}$ l, where l is the edge length of the unit cell. Thus

$$4r = \sqrt{3} \, (l)$$

or

$$l = \frac{4r}{\sqrt{3}}$$

Because the unit cell contains two spheres, the volume occupied by the spheres is

$$V_{spheres} = 2 \times \frac{4}{3}\pi r^3$$

and

$$V_{cell} = l^3 = \left[\frac{4r}{\sqrt{3}}\right]^3 = \frac{64r^3}{3\sqrt{3}}$$

Finally, to obtain the percent volume of the cell occupied, divide $V_{spheres}$ by V_{cell}:

$$\text{Percent V} = \frac{V_{spheres}}{V_{cell}} \times 100\% = \frac{2\left[\dfrac{4\pi r^3}{3}\right]}{\dfrac{64r^3}{3\sqrt{3}}} \times 100\% = \frac{\pi\sqrt{3}}{8} \times 100\%$$

$$= 68.01 = 68 \text{ percent}$$

11.111 a. The boiling point increases as the size (number of electrons in the atom or molecule) increases. The London forces or dispersion forces increase.

  b. Hydrogen bonding occurs between the H–F molecules, and is much stronger than the London forces.

  c. In addition to the dispersion forces, the hydrogen halides are polar (have dipole moments) so there are dipole-dipole interactions.

11.113 a. Diamond and silicon carbide are giant molecules with strong covalent bonds between all the atoms. Graphite is a layered structure, and the forces holding the layers together are weak dispersion forces.

  b. Silicon dioxide is a giant molecule with an infinite array of O–S–O bonds. Each silicon is bonded to four oxygen atoms in a covalent network solid. Carbon dioxide is a discrete, nonpolar, molecule.

11.115 a. CO_2 consists of discrete non-polar molecules that are held together in the solid by weak dispersion forces. SiO_2 is a giant molecule with all the atoms held together by strong covalent bonds.

b. HF(l) has extensive hydrogen bonding among the molecules. HCl(l) boils much lower because it doesn't have H-bonding.

c. SiF_4 is a larger molecule (it has more electrons than CF_4), so it has stronger dispersion forces and a higher boiling point than CF_4. Both molecules are tetrahadrally symmetrical and, therefore, non-polar.

■ Solutions to Cumulative-Skills Problems

11.117 Use the ideal gas law to calculate n, the number of moles of N_2:

$$N_2:\ n = \frac{PV}{RT} = \frac{(745/760\ atm)(5.40\ L)}{(0.082057\ L \cdot atm/K \cdot mol)(293\ K)} = 0.220\underline{1}\ mol$$

$$C_3H_8O:\ n = 0.6149\ g\ C_3H_8O \times \frac{1\ mol\ C_3H_8O}{60.094\ g\ C_3H_8O} = 0.01023\underline{2}\ mol$$

$$X_{C_3H_8O} = \frac{0.010232\ mol\ C_3H_8O}{(0.010232\ mol + 0.2201\ mol)} = 0.044\underline{4}2\ mole\ fraction$$

Partial P = 0.04442 x 745 mmHg = 33.\underline{0}93 mmHg = 33.1 mmHg

Vapor pressure of C_3H_8O = 33.1 mmHg

11.119 Calculate the moles of HCN in 10.0 mL of the solution (density = 0.687 g HCN/mL HCN):

$$10.0\ mL\ HCN \times \frac{0.687\ g\ HCN}{1\ mL\ HCN} \times \frac{1\ mol\ HCN}{27.03\ g\ HCN} = 0.254\underline{1}\ mol\ HCN$$

0.2541 mol HCN(l)   →   0.2541 mol HCN(g)

($\Delta H_f° = 105\ kJ/mol$)   ($\Delta H_f° = 135\ kJ/mol$)

$\Delta H° = 0.2541\ mol \times [135\ kJ/mol - 105\ kJ/mol] = 7.\underline{6}23 = 7.6\ kJ$

11.121 First, convert the mass to moles; then, multiply by the standard heat of formation to obtain the heat absorbed in vaporizing this mass:

$$12.5 \text{ g } P_4 \times \frac{1 \text{ mol } P_4}{123.88 \text{ g } P_4} = 0.10\underline{0}9 \text{ mol } P_4$$

$$0.1009 \text{ mol } P_4 \times \frac{95.4 \text{ J}}{°C \cdot \text{mol } P_4} \times (44.1 °C - 25.0 °C)$$

$$= 18\underline{3}.86 \text{ J} = 0.18\underline{3}86 \text{ kJ}$$

$$2.63 \text{ kJ/mol } P_4 \times 0.1009 \text{ mol } P_4 = 0.26\underline{5}37 \text{ kJ}$$

Total heat = 0.26537 kJ + 0.18386 kJ = 0.44\underline{9}23 = 0.449 kJ = 449 J

11.123 Use the ideal gas law to calculate the total number of moles of monomer and dimer:

$$n = \frac{PV}{RT} = \frac{(436/760)\text{atm} \times 1.000 \text{ L}}{[0.082057 \text{ L} \cdot \text{atm}/(K \cdot \text{mol})] \times 373.75 \text{ K}}$$

$$= 0.018\underline{7}057 \text{ mol monomer } + \text{ dimer}$$

$$(0.018\underline{7}057 \text{ mol monomer } + \text{ dimer}) \times \frac{0.630 \text{ mol dimer}}{1 \text{ mol dimer } + \text{ monomer}}$$

$$= 0.011\underline{7}8 \text{ mol dimer}$$

0.0187057 mol both - 0.01178 mol dimer = 0.006\underline{9}2 mol monomer

$$\text{Mass dimer} = 0.011\underline{7}8 \text{ mol dimer} \times \frac{120.1 \text{ g dimer}}{1 \text{ mol dimer}} = 1.4\underline{1}4 \text{ g dimer}$$

$$\text{Mass monomer} = 0.00692 \text{ mol monomer} \times \frac{60.05 \text{ g monomer}}{1 \text{ mol monomer}}$$

$$= 0.4\underline{1}55 \text{ g monomer}$$

$$\text{Density} = \frac{1.414 \text{ g} + 0.4155 \text{ g}}{1.000 \text{ L}} = 1.8\underline{2}9 = 1.83 \text{ g/L vapor}$$

12. SOLUTIONS

■ Solutions to Exercises

Note on significant figures: If the final answer to a solution needs to be rounded off, it is given first with one nonsignificant figure, and the last significant figure is underlined. The final answer is then rounded to the correct number of significant figures. In multiple-step problems, intermediate answers are given with at least one nonsignificant figure; however, only the final answer has been rounded off.

12.1   An example of a solid solution prepared from a liquid and a solid is a dental filling made of liquid mercury and solid silver.

12.2   The C_4H_9OH molecules will be more soluble in water because their -OH ends can form hydrogen bonds with water.

12.3   The Na^+ ion has a larger energy of hydration because its ionic radius is smaller, giving Na^+ a more concentrated electric field than K^+.

12.4   Write Henry's law ($S = k_HP$) for 159 mmHg (P_2), and divide it by Henry's law for one atm, or 760 mmHg (P_1). Then, substitute the experimental values of P_1, P_2, and S_1 to solve for S_2.

$$\frac{S_2}{S_1} = \frac{k_HS_2}{k_HP_1} = \frac{P_2}{P_1}$$

(continued)

Solving for S_2 gives

$$S_2 = \frac{P_2 S_1}{P_1} = \frac{(159 \text{ mmHg})(0.0404 \text{ g O}_2/\text{L})}{760 \text{ mmHg}} = 8.4\underline{5}2 \times 10^{-3}$$

$$= 8.45 \times 10^{-3} \text{ g O}_2/\text{L}$$

12.5   The mass of HCl in 20.2 percent HCl (0.202 = fraction of HCl) is

$$0.202 \times 35.0 \text{ g} = 7.0\underline{7}0 = 7.07 \text{ g HCl}$$

The mass of H_2O in 20.2 percent HCl is

$$35.0 \text{ g solution} - 7.07 \text{ g HCl} = 27.\underline{9}3 = 27.9 \text{ g H}_2\text{O}$$

12.6   Calculate the moles of toluene using its molar mass of 92.14 g/mol:

$$35.6 \text{ g toluene} \times \frac{1 \text{ mol toluene}}{92.14 \text{ g toluene}} = 0.386\underline{3} \text{ mol toluene}$$

To calculate molality, divide the moles of toluene by the mass in kg of the solvent (C_6H_6):

$$\text{Molality} \times \frac{0.3863 \text{ mol toluene}}{0.125 \text{ kg solvent}} = 3.09\underline{0}4 = 3.09 \text{ m toluene}$$

12.7   The number of moles of toluene = 0.3863 (previous exercise); the number of moles of benzene is

$$125 \text{ g benzene} \times \frac{1 \text{ mol benzene}}{78.11 \text{ g benzene}} = 1.6\underline{0}03 \text{ mol benzene}$$

The total number of moles is 1.6003 + 0.3863 = 1.9\underline{8}66, and the mole fractions are

$$\text{Mol fraction benzene} \times \frac{1.6003 \text{ mol benzene}}{1.9866 \text{ mol}} = 0.805\underline{5}4 = 0.806$$

$$\text{Mol fraction toluene} \times \frac{0.3863 \text{ mol toluene}}{1.9866 \text{ mol}} = 0.194\underline{4} = 0.194$$

The sum of the mole fractions = 1.000.

12.8   This solution contains 0.120 moles of methanol dissolved in 1.00 kg of ethanol. The number of moles in 1.00 kg of ethanol is

$$1.00 \times 10^3 \text{ g } C_2H_5OH \times \frac{1 \text{ mol } C_2H_5OH}{46.07 \text{ g } C_2H_5OH} = 21.\underline{7}06 \text{ mol } C_2H_5OH$$

The total number of moles is 21.706 + 0.120 = 21.\underline{8}26, and the mole fractions are

$$\text{Mol fraction } C_2H_5OH = \frac{21.706 \text{ mol } C_2H_5OH}{21.826 \text{ mol}} = 0.99\underline{4}501 = 0.995$$

$$\text{Mol fraction } CH_3OH = \frac{0.120 \text{ mol } CH_3OH}{21.826 \text{ mol}} = 0.00549\underline{8} = 0.00550$$

The sum of the mole fractions is 1.000.

12.9   One mole of solution contains 0.250 moles methanol and 0.750 moles ethanol. The mass of this amount of ethanol, the solvent, is

$$0.750 \text{ mol } C_2H_5OH \times \frac{46.07 \text{ g } C_2H_5OH}{1 \text{ mol } C_2H_5OH} = 34.\underline{5}5 \text{ g } C_2H_5OH \ (0.03455 \text{ kg})$$

The molality of methanol in the ethanol solvent is

$$\frac{0.250 \text{ mol } CH_3OH}{0.03455 \text{ kg } C_2H_5OH} = 7.2\underline{3}58 = 7.24 \text{ m } CH_3OH$$

12.10   Assume an amount of solution contains one kilogram of water. The mass of urea in this mass is

$$3.42 \text{ mol urea} \times \frac{60.05 \text{ g urea}}{1 \text{ mol urea}} = 20\underline{5}.4 \text{ g urea}$$

The total mass of solution is 205.4 + 1000.0 g = 120\underline{5}.4 g. The volume and molarity are

$$\text{Volume of solution} = 120\underline{5}.4 \text{ g} \times \frac{1 \text{ mL}}{1.045 \text{ g}} = 115\underline{3}.49 \text{ mL} = 1.15349 \text{ L}$$

$$\text{Molarity} = \frac{3.42 \text{ mol urea}}{1.15349 \text{ L solution}} = 2.9\underline{6}49 \text{ mol/L} = 2.96 \text{ M}$$

12.11 Assume a volume equal to 1.000 L of solution. Then

$$\text{Mass of solution} = 1.029 \text{ g/mL} \times (1.000 \times 10^3 \text{ mL}) = 1029 \text{ g}$$

$$\text{Mass of urea} = 2.00 \text{ mol urea} \times \frac{60.05 \text{ g urea}}{1 \text{ mol urea}} = 120.1 \text{ g urea}$$

$$\text{Mass of water} = (1029 - 120.1) \text{ g} = 90\underline{8}.9 \text{ g water} \ (0.90\underline{89} \text{ kg})$$

$$\text{Molality} = \frac{2.00 \text{ mol urea}}{0.9089 \text{ kg solvent}} = 2.2\underline{0}04 = 2.20 \text{ m urea}$$

12.12 Calculate the moles of naphthalene and moles of chloroform:

$$0.515 \text{ g } C_{10}H_8 \times \frac{1 \text{ mol } C_{10}H_8}{128.17 \text{ g } C_{10}H_8} = 0.0040\underline{18} \text{ mol } C_{10}H_8$$

$$60.8 \text{ g } CHCl_3 \times \frac{1 \text{ mol } CHCl_3}{119.38 \text{ g } CHCl_3} = 0.50\underline{9}29 \text{ mol } CHCl_3$$

The total number of moles is $0.004018 + 0.50929 = 0.51\underline{33}$ mol, and the mole fraction of chloroform is

$$\text{Mol fraction } CHCl_3 = \frac{0.50929 \text{ mol } CHCl_3}{0.5133 \text{ mol}} = 0.99\underline{21}$$

$$\text{Mol fraction } C_{10}H_8 = \frac{0.004018 \text{ mol } C_{10}H_8}{0.5133 \text{ mol}} = 0.0078\,\underline{2}8$$

The vapor-pressure lowering is

$$\Delta P = P° X_{C_{10}H_8} = (156 \text{ mmHg})(0.007828) = 1.2\underline{2}1 = 1.22 \text{ mmHg}$$

Use Raoult's law to calculate the vapor-pressure of chloroform:

$$P = P° X_{CHCl_3} = (156 \text{ mmHg})(0.9921) = 15\underline{4}.7 = 155 \text{ mmHg}$$

12.13 Solve for c_m in the freezing-point-depression equation ($\Delta T = K_f c_m$; K_f in Table 12.3):

$$C_m = \frac{\Delta T}{K_f} = \frac{0.150 \text{ °C}}{1.858 \text{ °C/m}} = 0.80\underline{7}3 \text{ m}$$

(continued)

Use the molal concentration to solve for the mass of ethylene glycol:

$$\frac{0.08073 \text{ m glycol}}{1 \text{ kg solvent}} \times 0.0378 \text{ kg solvent} = 0.003051 \text{ mol glycol}$$

$$0.003051 \text{ mol glycol} \times \frac{62.1 \text{ g glycol}}{1 \text{ mol glycol}} = 1.894 \times 10^{-1} = 1.89 \times 10^{-1} \text{ g glycol}$$

12.14 Calculate the moles of ascorbic acid (vit. C) from the molality, and then divide the mass of 0.930 g by the number of moles to obtain the molar mass:

$$\frac{0.0555 \text{ mol vit. C}}{1 \text{ kg } H_2O} \times 0.0950 \text{ kg } H_2O = 0.005272 \text{ mol vit C}$$

$$\frac{0.930 \text{ g vit. C}}{0.005272 \text{ mol vit. C}} = 176.4 = 176 \text{ g/mol}$$

The molecular weight of ascorbic acid, or vitamin C, is 176 amu.

12.15 The molal concentration of white phosphorus is

$$c_m = \frac{\Delta T}{K_b} = \frac{0.159 \text{ °C}}{2.40 \text{ °C/m}} = 0.06625 \text{ m}$$

The number of moles of white phosphorus (P_x) present in this solution is

$$\frac{0.06625 \text{ mol } P_x}{1 \text{ kg } CS_2} \times 0.0250 \text{ kg } CS_2 = 0.001656 \text{ mol } P_x$$

The molar mass of white phosphorus equals the mass divided by moles:

$$0.205 \text{ g} \div 0.001656 \text{ mol} = 123.77 = 124 \text{ g/mol}$$

Thus, the molecular weight of P_x is 124 amu. The number of P atoms in the molecule of white phosphorus is obtained by dividing the molecular weight by the atomic weight of P:

$$\frac{123.77 \text{ amu } P_x}{30.97 \text{ amu } P} = 3.9965 = 4.00$$

Hence, the molecular formula is P_4 ($x = 4$).

12.16   The number of moles of sucrose is

$$5.0 \text{ g sucrose } \times \frac{1 \text{ mol sucrose}}{342.3 \text{ g sucrose}} = 0.014\underline{6} \text{ mol sucrose}$$

The molarity of the solution is

$$\frac{0.0146 \text{ mol sucrose}}{0.100 \text{ L}} = 0.14\underline{6} \text{ M sucrose}$$

The osmotic pressure, π, is equal to MRT and is calculated as follows:

$$\frac{0.146 \text{ mol sucrose}}{1 \text{ L}} \times \frac{0.0821 \text{ L} \cdot \text{atm}}{\text{K} \cdot \text{mol}} \times 293 \text{ K} = 3.\underline{5}1 = 3.5 \text{ atm}$$

12.17   The number of ions from each formula unit is i. Here,

$$i = 1 + 2 = 3$$

The boiling-point elevation is

$$\Delta T_b = K_b c_m = 3 \times \frac{0.512 \text{ °C}}{m} \times 0.050 \text{ m} = 0.07\underline{6}81 = 0.077 \text{ °C}$$

The boiling point of aqueous $MgCl_2$ is 100.077 °C.

12.18 $AlCl_3$ would be most effective in coagulating colloidal sulfur because of the greater magnitude of charge on the Al ion (+3).

■ Answers to Review Questions

12.1   An example of a gaseous solution is air, in which nitrogen (78 percent) acts as a solvent for a gas such as oxygen (21 percent). Recall that the solvent is the component present in greater amount. An example of a liquid solution containing a gas is any carbonated beverage in which water acts as the solvent for carbon dioxide gas. Ethanol in water is an example of a liquid-liquid solution. An example of a solid solution is any gold-silver alloy.

12.2   The two factors that explain differences in solubilities are (1) the natural tendency of substances to mix together, or the natural tendency of substances to become disordered, and (2) the relative forces of attraction between solute species (or solvent species) compared to that between the solute and solvent species. The strongest interactions are always achieved.

12.3   This is not a case of "like dissolves like." There are strong hydrogen-bonding forces between the water molecules. For the octane to mix with water, hydrogen bonds must be broken and replaced by the much weaker London forces between water and octane. Thus, octane does not dissolve in water because the maximum forces of attraction among molecules are obtained if it does not dissolve.

12.4   In most cases, the wide differences in solubility can be explained in terms of the different energies of attraction between ions in the crystal and between ions and water. Hydration energy is used to measure the attraction of ions for water molecules, and lattice energy is used to measure the attraction of positive ions for negative ions in the crystal lattice. An ionic substance is soluble when the hydration energy is much larger than the lattice energy. An ionic substance is insoluble when the lattice energy is much larger than the hydration energy.

12.5   A sodium chloride crystal dissolves in water because of two factors. The positive Na^+ ion is strongly attracted to the oxygen (negative end of the water dipole) and dissolves as $Na^+(aq)$. The negative Cl^- ion is strongly attracted to the hydrogens (positive end of the water dipole) and dissolves as $Cl^-(aq)$.

12.6   When the temperature (energy of a solution) is increased, the solubility of an ionic compound usually increases. A number of salts are exceptions to this rule, particularly a number of calcium salts such as calcium acetate, calcium sulfate, and calcium hydroxide (although the solubilities of calcium bromide, calcium chloride, calcium fluoride, and calcium iodide all increase with temperature).

12.7   Calcium chloride is an example of a salt that releases heat when it dissolves (exothermic heat of solution). Ammonium nitrate is an example of a salt that absorbs heat when it dissolves (endothermic heat of solution).

12.8   As the temperature of the solution is increased by heating, the concentration of the dissolved gas would decrease.

12.9   A carbonated beverage must be stored in a closed container because carbonated beverages must contain more carbon dioxide than is soluble in water at atmospheric pressure. It is possible to add carbon dioxide under pressure to a closed container before it is sealed and increase the solubility of carbon dioxide. This is an illustration of Le Chatelier's principle, which states that the equilibrium between gaseous and dissolved carbon dioxide is shifted in favor of the dissolved carbon dioxide by an increase in pressure.

12.10   According to Le Chatelier's principle, a gas is more soluble in a liquid at higher pressures because, when the gas dissolves in the liquid, the system decreases in volume, tending to decrease the applied pressure. However, when a solid dissolves in a liquid, there is very little volume change. Thus, pressure has very little effect on the solubility of a solid in a liquid.

12.11   The four ways to express the concentration of a solute in a solution are (1) molarity, which is moles per liter; (2) mass percentage of solute, which is the percentage by mass of solute contained in a given mass of solution; (3) molality, which is the moles of solute per kilogram of solvent; and (4) mole fraction, which is the moles of the component substance divided by the total moles of solution.

12.12   The vapor pressure of the solvent of the dilute solution is larger than the vapor pressure of the solvent of the more concentrated solution (Raoult's law). Thus, the more dilute solution loses solvent and becomes more concentrated while the solvent molecules in the vapor state condense into the concentrated solution making it more dilute. After sufficient time has passed, the vapor pressure of the solvent in the closed container will reach a steady value (equilibrium), at which time the concentration of solute in the two solutions will be the same.

12.13   In fractional distillation, the vapor that first appears over a solution will have a greater mole fraction of the more volatile component. If a portion of this is vaporized and condensed, the liquid will be still richer in the more volatile component. After successive distillation stages, eventually the more volatile component will be obtained in pure form (Figure 12.19).

12.14   The boiling point of the solution is higher because the nonvolatile solute lowers the vapor pressure of the solvent. Thus, the temperature must be increased to a value greater than the boiling point of the pure solvent to achieve a vapor pressure equal to atmospheric pressure.

12.15   One application is the use of ethylene glycol in automobile radiators as antifreeze; the glycol-water mixture usually has a freezing point well below the average low temperature during the winter. A second application is spreading sodium chloride on icy roads in the winter to melt the ice. The ice usually melts because, at equilibrium, a concentrated solution of NaCl usually freezes at a temperature below that of the roads.

12.16   If a pressure greater than the osmotic pressure of the ocean water is applied, the natural osmotic flow can be reversed. Then, the water solvent flows from the ocean water through a membrane to a more dilute solution or to pure water, leaving behind the salt and other ionic compounds from the ocean in a more concentrated solution.

12.17   Part of the light from the sun is scattered in the direction of an observer by fine particles in the clouds (Tyndall effect) rather than being completely absorbed by the clouds. The scattered light becomes visible against the darker background of dense clouds.

12.18   The examples are: fog is an aerosol, whipped cream is a foam, mayonnaise is an emulsion, solid silver chloride dispersed in water is a sol, and fruit jelly is a gel.

12.19   The polar -OH group on glycerol allows it to interact (hydrogen bond) with the polar water molecules, which means it is like water and, therefore, will dissolve in water. Benzene is a nonpolar molecule, which indicates that it is not "like" water and, therefore, will not dissolve in water.

12.20   Soap removes oil from a fabric by absorbing the oil into the hydrophobic centers of the soap micelles and off the surface of the fabric. Rinsing removes the micelles from contact with the fabric and leaves only water on the fabric, which can then be dried.

■ Solutions to Practice Problems

Note on significant figures: If the final answer to a solution needs to be rounded off, it is given first with one nonsignificant figure, and the last significant figure is underlined. The final answer is then rounded to the correct number of significant figures. In multiple-step problems, intermediate answers are given with at least one nonsignificant figure; however, only the final answer has been rounded off. Atomic weights are rounded to two decimal places, except for that of hydrogen.

12.31   An example of a liquid solution prepared by dissolving a gas in a liquid is household ammonia, which consists of ammonia (NH_3) gas dissolved in water.

12.33   Boric acid would be more soluble in ethanol because this acid is polar and is more soluble in a more polar solvent. It can also hydrogen-bond to ethanol but not to benzene.

12.35   The order of increasing solubility is $H_2O < CH_2OHCH_2OH < C_{10}H_{22}$. The solubility in nonpolar hexane increases with the decreasing polarity of the solute.

12.37   The Al^{3+} ion has both a greater charge and a smaller ionic radius than Mg^{2+}, so Al^{3+} should have a greater energy of hydration.

12.39   The order is $Ba(IO_3)_2 < Sr(IO_3)_2 < Ca(IO_3)_2 < Mg(IO_3)_2$. The iodate ion is fairly large, so the lattice energy for all these iodates should change to a smaller degree than the hydration energy of the cations. Therefore, solubility should increase with decreasing cation radius.

12.41 Using Henry's law, let S_1 = the solubility at 1.00 atm (P_1), and S_2 = the solubility at 5.50 atm (P_2).

$$S_2 = \frac{P_2 S_1}{P_1} = \frac{(5.50 \text{ atm})(0.161 \text{ g}/100 \text{ mL})}{1.00 \text{ atm}} = 0.885\underline{5}$$

$$= 0.886 \text{ g}/100 \text{ mL}$$

12.43 First, calculate the mass of KI in the solution; then calculate the mass of water needed.

$$\text{Mass KI} = 72.5 \text{ g} \times \frac{5.00 \text{ g KI}}{100 \text{ g solution}} = 3.6\underline{2}50 = 3.63 \text{ g KI}$$

$$\text{Mass H}_2\text{O} = 72.5 \text{ g soln} - 3.6250 \text{ g KI} = 68.\underline{8}75 = 68.9 \text{ g H}_2\text{O}$$

Dissolve 3.63 g KI in 68.9 g of water.

12.45 Multiply the mass of KI by 100 g of solution per 5.00 g KI (reciprocal of percentage).

$$0.258 \text{ g KI} \times \frac{100 \text{ g solution}}{5.00 \text{ g KI}} = 5.1\underline{6}0 = 5.16 \text{ g solution}$$

12.47 Convert mass of vanillin ($C_8H_8O_3$, molar mass 152.14 g/mol) to moles, convert mg of ether to kg, and divide for molality.

$$0.0391 \text{ g vanillin} \times \frac{1 \text{ mol vanillin}}{152.14 \text{ g vanillin}} = 2.5\underline{7}0 \times 10^{-4} \text{ mol vanillin}$$

$$168.5 \text{ mg ether} \times 1 \text{ kg}/10^6 \text{ mg} = 168.5 \times 10^{-6} \text{ kg ether}$$

$$\text{Molality} = \frac{2.570 \times 10^{-4} \text{ mol vanillin}}{168.5 \times 10^{-6} \text{ kg ether}} = 1.5\underline{2}52 = 1.53 \text{ m vanillin}$$

12.49 Convert mass of fructose ($C_6H_{12}O_6$, molar mass 180.16 g/mol) to moles, and then multiply by one kg H_2O per 0.125 mol fructose (the reciprocal of molality).

$$1.75 \text{ g fructose} \times \frac{1 \text{ mol fructose}}{180.16 \text{ g fructose}} \times \frac{1 \text{ kg H}_2\text{O}}{0.125 \text{ mol fructose}}$$

$$= 0.077\underline{7}0 \text{ kg} \quad (77.7 \text{ g H}_2\text{O})$$

12.51 Convert masses to moles, and then calculate the mole fractions.

$$65.0 \text{ g alc.} \times \frac{1 \text{ mol alc.}}{60.09 \text{ g alc.}} = 1.0817 \text{ mol alc.}$$

$$35.0 \text{ g H}_2\text{O} \times \frac{1 \text{ mol H}_2\text{O}}{18.02 \text{ g H}_2\text{O}} = 1.9422 = 1.94 \text{ mol H}_2\text{O}$$

$$\text{Mol fraction alc.} = \frac{\text{mol alc.}}{\text{total mol}} = \frac{1.0817 \text{ mol}}{3.0239 \text{ mol}} = 0.3577 = 0.358$$

$$\text{Mol fraction H}_2\text{O} = \frac{\text{mol H}_2\text{O}}{\text{total mol}} = \frac{1.9422 \text{ mol}}{3.0239 \text{ mol}} = 0.6422 = 0.642$$

12.53 In the solution, for every 0.650 mol of NaClO there is 1.00 kg, or 1.00×10^3 g, H_2O, so

$$1.00 \times 10^3 \text{ g H}_2\text{O} \times \frac{1 \text{ mol H}_2\text{O}}{18.02 \text{ g H}_2\text{O}} = 55.49 \text{ mol H}_2\text{O}$$

Total mol = 55.49 mol H_2O + 0.650 mol NaClO = 56.14 mol

$$\text{Mol fraction NaClO} = \frac{\text{mol NaClO}}{\text{total mol}} = \frac{0.650 \text{ mol}}{56.14 \text{ mol}} = 0.01157 = 0.0116$$

12.55 The total moles of solution = 3.31 mol H_2O + 1.00 mol HCl = 4.31 mol.

$$\text{Mol fraction HCl} = \frac{1 \text{ mol HCl}}{4.31 \text{ mol}} = 0.23201 = 0.232$$

$$3.31 \text{ mol H}_2\text{O} \times \frac{18.02 \text{ g H}_2\text{O}}{1 \text{ mol H}_2\text{O}} \times \frac{1 \text{ kg H}_2\text{O}}{10^3 \text{ g H}_2\text{O}} = 5.965 \times 10^{-2} \text{ kg H}_2\text{O}$$

$$\text{Molality} = \frac{1.00 \text{ mol HCl}}{5.965 \times 10^{-2} \text{ kg H}_2\text{O}} = 16.76 = 16.8 \text{ m}$$

12.57　The mass of 1.000 L of solution is 1.022 kg. In the solution, there are 0.580 moles of $H_2C_2O_4$ (OA) for every 1.0000 kg of water. Convert this number of moles to mass.

$$0.580 \text{ mol OA} \times \frac{90.04 \text{ g OA}}{1 \text{ mol OA}} \times \frac{1 \text{ kg OA}}{10^3 \text{ g OA}} = 0.05222 \text{ kg OA}$$

The total mass of the solution containing 1.000 kg H_2O and 0.580 moles of OA is calculated as follows:

$$\text{Mass} = 1.0000 \text{ kg } H_2O + 0.05222 \text{ kg OA} = 1.05222 \text{ kg}$$

Use this to relate the mass of 1.000 L (1.022 kg) of solution to the amount of solute:

$$1.022 \text{ kg solution} \times \frac{0.580 \text{ mol OA}}{1.05222 \text{ kg solution}} = 0.5633 \text{ mol OA}$$

$$\text{Molarity} = \frac{0.5633 \text{ mol OA}}{1.00 \text{ L solution}} = 0.5633 = 0.563 \text{ M}$$

12.59　In 1.000 L of vinegar, there is 0.763 mole of acetic acid. The total mass of the 1.000 L solution is 1.004 kg. Start by calculating the mass of acetic acid (AA) in the solution.

$$0.763 \text{ mol AA} \times \frac{60.05 \text{ g AA}}{1 \text{ mol AA}} = 45.82 \text{ g AA} \quad (0.04582 \text{ kg AA})$$

The mass of water may be found by difference:

$$\text{Mass } H_2O = 1.004 \text{ kg soln} - 0.04582 \text{ kg AA} = 0.9582 \text{ kg } H_2O$$

$$\text{Molality} = \frac{0.763 \text{ mol AA}}{0.9582 \text{ kg } H_2O} = 0.7962 = 0.796 \text{ m AA}$$

12.61　To find the mole fraction of sucrose, first find the amounts of both sucrose (suc.) and water:

$$20.2 \text{ g sucrose} \times \frac{1 \text{ mol sucrose}}{342.30 \text{ g sucrose}} = 0.05901 \text{ mol sucrose}$$

$$70.1 \text{ g } H_2O \times \frac{1 \text{ mol } H_2O}{18.02 \text{ g } H_2O} = 3.890 \text{ mol } H_2O$$

$$X_{sucrose} = \frac{0.05901 \text{ mol sucrose}}{(3.890 + 0.05901) \text{ mol}} = 0.01494$$

(continued)

From Raoult's law, the vapor pressure (P) and lowering (ΔP) are

$$P = P^\circ_{H_2O} \, X_{H_2O} = P^\circ_{H_2O} \, (1 - X_{suc.}) = (42.2 \text{ mmHg})(1 - 0.01494)$$

$$= 41.\underline{5}69 = 41.6 \text{ mmHg}$$

$$\Delta P = P^\circ_{H_2O} \, X_{suc.} = (42.2 \text{ mmHg})(0.01494) = 0.63\underline{0}6 = 0.631 \text{ mmHg}$$

12.63   Find the molality of glycerol (gly.) in the solution first:

$$0.150 \text{ g gly.} \times \frac{1 \text{ mol gly.}}{92.095 \text{ g gly.}} = 0.0016\underline{2}8 \text{ mol gly.}$$

$$\text{Molality} = \frac{0.001628 \text{ mol gly.}}{0.0200 \text{ kg solvent}} = 0.081\underline{4}37 \text{ m}$$

Substitute K_b = 0.512 °C/m and K_f = 1.858 °C/m (Table 12.3) into equations for ΔT_b and ΔT_f:

$$\Delta T_b = K_b m = 0.512 \text{ °C/m} \times 0.081\underline{4}37 \text{ m} = 0.041\underline{6}9 \text{ °C}$$

$$T_b = 100.000 + 0.04169 = 100.04\underline{1}69 = 100.042 \text{ °C}$$

$$\Delta T_f = K_f m = 1.858 \text{ °C/m} \times 0.081\underline{4}37 \text{ m} = 0.151\underline{3}1 \text{ °C}$$

$$T_f = 0.000 - 0.15\underline{1}31 = -0.15\underline{1}31 = -0.151 \text{ °C}$$

12.65   Calculate ΔT_f, the freezing-point depression, and, using K_f = 1.858 °C/m (Table 12.3), the molality, c_m.

$$\Delta T_f = 0.000 \text{ °C} - (-0.086 \text{ °C}) = 0.086 \text{ °C}$$

$$c_m = \frac{\Delta T_f}{K_f} = \frac{0.086 \text{ °C}}{1.858 \text{ °C/m}} = 4.\underline{6}28 \times 10^{-2} = 4.6 \times 10^{-2} \text{ m}$$

12.67 Find the moles of unknown solute from the definition of molality:

$$\text{Mol}_{solute} = m \times \text{kg solvent} = \frac{0.0698 \text{ mol}}{1 \text{ kg solvent}} \times 0.002135 \text{ kg solvent}$$

$$= 1.4\underline{9}0 \times 10^{-4} \text{ mol}$$

$$\text{Molar mass} = \frac{0.0182 \text{ g}}{1.490 \times 10^{-4} \text{ mol}} = 12\underline{2}.1 = 122 \text{ g/mol}$$

The molecular weight is 122 amu.

12.69 Calculate ΔT_f, the freezing-point depression, and, using $K_f = 1.858 \text{ °C/m}$ (Table 12.3), the molality, c_m.

$$\Delta T_f = 26.84 \text{ °C} - 25.70 \text{ °C} = 1.14 \text{ °C}$$

$$c_m = \frac{\Delta T_f}{K_f} = \frac{1.14 \text{ °C}}{8.00 \text{ °C/m}} = 0.14\underline{2}5 \text{ m}$$

Find the moles of solute by rearranging the definition of molality:

$$\text{Mol} = m \times \text{kg solvent} = \frac{0.1425 \text{ mol}}{1 \text{ kg solvent}} \times 103 \times 10^{-6} \text{ kg solvent}$$

$$= 1.4\underline{6}7 \times 10^{-5} \text{ mol}$$

$$\text{Molar mass} = \frac{2.39 \times 10^{-3} \text{ g}}{1.467 \times 10^{-5} \text{ mol}} = 16\underline{2}.9 = 163 \text{ g/mol}$$

The molecular weight is 163 amu.

12.71 Use the equation for osmotic pressure (π) to solve for the molarity of the solution.

$$M = \frac{\pi}{RT} = \frac{1.47 \text{ mmHg} \times \left(\dfrac{1 \text{ atm}}{760 \text{ mmHg}}\right)}{(0.0821 \text{ L} \cdot \text{atm/K} \cdot \text{mol})(21 + 273)\text{K}} = 8.0\underline{1}3 \times 10^{-5} \text{ mol/L}$$

Now, find the number of moles in 106 mL (0.106 L) using the molarity.

$$0.106 \text{ L} \times \frac{8.013 \times 10^{-5} \text{ mol}}{1 \text{ L}} = 8.4\underline{9}4 \times 10^{-6} \text{ mol}$$

(continued)

$$\text{Molar mass} = \frac{0.582 \text{ g}}{8.494 \times 10^{-6} \text{ mol}} = 6.851 \times 10^4 = 6.85 \times 10^4 \text{ g/mol}$$

The molecular weight is 6.85×10^4 amu.

12.73   Begin by noting that $i = 3$. Then, calculate ΔT_f from the product of $iK_f c_m$:

$$3 \times \frac{1.858 \text{ °C}}{m} \times 0.0075 \text{ m} = 0.0418 \text{ °C}$$

The freezing point = 0.000 °C - 0.0418 °C = -0.0418 = -0.042 °C.

12.75   Begin by calculating the molarity of $Cr(NH_3)_5Cl_3$.

$$1.40 \times 10^{-2} \text{ g } Cr(NH_3)_5Cl_3 \times \frac{1 \text{ mol } Cr(NH_3)_5Cl_3}{243.5 \text{ g } Cr(NH_3)_5Cl_3}$$

$$= 5.749 \times 10^{-5} \text{ mol } Cr(NH_3)_5Cl_3$$

$$\text{Molarity} = \frac{5.749 \times 10^{-5} \text{ mol } Cr(NH_3)_5Cl_3}{0.0250 \text{ L}} = 0.002299 \text{ M}$$

Now find the hypothetical osmotic pressure, assuming $Cr(NH_3)_5Cl_3$ does not ionize:

$$\pi = MRT = (2.30 \times 10^{-3} \text{ M}) \times \frac{0.0821 \text{ L} \cdot \text{atm}}{K \cdot \text{mol}} \times 298 \text{ K} \times \frac{760 \text{ mmHg}}{1 \text{ atm}}$$

$$= 42.77 \text{ mmHg}$$

The measured osmotic pressure is greater than the hypothetical osmotic pressure. The number of ions formed per formula unit equals the ratio of the measured pressure to the hypothetical pressure:

$$i = \frac{119 \text{ mmHg}}{42.77 \text{ mmHg}} = 2.782 \cong 3 \text{ ions/formula unit}$$

12.77 a. Aerosol (liquid water in air)

   b. Sol [solid $Mg(OH)_2$ in liquid water]

   c. Foam (air in liquid soap solution)

   d. Sol (solid silt in liquid water)

12.79   Because the As_2S_3 particles are negatively charged, the effective coagulation requires a highly charged cation, so $Al_2(SO_4)_3$ is the best choice.

■ Solutions to General Problems

12.81   Using Henry's law [$S_2 = S_1 \times (P_2/P_1)$, where P_1 = 1.00 atm], find the solubility of each gas at P_2, its partial pressure. For N_2, P_2 = 0.800 × 1.00 atm = 0.800 atm; for O_2, P_2 = 0.200 atm.

$$N_2: \ S_2 \times \frac{P_2}{P_1} = (0.0175 \ \text{g/L} \ H_2O) \times \frac{0.800 \ \text{atm}}{1.00 \ \text{atm}} = 0.01400 \ \text{g/L} \ H_2O$$

$$O_2: \ S_2 \times \frac{P_2}{P_1} = (0.0393 \ \text{g/L} \ H_2O) \times \frac{0.200 \ \text{atm}}{1.00 \ \text{atm}} = 0.007860 \ \text{g/L} \ H_2O$$

In 1.00 L of the water, there are 0.0140 g of N_2 and 0.00786 g of O_2. If the water is heated to drive off both dissolved gases, the gas mixture that is expelled will contain 0.0140 g of N_2 and 0.00786 g of O_2.

Convert both masses to moles using the molar masses:

$$0.0140 \ \text{g} \ N_2 \times \frac{1 \ \text{mol} \ N_2}{28.01 \ \text{g} \ N_2} = 4.998 \times 10^{-4} \ \text{mol}$$

$$0.00786 \ \text{g} \ O_2 \times \frac{1 \ \text{mol} \ O_2}{32.00 \ \text{g} \ O_2} = 2.456 \times 10^{-4} \ \text{mol}$$

Now calculate the mole fractions of each gas:

$$X_{N_2} = \frac{\text{mol} \ N_2}{\text{total mol}} = \frac{4.998 \times 10^{-4} \ \text{mol}}{(4.998 + 2.456) \times 10^{-4} \ \text{mol}} = 0.67051 = 0.671$$

$$X_{O_2} = 1 - X_{N_2} = 1 - 0.67051 = 0.32949 = 0.329$$

12.83 Assume a volume of 1.000 L whose mass is then 1.024 kg. Use the percent composition given to find the mass of each of the components of the solution.

$$1.024 \text{ kg soln } \times \frac{8.50 \text{ kg NH}_4\text{Cl}}{100.00 \text{ kg soln}} = 0.08704 \text{ kg NH}_4\text{Cl}$$

Mass of H_2O = 1.024 kg soln - 0.08704 kg NH_4Cl = 0.9370 kg H_2O

Convert mass of NH_4Cl and water to moles:

$$87.04 \text{ g NH}_4\text{Cl} \times \frac{1 \text{ mol NH}_4\text{Cl}}{53.49 \text{ g NH}_4\text{Cl}} = 1.627 \text{ mol NH}_4\text{Cl}$$

$$937.0 \text{ g H}_2\text{O} \times \frac{1 \text{ mol H}_2\text{O}}{18.015 \text{ g H}_2\text{O}} = 52.01 \text{ mol H}_2\text{O}$$

$$\text{Molarity} = \frac{\text{mol NH}_4\text{Cl}}{\text{L solution}} = \frac{1.627 \text{ mol}}{1.00 \text{ L}} = 1.627 = 1.63 \text{ M}$$

$$\text{Molality} = \frac{\text{mol NH}_4\text{Cl}}{\text{kg H}_2\text{O}} = \frac{1.627 \text{ mol}}{0.9370 \text{ kg H}_2\text{O}} = 1.736 = 1.74 \text{ m}$$

$$X_{\text{NH}_4\text{Cl}} = \frac{\text{mol NH}_4\text{Cl}}{\text{total moles}} = \frac{1.627 \text{ mol}}{(52.01 + 1.627) \text{ mol}} = 0.03033 = 0.0303$$

12.85 In 1.00 mol of gas mixture, there are 0.51 mol of propane (pro.) and 0.49 mol of butane (but.). First, calculate the masses of these components.

$$0.51 \text{ mol pro. } \times \frac{44.10 \text{ g pro.}}{1 \text{ mol pro.}} = 22.491 \text{ g pro.}$$

$$0.49 \text{ mol but. } \times \frac{58.12 \text{ g but.}}{1 \text{ mol but.}} = 28.478 \text{ g but.}$$

The mass of 1.00 mol of gas mixture is the sum of the masses of the two components:

22.491 g pro. + 28.478 g but. = 50.969 g mixture

Therefore, in 50.969 g of the mixture, there are 22.491 g of propane and 28.478 g of butane. For a sample with a mass of 55 g:

$$55 \text{ g mixture } \times \frac{22.491 \text{ g pro.}}{50.969 \text{ g mixture}} = 24.26 = 24 \text{ g pro.}$$

$$55 \text{ g mixture } \times \frac{28.478 \text{ g pro.}}{50.969 \text{ g mixture}} = 30.73 = 31 \text{ g but.}$$

12.87 $P_{ED} = P°_{ED}(X_{ED}) = 173$ mmHg $(0.25) = 4\underline{3}.25$ mmHg

$P_{PD} = P°_{PD}(E_{PD}) = 127$ mmHg $(0.75) = 9\underline{5}.25$ mmHg

$P = P_{ED} + P_{PD} = 43.25 + 95.25 = 13\underline{8}.50 = 139$ mmHg

12.89 Calculate the moles of $KAl(SO_4)_2 \bullet 12H_2O$ using its molar mass of 474.4 g/mol, and use this to calculate the three concentrations.

a. The moles of $KAl(SO_4)_2 \bullet 12H_2O$ are calculated below, using the abbreviation of "Hyd" for the formula of $KAl(SO_4)_2 \bullet 12H_2O$.

$$\text{mol Hyd.} = 0.1186 \text{ g Hyd.} \times \frac{1 \text{ mol Hyd}}{474.4 \text{ g Hyd}} = 0.0002500\underline{0} = 0.0002500 \text{ mol}$$

Note that one mol of $KAl(SO_4)_2 \bullet 12H_2O$ contains one mole of $KAl(SO_4)_2$, so calculating the molarity of $KAl(SO_4)_2$ can be performed using the moles of $KAl(SO_4)_2 \bullet 12H_2O$.

$$\frac{\text{mol}}{L} = \frac{0.0002500 \text{ mol Hyd}}{1.000 \text{ L soln.}} \times \frac{1 \text{ mol } KAl(SO_4)_2}{1 \text{ mol Hyd}} = 0.0002500 \text{ M } KAl(SO_4)_2$$

b. The molarity of the SO_4^{2-} ion will be twice that of the $KAl(SO_4)_2$.

$$\frac{\text{mol } SO_4^{2-}}{L} = 0.0002500 \text{ M } KAl(SO_4)_2 \times \frac{2 \text{ mol } SO_4^{2-}}{1 \text{ mol } KAl(SO_4)_2} = 0.0005000 \text{ M}$$

c. Since the density of the solution is 1.00 g/mL, the mass of 1.000 L of solution is 1000 g, or 1.000 kg. Since molality is moles per 1.000 kg of solvent, the molality of $KAl(SO_4)_2$ equals 0.0002500 moles divided by 1.000 kg or 0.0002500 m (the same as the molarity).

12.91 In 1.00 kg of a saturated solution of urea, there are 0.41 kg of urea (a molecular solute) and 0.59 kg of water. First, convert the mass of urea to moles.

$$0.41 \times 10^3 \text{ g urea} \times \frac{1 \text{ mol urea}}{60.06 \text{ g urea}} = 6.\underline{8}26 \text{ mol urea}$$

Then, find the molality of the urea in the solution:

$$\text{Molality} = \frac{\text{mol urea}}{\text{kg } H_2O} \times \frac{6.826 \text{ mol urea}}{0.59 \text{ kg}} = 1\underline{1}.57 \text{ m}$$

$$\Delta T_f = K_f c_m = (1.858 \text{ °C/m})(11.57 \text{ m}) = 2\underline{1}.4 \text{ °C}$$

$$T_f = 0.0 \text{ °C} - 21.4 \text{ °C} = -2\underline{1}.4 = -21 \text{ °C}$$

12.93 $M = \dfrac{\pi}{RT} = \dfrac{7.7 \text{ atm}}{(0.0821 \text{ L} \cdot \text{atm/K} \cdot \text{mol})(37 + 273)K} = 0.3\underline{0}2 = 0.30 \text{ mol/L}$

12.95   Consider the equation $\Delta T_f = iK_f c_m$. For $CaCl_2$, i = 3; for glucose i = 1. Because $c_m = 0.10$ for both solutions, the product of i and c_m will be larger for $CaCl_2$, as will ΔT_f. The solution of $CaCl_2$ will thus have the lower freezing point.

12.97   Assume there is 1.000 L of the solution, which will contain eighteen mol H_2SO_4, molar mass 98.09 g/mol. The mass of the solution is

$$\text{Mass solution} = \dfrac{18 \text{ mol} \times 98.09 \text{ g/mol}}{0.98} = 18\underline{0}2 \text{ g}$$

Thus, the density of the solution is

$$d = \dfrac{1802 \text{ g}}{1000 \text{ mL}} = 1.\underline{8}02 = 1.8 \text{ g/mL}$$

The mass of water in the solution is

$$\text{mass } H_2O = 1802 \text{ g} \times 0.02 = \underline{3}6.0 \text{ g} = 0.0\underline{3}60 \text{ kg}$$

Thus, the molality of the solution is

$$m = \dfrac{18 \text{ mol } H_2SO_4}{0.0360 \text{ Kg}} = \underline{5}00 = 5 \times 10^2 \text{ m}$$

12.99   Use the freezing point depression equation to find the molality of the solution. The freezing point of pure cyclohexane is 6.55 °C, and $K_f = 20.2$ °C/m. Thus

$$m = \dfrac{\Delta T_f}{K_f} = \dfrac{(6.55 - 5.28) \text{ °C}}{20.2 \text{ °C/m}} = 0.0628\underline{7} \text{ m}$$

The moles of the compound are

$$\text{mol compound} = \dfrac{0.06287 \text{ mol}}{1 \text{ kg solvent}} \times 0.00538 \text{ kg} = 3.3\underline{8}2 \times 10^{-4} \text{ mol}$$

The molar mass of the compound is

$$\text{molar mass} = \dfrac{0.125 \text{ g}}{3.382 \times 10^{-4} \text{ mol}} = 36\underline{9}.6 \text{ g/mol}$$

(continued)

The moles of the elements in 100 g of the compound are

$$\text{mol Mn} = 28.17 \text{ g Mn} \times \frac{1 \text{ mol}}{54.94 \text{ g Mn}} = 0.51274 \text{ mol}$$

$$\text{mol C} = 30.80 \text{ g C} \times \frac{1 \text{ mol}}{12.01 \text{ g C}} = 2.5645 \text{ mol}$$

$$\text{mol O} = 41.03 \text{ g O} \times \frac{1 \text{ mol}}{16.00 \text{ g O}} = 2.5644 \text{ mol}$$

This gives mole ratios of one mol Mn to five mol C to five mol O. Therefore, the empirical formula of the compound is MnC_5O_5. The weight of this formula unit is approximately 195 amu. Since the molar mass of the compound is 370. g/mol, the value of n is

$$n = \frac{370 \text{ g/mol}}{195 \text{ g/unit}} = 2.00, \text{ or } 2$$

Therefore, the formula of the compound is $Mn_2C_{10}O_{10}$.

12.101 a. Use the freezing point depression equation to find the molality of the solution. The freezing point of pure water is 0 °C and K_f = 1.858 °C/m. Thus

$$m = \frac{\Delta T_f}{K_f} = \frac{[0 - (-2.2)] \text{ °C}}{1.858 \text{ °C/m}} = 1.184 \text{ m}$$

The moles of the compound are

$$\text{mol compound} = \frac{1.184 \text{ mol}}{1 \text{ kg solvent}} \times 0.100 \text{ kg} = 0.1184 \text{ mol}$$

The molar mass of the compound is

$$\text{molar mass} = \frac{18.0 \text{ g}}{0.1184 \text{ mol}} = 152.0 \text{ g/mol}$$

(continued)

The moles of the elements in 100 g of the compound are

$$\text{mol C} = 48.64 \text{ g C} \times \frac{1 \text{ mol}}{12.01 \text{ g C}} = 4.0500 \text{ mol}$$

$$\text{mol H} = 8.16 \text{ g H} \times \frac{1 \text{ mol}}{1.008 \text{ g H}} = 8.095 \text{ mol}$$

$$\text{mol O} = 43.20 \text{ g O} \times \frac{1 \text{ mol}}{16.00 \text{ g O}} = 2.7000 \text{ mol}$$

This gives mole ratios of 1.5 mol C to three mol H to one mol O. Multiplying by two gives ratios of three mol C to six mol H to two mol O. Therefore, the empirical formula of the compound is $C_3H_6O_2$. The weight of this formula unit is approximately 74 amu. Since the molar mass of the compound is 152.0. g/mol, the value of n is

$$n = \frac{152.0 \text{ g/mol}}{74 \text{ g/unit}} = 2.05, \text{ or } 2$$

Therefore, the formula of the compound is $C_6H_{12}O_4$.

b. The molar mass of the compound is (to the nearest tenth of a gram):

$$6(12.01) + 12(1.008) + 4(16.00) = 148.156 = 148.2 \text{ g/mol}$$

12.103 Use the freezing point depression equation to find the molality of the solution. The freezing point of pure water is 0 °C, and K_f = 1.86 °C/m. Thus

$$m = \frac{\Delta T_f}{K_f} = \frac{[0 - (-2.3)] \text{ °C}}{1.86 \text{ °C/m}} = 1.24 \text{ m}$$

Since this is a relatively dilute solution, we can assume the molarity and molality of the fish blood are approximately equal. The calculated molarity is for the total number of particles, assuming they behave ideally.

$$\pi = MRT = 1.24 \text{ mol/L} \times 0.08206 \text{ L•atm/K•mol} \times 298.2 \text{ K}$$

$$= 30.3 = 30. \text{ atm}$$

■ Solutions to Cumulative-Skills Problems

12.105 First, determine the initial moles of each ion present in the solution before any reaction has occurred. There are five ions present: Na^+, 2 x 0.375 = 0.750 mol; CO_3^{2-}, 0.375 mol; Ca^{2+}, 0.125 mol; Ag^+, 0.200 mol; and NO_3^-, 0.200 + 2 x 0.125 = 0.450 mol. Two precipitates form, $CaCO_3$ and Ag_2CO_3. The balanced net ionic equations for their formation are

$$Ca^{2+}(aq) + CO_3^{2-}(aq) \rightarrow CaCO_3(s)$$

$$2\, Ag^+(aq) + CO_3^{2-}(aq) \rightarrow Ag_2CO_3(s)$$

There is sufficient CO_3^{2-} to precipitate all of the Ca^{2+} and all of the Ag^+ ions. The moles of excess CO_3^{2-} are calculated as follows: 0.375 - 0.125 - 1/2 x 0.200 = 0.150 mol remaining. Since the volume is 2.000 L, the molarities of the various ions are

M of CO_3^{2-} left = 0.150 mol ÷ 2.000 L = 0.0750 M

M of NO_3^- left = 0.450 mol ÷ 2.000 L = 0.225 M

M of Na^+ left = 0.750 mol ÷ 2.000 L = 0.375 M

12.107

$Na^+(g) + Cl^-(g)$	$\rightarrow$	$NaCl(s)$	ΔH = -787 kJ/mol
$NaCl(s)$	$\rightarrow$	$Na^+(aq) + Cl^-(aq)$	ΔH = + 4 kJ/mol

$$Na^+(g) + Cl^-(g) \rightarrow Na^+(aq) + Cl^-(aq) \qquad \Delta H = -783 \text{ kJ/mol}$$

The heat of hydration of Na^+ is

$Na^+(g) + Cl^-(g)$	$\rightarrow$	$Na^+(aq) + Cl^-(aq)$	ΔH = -783 kJ/mol
$Cl^-(aq)$	$\rightarrow$	$Cl^-(g)$	ΔH = +338 kJ/mol

$$Na^+(g) \rightarrow Na^+(aq) \qquad \Delta H = -445 \text{ kJ/mol}$$

12.109

$$15.0 \text{ g MgSO}_4\text{•7H}_2\text{O} \times \frac{1 \text{ mol}}{246.5 \text{ g}} = 0.060854 \text{ mol MgSO}_4\text{•7H}_2\text{O}$$

$$0.060854 \text{ mol MgSO}_4 \text{•7H}_2\text{O} \times \frac{7 \text{ mol H}_2\text{O}}{1 \text{ mol hydrate}} \times \frac{18.0 \text{ g H}_2\text{O}}{1 \text{ mol H}_2\text{O}} = 7.667 \text{ g H}_2\text{O}$$

$$\text{kg H}_2\text{O} = (100.0 \text{ g H}_2\text{O} + 7.667 \text{ g H}_2\text{O}) \times \frac{1 \text{ kg H}_2\text{O}}{1000 \text{ g H}_2\text{O}} = 0.10766 \text{ kg H}_2\text{O}$$

$$m = \frac{0.060854 \text{ mol MgSO}_4}{0.10766 \text{ kg H}_2\text{O}} = 0.5652 \text{ mol/kg} = 0.565 \text{ m}$$

12.111 $15.0 \text{ g CuSO}_4 \cdot 5\text{H}_2\text{O} \times \dfrac{1 \text{ mol}}{249.7 \text{ g}} = 0.06007 \text{ mol CuSO}_4 \cdot 5\text{H}_2\text{O, or CuSO}_4$

$100 \text{ g soln} \times \dfrac{1 \text{ mL soln}}{1.167 \text{ g soln}} \times \dfrac{1 \text{ L}}{1000 \text{ mL}} = 0.085689 \text{ L}$

$M = \dfrac{0.06007 \text{ mol CuSO}_4}{0.085689 \text{ L}} = 0.70102 = 0.701 \text{ mol/L}$

12.113 $0.159 \text{ °C} \times \dfrac{m}{1.858 \text{ °C}} = 0.08557 \text{ m} = \dfrac{0.08557 \text{ mol AA} + \text{H}^+}{1000 \text{ g (or 1000 mL) H}_2\text{O}}$

Note that 0.0830 mol AA + mol H^+ = 0.08557 mol/L (AA + H^+).

Mol H^+ = 0.08557 - 0.0830 = 0.00257 mol

$\text{Percent dissoc.} = \dfrac{\text{mol H}^+}{\text{mol AA}} \times 100\% = \dfrac{0.00257 \text{ mol}}{0.0830 \text{ mol}} \times 100\%$

$= 3.09 = 3.1 \text{ percent}$

12.115 Calculate the empirical formula first, using the masses of C, O, and H in 1.000 g:

$1.434 \text{ g CO}_2 \times \dfrac{12.01 \text{ g C}}{44.01 \text{ g CO}_2} = 0.39132 \text{ g C}$

$0.783 \text{ g H}_2\text{O} \times \dfrac{2.016 \text{ g H}}{18.016 \text{ g H}_2\text{O}} = 0.08761 \text{ g H}$

g O = 1.000 g - 0.39132 g - 0.08761 g = 0.5211 g O

Mol C = 0.39132 g C × 1 mol/12.01 g = 0.03258 mol C (lowest integer = 1)

Mol H = 0.08761 g H × 1 mol/1.008 g = 0.08691 mol H (lowest integer = 8/3)

Mol O = 0.5211 g O × 1 mol/16.00 g = 0.03257 mol O (lowest integer = 1)

(continued)

Therefore, the empirical formula is $C_3H_8O_3$. The formula weight from the freezing point is calculated by first finding the molality:

$$0.0894 \text{ °C} \times (m/1.858 \text{ °C}) = 0.04811 = (0.04811 \text{ mol}/1000 \text{ g } H_2O)$$

$$(0.04811 \text{ mol}/1000 \text{ g } H_2O) \times 25.0 \text{ g } H_2O = 0.001203 \text{ mol (in 25.0 g } H_2O)$$

$$\text{Molar mass} = M_m = 0.1107 \text{ g}/0.001203 \text{ mol} = 92.02 \text{ g/mol}$$

Because this is also the formula weight, the molecular formula is also $C_3H_8O_3$.

13. MATERIALS OF TECHNOLOGY

■ Answers to Review Questions

13.1   An alloy is a material with metallic properties that is either a compound or a mixture. If the alloy is a mixture, it may be homogeneous (a solution) or heterogeneous. Gold jewelry is made from an alloy that is a solid solution of gold containing some silver.

13.2   A metal is a material that is lustrous (shiny), has high electrical and heat conductivities, and is malleable and ductile.

13.3   A rock is a naturally occurring solid material composed of one or more minerals. A mineral is a naturally occurring inorganic solid substance or solid solution with a definite crystalline structure. An ore is a rock or mineral from which a metal or nonmetal can be economically produced. Bauxite, the principle ore of aluminum, is a rock.

13.4   The usual compounds from which metals are obtained are oxides (and hydroxides), sulfides, and carbonates (Table 13.2).

13.5   The basic steps in the production of a pure metal from a natural source are:
(1) Preliminary treatment: separating the metal-containing mineral from the less desirable parts of the ore. The mineral may also be transformed by chemical reaction to a metal compound that is more easily reduced to the free metal; (2) Reduction: The metal compound is reduced to the free metal by electrolysis or chemical reduction; (3) Refining: The free metal is purified.

(1) Preliminary treatment: Aluminum oxide is obtained from bauxite by the Bayer process. Bauxite contains aluminum hydroxide, aluminum oxide hydroxide, and other worthless constituents. It is mixed with hot, aqueous sodium hydroxide solution, which dissolves the amphoteric aluminum minerals along with some silicates. When the solution is cooled, $Al(OH)_3$ precipitates, leaving the silicates behind.

(continued)

The aluminum hydroxide is finally calcined (heated strongly in a furnace) to produce purified aluminum oxide, Al_2O_3. (2) Reduction: Aluminum is then obtained by reduction using the Hall-Héroult process, which is the electrolysis of a molten mixture of aluminum oxide in cryolyte (Na_3AlF_6). (3) Refining: The aluminum can be further refined and purified.

13.6   Flotation is a physical method of separating a mineral from gangue that depends on differences in their wettabilities by a liquid solution. Flotation agents coat the mineral particles selectively, and during the process air bubbles cause them to float to the surface in the froth. The gangue settles to the bottom.

13.7   Zinc is obtained from its sulfide ore by roasting, which forms zinc oxide. It is then heated with coke in a blast furnace where it is reduced to zinc metal.

13.8   Tungsten reacts with carbon to produce tungsten carbide, WC, which is unreactive. Instead, the metal is prepared from tungsten (VI) oxide, obtained from the processing of tungsten ore. The oxide is reduced by heating it in a stream of hydrogen gas.

13.9   (1) Iron oxide is reduced to iron using coke and limestone. The molten iron flows to the bottom of the blast furnace while carbon dioxide gas escapes. Impurities in the iron react with the calcium oxide from the limestone and produce slag, which floats to the top. (2) The Mond process is a chemical procedure that depends on the formation, and later decomposition, of a volatile compound of the metal (nickel tetracarbonyl). The compound decomposes over pellets of pure nickel heated to 230 °C. (3) Copper is purified by electrolysis using a pure copper negative electrode. Electrons from the impure copper positive electrode flow to the positive pole of the battery leaving copper(II) ions behind in the solution. These replace the ones that plate out on the pure copper electrode, enlarging it.

13.10   The Dow process use seashells ($CaCO_3$) as a source of the base. When heated, calcium carbonate decomposes to calcium oxide (CaO) and carbon dioxide (CO_2).

$$CaCO_3(s) \xrightarrow{\Delta} CaO(s) + CO_2(g)$$

The calcium oxide reacts with water to produce calcium hydroxide, the base.

$$CaO(s) + H_2O(l) \longrightarrow Ca(OH)_2(aq)$$

The calcium hydroxide reacts with the magnesium ion in seawater to form a magnesium hydroxide precipitate.

$$Ca(OH)_2(aq) + Mg^{2+} \longrightarrow Mg(OH)_2(s) + Ca^{2+}(aq)$$

The magnesium hydroxide is then treated with hydrochloric acid (HCl) to yield magnesium chloride.

$$Mg(OH)_2(s) + 2HCl(aq) \longrightarrow MgCl_2(aq) + 2H_2O(l)$$

The dry magnesium chloride is melted and electrolyzed at 700 °C to yield the metal.

$$MgCl_2(l) \rightarrow Mg(l) + Cl_2(g)$$

13.11   The flowchart for the preparation of aluminum from its ore (bauxite) is:

   (1) Preliminary treatment (The Bayer Process): Mix the ore with hot, NaOH(aq). Cool and filter the Al(OH)$_3$ precipitate. Heat the precipitate in a furnace to convert to Al$_2$O$_3$.

   (2) Reduction (Hall-Héroult Process): Melt a mixture of Al$_2$O$_3$ with cryolyte produced from Al(OH)$_3$ by the following reaction.

$$Al(OH)_3(s) + 3NaOH(aq) + 6HF(aq) \longrightarrow Na_3AlF_6(aq) + 6H_2O(l)$$

The electrolytic cell has carbon electrodes and a molten cryolyte electrolyte, at about 1000 °C, into which some aluminum oxide is dissolved. The reaction is

$$2Al_2O_3(s) + 3C(sq) \xrightarrow{\text{electrolysis}} 4Al(l) + 3CO_2(g)$$

13.12   In a metal, the outer orbits of an enormous number of metal atoms overlap to form an enormous number of molecular orbitals that are delocalized over the metal. As a result, a large number of energy levels are crowded together into bands. These bands are half-filled with electrons. When a voltage is applied to the metal crystal, electrons are excited to the unoccupied orbitals and move toward the positive pole of the voltage source. This is electrical conductivity.

13.13   An energy band is formed in a metal as metal atoms are brought together. When two metal atoms approach each other, their outer orbitals overlap to form two molecular orbitals. When a third atom is brought to this diatomic metal molecule, the three outer orbitals all overlap, forming a larger delocalized outer orbital. When a large number, N (close to Avogadro's number), of atoms are brought together, the atoms will form N molecular orbitals delocalized over the entire crystalline metal molecule. This is a band of electrons of continuous energies.

13.14   The three allotropes of carbon are diamond, graphite, and fullerenes. Diamond is a covalent network solid in which each carbon is tetrahedrally (sp^3) bonded to four other carbon atoms. Graphite has a layer structure with each layer being attracted to one another by London forces. Within a layer, each carbon atom is covalently bonded to three other atoms, giving a flat layer of carbon-atom hexagons. The bonding is sp^2 with delocalized π bonds. Fullerenes are molecules with a closed cage of carbon atoms arranged in pentagons and hexagons, like a soccer ball, or graphite-like sheets rolled into tubes.

13.15   In 1955, General Electric synthesized diamonds using a transition-metal catalyst at temperatures of about 2000 °C and a pressure of about 100,000 atm. These synthetic diamonds are very small and are used as abrasives or as grinding material on drill bits and cutting wheels. More recently, chemical vapor deposition has been used to produce diamond films. In this method, a mixture of methane and hydrogen is passed over a hot filament to produce reactive species that deposit diamond onto a solid substrate. This can be used as a coating on cutting tools or as a base for microelectronic parts that generate heat. Also, the diamond can be doped with a material like boron, and will then behave like a semiconductor.

13.16 Natural graphite is still used in pencils, but modern uses are more varied. Graphite is one of the most studied materials for use as the negative electrode in batteries. It is also used in the construction of artificial heart valves.

13.17 Atoms can be trapped inside a buckminsterfullerene, so it can act as a special storage device. With a potassium atom inside, they show the properties of a superconductor at about 18 K. They also show catalytic activity. Nanotubes can be used to construct molecular-scale tweezers and possibly computer memory devices.

13.18 An n-type semiconductor is silicon that has been doped with atoms with more valence electrons than the silicon, like phosphorus. The extra electrons in phosphorus-doped silicon are free to conduct an electric current by carrying negative charges (electrons). A p-type semiconductor is silicon that has been doped with atoms with less valence electrons than silicon, such as boron. This creates positively charged holes. As the electrons move to fill these holes, the holes move and become a conductor.

13.19 Elementary silicon is obtained by reducing quartz sand (SiO_2) with coke (C) in an electric furnace at 3000 °C. The reaction is

$$SiO_2(l) + 2C(s) \longrightarrow Si(l) + 2CO(g)$$

The impure silicon is converted into silicon tetrachloride ($SiCl_4$), which is purified by distillation (b.p. 58 °C).

$$Si(s) + 2Cl_2(g) \longrightarrow SiCl_4(g)$$

The purified $SiCl_4$ is reduced by passing the vapor with hydrogen through a hot tube, where pure silicon crystallizes on the tube. In the final step, this polycrystalline silicon is fashioned into a rod that consists of a single crystal from which wafers are cut.

13.20 When cut to precise dimensions, the quartz crystal responds most strongly to a certain vibrational frequency. Such crystals are used to control the frequency of an alternating electric current. When the alternating current frequency deviates from the natural frequency of the crystal, a feedback mechanism adjusts the alternating current frequency. These crystals are used to control radio and television frequencies as well as clocks.

13.21 The condensation reaction between two silicic acid molecules is represented by

13.22   Spodumene, $LiAl(SiO_3)_2$, is a long-chain silicate mineral. Such a mineral consists of silicate chains in which the SiO_4 tetrahedra are linked to form anions that are long chains or sheets.

13.23   An aluminosilicate mineral is a mineral consisting of silicate sheets or three-dimensional networks in which some of the SiO_4 tetrahedra of the silicate structure have been replaced by AlO_4 tetrahedra. Orthoclase, $KAlSi_3O_8$, which occurs in granite, is an example of a three-dimensional network aluminosilicate.

13.24   Ceramics are nonmetallic, inorganic solids that are hard and brittle and usually produced at elevated temperatures. Clay ceramics are fired, and consist of particles suspended in a glass matrix. It can be earthenware or porcelain and can be made into vases and pottery.

13.25   The newer non-clay ceramics, or glass-ceramics, are used to make stovetop pots. Silicon carbide is used as an abrasive in grinding wheels and cutting tools. Silicon nitride is used for special cutting tools and for high-temperature engine components. Other uses for glass-ceramics include medical implants and bone replacements.

13.26   Ordinary glass is a silicate produced by fusing silica (SiO_2) with other oxides (or substances such as carbonates that yield oxides when heated). Common glass (soda-lime glass) contains sodium and calcium oxides in addition to silica. It is a supercooled liquid whose viscosity is so high that it has the properties of a solid.

13.27   Glass-ceramic is a material that is made by adding materials to a glass to introduce many crystal nuclei into it. An example is the glass-ceramic stovetop pots that are made from a lithium aluminosilicate glass to which titanium and zirconium oxides is added. It has a very low thermal expansion, so it doesn't easily break when heated, and it remains transparent.

13.28   A composite is a material constructed of two or more different kinds of other materials. Bone is a natural composite of calcium phosphate and protein collagen. A commercial composite is epoxy plastic reinforced with carbon fibers and used in aircraft parts, golf clubs, fishing rods, and so forth.

■ Solutions to Practice Problems

Note on significant figures: If the final answer to a solution needs to be rounded off, it is given first with one nonsignificant figure, and the last significant figure is underlined. The final answer is then rounded to the correct number of significant figures. In multiple-step problems, intermediate answers are given with at least one nonsignificant figure; however, only the final answer has been rounded off.

13.37 For copper, the impure lead metal serves as the anode, and the pure lead serves as the cathode. During electrolysis, lead(II) ions leave the anode and deposit on the cathode; the electrolyte of $PbSiF_6$ may be considered the form of lead(II) that reacts at the cathode.

Anode reaction: $Pb(s) \rightarrow Pb^{2+}(aq) + 2e^-$

Cathode reaction: $PbSiF_6(aq) + 2e^- \rightarrow Pb(s) + SiF_6^{2-}(aq)$

13.39 $Fe_2O_3(s) + 3H_2(g) \rightarrow 2Fe(s) + 3H_2O(g)$

13.41 In the reaction (Problem 13.39), three mol of H_2 are used to form two mol of iron. Using the respective atomic masses of 55.85 g/mol for Fe and 2.016 g/mol for H_2, you can calculate the mass of Fe as follows:

$$2.00 \times 10^3 \text{ g } H_2 \times \frac{1 \text{ mol } H_2}{2.016 \text{ g } H_2} \times \frac{2 \text{ mol Fe}}{3 \text{ mol } H_2} \times \frac{55.85 \text{ g Fe}}{1 \text{ mol Fe}}$$

$$= 3.6\underline{9}3 \times 10^4 = 3.69 \times 10^4 \text{ g} = 36.9 \text{ kg Fe}$$

13.43 The equation with $\Delta H°_f$'s recorded beneath each substance is

$$PbS(s) + 3/2O_2(g) \rightarrow PbO(s) + SO_2(g)$$

| -98.3 | 0 | -219.0 | -296.8 (kJ) |

$\Delta H° = [(-219.0) + (-296.8) - (-98.3)] \text{ kJ} = -417.5 \text{ kJ (exothermic)}$

13.45   The energy levels in potassium metal are

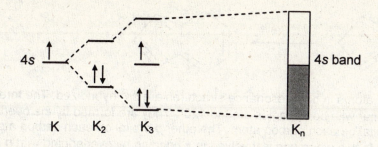

13.47   The number of sodium atoms in a 1.00-mg crystal of sodium is

$$1.00 \times 10^{-3} \text{ g Na} \times \frac{1 \text{ mol Na}}{22.99 \text{ g Na}} \times \frac{6.02 \times 10^{23} \text{ Na atoms}}{1 \text{ mol Na}}$$

$$= 2.6\underline{1}8 \times 10^{19} \text{ Na atoms}$$

Each sodium atom contributes an energy level for the $3s$ orbital to the valence band, so the number of energy levels in the valence band is 2.62×10^{19}.

13.49   The density of diamond is 3.5155 g/cm^3, and the density of graphite is 2.2670 g/cm^3 (*CRC Handbook of Chemistry and Physics*, 71st ed, p 4-160). Therefore, diamond is more dense than graphite. According to Le Chatelier's principle, as the pressure increases, the volume decreases, and higher densities are favored. Therefore, at higher pressures, diamond is the more stable allotrope.

13.51   Assume that 1.00 mg of diamond is 1.00 mg (1.00×10^{-3} g) of carbon. Convert this to moles of methane

$$1.00 \times 10^{-3} \text{ g C} \times \frac{1 \text{ mol C}}{12.01 \text{ g C}} \times \frac{1 \text{ mol CH}_4}{1 \text{ mol C}} = 8.3\underline{2}63 \times 10^{-5} \text{ mol CH}_4$$

At STP, one mole of a gas occupies 22.4 L, so the volume of methane is

$$V = 8.3263 \times 10^{-5} \text{ mol} \times 22.4 \text{ L/mol} = 1.8\underline{6}51 \times 10^{-3} \text{ L} = 1.87 \text{ mL}$$

13.53   A portion of the structure of a buckminsterfullerene molecule can be represented by

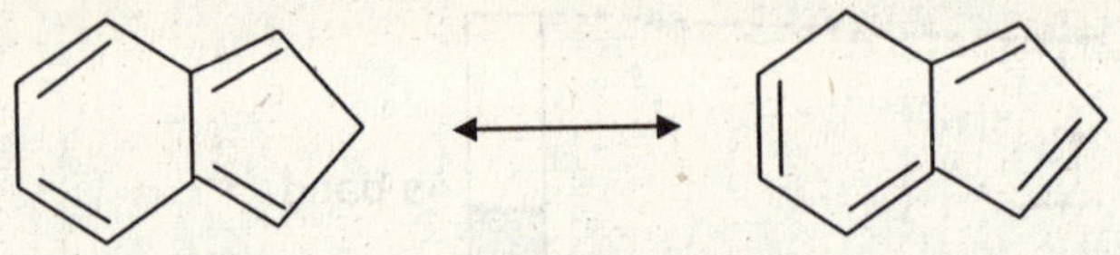

All the carbon atoms in both resonance structures are sp^2 hybridized. The three sigma bonds are planar with bond angles around 120°. They are formed by the overlap of an sp^2 hybrid orbital on each carbon atom. The other p orbital on each carbon atom is perpendicular to this plane and is involved in π bonding by overlapping with p orbitals from the other neighboring carbon atoms.

13.55   The n-type semiconductors have a doping agent with an additional electron. This would include Ge (four valence electrons) doped with As (five valence electrons) and doped with P (five valence electrons). Thus the answer is b and d.

13.57   The diamond-like structure of silicon implies that each silicon atom is tetrahedrally bonded to four other silicon atoms. The hybridization is sp^3. Each bond is formed by the overlap of an sp^3 hybrid orbital on each silicon atom to make a sigma bond.

13.59   The molar mass of silicon dioxide (SiO_2) is 60.09 g/mol. Thus,

$$5.00 \text{ kg Si} \times \frac{1 \text{ mol Si}}{28.09 \text{ g Si}} \times \frac{1 \text{ mol SiO}_2}{1 \text{ mol Si}} \times \frac{60.09 \text{ g SiO}_2}{1 \text{ mol SiO}_2}$$

$$= 10.\underline{6}959 = 10.7 \text{ kg SiO}_2$$

13.61   Spodumene, $LiAl(SiO_3)_2$ has the following structure:

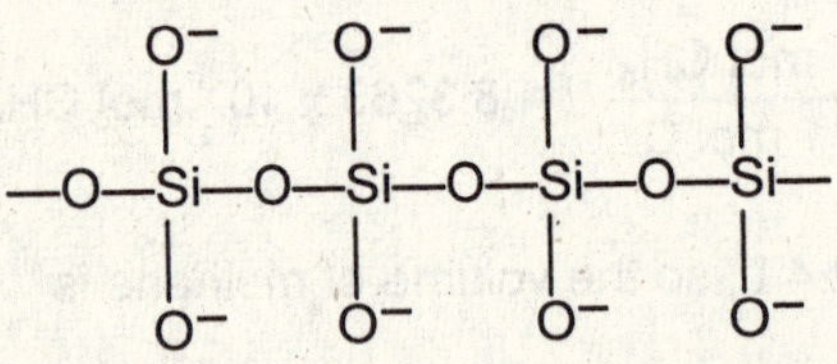

The drawing represents two of the sets of tetrahedra forming a portion of a long chain. There is a net -8 charge on the fragment, which includes two sets of two SiO_3 units. Thus, the $(SiO_3)_2$ unit has a 4- charge. This would have to be balanced by one Li^+ and one Al^{3+} per two SiO_3. Thus, the empirical formula is $LiAl(SiO_3)_2$.

13.63 The roasting of sphalerite to give the oxide is

$$2ZnS(s) + 3O_2(g) \rightarrow 2ZnO(s) + 2SO_2(g)$$

The reduction with carbon is

$$ZnO(s) + C(s) \rightarrow Zn(g) + CO(g)$$

13.65 Sphalerite, ZnS, has the molar mass 97.45 g/mol. Thus

$$1.00 \text{ metric ton} \times 0.870 \text{ ZnS} \times \frac{1 \text{ mol ZnS}}{97.45 \text{ g ZnS}} \times \frac{1 \text{ mol Zn}}{1 \text{ mol ZnS}} \times \frac{65.38 \text{ g Zn}}{1 \text{ mol Zn}}$$

$$= 0.58\underline{3}69 = 0.584 \text{ metric ton Zn}$$

13.67 Graphite is a covalent network solid with a layered structure. The layers are attracted to one another by London forces. Within each layer, each carbon is covalently bonded to three other carbon atoms, in hexagons, with sp^2 hybridization and delocalized π bonds. Fullerenes are molecules consisting of a closed cage of carbon atoms arranged in pentagons and hexagons. Other fullerenes consist of graphite-like sheets of carbon atoms rolled into tubes.

They are both similar in that they comprise only carbon atoms with sp^2 hybridization, each attached to three other carbon atoms and also with delocalized π bonds. They are different in that fullerenes can have pentagons and hexagons but graphite has only hexagons. Also, fullerenes are distinct molecules while graphite is a covalent network solid.

13.69 Write the reaction with the $\Delta H°_f$' below

	SiO$_2$(s)	+	2C(s)	→	Si(s)	+	2CO(g)
$\Delta H°_f$:	-910.9		0		0		2(-110.5) kJ

$$\Delta H° = \Sigma n\Delta H°_f(\text{products}) - \Sigma m\Delta H°_f(\text{reactants})$$

$$= [2(-110.5) - (-910.9)] \text{ kJ} = 689.9 \text{ kJ}$$

14. RATES OF REACTION

■ Solutions to Exercises

Note on significant figures: If the final answer to a solution needs to be rounded off, it is given first with one nonsignificant figure, and the last significant figure is underlined. The final answer is then rounded to the correct number of significant figures. In multiple-step problems, intermediate answers are given with at least one nonsignificant figure; however, only the final answer has been rounded off.

14.1 Rate of formation of NO_2F = $\Delta[NO_2F]/\Delta t$. Rate of reaction of NO_2 = $-\Delta[NO_2]/\Delta t$. Divide each rate by the coefficient of the corresponding substance in the equation:

$$1/2\frac{\Delta[NO_2F]}{\Delta t} = -1/2\frac{\Delta[NO_2]}{\Delta t}; \quad \text{or} \quad \frac{\Delta[NO_2F]}{\Delta t} = -\frac{\Delta[NO_2]}{\Delta t}$$

14.2 $\text{Rate} = -\dfrac{\Delta[I^-]}{\Delta t} = -\dfrac{\Delta[0.00101\ M\ -\ 0.00169\ M]}{8.00\ s\ -\ 2.00\ s} = 1.\underline{1}3 \times 10^{-4} = 1.1 \times 10^{-4}\ M/s$

14.3 The order with respect to CO is zero, and with respect to NO_2 is two. The overall order is two, the sum of the exponents in the rate law.

14.4 By comparing experiments 1 and 2, you see the rate is quadrupled when the $[NO_2]$ is doubled. Thus, the reaction is second order in NO_2, and the rate law is

$$\text{Rate} = k[NO_2]^2$$

(continued)

The rate constant may be found by substituting experimental values into the rate-law expression. Based on values from Experiment 1,

$$k = \frac{rate}{[NO_2]^2} = \frac{7.1 \times 10 \; mol/(L \bullet s)}{(0.010 \; mol/L)^2} = 0.7\underline{1}0 = 0.71 \; L/(mol \bullet s)$$

14.5 a. For $[N_2O_5]$ after 6.00×10^2 s, use the first-order rate law, and solve for the concentration at time t:

$$\ln \frac{[N_2O_5]_t}{[1.65 \times 10^{-2} \; M]} = -(4.80 \times 10^{-4} \; /s)(6.00 \times 10^2 \; s) = -0.288\underline{0}$$

Take the antilog of both sides to get

$$\frac{[N_2O_5]_t}{[1.65 \times 10^{-2} \; M]} = e^{-0.2880} = 0.749\underline{7}6$$

Hence,

$$[N_2O_5]_t = 1.65 \times 10^{-2} \; M \times 0.74976 = 0.012\underline{3}7 = 0.0124 \; mol/L$$

b. Substitute the values into the rate equation to get

$$\ln \frac{10.0\%}{100.0\%} = -(4.80 \times 10^{-4}/s) \times t$$

Taking the log on the left side gives $\ln(0.100) = -2.302\underline{5}8$. Hence,

$$-2.30\underline{2}58 = -(4.80 \times 10^{-4} \; /s) \times t$$

Or,

$$t = \frac{2.30258}{4.80 \times 10^{-4}/s} = 47\underline{9}7 = 4.80 \times 10^3 \; s \; (80.0 \; min)$$

14.6 Substitute k = 9.2/s into the equation relating K and $t_{1/2}$.

$$t_{1/2} = \frac{0.693}{k} = \frac{0.693}{9.2/s} = 0.07\underline{5}3 = 0.075 \; s$$

By definition, the half-life is the amount of time it takes to decrease the amount of substance present by one-half. Thus, it takes 0.07\underline{5}3 s for concentration to decrease by 50 percent and another 0.07\underline{5}3 s for the concentration to decrease by 50 percent of the remaining 50 percent (to 25 percent left), for a total of 0.15\underline{0}6, or 0.151, s.

14.7   Solve for E_a by substituting the given values into the two-temperature Arrhenius equation:

$$\ln \frac{2.14 \times 10^{-2}}{1.05 \times 10^{-3}} = \frac{E_a}{8.31 \ J/(mol \cdot K)} \left(\frac{1}{759 \ K} - \frac{1}{836 \ K} \right)$$

$$3.01\underline{4}6 = \frac{E_a}{8.31 \ J/(mol \cdot K)} \ (1.2\underline{1}35 \times 10^{-4} \ /K)$$

$$E_a = \frac{3.0146 \cdot \times \ 8.31 \ J/mol}{1.2135 \times 10^{-4}} = 2.0\underline{6}4 \times 10^5 = 2.06 \times 10^5 \ J/mol$$

Solve for the rate constant, k_2, at 865 K by using the same equation and using $E_a = 2.064 \times 10^5$ J/mol:

$$\ln \frac{k_2}{2.14 \times 10^{-2}/(M^{1/2} \cdot s)} = \frac{2.064 \times 10^5 \ J/mol}{8.31 \ J/(mol \cdot K)} \left(\frac{1}{836 \ K} - \frac{1}{865 \ K} \right) = 0.99\underline{6}23$$

Taking antilogarithms,

$$\frac{k_2}{2.14 \times 10^{-2}/(M^{1/2} \cdot s)} = e^{0.99623} = 2.7\underline{0}80$$

$$k_2 = 2.7080 \times (2.14 \times 10^{-2}) /(M^{1/2} \cdot s) = 5.7\underline{9}52 \times 10^{-2} = 5.80 \times 10^{-2} \ (M^{1/2} \cdot s)$$

14.8   The net chemical equation is the overall sum of the two elementary reactions:

$$H_2O_2 + I^- \quad \rightarrow \quad H_2O + IO^-$$
$$H_2O_2 + IO^- \quad \rightarrow \quad H_2O + O_2 + I^-$$
$$\overline{}$$
$$2H_2O_2 \quad \rightarrow \quad 2H_2O + O_2$$

The IO^- is an intermediate; I^- is a catalyst. Neither appears in the net equation.

14.9   The reaction is bimolecular because it is an elementary reaction that involves two molecules.

14.10   For $NO_2 + NO_2 \rightarrow N_2O_4$, the rate law is

$$Rate = k[NO_2]^2$$

(The rate must be proportional to the concentration of both reactant molecules.)

14.11   The first step is the slow, rate-determining step. Therefore, the rate law predicted by the mechanism given is

$$\text{Rate} = k_1 [H_2O_2][I^-]$$

14.12   According to the rate-determining (slow) step, the rate law is

$$\text{Rate} = k_2[NO_3][NO]$$

Eliminate NO_3 from the rate law by looking at the first step, which is fast and reaches equilibrium. At equilibrium, the forward rate and the reverse rate are equal.

$$k_1[NO][O_2] = k_{-1}[NO_3]$$

Therefore, $[NO_3] = (k_1/k_{-1})[NO][O_2]$, so

$$\text{Rate} = \frac{k_2 k_1}{k_{-1}} [NO]^2[O_2] = k[NO]^2[O_2]$$

Where $k_2(k_1/k_{-1})$ has been replaced by k, which represents the experimentally observed rate constant.

■ Answers to Review Questions

14.1   The four variables that can affect rate are (1) the concentrations of the reactants, although in some cases a particular reactant's concentration does not affect the rate; (2) the presence and concentration of a catalyst; (3) the temperature of the reaction; and (4) the surface area of any solid reactant or solid catalyst.

14.2   The rate of reaction of HBr can be defined as the decrease in HBr concentration (or the increase in Br_2 product formed) over the time interval, Δt:

$$\text{Rate} = -1/4 \frac{\Delta[HBr]}{\Delta t} = 1/2 \frac{\Delta[Br_2]}{\Delta t} \quad \text{or} \quad -\frac{\Delta[HBr]}{\Delta t} = 2 \frac{\Delta[Br_2]}{\Delta t}$$

14.3   Two physical properties used to determine the rate are color, or absorption of electromagnetic radiation, and pressure. If a reactant or product is colored, or absorbs a different type of electromagnetic radiation than the other species, then measurement of the change in color (change in absorption of electromagnetic radiation) may be used to determine the rate. If a gas reaction involves a change in the number of gaseous molecules, measurement of the pressure change may be used to determine the rate.

14.4   Use the general example of a rate law in Section 14.3, where A and B react to give D and E with C as a catalyst:

$$Rate = k \, [A]^m \, [B]^n \, [C]^p$$

Note that the exponents m, n, and p are the orders of the individual reactants and catalyst. Assuming that m and n are positive numbers, the rate law predicts that increasing the concentrations of A and/or B will increase the rate. In addition, the rate will be increased by increasing the surface of the solid catalyst (making it as finely divided as possible, etc.). Finally, increasing the temperature will increase the rate constant, k, and increase the rate.

14.5   An example that illustrates that exponents have no relationship to coefficients is the reaction of nitric oxide and hydrogen from Example 14.12 in the text.

$$2NO + 2H_2 \rightarrow N_2 + 2H_2O.$$

The experimental rate law given there is

$$Rate = k[NO]^2[H_2]$$

Thus, the exponent for hydrogen is one, not two like the coefficient, and the overall order is three, not four like the sum of the coefficients.

14.6   The rate law for this reaction of iodide ion, arsenic acid, and hydrogen ion is

$$Rate = k[I^-][H_3AsO_4] \, [H^+]$$

The overall order is $1 + 1 + 1 = 3$ (third order).

14.7   Use m to symbolize the reaction order as is done in the text. Then from the table for m and the change in rate in the text, m is two when the rate is quadrupled (increased fourfold). Using the equation in the text gives the same result:

$$2^m = \text{new rate/old rate} = 4/1; \text{ thus, } m = 2$$

14.8   Use m to symbolize the reaction order as is done in the text. The table for m and the change in rate in the text cannot be used in this case. When m = 0.5, the new rate should be found using the equation in the text:

$$2^{0.50} = \sqrt{2} = 1.41 = \text{new rate/old rate}$$

Thus, the new rate is 1.41 times the old rate.

14.9 Use the half-life concept to answer the question without an equation. If the half-life for the reaction of A(g) is 25 s, then the time for A(g) to decrease to 1/4 the initial value is two half-lives or 2 x 25 = 50 s. The time for A(g) to decrease to 1/8 the initial value is three half-lives, or 3 x 25 = 75 s.

14.10 The half-life equation for a first order reaction is $t_{1/2} = 0.693/k$, The half-life for the reaction is constant, independent of the reactant concentration. For a second order reaction, the half-life equation is $t_{1/2} = 1/(k[A]_o)$, and it depends on the initial reactant concentration.

14.11 According to transition-state theory, the two factors that determine whether a collision results in reaction or not are (1) the molecules must collide with the proper orientation to form the activated complex, and (2) the activated complex formed must have a kinetic energy greater than the activation energy.

14.12 The potential-energy diagram for the exothermic reaction of A and B to give activated complex $AB^{\ddagger}$ and products C and D is given below.

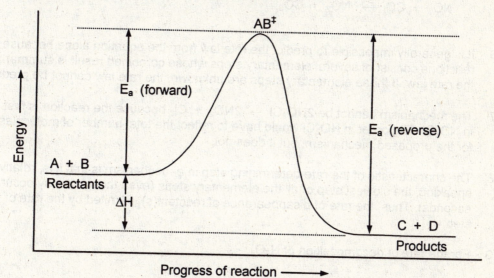

14.13 The activated complex for the reaction of NO_2 with NO_3 to give NO, NO_2, and O_2 has the structure below (dashed lines = bonds about to form or break):

O—N---O---O---N(—O)$_2$

14.14 The Arrhenius equation expressed with the base e is

$$k = A\, e^{-E_a/RT}$$

The A term is the frequency factor and is equal to the product of p and Z from collision theory. The term p is the fraction of collisions with properly oriented reactant molecules, and Z is the frequency of collisions. Thus, A is the number of collisions with the molecules properly oriented. The E_a term is the activation energy, the minimum energy of collision required for two molecules to react. The R term is the gas constant, and T is the absolute temperature.

14.15 In the reaction of $NO_2(g)$ with $CO(g)$, an example of an intermediate is the temporary formation of NO_3 from the reaction of two NO_2 molecules in the first step:

$$NO_2 + NO_2 \rightarrow NO_3 + NO$$

$$NO_3 + CO \rightarrow NO_2 + CO_2$$

14.16 It is generally impossible to predict the rate law from the equation alone because most reactions consist of several elementary steps whose combined result is summarized in the rate law. If these elementary steps are unknown, the rate law cannot be predicted.

14.17 The mechanism cannot be $2NO_2Cl \rightarrow 2NO_2 + Cl_2$ because the reaction is first order in NO_2Cl. The order in NO_2Cl would have to reflect the total number of molecules (two) for the proposed mechanism, but it does not.

14.18 The characteristic of the rate-determining step in a mechanism is that it is, relatively speaking, the slowest step of all the elementary steps (even though it may occur in seconds). Thus, the rate of disappearance of reactant(s) is limited by the rate of this step.

14.19 For the rate of decomposition of N_2O_4,

$$Rate = k_1[N_2O_4]$$

For the rate of formation of N_2O_4,

$$Rate = k_{-1}[NO_2]^2$$

At equilibrium the rates are equal, so

$$k_1[N_2O_4] = k_{-1}[NO_2]^2$$

$$[N_2O_4] = (k_{-1}/k_1)\,[NO_2]^2$$

14.20 A catalyst operates by providing a pathway (mechanism) that occurs faster than the uncatalyzed pathway (mechanism) of the reaction. The catalyst is not consumed because after reacting in an early step, it is regenerated in a later step.

14.21 In physical adsorption, molecules adhere to a surface through *weak* intermediate forces, whereas in chemisorption the molecules adhere to the surface by *stronger* chemical bonding.

14.22 In the first step of catalytic hydrogenation of ethylene, the ethylene and hydrogen molecules diffuse to the catalyst surface and undergo chemisorption. Then the *pi* electrons of ethylene form temporary bonds to the metal catalyst, and the hydrogen molecule breaks into two hydrogen atoms. The hydrogen atoms next migrate to an ethylene held in position on the metal catalyst surface, forming ethane. Finally, because it cannot bond to the catalyst, the ethane diffuses away from the surface.

■ Solutions to Practice Problems

Note on significant figures: If the final answer to a solution needs to be rounded off, it is given first with one nonsignificant figure, and the last significant figure is underlined. The final answer is then rounded to the correct number of significant figures. In multiple-step problems, intermediate answers are given with at least one nonsignificant figure; however, only the final answer has been rounded off.

14.33 For the reaction $2NO_2 \rightarrow 2NO + O_2$, the rate of decomposition of NO_2 and the rate of formation of O_2 are, respectively,

Rate $= -\Delta[NO_2]/\Delta t$

Rate $= \Delta[O_2]/\Delta t$

To relate the two rates, divide each rate by the coefficient of the corresponding substance in the chemical equation and equate them.

$$-1/2 \frac{\Delta[NO_2]}{\Delta t} = \frac{\Delta[O_2]}{\Delta t}$$

14.35 For the reaction $5Br^- + BrO_3^- + 6H^+ \rightarrow 3Br_2 + 3H_2O$, the rate of decomposition of Br^- and the rate of decomposition of BrO_3^- are, respectively,

$$Rate = -\Delta[Br^-]/\Delta t$$

$$Rate = -\Delta[BrO_3^-]/\Delta t$$

To relate the two rates, divide each rate by the coefficient of the corresponding substance in the chemical equation and equate them.

$$\frac{1}{5}\frac{\Delta[Br^-]}{\Delta t} = \frac{\Delta[BrO_3^-]}{\Delta t}$$

14.37 $Rate = -\dfrac{\Delta[NH_4NO_2]}{\Delta t} = -\dfrac{[0.0432\ M - 0.500\ M]}{[3.00\ hr - 0.00\ hr]} = 2.27 \times 10^{-2} = 2.3 \times 10^{-2}\ M/hr$

14.39 $Rate = -\dfrac{\Delta[Azo.]}{\Delta t} = -\dfrac{[0.0101\ M - 0.0150\ M]}{7.00\ min - 0.00\ min} \times \dfrac{1\ min}{60\ sec} = 1.16 \times 10^{-5}$

$$= 1.2 \times 10^{-5}\ M/s$$

14.41 If the rate law is rate $= k[H_2S][Cl_2]$, the order with respect to H_2S is 1 (first-order), and the order with respect to Cl_2 is also one (first-order). The overall order is $1 + 1 = 2$, second-order.

14.43 If the rate law is rate $= k[MnO_4^-][H_2C_2O_4]$, the order with respect to MnO_4^- is one (first-order), the order with respect to $H_2C_2O_4$ is one (first-order), and the order with respect to H^+ is zero. The overall order is two, second-order.

14.45 The reaction rate doubles when the concentration of CH_3NNCH_3 is doubled, so the reaction is first-order in azomethane. The rate equation should have the form

$$Rate = k[CH_3NNCH_3]$$

Substituting values for the rate and concentration yields a value for k:

$$k = \frac{rate}{[Azo.]} = \frac{2.8 \times 10^{-6}\ M/s}{1.13 \times 10^{-2}\ M} = 2.47 \times 10^{-4} = 2.5 \times 10^{-4}/s$$

14.47 Doubling [NO] quadruples the rate, so the reaction is second-order in NO. Doubling [H$_2$] doubles the rate, so the reaction is first-order in H$_2$. The rate law should have the form

$$Rate = k[NO]^2[H_2]$$

Substituting values for the rate and concentrations yields a value for k:

$$k = \frac{rate}{[NO]^2[H_2]} = \frac{2.6 \times 10^{-5}\ M/s}{[6.4 \times 10^{-3}\ M]^2[2.2 \times 10^{-3}\ M]} = 2.\underline{8}8 \times 10^2 = 2.9 \times 10^2\ M^2 s$$

14.49 By comparing experiments 1 and 2, you see that tripling [ClO$_2$] increases the rate ninefold; that is, $3^m = 9$, so m = 2 (and the reaction is second-order in ClO$_2$). From experiments 2 and 3, you see that tripling [OH$^-$] triples the rate, so the reaction is first-order in OH$^-$. The rate law is

$$Rate = k[ClO_2]^2[OH^-]$$

Substituting values for the rate and concentrations yields a value for k:

$$k = \frac{rate}{[ClO_2]^2[OH^-]} = \frac{0.0248\ M/s}{[0.060\ M]^2[0.030\ M]} = 2.\underline{2}9 \times 10^2 = 2.3 \times 10^2 / M^2 s$$

14.51 Let [SO$_2$Cl$_2$]$_0$ = 0.0248 M and [SO$_2$Cl$_2$]$_t$ = the concentration after 2.0 hr. Substituting these and k = 2.2 × 10^{-5}/s into the first-order rate equation gives

$$\ln \frac{[SO_2Cl_2]_t}{[0.0248\ M]} = -(2.2 \times 10^{-5}/s)\left(2.0\ hr \times \frac{3600\ s}{1\ hr}\right) = -0.1\underline{5}84$$

Taking the antilog of both sides gives

$$\frac{[SO_2Cl_2]_t}{[0.0248\ M]} = e^{-0.1584} = 0.8\underline{5}35$$

Solving for [SO$_2$Cl$_2$]$_t$ gives

$$[SO_2Cl_2]_t = 0.8\underline{5}25 \times [0.0248\ M] = 0.02\underline{1}16 = 0.021 = 2.1 \times 10^{-2}\ M$$

14.53 The second-order integrated rate law is

$$\frac{1}{[A]_t} = kt + \frac{1}{[A]_o}$$

Using a rate constant value of 0.225 L/(mol·s) and an initial concentration of 0.293 mol/L, after 35.4 s the concentration of A is

$$\frac{1}{[A]_t} = 0.225 \text{ L/(mol·s)} \times 35.4 \text{ s} + \frac{1}{0.293 \text{ mol/L}}$$

$$\frac{1}{[A]_t} = 7.9\underline{6}5 \text{ L/mol} + 3.4\underline{1}2 \text{ L/mol} = 11.3\underline{7}7 \text{ L/mol}$$

$$[A]_t = \frac{1}{11.377 \text{ L/mol}} = 0.08788\underline{9} = 0.08789 \text{ M}$$

14.55 First, find the rate constant, k, by substituting experimental values into the first-order rate equation. Let $[Et. Cl.]_o = 0.00100$ M, $[Et. Cl.]_t = 0.00067$ M, and t = 155 s. Solving for k yields

$$k = - \frac{\ln \dfrac{[0.00067 \text{ M}]_t}{[0.00100 \text{ M}]_o}}{155 \text{ s}} = 2.5\underline{8}3 \times 10^{-3}/\text{s}$$

Now let $[Et. Cl.]_t$ = the concentration after 256 s, $[Et. Cl.]_o$ again = 0.00100 M, and use the value of k of $1.5\underline{6}4 \times 10^{-3}$/s to calculate $[Et. Cl.]_t$.

$$\ln \frac{[Et. Cl.]_t}{[0.00100 \text{ M}]} = -(2.584 \times 10^{-3}/\text{s})(256 \text{ s}) = -0.6\underline{6}14$$

Converting both sides to antilogs gives

$$\frac{[Et. Cl.]_t}{[0.00100 \text{ M}]} = 0.5\underline{1}61$$

$$[Et. Cl.]_t = 0.5161 \times [0.00100 \text{ M}] = 5.\underline{1}6 \times 10^{-4} = 5.2 \times 10^{-4} \text{ M}$$

14.57 For a first-order reaction, divide 0.693 by the rate constant to find the half-life:

$$t_{1/2} = 0.693/(6.3 \times 10^{-4}/s) = 1.\underline{1}0 \times 10^3 = 1.1 \times 10^3 \text{ s (1}\underline{8}.3 \text{ min)}$$

Now, use the half-life to determine the concentrations. When the concentration decreases to 50.0 percent of its initial value, this is equal to the half-life.

$$t_{50.0\% \text{ left}} = t_{1/2} = 1.10 \times 10^3 \text{ s}$$

When the concentration decreases to 25.0 percent of its initial value, this is equal to two half-lives.

$$t_{25.0\% \text{ left}} = t_{1/4 \text{ left}} = 2 \times t_{1/2} = 2 \times (1.10 \times 10^3 \text{ s}) = 2.\underline{2}0 \times 10^3 \text{ s (37 min)}$$

14.59 For a first-order reaction, divide 0.693 by the rate constant to find the half-life:

$$t_{1/2} = 0.693/(2.0 \times 10^{-6}/s) = 3.\underline{4}65 \times 10^5 \text{ s (9}\underline{6}.25 \text{ or 96 hr)}$$

$$t_{25\% \text{ left}} = t_{1/4 \text{ left}} = 2 \times t_{1/2} = 2 \times 96.25 \text{ hr} = 19\underline{2}.5 = 1.9 \times 10^2 \text{ hr}$$

$$t_{12.5\% \text{ left}} = t_{1/8 \text{ left}} = 3 \times t_{1/2} = 3 \times 96.25 \text{ hr} = 28\underline{8}.75 = 2.9 \times 10^2 \text{ hr}$$

$$t_{6.25\% \text{ left}} = t_{1/16 \text{ left}} = 4 \times t_{1/2} = 4 \times 96.25 \text{ hr} = 3\underline{8}5.0 = 3.9 \times 10^2 \text{ hr}$$

$$t_{3.125\% \text{ left}} = t_{1/32 \text{ left}} = 5 \times t_{1/2} = 5 \times 96.25 \text{ hr} = 4\underline{8}1.25 = 4.8 \times 10^2 \text{ hr}$$

14.61 The half-life for a second-order reaction is

$$t_{1/2} = \frac{1}{k[A]_o}$$

Using a rate constant of 0.413 L/(mol•s) and an initial A concentration of 5.25×10^{-3} mol/L, the half-life is

$$t_{1/2} = \frac{1}{(0.413 \text{ L/(mol} \cdot \text{s)}(5.25 \times 10^{-3} \text{ mol/L)}} = 46\underline{1}.2 = 461 \text{ s}$$

14.63 Use the first-order rate equation, and solve for time, t. Let $[Cr^{3+}]_o$ = 100.0 percent; then the concentration at time t, $[Cr^{3+}]_t$, = (100.0% - 85.0%, or 15.0%). Use $k = 2.0 \times 10^{-6}/s$.

$$\ln \frac{[15.0]}{[100.0]} = -(2.0 \times 10^{-6}/s) \, t$$

$$t = -\frac{\ln(0.150)}{2.0 \times 10^{-6}/s} = 9.\underline{4}8 \times 10^5 \text{ s, or } 2.6 \times 10^2 \text{ hr}$$

14.65 The rate law for a zero-order reaction is

$$[A] = -kt + [A]_0.$$

Using a rate constant of 8.1×10^{-2} mol/(L•s) and an initial concentration of 0.10 M, the time it would take for the concentration to change to 1.0×10^{-2} M is

$$1.0 \times 10^{-2} \text{ M} = -8.1 \times 10^{-2} \text{ mol/(L•s)} \times t + 0.10 \text{ M}$$

$$t = \frac{0.10 \text{ M} - 1.0 \times 10^{-2} \text{ M}}{8.1 \times 10^{-2} \text{ M/s}} = 1.\underline{11} = 1.1 \text{ s}$$

14.67 For the first-order plot, follow Figure 14.9, and plot ln $[ClO_2]$ versus the time in seconds. The data used for plotting are

t, sec	$[ClO_2]$, M	ln $[ClO_2]$
0.00	4.77×10^{-4}	-7.647
1.00	4.31×10^{-4}	-7.749
2.00	3.91×10^{-4}	-7.846
3.00	3.53×10^{-4}	-7.949
5.00	2.89×10^{-4}	-8.149
10.00	1.76×10^{-4}	-8.645
30.00	2.4×10^{-5}	-10.63
50.00	3.2×10^{-5}	-12.65

The plot yields an approximate straight line, demonstrating the reaction is first order in $[ClO_2]$. The slope of the line may be calculated from the difference between the last point and the first point:

$$\text{Slope} = \frac{[(-12.65) - (-7.647)]}{[50.00 - 0.00]\text{s}} = -0.10\underline{0}1/\text{s}$$

Just as the slope, m, was obtained for the plot in Figure 14.9, you can also equate m to -k, and calculate k as follows:

$$k = -\text{slope} = 0.10\underline{0}1 = 0.100/\text{s}$$

14.69 The potential-energy diagram is below. Because the activation energy for the forward reaction is +10 kJ, and $\Delta H° = -200$ kJ, the activation energy for the reverse reaction is +210 kJ.

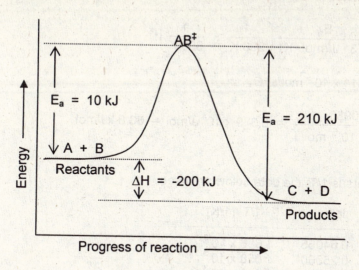

14.71 Solve the two-temperature Arrhenius equation for E_a by substituting $T_1 = 308$ K (from 35°C), $k_1 = 1.4 \times 10^{-4}$/s, $T_2 = 318$ K (from 45 °C), and $k_2 = 5.0 \times 10^{-4}$/s:

$$\ln \frac{5.0 \times 10^{-4}}{1.4 \times 10^{-4}} = \frac{E_a}{8.31 \text{ J/(mol·K)}} \left(\frac{1}{308 \text{ K}} - \frac{1}{318 \text{ K}} \right)$$

Rearranging E_a to the left side and calculating [1/308 - 1/318] gives

$$E_a = \frac{8.31 \text{ J/K} \times \ln(3.5714)}{1.0209 \times 10^{-4}\text{/K}} = 1.\underline{0}37 \times 10^5 = 1.0 \times 10^5 \text{ J/mol}$$

To find the rate at 55 °C (328 K), use the first equation. Let k_2 in the numerator be the unknown and solve:

$$\ln \frac{k_2}{1.4 \times 10^{-4}} = \frac{1.037 \times 10^5 \text{ J/mol}}{8.31 \text{ J/(mol·K)}} \left(\frac{1}{308 \text{ K}} - \frac{1}{328 \text{ K}} \right) = 2.\underline{4}68$$

$$\frac{k_2}{1.4 \times 10^{-4}} = 1\underline{1}.80$$

$$k_2 = 11.80 \times (1.4 \times 10^{-4}\text{/s}) = 1.\underline{6}52 \times 10^{-3} = 1.7 \times 10^{-3}\text{/s}$$

14.73 Because the rate constant is proportional to the rate of a reaction, tripling the rate at 25 °C also means that the rate constant at 25 °C is tripled. Thus, $k_{35} = 3k_{25}$, and the latter can be substituted for k_{25} in the Arrhenius equation:

$$\ln \frac{3\,k_{25}}{k_{25}} = \frac{E_a}{8.31\ \text{J/(mol}\cdot\text{K)}} \left(\frac{1}{298\ \text{K}} - \frac{1}{308\ \text{K}} \right)$$

$$1.0986 = (1.311 \times 10^{-5}\ \text{mol/J})\ E_a$$

$$E_a = \frac{1.0986}{1.311 \times 10^{-5}\ \text{mol/J}} = 8.379 \times 10^4\ \text{J/mol} = 83.8\ \text{kJ/mol}$$

14.75 For plotting ln k versus 1/T, the data below are used:

k	ln k	1/T (1/K)
0.527	-0.64055	1.686×10^{-3}
0.776	-0.25360	1.658×10^{-3}
1.121	0.11422	1.631×10^{-3}
1.607	0.47436	1.605×10^{-3}

The plot yields an approximate straight line. The slope of the line is calculated from the difference between the last and the first points:

$$\text{Slope} = \frac{(0.47436) - (-0.64055)}{[1.605 \times 10^{-3} - 1.686 \times 10^{-3}]/\text{K}} = -13764\ \text{K}$$

Because the slope $= -E_a/R$, you can solve for E_a using R = 8.31 J/K:

$$\frac{-E_a}{8.31\ \text{J/ (K}\cdot\text{mol)}} = -13764\ \text{K}$$

$$E_a = 5976\ \text{K} \times 8.31\ \text{J/K} = 1.143 \times 10^5\ \text{J/mol, or } 1.1 \times 10^2\ \text{kJ/mol}$$

14.77 The $NOCl_2$ is a reaction intermediate that is produced in the first reaction and consumed in the second. The overall reaction is the sum of the two elementary reactions:

$$NO + Cl_2 \rightarrow NOCl_2$$

$$NOCl_2 + NO \rightarrow 2NOCl$$

$$\overline{}$$

$$2NO + Cl_2 \rightarrow 2NOCl$$

14.79 a. Bimolecular   b. Bimolecular   c. Unimolecular   d. Termolecular

14.81 a. Only O_3 occurs on the left side of the equation, so the rate law is

$$Rate = k[O_3]$$

    b. Both $NOCl_2$ and NO occur on the left side of the equation, so the rate law is

$$Rate = k[NOCl_2][NO]$$

14.83   Step 1 of the isomerization of cyclopropane, C_3H_6, is slow, so the rate law for the overall reaction will be the rate law for this step, with $k_1 = k$, the overall rate constant:

$$Rate = k[C_3H_6]^2$$

14.85   Step 2 of this reaction is slow, so the rate law for the overall reaction would appear to be the rate law for this step:

$$Rate = k_2[I]^2[H_2]$$

However, the rate law includes an intermediate, the I atom, and cannot be used unless the intermediate is eliminated. This can only be done using an equation for step 1. At equilibrium, you can write the following equality for step 1:

$$k_1[I_2] = k_{-1}[I]^2$$

Rearranging and then substituting for the $[I]^2$ term yields

$$[I]^2 = [I_2]\frac{k_1}{k_{-1}}$$

$$Rate = k_2(k_1/k_{-1})[I_2][H_2] = k[I_2][H_2] \quad (k = \text{the overall rate constant})$$

14.87   The Br^- ion is the catalyst. It is consumed in the first step and regenerated in the second step. It speeds up the reaction by providing a pathway with a lower activation energy than that of a reaction pathway involving no Br^-. The overall reaction is obtained by adding the two steps together:

$$2H_2O_2 \rightarrow 2H_2O + O_2$$

Bromide ion is added to the mixture to give the catalytic activity, and BrO^- is an intermediate.

■ Solutions to General Problems

14.89 All rates of reaction are calculated by dividing the decrease in concentration by the difference in times; hence, only the setup for the first rate (after 10 minutes) is given below. This setup is

$$\text{Rate (10 min)} = -\frac{(1.29 - 1.50) \times 10^{-2} \text{ M}}{(10 - 0) \text{ min}} \times \frac{1 \text{ min}}{60 \text{ s}}$$

$$= 3.5 \times 10^{-6} \text{ M/s}$$

A summary of the times and rates is given in the table.

Time, min	Rate
10	$3.\underline{50} \times 10^{-6} = 3.5 \times 10^{-6}$ M/s
20	$3.\underline{17} \times 10^{-6} = 3.2 \times 10^{-6}$ M/s
30	$2.\underline{50} \times 10^{-6} = 2.5 \times 10^{-6}$ M/s

14.91 The calculation of the average concentration and the division of the rate by the average concentration are the same for all three time intervals. Thus, only the setup for the first interval is given:

$$k_{10 \text{ min}} = \frac{\text{rate}}{\text{avg. conc.}} = \frac{3.50 \times 10^{-6} \text{ M/s}}{\left[\dfrac{(1.50 + 1.29) \times 10^{-2} \text{ M}}{2}\right]} = 2.5\underline{0}8 \times 10^{-4} / \text{s}$$

A summary of the times, rate constants, and average rate constant is given in the table.

Time	Rate	k
10 min	3.50×10^{-6} M/s	$2.\underline{5}08 \times 10^{-4}$/s
20 min	3.17×10^{-6} M/s	$2.\underline{6}52 \times 10^{-4}$/s
30 min	2.50×10^{-6} M/s	$2.\underline{4}39 \times 10^{-4}$/s
-----	---- average k	$2.\underline{5}33 \times 10^{-4} = 2.5 \times 10^{-4}$/s

14.93 Use the first-order rate equation $k = 1.26 \times 10^{-4}$/s, the initial methyl acetate $[MA]_o = 100$ percent, and $[MA]_t = (100\% - 65\%, \text{ or } 35\%)$.

$$t = -\frac{\ln \dfrac{35\%}{100\%}}{1.26 \times 10^{-4}/\text{s}} = 8.3\underline{3}19 \times 10^{3} = 8.33 \times 10^{3} \text{ s}$$

14.95   Use $k = 1.26 \times 10^{-4}/s$, and substitute into the $t_{1/2}$ equation:

$$t_{1/2} = \frac{0.693}{k} = \frac{0.693}{1.26 \times 10^{-4}/s} = 5.5\underline{0}0 \times 10^3 = 5.50 \times 10^3 \text{ s} \quad (1.53 \text{ hr})$$

14.97   First, find the rate constant from the first-order rate equation, substituting the initial concentration of $[comp.]_o = 0.0350$ M, and the $[comp.]_t = 0.0250$ M.

$$\ln \frac{0.0250 \text{ M}}{0.0350 \text{ M}} = -k(65 \text{ s})$$

Rearranging and solving for k gives

$$k = \frac{\left(\ln \dfrac{0.0250 \text{ M}}{0.0350 \text{ M}} \right)}{65 \text{ s}} = 5.\underline{1}7 \times 10^{-3}/s$$

Now, arrange the first-order rate equation to solve for $[comp.]_t$; substitute the above value of k, again using $[comp.]_o = 0.0350$ M.

$$\ln \frac{[comp.]_t}{0.0350 \text{ M}} = -(5.18 \times 10^{-3}/s)(88 \text{ s}) = -0.4\underline{5}58$$

Taking the antilog of both sides gives

$$\frac{[comp.]_t}{0.0350 \text{ M}} = e^{-0.4558} = 0.6\underline{3}39$$

$$[comp.]_t = 0.6\underline{3}39 \times [0.0350 \text{ M}]_o = 0.02\underline{2}1 = 0.022 \text{ M}$$

14.99 a. The rate constant for a second-order reaction is related to the half-life by

$$k = \frac{1}{t_{1/2}[A]_o}$$

Using a half-life of 5.92×10^{-2} s and an initial A concentration of 0.50 mol/L, the rate constant is

$$k = \frac{1}{(5.92 \times 10^{-2} \text{ s})(0.50 \text{ mol/L})} = 33.\underline{7}8 = 34 \text{ L/(mol•s)}$$

(continued)

b. The second-order integrated rate law is

$$\frac{1}{[A]_t} = kt + \frac{1}{[A]_o}$$

Using an initial concentration of C_4H_8 (A) of 0.010 M, after 3.6×10^2 s, the concentration will be

$$\frac{1}{[A]_t} = 33.78 \text{ L/(mol•s)}(3.6 \times 10^2 \text{ s}) + \frac{1}{0.010 \text{ mol/L}}$$

$$\frac{1}{[A]_t} = 12{,}162 \text{ L/mol} + 100.0 \text{ L/mol} = 12{,}262 \text{ L/mol}$$

$$[A]_t = \frac{1}{12{,}262 \text{ L/mol}} = 8.155 \times 10^{-5} = 8.2 \times 10^{-5} \text{ M}$$

14.101   The second-order integrated rate law is

$$\frac{1}{[A]_t} = kt + \frac{1}{[A]_o}$$

The starting concentration of NO_2 (A) is 0.050 M. Using a rate constant of 0.775 L/(mol•s), after 2.5×10^2 s the concentration of NO_2 will be

$$\frac{1}{[A]_t} = 0.775 \text{ L/(mol•s)}(2.5 \times 10^2 \text{ s}) + \frac{1}{0.050 \text{ mol/L}}$$

$$\frac{1}{[A]_t} = 193 \text{ L/mol} + 20.0 \text{ L/mol} = 213. \text{ L/mol}$$

$$[A]_t = \frac{1}{213 \text{ L/mol}} = 4.67 \times 10^{-3} = 4.7 \times 10^{-3} \text{ M}$$

The half life is

$$t_{1/2} = \frac{1}{k[A]_o}$$

Using a rate constant of 0.775 L/(mol•s) and an initial A concentration of 0.10 mol/L, the half-life is

$$t_{1/2} = \frac{1}{(0.775 \text{ L/(mol • s)})(0.050 \text{ mol/L})} = 25.80 = 26 \text{ s}$$

14.103 The ln [CH_3NNCH_3] and time data for the plot are tabulated below.

t. min	ln [CH_3NNCH_3]
0	-4.1997
10	-4.3505
20	-4.5098
30	-4.6564

From the graph the slope, m, is calculated:

$$m = \frac{(-4.6564) - (-4.1997)}{(30 - 0) \text{ min}}$$

$$= -0.01522 \text{ /min}$$

Because the slope also = -k, this gives

$$k = -(-0.01522/\text{min})(1 \text{ min}/60 \text{ s}) = 2.\underline{5}3 \times 10^{-4} = 2.5 \times 10^{-4}/\text{s}$$

14.105 The rate law for a zero-order reaction is

$$[A] = -kt + [A]_o.$$

Using a rate constant of 3.7×10^{-6} mol/(L•s) and an initial concentration of 5.0×10^{-4} M, the time it would take for the concentration to drop to 5.0×10^{-5} M is

$$5.0 \times 10^{-5} \text{ M} = -8.1 \times 10^{-2} \text{ mol/(L•s)} \times t + 5.0 \times 10^{-4} \text{ M}$$

$$t = \frac{(5.0 \times 10^{-4} \text{ M}) - (5.0 \times 10^{-5} \text{ M})}{3.7 \times 10^{-6} \text{ M/s}} = 12\underline{1}.6 = 1.2 \times 10^{2} \text{ s}$$

The half-life for a zero-order reaction is

$$t_{1/2} = \frac{[A]_o}{2k} = \frac{5.0 \times 10^{-4} \text{ M}}{2(3.7 \times 10^{-6} \text{ M/s})} = 6\underline{7}.57 = 68 \text{ s}$$

14.107 Rearrange the two-temperature Arrhenius equation to solve for E_a in J, using $k_1 = 0.498$ M/s at $T_1 = 592$ K (319 °C) and $k_2 = 1.81$ M/s at 627 K (354 °C). Assume $(1/T_1 - 1/T_2)$ has three significant figures.

$$E_a = \frac{(8.31 \text{ J/mol} \cdot \text{K}) \ln\left(\dfrac{1.81}{0.498}\right)}{\left(\dfrac{1}{592 \text{ K}} - \dfrac{1}{627 \text{ K}}\right)} = 1.1\underline{3}7 \times 10^5 \text{ J/mol} = 114 \text{ kJ/mol}$$

To obtain A, rearrange the ln form of the one-temperature Arrhenius equation; substitute the value of E_a obtained above, and use $k_1 = 0.498$ M/s at $T_1 = 592$ K.

$$\ln A = \ln 0.498 + \frac{1.137 \times 10^5 \text{ J/mol}}{8.31 \text{ J/(mol} \cdot \text{K)} \times 592 \text{ K}} = 22.4\underline{2}0$$

$$A = \underline{5}.46 \times 10^9 = 5 \times 10^9$$

To obtain k at 420°C (693 K), also use the ln form of the one-temperature Arrhenius equation:

$$\ln k = 22.420 - \frac{1.137 \times 10^5 \text{ J/mol}}{8.31 \text{ J/(mol} \cdot \text{K)} \times 693 \text{K}} = 2.\underline{6}71$$

$$k = e^{2.671} = 1\underline{4}.45 = 14 \text{ M/s}$$

14.109 If the reaction occurs in one step, the coefficients of NO_2 and CO in this elementary reaction are each one, so the rate law should be

$$\text{Rate} = k[NO_2][CO]$$

14.111 The slow step determines the observed rate, so the overall rate constant, k, should be equal to the rate constant for the first step, and the rate law should be

$$\text{Rate} = k[NO_2Br]$$

14.113 The slow step determines the observed rate; assuming k_2 is the rate constant for the second step, the rate law would appear to be

$$\text{Rate} = k_2[NH_3][HOCN]$$

However, this rate law includes two intermediate substances that are neither reactants nor products. The rate law cannot be used unless both are eliminated.

(continued)

This can only be done using an equation from step 1. At equilibrium in step 1, you can write the following equality, assuming k_1 and k_{-1} are the rate constants for the forward and back reactions, respectively:

$$k_1[NH_4^+][OCN^-] = k_{-1}[NH_3][HOCN]$$

Rearranging and then substituting for the $[NH_3][HOCN]$ product gives

$$[NH_3][HOCN] = (k_1/k_{-1})[NH_4^+][OCN^-]$$

$$Rate = k_2(k_1/k_{-1})[NH_4^+][OCN^-] = k[NH_4^+][OCN^-] \quad (k = \text{overall rate constant})$$

14.115 a. The reaction is first order in O_2 because the rate doubled with a doubling of the oxygen concentration. The reaction is second order in NO because the rate increased by a factor of eight when both the NO and O_2 concentrations were doubled.

$$Rate = k[NO]^2[O_2]$$

b. The initial rate of the reaction for Experiment 4 can be determined by first calculating the value of the rate constant using Experiment 1 for the data.

$$0.80 \times 10^{-2} \text{ M/s} = k[4.5 \times 10^{-2} \text{ M}]^2[2.2 \times 10^{-2} \text{ M}]$$

Solving for the rate constant gives

$$k = 1.\underline{7}96 \times 10^2 \text{ M}^{-2}\text{s}^{-1}$$

Now, use the data in Experiment 4 and the rate constant to determine the initial rate of the reaction.

$$Rate = 1.796 \times 10^2 \text{ M}^{-2}\text{s}^{-1} [3.8 \times 10^{-1} \text{ M}]^2[4.6 \times 10^{-3} \text{ M}]$$

$$Rate = 0.1\underline{1}9 = 0.12 \text{ mol/L•s}$$

14.117 a. i) Rate will decrease because OH^- will react with H_3O^+ and lower its concentration.

ii) Rate will decrease because the dilution with water will decrease both the H_3O^+ and CH_3CSNH_2 concentrations.

b. i) The catalyst will provide another pathway, and k will increase because E_a will be smaller.

ii) The rate constant changes with temperature, and it will decrease with a decrease in temperature as fewer molecules will have enough energy to react.

14.119 a. The diagram:

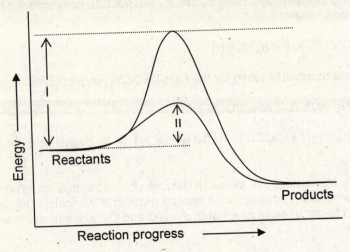

b. In the diagram, I represents the activation energy for the uncatalyzed reaction, and II represents the activation energy for the catalyzed reaction. A catalyst provides another pathway for a chemical reaction and, with a lower activation energy, more molecules have enough energy to react, so the reaction will be faster.

14.121 a. The rate of a chemical reaction is the change in the concentration of a reactant or product with time. For a reactant,

$$-\Delta c/\Delta t \text{ or } -d[c]/dt.$$

b. The rate changes because the concentration of the reactant has changed.

$$Rate = k[A]^m$$

c. $Rate = k[A]^m[B]^n$

The rate will equal k when the reactants all have 1.00 M concentrations.

■ Solutions to Cumulative-Skills Problems

14.123 The balanced equation is:

$$2N_2O_5 \quad \rightarrow \quad 4NO_2 + O_2.$$

Note that the moles of O_2 formed will be one-half that of the moles of N_2O_5 decomposed. Now use the integrated form of the first-order rate law to calculate the fraction of the 1.00 mol N_2O_5 decomposing in 20.0 hr, or 1200 min.

$$\ln \frac{[N_2O_5]_t}{[N_2O_5]_o} = -kt = -(6.2 \times 10^{-4}/min)(1200\ min) = -0.744$$

$$\frac{[N_2O_5]_t}{[N_2O_5]_o} = e^{-0.744} = 0.4752$$

$$\frac{\ln[A]}{\ln[1]} =$$

Fraction of N_2O_5 decomposed = 1.000 - 0.4752 = 0.5248

Since there is 1.00 mol of N_2O_5 present at the start, the moles of N_2O_5 decomposed is 0.5248 mol. Thus,

$$\text{Mol } O_2 \text{ formed} = (1\ O_2/2\ N_2O_5) \times 0.5248\ \text{mol } N_2O_5 = 0.26237\ \text{mol } O_2$$

Now, use the ideal gas law to calculate the volume of this number of moles of O_2 gas at 45 °C and 780 mmHg.

$$V = \frac{nRT}{P} = \frac{(0.26237\ mol)(0.08206\ L \cdot atm/(K \cdot mol)(318\ K)}{(770/760)\ atm}$$

$$= 6.758 = 6.8\ L$$

14.125 Using the first-order rate law, the initial rate of decomposition is given by

$$\text{Rate} = k[H_2O_2] = (7.40 \times 10^{-4}/s) \times (1.50\ M\ H_2O_2) = 1.110 \times 10^{-3}\ M\ H_2O_2/s$$

The heat liberated per second per mol of H_2O_2 can be found by first calculating the standard enthalpy of the decomposition of one mol of H_2O_2:

	$H_2O_2(aq)$	$\rightarrow$	$H_2O(l)$	+	$1/2\ O_2(g)$
ΔH°_f =	-191.2 kJ/mol		-285.84 kJ/mol		0 kJ/mol

(continued)

For the reaction, the standard enthalpy change is

$$\Delta H° = -285.84 \text{ kJ/mol} - (-191.2 \text{ kJ/mol}) = -94.\underline{6}4 \text{ kJ/mol } H_2O_2$$

The heat liberated per second is:

$$\frac{94.64 \text{ kJ}}{\text{mol } H_2O_2} \times \frac{1.110 \times 10^{-3} \text{ mol } H_2O_2}{L \cdot s} \times 2.00 \text{ L} = 0.21\underline{0}10 = 0.210 \text{ kJ/s}$$

14.127 Use the ideal gas law (P/RT = n/V) to calculate the mol/L of each gas:

$$[O_2] = \frac{(345/760) \text{ atm}}{0.082057 \text{ L} \cdot \text{atm}/(K \cdot \text{mol}) \times 612K} = 0.00903\underline{9} \text{ mol } O_2/L$$

$$[NO] = \frac{(155/760) \text{ atm}}{0.082057 \text{ L} \cdot \text{atm}/(K \cdot \text{mol}) \times 612K} = 0.00406\underline{1} \text{ mol NO/L}$$

The rate of decrease of NO is

$$\frac{1.16 \times 10^{-5} \text{ L}^2}{\text{mol}^2 \cdot s} \times \left(\frac{4.061 \times 10^{-3} \text{ mol}}{1 \text{ L}}\right)^2 \times \frac{9.039 \times 10^{-3} \text{ mol}}{1 \text{ L}}$$

$$= 1.7\underline{2}9 \times 10^{-12} \text{ mol}/(L \cdot s)$$

The rate of decrease in atm/s is found by multiplying by RT:

$$\frac{1.729 \times 10^{-12} \text{ mol}}{L \cdot s} \times \frac{0.082057 \text{ L} \cdot \text{atm}}{K \cdot \text{mol}} \times 612 \text{ K} = 8.6\underline{8}2 \times 10^{-11} \text{ atm/s}$$

The rate of decrease in mmHg/s is

$$8.682 \times 10^{-11} \text{ atm/s} \times (760 \text{ mmHg/atm}) = 6.598 \times 10^{-8} = 6.60 \times 10^{-8} \text{ mmHg/s}$$

15. CHEMICAL EQUILIBRIUM

■ Solutions to Exercises

Note on significant figures: If the final answer to a solution needs to be rounded off, it is given first with one nonsignificant figure, and the last significant figure is underlined. The final answer is then rounded to the correct number of significant figures. In multiple-step problems, intermediate answers are given with at least one nonsignificant figure; however, only the final answer has been rounded off.

15.1 Use the table approach, and give the starting, change, and equilibrium number of moles of each.

Amt. (mol)	$CO(g)$ +	$H_2O(g)$	$\rightleftharpoons$	$CO_2(g)$ +	$H_2(g)$
Starting	1.00	1.00		0	0
Change	-x	-x		+x	+x
Equilibrium	(1.00 - x)	(1.00 - x)		x	x = 0.43

Because we are given that x = 0.43 in the statement of the problem, we can use that to calculate the equilibrium amounts of the reactants and products:

Equilibrium amount CO = 1.00 - 0.43 = 0.57 mol

Equilibrium amount H_2O = 1.00 - 0.43 = 0.57 mol

Equilibrium amount CO_2 = x = 0.43 mol

Equilibrium amount H_2 = x = 0.43 mol

15.2 For the equation $2NO_2 + 7H_2 \rightarrow 2NH_3 + 4H_2O$, the expression for the equilibrium constant, K_c, is

$$K_c = \frac{[NH_3]^2 [H_2O]^4}{[NO_2]^2 [H_2]^7}$$

Notice that each concentration term is raised to a power equal to that of its coefficient in the chemical equation.

For the equation $NO_2 + 7/2H_2 \rightarrow NH_3 + 2H_2O$, the expression for the equilibrium constant, K_c, is

$$K_c = \frac{[NH_3][H_2O]^2}{[NO_2][H_2]^{7/2}}$$

Note the correspondence between the power and coefficient for each molecule.

15.3 The chemical equation for the reaction is

$$CO(g) + H_2O(g) \rightleftharpoons H_2(g) + CO_2(g)$$

The expression for the equilibrium constant for this reaction is

$$K_c = \frac{[CO_2][H_2]}{[CO][H_2O]}$$

We obtain the concentration of each substance by dividing the moles of substance by its volume. The equilibrium concentrations are as follows: $[CO] = 0.057$ M, $[H_2O] = 0.057$ M, $[CO_2] = 0.043$ M, and $[H_2] = 0.043$ M. Substituting these values into the equation for the equilibrium constant gives

$$K_c = \frac{(0.043)(0.043)}{(0.057)(0.057)} = 0.5\underline{6}9 = 0.57$$

15.4 Use the table approach, and give the starting, change, and equilibrium concentrations of each by dividing moles by volume in liters.

Conc. (M)	$2H_2S(g)$	$\rightleftharpoons$	$2H_2(g)$	+	$S_2(g)$
Starting	0.0100		0		0
Change	-2x		+2x (= 0.00285)		+x
Equilibrium	0.0100 - 2x		2x		x

Because the problem states that 0.00285 M H_2 was formed, we can use the 0.00285 M to calculate the other concentrations. The S_2 molarity should be one-half that, or 0.001425 M, and the H_2S molarity should be 0.0100 - 0.00285, or 0.00715 M. Substituting into the equilibrium expression gives

$$K_c = \frac{[H_2]^2[S_2]}{[H_2S]^2} = \frac{(0.00285)^2(0.001425)}{(0.00715)^2} = 2.26 \times 10^{-4} = 2.3 \times 10^{-4}$$

15.5 Use the expression that relates K_c to K_p:

$$K_p = K_c(RT)^{\Delta n}$$

The Δn term is the sum of the coefficients of the gaseous products minus the sum of the coefficients of the gaseous reactants. In this case, $\Delta n = (2 - 1) = 1$, and K_p is

$$K_p = (3.26 \times 10^{-2})(0.0821 \times 464)^1 = 1.241 = 1.24$$

15.6 For a heterogeneous equilibrium, the concentration terms for liquids and solids are omitted because such concentrations are constant at a given temperature and are incorporated into the measured value of K_c. For this case, K_c is defined

$$K_c = \frac{[Ni(CO)_4]}{[CO]^4}$$

15.7 Because the equilibrium constant is very large (> 10^4), the equilibrium mixture will contain mostly products. Rearrange the K_c expression to solve for $[NO_2]$.

$$[NO_2] = \sqrt{K_c[O_2][NO]^2} = \sqrt{(4.0 \times 10^{13})(2.0 \times 10^{-6})(2.0 \times 10^{-6})^2} = 1.78 \times 10^{-2}$$

$$= 1.8 \times 10^{-2} = 0.018\ M$$

15.8   First, divide moles by volume in liters to convert to molar concentrations, giving 0.00015 M CO_2 and 0.010 M CO. Substitute these values into the reaction quotient and calculate Q.

$$Q = \frac{[CO]^2}{[CO_2]} = \frac{(0.010)^2}{(0.00015)} = 0.6\underline{6}6 = 0.67$$

Because Q = 0.67 and is less than K_c, the reaction will go to the right, forming more CO.

15.9   Rearrange the K_c expression, and substitute for K_c (= 0.0415) and the given moles per 1.00 L to solve for moles per 1.00 L of PCl_5.

$$[PCl_5] = \frac{[PCl_3][Cl_2]}{K_c} = \frac{(0.020)(0.020)}{0.0415} = 9.\underline{6}3 \times 10^{-3}$$

$$= 9.6 \times 10^{-3} \text{ mol/ } 1.00 \text{ L}$$

Because the volume is 1.00 L, the moles of PCl_5 = 0.0096 mol.

15.10   Use the table approach, and give the starting, change, and equilibrium number of moles of each.

Amt. (mol)	$H_2(g)$	+	$I_2(g)$	$\rightleftharpoons$	2HI(g)
Starting	0.500		0.500		0
Change	-x		-x		+2x
Equilibrium	0.500 - x		0.500 - x		2x

Substitute the equilibrium concentrations into the expression for K_c (= 49.7).

$$K_c = \frac{[HI]^2}{[H_2][I_2]}; \; 49.7 = \frac{(2x)^2}{(0.500 - x)(0.500 - x)} = \frac{(2x)^2}{(0.500 - x)^2}$$

Taking the square root of both sides of the right-hand equation and solving for x gives:

$$\pm 7.05 = \frac{2x}{(0.500 - x)}, \text{ or } \pm 7.05(0.500 - x) = 2x$$

(continued)

Using the positive root, x = 0.390.

Using the negative root, x = 0.698 (this must be rejected because 0.698 is greater than the 0.500 starting number of moles).

Substituting x = 0.390 mol into the last line of the table to solve for equilibrium concentrations gives these amounts: 0.11 mol H_2, 0.11 mol I_2, and 0.78 mol HI.

15.11 Use the table approach for starting, change, and equilibrium concentrations of each species.

Conc. (M)	$PCl_5(g)$	⇌	$PCl_3(g)$	+	$Cl_2(g)$
Starting	1.00		0		0
Change	-x		+x		+x
Equilibrium	1.00 - x		x		x

Substitute the equilibrium concentration expressions from the table into the equilibrium equation, and solve for x using the quadratic formula.

$[PCl_3][Cl_2] = K_c \times [PCl_5] = 0.0211(1.00 - x) = x^2$

$x^2 + 0.0211x - 0.0211 = 0$

$$x = \frac{-0.0211 \pm \sqrt{(0.0211)^2 - 4(-0.0211)}}{2} = \frac{-0.0211 \pm 0.2913}{2}$$

x = -0.1562 (impossible; reject), or x = 0.13509 = 0.135 M (logical)

Solve for the equilibrium concentrations using x = 0.135 M: $[PCl_5]$ = 0.86 M, $[Cl_2]$ = 0.135 M, and $[PCl_3]$ = 0.135 M.

15.12 a. Increasing the pressure will cause a net reaction to occur from right to left, and more $CaCO_3$ will form.

b. Increasing the concentration of hydrogen will cause a net reaction to occur from right to left, forming more Fe and H_2O.

15.13 a. Because there are equal numbers of moles of gas on each side of the equation, increasing the pressure will not increase the amount of product.

b. Because the reaction increases the number of moles of gas, increasing the pressure will decrease the amount of product.

c. Because the reaction decreases the number of moles of gas, increasing the pressure will increase the amount of product.

15.14 Because this is an endothermic reaction and absorbs heat, high temperatures will be more favorable to the production of carbon monoxide.

15.15 Because this is an endothermic reaction and absorbs heat, high temperatures will give the best yield of carbon monoxide. Because the reaction increases the number of moles of gas, decreasing the pressure will also increase the yield.

■ Answers to Review Questions

15.1 A reasonable graph showing the decrease in concentration of $N_2O_4(g)$ and the increase in concentration of $NO_2(g)$ is shown below:

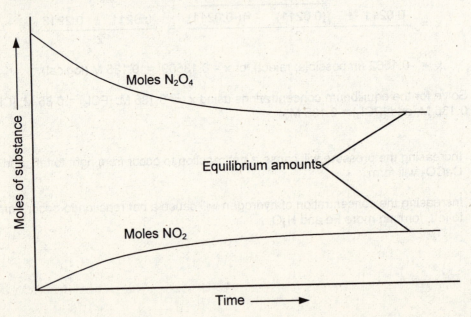

(continued)

At first, the concentration of N_2O_4 is large, and the rate of the forward reaction is large, but then as the concentration of N_2O_4 decreases, the rate of the forward reaction decreases. In contrast, the concentration of NO_2 builds up from zero to a low concentration. Thus, the initial rate of the reverse reaction is zero, but it steadily increases as the concentration of NO_2 increases. Eventually, the two rates become equal when the reaction reaches equilibrium. This is a dynamic equilibrium because both the forward and reverse reactions are occurring at all times even though there is no net change in concentration at equilibrium.

15.2   The 1.0 mol of $H_2(g)$ and 1.0 mol of $I_2(g)$ in the first mixture reach equilibrium when the amounts of reactants decrease to 0.50 mol each and when the amount of product increases to 1.0 mol. The total number of moles of the reactants at the start is 2.0 mol, which is the same number of moles as in the second mixture, the 2.0 mol of HI that is to be allowed to come to equilibrium. The second mixture should produce the same number of moles of H_2, I_2, and HI at equilibrium because, if the total number of moles is constant, it should not matter from which direction an equilibrium is approached.

15.3   The equilibrium constant for a gaseous reaction can be written using partial pressures instead of concentrations because all the reactants and products are in the same vessel. Therefore, at constant temperature, the pressure, P, is proportional to the concentration, n/V. (The ideal gas law says that $P = [n/V]RT$).

15.4   The addition of reactions 1 and 2 yields reaction 3:

(Reaction 1) $HCN + OH^- \rightleftharpoons CN^- + H_2O$

(Reaction 2) $H_2O \rightleftharpoons H^+ + OH^-$

(Reaction 3) $HCN \rightleftharpoons H^+ + CN^-$

The rule states that, if a given equation can be obtained from the sum of other equations, the equilibrium constant for the given equation equals the product of the other equilibrium constants. Thus, K for reaction 3 is

$K = K_1 \times K_2 = (4.9 \times 10^4) \times (1.0 \times 10^{-14}) = 4.9 \times 10^{-10}$

15.5   a. Homogeneous equilibrium. All substances are gases and thus exist in one phase, a mixture of gases.

b. Heterogeneous equilibrium. The two copper compounds are solids, but the other substances are gases. This fulfills the definition of a heterogeneous equilibrium.

c. Homogeneous equilibrium. All substances are gases and thus exist in one phase, a mixture of gases.

d. Heterogeneous equilibrium. The two copper-containing substances are solids, but the other substances are gases. This fulfills the definition of a heterogeneous equilibrium.

15.6   Pure liquids and solids can be ignored in an equilibrium-constant expression because their concentrations do not change. (If a solid is present, it has not dissolved in any gas or solution present.) In effect, concentrations of liquids and solids are incorporated into the value of K_c, as discussed in the chapter.

15.7   A qualitative interpretation of the equilibrium constant involves using the magnitude of the equilibrium constant to predict the relative amounts of reactants and products at equilibrium. If K_c is around one, the equilibrium mixture contains appreciable amounts (same order of magnitude) of reactants and products. If K_c is large, the equilibrium mixture is mostly products. If K_c is small, the equilibrium mixture is mostly reactants. The type of reaction governs what "large" and "small" values are; but for some types of reactions, "large" might be no less than 10^2 to 10^4, whereas "small" might be no more than 10^{-4} to 10^{-2}.

15.8   The reaction quotient, Q_c, is an expression that has the same form as the equilibrium-constant expression but whose concentrations are not necessarily equilibrium concentrations. It is useful in determining whether a reaction mixture is at equilibrium or, if not, what direction the reaction will go as it approaches equilibrium.

15.9   The ways in which the equilibrium composition of a mixture can be altered are (1) changing concentrations by removing some of the products and/or some of the reactants, (2) changing the partial pressure of a gaseous reactant and/or a gaseous product, and (3) changing the temperature of the reaction mixture. (Adding a catalyst cannot alter the equilibrium concentration, but can affect only the rate of reaction.)

15.10   The role of the platinum is that of a catalyst; it provides conditions (a surface) suitable for speeding up the attainment of equilibrium. The platinum has no effect on the equilibrium composition of the mixture even though it greatly increases the rate of reaction.

15.11   In some cases, a catalyst can affect the product in a reaction because it affects only the rate of one reaction out of several reactions that are possible. If two reactions are possible, and the uncatalyzed rate of one is much slower but is the only reaction that is catalyzed, then the products with and without a catalyst will be different. The Ostwald process is a good example. In the absence of a catalyst, NH_3 burns in O_2 to form only N_2 and H_2O, even though it is possible for NH_3 to react to form NO and H_2O. Ostwald found that adding a platinum catalyst favors the formation of the NO and H_2O almost to the exclusion of the N_2 and H_2O.

15.12   Four ways in which the yield of ammonia can be improved are (1) removing the gaseous NH_3 from the equilibrium by liquefying it, (2) increasing the nitrogen or hydrogen concentration, (3) increasing the total pressure on the mixture (the moles of gas decrease), and (4) lowering the temperature ($\Delta H°$ is negative, so heat is evolved). Each causes a shift to the right in accordance with Le Chatelier's principle.

■ Solutions to Practice Problems

Note on significant figures: If the final answer to a solution needs to be rounded off, it is given first with one nonsignificant figure, and the last significant figure is underlined. The final answer is then rounded to the correct number of significant figures. In multiple-step problems, intermediate answers are given with at least one nonsignificant figure; however, only the final answer has been rounded off.

15.23 Use the table approach, and give the starting, change, and equilibrium number of moles of each.

Amt. (mol)	$PCl_5(g)$	$\rightleftharpoons$	$PCl_3(g)$	+	$Cl_2(g)$
Starting	2.500		0		0
Change	-x		+x		+x
Equilibrium	2.500 - x		x = 0.338		x

Since the equilibrium amount of PCl_3 is given in the problem, this tells you x = 0.338. The equilibrium amounts for the other substances can now be determined.

Equilibrium amount Cl_2 = x = 0.338 mol

Equilibrium amount PCl_5 = 2.500 - x = 2.500 - 0.338 = 2.162 mol

Therefore, the amounts of the substances in the equilibrium mixture are 2.162 mol PCl_5, 0.338 mol PCl_3, and 0.338 mol Cl_2.

15.25 Use the table approach, and give the starting, change, and equilibrium number of moles of each.

Amt. (mol)	$N_2(g)$	+	$3H_2(g)$	$\rightleftharpoons$	$2NH_3(g)$
Starting	0.600		1.800		0
Change	-x		-3x		+2x
Equilibrium	0.600 - x		1.800 - 3x		2x = 0.048

Since the equilibrium amount of NH_3 is given in the problem, this tells you x = 0.024. The equilibrium amounts for the other substances can now be determined.

Equilibrium amount N_2 = 0.600 - 0.024 = 0.576 mol

Equilibrium amount H_2 = 1.800 - 3 x (0.024) = 1.728 mol

Therefore, the amounts of the substances in the equilibrium mixture are 0.576 mol N_2, 1.728 mol H_2, and 0.048 mol NH_3.

15.27 Use the table approach, and give the starting, change, and equilibrium number of moles of each.

Amt. (mol)	$2SO_2(g)$	$+$	$O_2(g)$	$\rightleftharpoons$	$2SO_3(g)$
Starting	0.0400		0.0200		0
Change	-2x		-x		+2x
Equilibrium	0.0400 - 2x		0.0200 - x		2x (= 0.0296)

Therefore, x = 0.0148, and the amounts of substances at equilibrium are 0.0104 mol SO_2, 0.0052 mol O_2, and 0.0296 mol SO_3.

15.29 a. $K_c = \dfrac{[NO_2][NO]}{[N_2O_3]}$

b. $K_c = \dfrac{[H_2]^2 [S_2]}{[H_2S]^2}$

c. $K_c = \dfrac{[NO_2]^2}{[NO]^2 [O_2]}$

d. $K_c = \dfrac{[P(NH_2)_3][HCl]^3}{[PCl_3][NH_3]^3}$

15.31 The reaction is $2H_2S(g) + 3O_2(g) \rightleftharpoons 2H_2O(g) + 2SO_2(g)$

15.33 When the reaction is halved, the equilibrium-constant expression is

$$K_c = \frac{[NO_2]^2 [O_2]^{1/2}}{[N_2O_5]}$$

When the reaction is then reversed, the equilibrium-constant expression becomes

$$K_c = \frac{[N_2O_5]}{[NO_2]^2 [O_2]^{1/2}}$$

15.35 Because $K_c = 1.84$ for $2HI \rightleftharpoons H_2 + I_2$, the value of K_c for $H_2 + I_2 \rightleftharpoons 2HI$ must be the reciprocal of K_c for the first reaction. Mathematically, this can be shown as follows:

Forward: $K_c = \dfrac{[H_2][I_2]}{[HI]^2} = 1.84$

(continued)

Reverse: $K_c = \dfrac{[HI]^2}{[H_2][I_2]} = \dfrac{1}{\dfrac{[H_2][I_2]}{[HI]^2}} = \dfrac{1}{K_c(f)}$

Thus, for the reverse reaction, K_c is calculated as follows:

$K_c = 1 \div 1.84 = 5.4\underline{3}4 \times 10^{-1} = 0.543$

15.37   First, calculate the molar concentrations of each of the compounds in the equations:

$[H_2] = 0.488$ mol $H_2 \div 6.00$ L $= 0.081\underline{3}3$ M

$[I_2] = 0.206$ mol $I_2 \div 6.00$ L $= 0.034\underline{3}3$ M

$[HI] = 2.250$ mol HI $\div 6.00$ L $= 0.375\underline{0}$ M

Now, substitute into the K_c expression:

$K_c = \dfrac{(0.3750)^2}{(0.08133)(0.03433)} = 50.\underline{3}6 = 50.4$

15.39   Substitute the following concentrations into the K_c expression:

$[SO_3] = 0.0296$ mol $\div 2.000$ L $= 0.0148\underline{0}$ M

$[SO_2] = 0.0104$ mol $\div 2.000$ L $= 0.00520\underline{0}$ M

$[O_2] = 0.0052$ mol $\div 2.000$ L $= 0.0026\underline{0}$ M

$K_c = \dfrac{(0.01480)^2}{(0.005200)^2(0.00260)} = 3.\underline{1}1 \times 10^3 = 3.1 \times 10^3$

15.41 For each mole of NOBr that reacts, (1.000 - 0.094 = 0.906) mol remains. Starting with 2.00 mol NOBr, 2 x 0.906 mol NOBr, or 1.812 mol NOBr, remain. Because the volume is 1.00 L, the concentration of NOBr at equilibrium is 1.812 M. Assemble a table of starting, change, and equilibrium concentrations:

Conc. (M)	$2NOBr(g)$	$\rightleftharpoons$	$2NO(g)$	+	$Br_2(g)$
Starting	2.00		0		0
Change	-2x		+2x		+x
Equilibrium	2.00 - 2x (= 1.812)		2x		x

Because 2.00 - 2x = 1.812, x = 0.094 M. Therefore, the equilibrium concentrations are [NOBr] = 1.812 M, [NO] = 0.188 M, and [Br$_2$] = 0.094 M.

$$K_c = \frac{[NO]^2[Br_2]}{[NOBr]^2} = \frac{(0.188 \text{ M})^2(0.094 \text{ M})}{(1.812 \text{ M})^2} = 1.01 \times 10^{-3} = 1.0 \times 10^{-3}$$

15.43 a. $K_p = \dfrac{P_{HBr}^2}{P_{H_2}\, P_{Br_2}}$ b. $K_p = \dfrac{P_{CH_4}\, P_{H_2S}^2}{P_{CS_2}\, P_{H_2}^4}$

c. $K_p = \dfrac{P_{H_2O}^2\, P_{Cl_2}^2}{P_{HCl}^4\, P_{O_2}}$ d. $K_p = \dfrac{P_{CH_3OH}}{P_{CO}\, P_{H_2}^2}$

15.45 There are three mol of gaseous product for every five mol of gaseous reactant, so Δn = 3 - 5 = -2. Using this, calculate K_p from K_c:

$$K_p = K_c(RT)^{\Delta n} = 0.28\,(0.0821 \times 1173)^{-2} = 3.\underline{0}19 \times 10^{-5} = 3.0 \times 10^{-5}$$

15.47 For each one mol of gaseous product, there are 1.5 mol of gaseous reactants; thus, Δn = 1 - 1.5 = -0.5. Using this, calculate K_c from K_p:

$$K_c = \frac{K_p}{(RT)^{\Delta n}} = \frac{6.55}{(0.0821 \times 900)^{-0.5}} = 6.55 \times (0.0821 \times 900)^{0.5}$$

$$= 56.\underline{3}03 = 56.3$$

15.49 a. $K_c = \dfrac{[CO]^2}{[CO_2]}$ b. $K_c = \dfrac{[CO_2]}{[CO]}$

c. $K_c = \dfrac{[CO_2]}{[SO_2][O_2]^{1/2}}$ d. $K_c = [Pb^{2+}][I^-]^2$

15.51 a. Not complete; K_c is very small (10^{-31}), indicating very little reaction.

b. Nearly complete; K_c is very large (10^{21}), indicating nearly complete reaction.

15.53 K_c is extremely small, indicating very little reaction at room temperature. Because the decomposition of HF yields equal amounts of H_2 and F_2, at equilibrium $[H_2] = [F_2]$. So, for the decomposition of HF,

$$K_c = \frac{[H_2][F_2]}{[HF]^2} = \frac{[H_2]^2}{[HF]^2}$$

$$[H_2] = K_c^{1/2}[HF] = (1.0 \times 10^{-95})^{1/2}(1.0\ M) = 3.\underline{1}6 \times 10^{-48} = 3.2 \times 10^{-48}\ mol/L$$

This result does agree with what is expected from the very small magnitude of K_c.

15.55   Calculate Q, the reaction quotient, and compare it to the equilibrium constant ($K_c = 3.07 \times 10^{-4}$). If Q is larger, the reaction will go to the left; smaller, the reaction will go to the right; equal, the reaction is at equilibrium. In all cases, Q is found by combining these terms:

$$Q = \frac{[NO]^2[Br_2]}{[NOBr]^2}$$

a. $Q = \dfrac{(0.0162)^2(0.0123)}{(0.0720)^2} = 6.2\underline{2}6 \times 10^{-4} = 6.23 \times 10^{-4}$

$Q > K_c$. The reaction should go to the left.

b. $Q = \dfrac{(0.0159)^2(0.0139)}{(0.121)^2} = 2.4\underline{0}0 \times 10^{-4} = 2.40 \times 10^{-4}$

$Q < K_c$. The reaction should go to the right.

(continued)

c. $Q = \dfrac{(0.0134)^2(0.0181)}{(0.103)^2} = 3.0\underline{6}3 \times 10^{-4} = 3.06 \times 10^{-4}$

$Q = K_c$. The reaction should be at equilibrium.

d. $Q = \dfrac{(0.0121)^2(0.0105)}{(0.0472)^2} = 6.9\underline{0}0 \times 10^{-4} = 6.90 \times 10^{-4}$

$Q > K_c$. The reaction should go to the left.

15.57 Calculate Q, the reaction quotient, and compare it to the equilibrium constant. If Q is larger, the reaction will go to the left, and vice versa. Q is found by combining these terms:

$$Q = \dfrac{[CH_3OH]}{[CO][H_2]^2} = \dfrac{(0.020)}{(0.010)(0.010)^2} = 20.0 \;(>10.5)$$

The reaction goes to the left.

15.59 Substitute into the expression for K_c and solve for $[COCl_2]$:

$$K_c = 1.23 \times 10^3 = \dfrac{[COCl_2]}{[CO][Cl_2]} \doteq \dfrac{[COCl_2]}{(0.012)(0.025)}$$

$[COCl_2] = (1.23 \times 10^3)(0.012)(0.025) = 0.3\underline{6}9 = 0.37 \; M$

15.61 Divide moles of substance by the volume of 5.0 L to obtain concentration. The starting concentrations are 3.0×10^{-4} M for both $[I_2]$ and $[Br_2]$. Assemble a table of starting, change, and equilibrium concentrations.

Conc. (M)	$I_2(g)$	+	$Br_2(g)$	$\rightleftharpoons$	$2\,IBr(g)$
Starting	3.0×10^{-4}		3.0×10^{-4}		0
Change	-x		-x		+2x
Equilibrium	$(3.0 \times 10^{-4}) - x$		$(3.0 \times 10^{-4}) - x$		2x

(continued)

Substituting into the equilibrium-constant expression gives

$$K_c = 1.2 \times 10^2 = \frac{[IBr]^2}{[I_2][Br_2]} = \frac{(2x)^2}{(3.0 \times 10^{-4} - x)(3.0 \times 10^{-4} - x)}$$

Taking the square root of both sides yields

$$10.95 = \frac{(2x)}{(3.0 \times 10^{-4} - x)}$$

Rearranging and simplifying the right side gives

$$(3.0 \times 10^{-4} - x) = \frac{(2x)}{10.95} = (0.182x)$$

$$x = 2.53 \times 10^{-4} \text{ M}$$

Thus, $[I_2] = [Br_2] = 4.70 \times 10^{-5} = 4.7 \times 10^{-5}$ M, and $[IBr] = 5.06 \times 10^{-4} = 5.1 \times 10^{-4}$ M.

15.63 Divide moles of substance by the volume of 1.25 L to obtain concentration. The starting concentration is 1.00 M for $[CO_2]$, but the concentration of carbon, a solid, is omitted. Assemble a table of starting, change, and equilibrium concentrations.

Conc. (M)	$CO_2(g)$	+	C(s)	$\rightleftharpoons$	2CO(g)
Starting	1.00				0
Change	-x				+2x
Equilibrium	1.00 - x				2x

Substituting into the equilibrium expression for K_c gives

$$K_c = 14.0 = \frac{[CO]^2}{[CO_2]} = \frac{(2x)^2}{(1.00 - x)}$$

Rearranging and solving for x yields

$$14.0 - 14.0x = 4x^2$$

$$4x^2 + 14.0x - 14.0 = 0 \quad \text{(quadratic equation)}$$

(continued)

Using the solution to the quadratic equation gives

$$x = \frac{-14.0 \pm \sqrt{(14.0)^2 - 4(4)(-14.0)}}{2(4)}$$

$x = -4.31$ (impossible; reject), or $x = 0.81\underline{17} = 0.812$ M (logical)

Thus, $[CO_2] = 0.19$ M, and $[CO] = 1.62$ M.

15.65 Divide moles of substance by the volume of 10.00 L to obtain concentration. The starting concentrations are 0.1000 M for [CO] and 0.3000 M for [H$_2$]. Assemble a table of starting, change, and equilibrium concentrations.

Conc. (M)	CO(g)	+	3H$_2$(g)	⇌	CH$_4$(g)	+	H$_2$O(g)
Starting	0.1000		0.3000		0		0
Change	-x		-3x		+x		+x
Equilibrium	0.1000 - x		0.3000 - 3x		x		x

Substituting into the equilibrium expression for K_c gives

$$K_c = 3.92 = \frac{[CH_4][H_2O]}{[CO][H_2]^3} = \frac{x^2}{(0.1000 - x)[3(0.1000 - x)]^3}$$

$$= \frac{x^2}{27(0.1000 - x)^4}$$

Multiplying both sides by 27 and taking the square root of both sides gives

$$10.29 = \frac{x}{(0.1000 - x)^2}$$

Or,

$$10.29x^2 - 3.058x + 0.1029 = 0$$

(continued)

Using the solution to the quadratic equation yields

$$x = \frac{3.058 \pm \sqrt{(-3.058)^2 - 4(10.29)(0.1029)}}{2(10.29)}$$

x = 0.2585 (can't be > 0.1000, so reject), or x = 0.03868 = 0.0387 M (use)

Thus, [CO] = 0.0613 M, [H₂] = 0.1839 M, [CH₄] = 0.0387 M, and [H₂O] = 0.0387 M.

15.67   Forward direction

15.69 a. A pressure increase has no effect because the number of moles of reactants equals that of products.

b. A pressure increase has no effect because the number of moles of reactants equals that of products.

c. A pressure increase causes the reaction to go to the left because the number of moles of reactants is less than that of products.

15.71   The fraction would not increase because an increase in temperature decreases the amounts of products of an exothermic reaction.

15.73   The value of $\Delta H°$ is calculated from the $\Delta H°_f$ values below each substance in the reaction:

$$2NO_2(g) \; + \; 7H_2(g) \;\; \rightleftharpoons \;\; 2NH_3(g) \; + \; 4H_2O(g)$$
$$2(33.2) \qquad\quad 7(0) \qquad\qquad\qquad 2(-45.9) \qquad 4(-241.8)$$

$\Delta H° = -967.2 + (-91.8) - 66.4 = -1125.4$ kJ/2 mol NO₂

The equilibrium constant will decrease with temperature because raising the temperature of an exothermic reaction will cause the reaction to go farther to the left.

15.75   Because the reaction is exothermic, the formation of products will be favored by low temperatures. Because there are more molecules of gaseous products than of gaseous reactants, the formation of products will be favored by low pressures.

■ Solutions to General Problems

15.77 Substitute the concentrations into the equilibrium expression to calculate K_c.

$$K_c = \frac{[CH_3OH]}{[CO][H_2]^2} = \frac{(0.015)}{(0.096)(0.191)^2} = 4.2\underline{8} = 4.3$$

15.79 Assume 100.00 g of gas: 90.55 g are CO, and 9.45 g are CO_2. The moles of each are

$$90.55 \text{ g CO} \times \frac{1 \text{ mol CO}}{28.01 \text{ g CO}} = 3.23\underline{2}8 \text{ mol COl}$$

$$9.45 \text{ g CO}_2 \times \frac{1 \text{ mol CO}_2}{44.01 \text{ g CO}_2} = 0.214\underline{7} \text{ mol CO}_2$$

Total moles of gas = (3.2328 + 0.2147) mol = 3.44$\underline{7}$5 mol. Use the ideal gas law to convert to the volume of gaseous solution:

$$V = \frac{nRT}{P} = \frac{(3.4475 \text{ mol})(0.082057 \text{ L} \cdot \text{atm/K} \cdot \text{mol})(850 + 273)K}{1.000 \text{ atm}} = 317.\underline{6}9 \text{ L}$$

The concentrations are

$$[CO] = \frac{3.2328 \text{ mol CO}}{317.69 \text{ L}} = 0.01017\underline{6} \text{ M}$$

$$[CO_2] = \frac{0.2147 \text{ mol CO}_2}{317.69 \text{ L}} = 6.7\underline{5}8 \times 10^{-4} \text{ M}$$

Find K_c by substituting into the equilibrium expression:

$$K_c = \frac{[CO]^2}{[CO_2]} = \frac{(0.010176)^2}{(6.758 \times 10^{-4})} = 0.153\underline{2} = 0.153$$

15.81 After calculating the concentrations after mixing, calculate Q, the reaction quotient, and compare it with K_c.

$$[N_2] = [H_2] = 1.00 \text{ mol} \div 2.00 \text{ L} = 0.500 \text{ M}$$

$$[NH_3] = 2.00 \text{ mol} \div 2.00 \text{ L} = 1.00 \text{ M}$$

$$Q = \frac{[NH_3]^2}{[N_2][H_2]^3} = \frac{(1.00)^2}{(0.500)(0.500)^3} = 16.\underline{0}0 = 16.0$$

Because Q is greater than K_c, the reaction will go in the reverse direction (to the left) to reach equilibrium.

15.83 To calculate the concentrations after mixing, assume the volume to be 1.00 L, symbolized as V. Because the volumes in the numerator and denominator cancel each other, they do not matter. Assume a 1.00-L volume and calculate Q, the reaction quotient, and compare it with K_c.

$$[CO] = [H_2O] = [CO_2] = [H_2] = 1.00 \text{ mol}/1.00 \text{ L}$$

$$Q = \frac{[CO_2][H_2]}{[CO][H_2O]} = \frac{(1.00 \text{ mol}/1.00 \text{ L})(1.00 \text{ mol}/1.00 \text{ L})}{(1.000 \text{ mol}/1.00 \text{ L})(1.00 \text{ mol}/1.00 \text{ L})} = 1.\underline{0}0$$

Because Q is greater than K_c, the reaction will go in the reverse direction (left) to reach equilibrium.

15.85 Assemble a table of starting, change, and equilibrium concentrations, letting 2x = the change in [HBr].

Conc. (M)	2HBr(g)	⇌	H$_2$(g)	+	Br$_2$(g)
Starting	0.010		0		0
Change	-2x		+x		+x
Equilibrium	0.010 - 2x		x		x

$$K_c = 0.016 = \frac{[H_2][Br_2]}{[HBr]^2} = \frac{(x)(x)}{(0.010 - 2x)^2}$$

(continued)

$$0.126 = \frac{(x)}{(0.010 - 2x)}$$

$$1.26 \times 10^{-3} - (2.52 \times 10^{-1})x = x$$

$$x = (1.26 \times 10^{-3}) \div 1.252 = 1.006 \times 10^{-3} = 1.0 \times 10^{-3} \text{ M}$$

Therefore, [HBr] = 0.008 M, or 0.008 mol; [H$_2$] = 0.0010 M, or 0.0010 mol; and [Br$_2$] = 0.0010 M, or 0.0010 mol.

15.87 The starting concentration of COCl$_2$ = 1.00 mol ÷ 25.00 L = 0.0400 M. Assemble a table of starting, change, and equilibrium concentrations.

Conc. (M)	COCl$_2$(g) $\rightleftharpoons$	Cl$_2$(g) +	CO(g)
Starting	0.0400	0	0
Change	-x	+x	+x
Equilibrium	0.0400 - x	x	x

Substituting into the equilibrium expression for K$_c$ gives

$$K_c = 8.05 \times 10^{-4} = \frac{[CO][Cl_2]}{[COCl_2]} = \frac{(x)(x)}{(0.0400 - x)}$$

Rearranging and solving for x yields

$$3.22 \times 10^{-5} - (8.05 \times 10^{-4})x - x^2 = 0$$

$$x^2 + (8.05 \times 10^{-4})x - 3.22 \times 10^{-5} = 0 \text{ (quadratic equation)}$$

Using the solution to the quadratic equation gives

$$x = \frac{-(8.05 \times 10^{-4}) \pm \sqrt{(8.05 \times 10^{-4})^2 - 4(1)(-3.22 \times 10^{-5})}}{2(1)}$$

$$x = -6.09 \times 10^{-3} \text{ (impossible; reject), or } x = 5.286 \times 10^{-3} \text{ (logical; use)}$$

Percent dissoc. = (change ÷ starting) × 100% = (0.005286 ÷ 0.0400) × 100%

$$= 13.21 = 13.2 \text{ percent}$$

15.89 Using 1.00 mol/10.00 L, or 0.100 M, and 4.00 mol/10.00 L, or 0.400 M, for the respective starting concentrations for CO and H_2, assemble a table of starting, change, and equilibrium concentrations.

Conc. (M)	CO(g)	+	$3H_2(g)$	$\rightleftharpoons$	$CH_2(g)$	+	$H_2O(g)$
Starting	0.100		0.400		0		0
Change	-x		-3x		+x		+x
Equilibrium	0.100 - x		0.400 -3 x		x		x

Substituting into the equilibrium expression for K_c gives

$$K_c = 3.92 = \frac{[CH_4][H_2O]}{[CO][H_2]^3} = \frac{x^2}{(0.100 - x)(0.400 - 3x)^3}$$

$$f(x) = \frac{x^2}{(0.100 - x)(0.400 - 3x)^3}$$

Because K_c is > 1 (> 50 percent reaction), choose x = 0.05 (about half of CO reacting), and use that for the first entry in the table of x, f(x), and interpretations:

x	f(x)	Interpretation
0.05	3.20	x > 0.05
0.06	8.45	x < 0.06
0.055	5.18	x < 0.055
0.0525	4.07	x < 0.0525 (but close)
0.052	3.87	f(x) of 3.87 $\cong$ 3.92

At equilibrium, concentrations and moles are CO: 0.048 M and 0.48 mol; H_2: 0.244 M and 2.44 mol; CH_4: 0.052 M and 0.52 mol; and H_2O: 0.052 M and 0.52 mol.

15.91 The dissociation is endothermic.

15.93 For $N_2 + 3H_2 \rightleftharpoons 2NH_3$, K_p is defined in terms of pressures as

$$K_p = \frac{P_{NH_3}^{\,2}}{P_{N_2} P_{H_2}^{\,3}}$$

(continued)

But by the ideal gas law, where $[i]$ = mol/L,

$$P_i = (n_iRT)/V, \text{ or } P_i = [i]RT$$

Substituting the right-hand equality into the K_p expression gives

$$K_p = \frac{[NH_3]^2(RT)^2}{[N_2](RT)[H_2]^3(RT)^3} = \frac{[NH_3]^2}{[N_2][H_2]^3}(RT)^{-2}$$

$$K_p = K_c(RT)^{-2}, \text{ or } K_c = K_p (RT)^2$$

15.95 a. The change in the number of moles of gas for the reaction is $\Delta n = 2 - 1 = 1$. Using this, calculate K_p from K_c:

$$K_p = K_c(RT)^{\Delta n} = 0.153 (0.08206 \times 1123)^1 = 14.\underline{0}99 = 14.1$$

b. Use the table approach, and give the starting, change, and equilibrium pressures in atm.

Press. (atm)	C(s) +	CO$_2$(g)	⇌	2CO(g)
Starting		1.50		0
Change		-x		+2x
Equilibrium		1.50 - x		2x

Substituting into the equilibrium-constant expression gives

$$K_p = 14.10 = \frac{P_{CO}^2}{P_{CO_2}} = \frac{(2x)^2}{(1.50 - x)}$$

Rearranging and solving for x gives a quadratic equation.

$$4x^2 + 14.10x - 21.15 = 0$$

Using the quadratic formula gives

$$x = \frac{-14.10 \pm \sqrt{(14.10)^2 - (4)(4)(-21.15)}}{2(4)}$$

$$x = 1.1\underline{3}5 \text{ (positive root)}$$

(continued)

Thus, at equilibrium, the pressures of CO and CO_2 are

$$P_{CO_2} = 1.50 - 1.1\underline{3}5 = 0.3\underline{6}5 = 0.37 \text{ atm}$$

$$P_{CO} = 2x = 2(1.1\underline{3}5) = 2.2\underline{7}0 = 2.27 \text{ atm}$$

c. Because the reaction is endothermic, the equilibrium will shift to the left, and the pressure of CO will decrease.

15.97 a. The molar mass of PCl_5 is 208.22 g/mol. Thus, the initial concentration of PCl_5 is

$$\frac{35.8 \text{ g } PCl_5 \times \dfrac{1 \text{ mol } PCl_5}{208.22 \text{ g } PCl_5}}{5.0 \text{ L}} = 0.03\underline{4}4 \text{ M}$$

Use the table approach, and give the starting, change, and equilibrium concentrations.

Conc. (M)	$PCl_3(g)$	+	$Cl_2(g)$	$\rightleftharpoons$	$PCl_5(g)$
Starting	0		0		0.03\underline{4}4
Change	x		x		- x
Equilibrium	x		x		0.03\underline{4}4 - x

Substituting into the equilibrium-constant expression gives

$$K_c = 4.1 = \frac{[PCl_5]}{[PCl_3][Cl_2]} = \frac{0.0344 - x}{x^2}$$

Rearranging and solving for x gives a quadratic equation.

$$4.1x^2 + x - 0.0344 = 0$$

Using the quadratic formula gives

$$x = \frac{-1 \pm \sqrt{(1)^2 - (4)(4.1)(-0.0344)}}{2(4.1)}$$

$$x = 0.03\underline{0}6 \text{ (positive root)}$$

Thus, at equilibrium, $[PCl_3] = [Cl_2] = x = 0.031$ M. The concentration of PCl_5 is

$$[PCl_5] = .03\underline{4}4 - x = 0.0344 - 0.0306 = 0.00\underline{3}8 = 0.004 \text{ M}.$$

(continued)

b. The fraction of PCl_5 decomposed is

$$\text{fraction decomposed} = \frac{0.0306 \text{ M}}{0.0344 \text{ M}} = 0.0889 = 0.89$$

c. There would be a greater pressure, so less PCl_5 would decompose in order to minimize the increase in pressure.

15.99 The initial moles of $SbCl_5$ (molar mass 299.01 g/mol) are

$$65.4 \text{ g SbCl}_5 \times \frac{1 \text{ mol SbCl}_5}{299.01 \text{ g SbCl}_5} = 0.2187 \text{ mol}$$

The initial pressure of $SbCl_5$ is

$$P = \frac{nRT}{V} = \frac{(0.2187 \text{ mol})(0.08206 \text{ L} \cdot \text{atm/K} \cdot \text{mol})(468K)}{5.00 \text{ L}} = 1.680 \text{ atm}$$

Use the table approach, and give the starting, change, and equilibrium pressures in atm.

Press. (atm)	$SbCl_5(g)$	$\rightleftharpoons$	$SbCl_3(g)$	+	$Cl_2(g)$
Starting	1.680		0		0
Change	-x		x		x
Equilibrium	(1.680)(0.642)		(1.680)(0.358)		(1.680)(0.358)

At equilibrium, 35.8 percent of the $SbCl_5$ is decomposed, so x = (1.680)(0.358) in this table. The equilibrium-constant expression is

$$K_p = \frac{P_{SbCl_3} \cdot P_{Cl_2}}{P_{SbCl_5}} = \frac{(1.680 \times 0.358)^2}{1.680 \times 0.642} = 0.3354 = 0.335$$

15.101 a. The initial concentration of SO_2Cl_2 (molar mass 134.97 g/mol) is

$$\frac{8.25 \text{ g } SO_2Cl_2 \times \dfrac{1 \text{ mol } SO_2Cl_2}{134.97 \text{ g } SO_2Cl_2}}{1.00 \text{ L}} = 0.061\underline{1}2 \text{ M}$$

Use the table approach, and give the starting, change, and equilibrium concentrations.

Conc. (M)	$SO_2Cl_2(g)$	$\rightleftharpoons$	$SO_2(g)$	+	$Cl_2(g)$
Starting	0.06112		0		0
Change	-x		x		x
Equilibrium	0.06112 - x		x		x

Substituting into the equilibrium-constant expression gives

$$K_c = \frac{[SO_2][Cl_2]}{[SO_2Cl_2]} = \frac{x^2}{(0.06112 - x)} = 0.045$$

Rearranging and solving for x gives a quadratic equation.

$$x^2 + 0.045x - 0.0027506 = 0$$

Using the quadratic formula gives

$$x = \frac{-(0.045) \pm \sqrt{(0.045)^2 - (4)(1)(-0.0027506)}}{2(1)}$$

x = 0.03\underline{4}56 (positive root)

The concentrations at equilibrium are $[SO_2]$ = $[Cl_2]$ = x = 0.034 M. For SO_2Cl_2,

$[SO_2Cl_2]$ = 0.06112 - x = 0.06112 - 0.03\underline{4}56 = 0.02\underline{6}55 = 0.027M.

b. The fraction of SO_2Cl_2 decomposed is

$$\text{Fraction decomposed} = \frac{0.03456 \text{ M}}{0.06112 \text{ M}} = 0.5\underline{6}5 = 0.57$$

c. This would shift the equilibrium to the left and decrease the fraction of SO_2Cl_2 that has decomposed.

15.103a. First, determine the initial concentration of the dimer assuming complete reaction. The reaction can be described as $2A \rightarrow D$. Therefore, the initial concentration of dimer is one-half of the concentration of monomer, or 2.0×10^{-4} M. Next, allow the dimer to dissociate into the monomer in equilibrium. Use the table approach, and give the starting, change, and equilibrium concentrations.

Conc. (M)	D(g)	$\rightleftharpoons$	2A(g)
Starting	2.0×10^{-4}		0
Change	-x		2x
Equilibrium	2.0×10^{-4} - x		2x

Substituting into the equilibrium-constant expression gives

$$K_c = \frac{[A]^2}{[D]} = \frac{(2x)^2}{(2.0 \times 10^{-4} - x)} = \frac{1}{3.2 \times 10^4} = 3.125 \times 10^{-5}$$

Rearranging and solving for x gives a quadratic equation.

$$4x^2 + (3.125 \times 10^{-5})x - (6.250 \times 10^{-9}) = 0$$

Using the quadratic formula gives

$$x = \frac{-(3.125 \times 10^{-5}) \pm \sqrt{(3.125 \times 10^{-5})^2 - (4)(4)(-6.250 \times 10^{-9})}}{2(4)}$$

$$x = 3.\underline{5}8 \times 10^{-5} \text{ (positive root)}$$

Thus, the concentrations at equilibrium are

$$[CH_3COOH] = 2x = 2(3.\underline{5}8 \times 10^{-5}) = 7.\underline{1}6 \times 10^{-5} = 7.2 \times 10^{-5} \text{ M}$$
$$[\text{Dimer}] = 2.0 \times 10^{-4} - 3.\underline{5}8 \times 10^{-5} = 1.\underline{6}4 \times 10^{-4} = 1.6 \times 10^{-4} \text{ M}$$

b. Some hydrogen bonding can occur that results in a more stable system. The proposed structure of the dimer is

c. An increase in the temperature would facilitate bond breaking and would decrease the amount of dimer. We could also use a Le Chatelier-type argument.

15.105 The molar mass of Br_2 is 159.82 g/mol. Thus, the initial concentration of Br_2 is

$$\frac{18.22 \text{ g } Br_2 \times \dfrac{1 \text{ mol } Br_2}{159.82 \text{ g } Br_2}}{1.00 \text{ L}} = 0.11\underline{4}0 \text{ M}$$

Use the table approach, and give the starting, change, and equilibrium concentrations.

Conc. (M)	2NO(g)	+	Br_2(g)	⇌	2NOBr(g)
Starting	0.112		0.11$\underline{4}$0		0
Change	-2x		-x		2x
Equilibrium	0.112 - 2x		0.1140 - x		2x (= 0.0824 M)
	= 0.029$\underline{6}$0		= 0.07$\underline{2}$80		= 0.0824

Substituting into the equilibrium-constant expression gives

$$K_c = \frac{[NOBr]^2}{[NO]^2[Br_2]} = \frac{(0.0824)^2}{(0.02960)^2(0.07280)} = 1\underline{0}6.4 = 1.1 \times 10^2$$

■ Solutions to Cumulative-Skills Problems

15.107 For $Sb_2S_3(s) + 3H_2(g) \rightleftharpoons 2Sb(s) + 3H_2S(g)$ [$+3Pb^{2+} \rightarrow 3PbS(s) + 6H^+$]:

Starting M of H_2(g) = 0.0100 mol ÷ 2.50 L = 0.00400 M H_2

1.029 g PbS ÷ 239.26 g PbS/mol H_2S = 4.30$\underline{0}$7 x 10^{-3} mol H_2S

4.3007 x 10^{-3} mol H_2S ÷ 2.50 L = [1.7203 x 10^{-3}] = M of H_2S

Conc. (M)	3H_2(g)	+	Sb_2S_3(s)	⇌	3H_2S(g)	+	2Sb(s)
Starting	0.00400				0		
Change	-0.0017203				+0.0017203		
Equilibrium	0.0022797				0.0017203		

Substituting into the equilibrium expression for K_c gives

$$K_c = \frac{[H_2S]^3}{[H_2]^3} = \frac{(0.0017203)^3}{(0.0022797)^3} = 0.42\underline{9}7 = 0.430$$

15.109 For $PCl_5(g) \rightleftharpoons PCl_3(g) + Cl_2(g)$:

Starting M of PCl_5 = 0.0100 mol ÷ 2.00 L = 0.00500 M

Conc. (M)	$PCl_5(g)$	$\rightleftharpoons$	$PCl_3(g)$	+	$Cl_2(g)$
Starting	0.00500		0		0
Change	-x		+x		+x
Equilibrium	0.00500 - x		x		x

Substituting into the equilibrium expression for K_c gives

$$K_c = 4.15 \times 10^{-2} = \frac{[PCl_3][Cl_2]}{[PCl_5]} = \frac{(x)(x)}{(0.00500 - x)}$$

$x^2 + (4.15 \times 10^{-2})x - 2.075 \times 10^{-4} = 0$ (quadratic)

Solving the quadratic equation gives x = -4.60 x 10^{-2} (impossible) and x = 4.51 x 10^{-3} M (use).

Total M of gas = 0.00451 + 0.00451 + 0.00049 = 0.009510 M

P = (n/V)RT = (0.009510 M)[0.082057 L•atm/(K•mol)](523 K)

= 0.4081 = 0.408 atm

16. ACIDS AND BASES

■ Solutions to Exercises

Note on significant figures: If the final answer to a solution needs to be rounded off, it is given first with one nonsignificant figure, and the last significant figure is underlined. The final answer is then rounded to the correct number of significant figures. In multiple-step problems, intermediate answers are given with at least one nonsignificant figure; however, only the final answer has been rounded off.

16.1 See labels below reaction:

$$H_2CO_3(aq) + CN^-(aq) \rightleftharpoons HCN(aq) + HCO_3^-(aq)$$

acid   base   acid   base

H_2CO_3 is the proton donor (Bronsted-Lowry acid) on the left, and HCN is the proton donor (Bronsted-Lowry acid) on the right. The CN^- and HCO_3^- ions are proton acceptors (Bronsted-Lowry bases). HCN is the conjugate acid of CN^-.

16.2 Part (a) involves molecules with all single bonds; part (b) does not, so bonds are drawn in.

a.

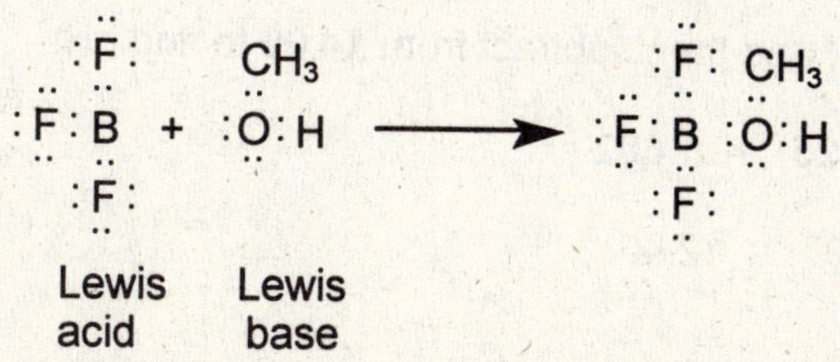

$$\ddot{:}\overset{\ddot{}}{F}\ddot{:} \quad CH_3$$
$$\ddot{:}\overset{\ddot{}}{F}\ddot{:}B + \ddot{:}\overset{\ddot{}}{O}H \longrightarrow \ddot{:}\overset{\ddot{}}{F}\ddot{:}B\ddot{:}\overset{\ddot{}}{O}H$$
$$\ddot{:}\overset{\ddot{}}{F}\ddot{:} \qquad \qquad \ddot{:}\overset{\ddot{}}{F}\ddot{:}$$

Lewis   Lewis
acid   base

(continued)

b.

$$: \overset{..}{\underset{..}{O}} :^{2-} \quad + \quad \overset{\overset{..}{O}}{\underset{\underset{..}{O}}{\overset{||}{C}}}^{} \quad \longrightarrow \quad \left[\overset{\overset{..}{O}:}{\underset{\underset{..}{O}:}{\overset{|}{:O-C}}} \right]^{2-}$$

Lewis   Lewis
base   acid

16.3   The $HC_2H_3O_2$ is a stronger acid than H_2S, and HS^- is a stronger base than the $C_2H_3O_2^-$ ion. The equilibrium favors the weaker acid and weaker base; therefore, the reactants are favored.

16.4 a. PH_3 b. HI   c. H_2SO_3 d. H_3AsO_4 e. HSO_4^-

16.5   A 0.125 M solution of $Ba(OH)_2$, a strong base, ionizes completely to yield 0.125 M Ba^{2+} ion and 2 x 0.125 M, or 0.250 M, OH^- ion. Use the K_w equation to calculate the $[H_3O^+]$.

$$[H_3O^+] = \frac{K_w}{[OH^-]} = \frac{1.0 \times 10^{-14}}{(0.250)} = 4.\underline{0}0 \times 10^{-14} = 4.0 \times 10^{-14} \text{ M}$$

16.6   Use the K_w equation to calculate the $[H_3O^+]$.

$$[H_3O^+] = \frac{K_w}{[OH^-]} = \frac{1.0 \times 10^{-14}}{1.0 \times 10^{-5}} = 1.\underline{0}0 \times 10^{-9} = 1.0 \times 10^{-9} \text{ M}$$

Since the $[H_3O^+]$ concentration is less than 1.0×10^{-7}, the solution is basic.

16.7   Calculate the negative log of the $[H_3O^+]$:

$$pH = -\log [H_3O^+] = -\log (0.045) = 1.3\underline{4}6 = 1.35$$

16.8   Calculate the pOH of 0.025 M OH^-, and then subtract from 14.00 to find pH:

$$pOH = -\log [OH^-] = -\log (0.025) = 1.6\underline{0}2$$

$$pH = 14.00 - 1.602 = 12.3\underline{9}7 = 12.40$$

16.9   Because pH = 3.16, by definition the log $[H_3O^+]$ = -3.16. Enter this on the calculator and convert to the antilog (number) of -3.16.

$$[H_3O^+] = \text{antilog} (-3.16) = 10^{-3.16} = 6.\underline{9}1 \times 10^{-4} = 6.9 \times 10^{-4} \text{ M}$$

16.10   Find the pOH by subtracting the pOH from 14.00. Then enter -3.40 on the calculator to convert to the antilog (number) corresponding to -3.40.

$$pOH = 14.00 - 10.6 = 3.\underline{4}0$$

$$[H_3O^+] = \text{antilog} (-3.40) = 10^{-3.40} = \underline{3}.98 \times 10^{-4} = 4 \times 10^{-4} \text{ M}$$

■ Answers to Review Questions

16.1   You can classify these acids using the information in Section 16.1. Also recall that all diatomic acids of Group VIIA halides are strong except for HF.

   a. Weak   b. Weak   c. Strong

   d. Strong   e. Weak   f. Weak

16.2   In Section 16.1, we are told that all neutralizations involving strong acids and bases evolve 55.90 kJ of heat per one mol H_3O^+. Thus, the thermochemical evidence for the Arrhenius concept is based on the fact that when one mole of any strong acid (one mol H_3O^+) is neutralized by one mole of any strong base (one mol OH^-), the heat of neutralization is always the same ($\Delta H° = -55.90$ kJ/mol).

16.3   A Bronsted-Lowry acid is a molecule or ion that donates an H_3O^+ ion (proton donor) to a base in a proton-transfer reaction. A Bronsted-Lowry base is a molecule or ion that accepts an H_3O^+ ion (proton acceptor) from an acid in a proton-transfer reaction. An example of an acid-base equation:

$$HF(aq) + NH_3(aq) \rightarrow NH_4^+(aq) + F^-(aq)$$
   acid   base   acid   base

16.4   The conjugate acid of a base is a species that differs from the base by only one H_3O^+. Consider the base, HSO_3^-. Its conjugate acid would be H_2SO_3 but not H_2SO_4. H_2SO_4 differs from HSO_3^- by one H and one O.

16.5   You can write the equations by considering that $H_2PO_3^-$ is both a Bronsted-Lowry acid and a Bronsted-Lowry base. The $H_2PO_3^-$ acts as a Bronsted-Lowry acid when it reacts with a base such as OH^- :

$$H_2PO_3^-(aq) \quad + \quad OH^-(aq) \quad \rightarrow \quad HPO_3^{2-}(aq) \quad + \quad H_2O(l)$$

The $H_2PO_3^-$ acts as a Bronsted-Lowry base when it reacts with an acid such as HF:

$$H_2PO_3^-(aq) \quad + \quad HCl(aq) \quad \rightarrow \quad H_3PO_3(aq) \quad + \quad Cl^-(aq)$$

16.6   The Bronsted-Lowry concept enlarges on the Arrhenius concept in the following ways: (1) It expands the concept of a base to include any species that accepts protons, not just the OH^- ion or compounds containing the OH^- ion. (2) It enlarges the concepts of acids and bases to include ions as well as molecules. (3) It enables us to write acid-base reactions in nonaqueous solutions as well as in aqueous solutions, whereas the Arrhenius concept applies only to aqueous solutions. (4) It allows some species to be considered as acids or bases depending on the other reactant with which they are mixed.

16.7   According to the Lewis concept, an acid is an electron-pair acceptor and a base is an electron-pair donor. An example is

$$Ag^+(aq) \quad + \quad 2(:NH_3) \quad \rightarrow \quad Ag(NH_3)_2^+(aq)$$
$$\quad\text{acid} \qquad\qquad\quad \text{base}$$

16.8   Recall that, the weaker the acid, the stronger it holds on to its proton(s). Thus, if a reaction mixture consists of a stronger acid and base and a weaker acid and base, the weaker acid side will always be favored because the proton(s) will bond more strongly to the weaker acid.

16.9   The two factors that determine the strength of an acid are (1) the polarity of the bond to which the H atom is attached, and (2) the strength of the bond, or how tightly the proton is held by the atom to which it is bonded. An increase in the polarity of the bond makes it easier to remove the proton, increasing the strength of the acid. An increase in the strength of the bond makes it more difficult to remove the proton, decreasing the strength of the acid. The strength of the bond depends in turn on the size of the atom, so larger atoms have weaker bonds, whereas smaller atoms have stronger bonds.

16.10   The self-ionization of water is the reaction of two water molecules in which a proton is transferred from one molecule to the other to form H_3O^+ and OH^- ions. At 25 °C, the K_w expression is $K_w = [H_3O^+][OH^-] = 1.0 \times 10^{-14}$.

16.11   The pH = $-\log [H_3O^+]$ of an aqueous solution. Measure pH by using electrodes and a pH meter, or by interpolating the pH from the color changes of a series of acid-base indicators.

16.12   A solution of pH 4 has a $[H_3O^+] = 1 \times 10^{-4}$ M and is more acidic than a solution of pH 5, which has a $[H_3O^+] = 1 \times 10^{-5}$ M.

16.13   For a neutral solution, $[H_3O^+] = [OH^-]$; thus, the $[H_3O^+]$ of a neutral solution at 37 °C is the square root of K_w at 37 °C:

$$[H_3O^+] = \sqrt{2.5 \times 10^{-14}} = 1.\underline{5}8 \times 10^{-7} \text{ M}$$

$$pH = -\log(1.58 \times 10^{-7}) = 6.8\underline{0}1 = 6.80$$

16.14   Because $pH + pOH = pK_w$ at any temperature,

$$pH + pOH = -\log(2.5 \times 10^{-14}) = 13.60$$

■ Solutions to Practice Problems

Note on significant figures: If the final answer to a solution needs to be rounded off, it is given first with one nonsignificant figure, and the last significant figure is underlined. The final answer is then rounded to the correct number of significant figures. In multiple-step problems, intermediate answers are given with at least one nonsignificant figure; however, only the final answer has been rounded off.

16.23   The reaction with labels of "acid" or "base" written below is as follows:

$$OH^-(aq) + HF(l) \rightleftharpoons F^-(aq) + H_2O(l)$$
    base   acid   base   acid

16.25 a. PO_4^{3-} b. HS^- c. NO_2^- d. $HAsO_4^{2-}$

16.27 a. $HClO$ b. AsH_4^+ c. H_3PO_4 d. $HTeO_3^-$

16.29   Each equation is given below with the labels for acid or base:

   a. $HSO_4^-(aq) + NH_3(aq) \rightleftharpoons SO_4^{2-}(aq) + NH_4^+(aq)$
      acid   base   base   acid

   The conjugate acid-base pairs are: HSO_4^-, SO_4^{2-}, and NH_4^+, NH_3.

(continued)

b. $HPO_4^{2-}(aq) + NH_4^+(aq) \rightleftharpoons H_2PO_4^-(aq) + NH_3(aq)$

    base   acid   acid   base

The conjugate acid-base pairs are: $H_2PO_4^-$, HPO_4^{2-}, and NH_4^+, NH_3.

c. $Al(H_2O)_6^{3+}(aq) + H_2O(l) \rightleftharpoons Al(H_2O)_5(OH)^{2+}(aq) + H_3O^+(aq)$

    acid   base   base   acid

The conjugate acid-base pairs are: $Al(H_2O)_6^{3+}$, $Al(H_2O)_5(OH)^{2+}$, and H_3O^+, H_2O.

d. $SO_3^{2-}(aq) + NH_4^+(aq) \rightleftharpoons HSO_3^-(aq) + NH_3(aq)$

    base   acid   acid   base

The conjugate acid-base pairs are: HSO_3^-, SO_3^{2-}, and NH_4^+, NH_3.

16.31   The reaction is

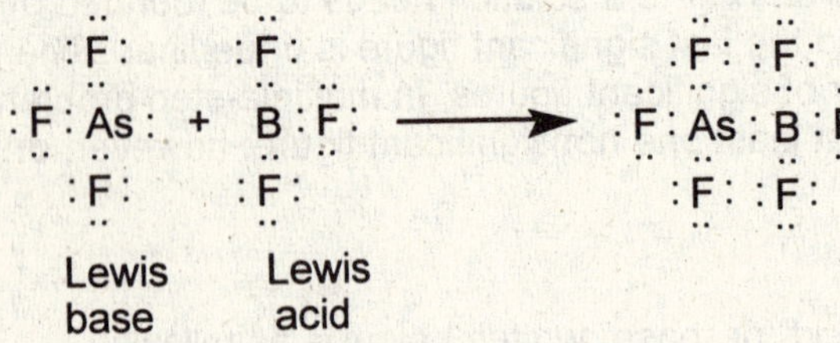

    Lewis   Lewis
    base   acid

16.33 a. The completed equation is

$$AlCl_3 + Cl^- \rightleftharpoons AlCl_4^-$$

The Lewis formula representation is

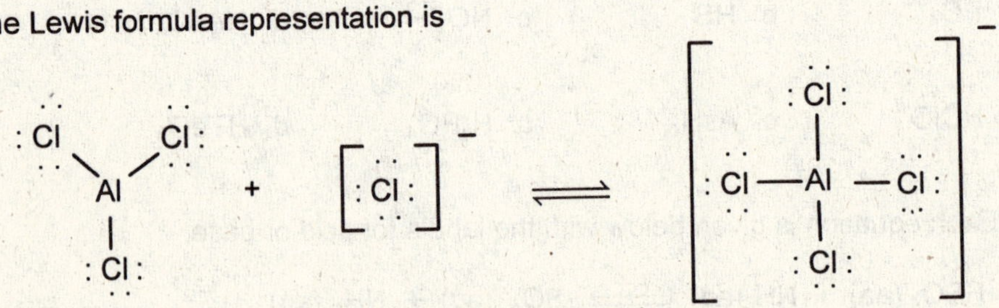

$AlCl_3$ is the electron-pair acceptor and is the acid. Cl^- is the electron pair donor and is the base.

                                                                       (continued)

b. The completed equation is

$$I^- + I_2 \rightleftharpoons I_3^-$$

The Lewis formula representation is

$$\left[:\overset{..}{\underset{..}{I}}:\right]^- \; + \; :\overset{..}{\underset{..}{I}} - \overset{..}{\underset{..}{I}}: \; \rightleftharpoons \; \left[:\overset{..}{\underset{..}{I}} - \overset{}{I} - \overset{..}{\underset{..}{I}}:\right]^-$$

I_2 is the electron-pair acceptor and is the acid. The I^- ion is the electron pair donor and is the base.

16.35 a. Each water molecule donates a pair of electrons to copper(II), making the water molecule a Lewis base and the Cu^{2+} ion a Lewis acid.

b. The AsH_3 donates a pair of electrons to the boron atom in BBr_3, making AsH_3 a Lewis base and the BBr_3 molecule a Lewis acid.

16.37 The equation is $H_2S + HOCH_2CH_2NH_2 \rightarrow HOCH_2CH_2NH_3^+ + HS^-$. The H_2S is a Lewis acid, and $HOCH_2CH_2NH_2$ is a Lewis base. The hydrogen ion from H_2S accepts a pair of electrons from the N atom in $HOCH_2CH_2NH_2$.

16.39 The reaction is $HSO_4^- + ClO^- \rightarrow HClO + SO_4^{2-}$. According to Table 16.2, $HClO$ is a weaker acid than HSO_4^-. Because the equilibrium for this type of reaction favors formation of the weaker acid (or weaker base), the reaction occurs to a significant extent.

16.41 a. NH_4^+ is a weaker acid than H_3PO_4, so the left-hand species are favored at equilibrium.

b. HCN is a weaker acid than H_2S, so the left-hand species are favored at equilibrium.

c. H_2O is a weaker acid than HCO_3^-, so the right-hand species are favored at equilibrium.

d. H_2O is a weaker acid than $Al(H_2O)_6^{3+}$, so the right-hand species are favored at equilibrium.

16.43 Trichloroacetic acid is the stronger acid because, in general, the equilibrium favors the formation of the weaker acid, which is formic acid in this case.

16.45 a. H_2S is stronger because acid strength decreases with increasing anion charge for polyprotic acid species.

b. H_2SO_3 is stronger because, for a series of oxoacids, acid strength increases with increasing electronegativity.

c. HBr is stronger because Br is more electronegative than Se. Within a period, acid strength increases as electronegativity increases.

d. HIO_4 is stronger because acid strength increases with the number of oxygen atoms bonded to the central atom.

e. H_2S is stronger because, within a group, acid strength increases with the increasing size of the central atom in binary acids.

16.47 a. $[H_3O^+] = 1.2$ M

$$[OH^-] = \frac{K_w}{[H_3O^+]} = \frac{1.0 \times 10^{-14}}{1.2} = 8.\underline{3}3 \times 10^{-15} = 8.3 \times 10^{-15} \text{ M}$$

b. $[OH^-] = 0.32$ M

$$[H_3O^+] = \frac{K_w}{[OH^-]} = \frac{1.0 \times 10^{-14}}{0.32} = 3.\underline{1}2 \times 10^{-14} = 3.1 \times 10^{-14} \text{ M}$$

c. $[OH^-] = 2 \times (0.085 \text{ M}) = 0.170$ M

$$[H_3O^+] = \frac{K_w}{[OH^-]} = \frac{1.0 \times 10^{-14}}{0.170} = 5.\underline{8}8 \times 10^{-14} = 5.9 \times 10^{-14} \text{ M}$$

d. $[H_3O^+] = 0.38$ M

$$[OH^-] = \frac{K_w}{[H_3O^+]} = \frac{1.0 \times 10^{-14}}{0.38} = 2.\underline{6}3 \times 10^{-14} = 2.6 \times 10^{-14} \text{ M}$$

16.49 The $[H_3O^+] = 0.050$ M (HCl is a strong acid); using K_w, the $[OH^-] = 2.0 \times 10^{-13}$ M.

16.51 Because the $Ba(OH)_2$ forms two OH^- per formula unit, the $[OH^-] = 2 \times 0.0085 = 0.017$ M.

$$[H_3O^+] = \frac{K_w}{[OH^-]} = \frac{1.0 \times 10^{-14}}{0.017} = 5.\underline{8}8 \times 10^{-13} = 5.9 \times 10^{-13} \text{ M}$$

16.53 a. 5×10^{-6} M H_3O^+ > 1.0×10^{-7}, so the solution is acidic.

b. Use K_w to determine $[H_3O^+]$

$$[H_3O^+] = \frac{K_w}{[OH^-]} = \frac{1.00 \times 10^{-14}}{5 \times 10^{-9}} = \underline{2}.0 \times 10^{-6} = 2 \times 10^{-6} \text{ M}$$

Since 2×10^{-6} M > 1.0×10^{-7}, the solution is acidic.

c. When $[OH^-]$ = 1.0×10^{-7} M, $[H_3O^+]$ = 1.0×10^{-7} M, and the solution is neutral.

d. 2×10^{-9} M H_3O^+ < 1.0×10^{-7}, so the solution is basic.

16.55   The $[H_3O^+]$ calculated below is > 1.0×10^{-7} M, so the solution is acidic.

$$[H_3O^+] = \frac{K_w}{[OH^-]} = \frac{1.00 \times 10^{-14}}{1.5 \times 10^{-9}} = 6.\underline{6}6 \times 10^{-6} = 6.7 \times 10^{-6} \text{ M}$$

16.57 a. pH 4.6, acidic solution   b. pH 7.0, neutral solution

c. pH 1.6, acidic solution   d. pH 10.5, basic solution

16.59 a. Acidic (3.5 < 7.0)   b. Neutral (7.0 = 7.0)

c. Basic (9.0 > 7.0)   d. Acidic (5.5 < 7.0)

16.61   Record the same number of places after the decimal point in the pH as the number of significant figures in the $[H_3O^+]$.

a. $-\log (1.0 \times 10^{-8})$ = 8.0$\underline{0}$0 = 8.00   b. $-\log (5.0 \times 10^{-12})$ = 11.3$\underline{0}$1 = 11.30

c. $-\log (7.5 \times 10^{-3})$ = 2.1$\underline{2}$4 = 2.12   d. $-\log (6.35 \times 10^{-9})$ = 8.19$\underline{7}$2 = 8.197

16.63   Record the same number of places after the decimal point in the pH as the number of significant figures in the $[H_3O^+]$.

$-\log (7.5 \times 10^{-3})$ = 2.1$\underline{2}$49 = 2.12

16.65 a. pOH = -log (5.25×10^{-9}) = 8.27$\underline{98}$; pH = 14.00 - 8.2798 = 5.7$\underline{2}$02 = 5.72

   b. pOH = -log (8.3×10^{-3}) = 2.0$\underline{8}$09; pH = 14.00 - 2.0809 = 11.9$\underline{1}$908 = 11.92

   c. pOH = -log (3.6×10^{-12}) = 11.44$\underline{36}$; pH = 14.00 - 11.4436 = 2.5$\underline{5}$63 = 2.56

   d. pOH = -log (2.1×10^{-8}) = 7.67$\underline{77}$; pH = 14.00 - 7.6777 = 6.3$\underline{2}$2 = 6.32

16.67 First, convert the $[OH^-]$ to $[H_3O^+]$ using the K_w equation. Then, find the pH, recording the same number of places after the decimal point in the pH as the number of significant figures in the $[H_3O^+]$.

$$[H_3O^+] = K_w \div [OH^-] = (1.0 \times 10^{-14}) \div (0.0040) = 2.\underline{5}0 \times 10^{-12} \text{ M}$$

$$pH = -log (2.50 \times 10^{-12}) = 11.6\underline{0}2 = 11.60$$

16.69 From the definition, pH = -log $[H_3O^+]$ and -pH = log $[H_3O^+]$, so enter the negative value of the pH on the calculator, and use the inverse and log keys (or 10^x) key to find the antilog of -pH:

$$log [H_3O^+] = -pH = -5.12$$

$$[H_3O^+] = \text{antilog} (-5.12) = 10^{-5.12} = 7.\underline{5}8 \times 10^{-6} = 7.6 \times 10^{-6} \text{ M}$$

16.71 From the definition, pH = -log $[H_3O^+]$ and -pH = log $[H_3O^+]$, so enter the negative value of the pH on the calculator, and use the inverse and log keys (or 10^x) key to find the antilog of -pH. Then, use the K_w equation to calculate $[OH^-]$ from $[H_3O^+]$.

$$log [H_3O^+] = -pH = -11.63$$

$$[H_3O^+] = \text{antilog} (-11.63) = 10^{-11.63} = 2.\underline{3}4 \times 10^{-12} \text{ M}$$

$$[OH^-] = K_w \div [H_3O^+] = (1.0 \times 10^{-14}) \div (2.34 \times 10^{-12}) = 4.\underline{2}7 \times 10^{-3}$$

$$= 4.3 \times 10^{-3} \text{ M}$$

16.73 First, calculate the molarity of the OH$^-$ ion from the mass of NaOH. Then, convert the [OH$^-$] to [H$_3$O$^+$] using the K$_w$ equation. Then, find the pH, recording the same number of places after the decimal point in the pH as the number of significant figures in the [H$_3$O$^+$].

$$\frac{5.80 \text{ g NaOH}}{1.00 \text{ L}} \times \frac{1 \text{ mol NaOH}}{40.01 \text{ g NaOH}} = \frac{0.1450 \text{ mol NaOH}}{1.00 \text{ L}} = 0.14\underline{5}0 \text{ M OH}^-$$

$$[H_3O^+] = K_w \div [OH^-] = (1.0 \times 10^{-14}) \div (0.1450) = 6.\underline{8}96 \times 10^{-14} \text{ M}$$

$$pH = -\log [H_3O^+] = -\log (6.896 \times 10^{-14}) = 13.1\underline{6}14 = 13.16$$

16.75 Figure 16.10 shows that the methyl-red indicator is yellow at pH values above about 5.5 (slightly past the midpoint of the range for methyl red). Bromthymol blue is yellow at pH values up to about 6.5 (slightly below the midpoint of the range for bromthymol blue). Therefore, the pH of the solution is between 5.5 and 6.5, and the solution is acidic.

■ Solutions to General Problems

16.77 a. BaO is a base; $BaO + H_2O \rightarrow Ba^{2+} + 2OH^-$

b. H$_2$S is an acid; $H_2S + H_2O \rightarrow H_3O^+ + HS^-$

c. CH$_3$NH$_2$ is a base; $CH_3NH_2 + H_2O \rightarrow CH_3NH_3^+ + OH^-$

d. SO$_2$ is an acid; $SO_2 + 2H_2O \rightarrow H_3O^+ + HSO_3^-$

16.79 a. $H_2O_2(aq) + S^{2-}(aq) \rightarrow HO_2^-(aq) + HS^-(aq)$

b. $HCO_3^-(aq) + OH^-(aq) \rightarrow CO_3^{2-}(aq) + H_2O(l)$

c. $NH_4^+(aq) + CN^-(aq) \rightarrow NH_3(aq) + HCN(aq)$

d. $H_2PO_4^-(aq) + OH^-(aq) \rightarrow HPO_4^{2-}(aq) + H_2O(l)$

16.81 a. The ClO⁻ ion is a Bronsted base, and water is a Bronsted acid. The complete chemical equation is $ClO^-(aq) + H_2O(l) \rightleftharpoons HClO(aq) + OH^-(aq)$. The equilibrium does not favor the products because ClO^- is a weaker base than OH^-. In Lewis language, a proton from H_2O acts as a Lewis acid by accepting a pair of electrons on the oxygen of ClO^-.

$$H^+ + \left[:\ddot{Cl}:\ddot{O}: \right]^- \longrightarrow :\ddot{Cl}:\ddot{O}:H$$

b. The NH_2^- ion is a Bronsted base, and NH_4^+ is a Bronsted acid. The complete chemical equation is $NH_4^+ + NH_2^- \rightleftharpoons 2NH_3$. The equilibrium favors the products because the reactants form the solvent, a weakly ionized molecule. In Lewis language, the proton from NH_4^+ acts as a Lewis acid by accepting a pair of electrons on the nitrogen of NH_2^-.

$$\left[\begin{array}{c} H \\ H:N:H \\ H \end{array} \right]^+ + \left[\begin{array}{c} :N:H \\ H \end{array} \right]^- \longrightarrow 2\ \begin{array}{c} H:N:H \\ H \end{array}$$

16.83 Table 16.2 shows that HNO_2 is a stronger acid than HF. Because an acid-base reaction normally goes in the direction of the weaker acid, the reaction is more likely to go in the direction written:

$$HNO_2 + F^- \rightleftharpoons HF + NO_2^-$$

16.85 The order is $H_2S < H_2Se < HBr$. H_2Se is stronger than H_2S because, within a group, acid strength increases with increasing size of the central atom in binary acids. HBr is a strong acid, whereas the others are weak acids.

16.87 The KOH is a strong base and is fully ionized in solution, so you can use its formula and molar concentration to determine the $[OH^-]$ of the solution. Therefore, the 0.25 M KOH contains 0.25 M OH^-. The $[H_3O^+]$ is obtained from the K_w expression:

$$K_w = 1.0 \times 10^{-14} = [H_3O^+] \times 0.25\ M\ OH^-$$

$$[H_3O^+] = \frac{1.0 \times 10^{-14}}{0.25} = 4.00 \times 10^{-14} = 4.0 \times 10^{-14}\ M$$

16.89 Enter the H_3O^+ concentration of 1.5×10^{-3} into the calculator, press the log key, and press the sign key to change the negative log to a positive log. This follows the negative log definition of pH. The number of decimal places of the pH should equal the significant figures in the H_3O^+.

$$pH = -\log [H_3O^+] = -\log (1.5 \times 10^{-3}) = 2.8\underline{2}3 = 2.82$$

16.91 Find the pOH from the pH using $pH + pOH = 14.00$. Then, calculate the $[OH^-]$ from the pOH by entering the pOH into the calculator, pressing the sign key to change the positive log to a negative log, and finding the antilog. On some calculators, the antilog is found by using the inverse of the log; on other calculators, the antilog is found using the 10^x key. The number of significant figures in the $[OH^-]$ should equal the number of decimal places in the pOH.

$$pOH = 14.00 - 3.15 = 10.85$$

$$[OH^-] = \text{antilog} (-10.85) = 10^{-10.85} = 1.\underline{4}2 \times 10^{-11} = 1.4 \times 10^{-11} \text{ M}$$

16.93 a. $H_2SO_4(aq) + 2NaHCO_3(aq) \rightarrow Na_2SO_4(aq) + 2CO_2(g) + 2H_2O(l)$ (molecular)

    $H_3O^+(aq) + HCO_3^-(aq) \rightarrow CO_2(g) + 2H_2O(l)$ (net ionic)

b. The total moles of H_3O^+ from the H_2SO_4 is

$$\text{mol } H_3O^+ = \frac{0.437 \text{ mol } H_2SO_4}{1 \text{ L}} \times 0.02500 \text{ L} \times \frac{2 \text{ mol } H_3O^+}{1 \text{ mol } H_2SO_4}$$

$$= 0.021\underline{8}5 \text{ mol } H_3O^+$$

The moles of H_3O^+ that reacted with the NaOH are given by

$$\text{mol } H_3O^+ = \frac{0.108 \text{ mol NaOH}}{1 \text{ L}} \times 0.0287 \text{ L} = 0.0031\underline{0}0 \text{ mol}$$

The moles of $NaHCO_3$ present in the original sample are equal to the moles of H_3O^+ that reacted with the HCO_3^-, which is given by

Total moles H_3O^+ - moles H_3O^+ reacted with the NaOH = moles HCO_3^-

$0.021\underline{8}5$ mol - 0.003823 mol = $0.018\underline{0}3$ = 0.0180 mol $NaHCO_3$

(continued)

c. The mass of $NaHCO_3$ (molar mass 84.01 g/mol) present in the original sample is

$$0.01803 \text{ mol NaHCO}_3 \times \frac{84.01 \text{ g NaHCO}_3}{1 \text{ mol NaHCO}_3} = 1.5\underline{1}4 \text{ g}$$

Thus, the percent $NaHCO_3$ in the original sample is given by

$$\text{Percent NaHCO}_3 = \frac{1.514 \text{ g}}{2.500 \text{ g}} \times 100\% = 60.\underline{5}6 = 60.6 \text{ percent}$$

The percent KCl in the original sample is

$$\text{Percent KCl} = 100 - 60.56 = 39.\underline{4}4 = 39.4 \text{ percent}$$

16.95 $HCO_3^-(aq) + H_2O(l) \rightleftharpoons H_3O^+(aq) + CO_3^{2-}(aq)$

$HCO_3^-(aq) + H_2O(l) \rightleftharpoons H_2CO_3(aq) + OH^-(aq)$

$HCO_3^-(aq) + Na^+(aq) + OH^-(aq) \rightleftharpoons Na^+(aq) + CO_3^{2-}(aq) + H_2O(l)$

$HCO_3^-(aq) + H^+(aq) + Cl^-(aq) \rightleftharpoons H_2O(l) + CO_2(g) + Cl^-(aq)$

16.97 $CaH_2(s) + 2H_2O(l) \rightarrow Ca(OH)_2(s) + H_2(g)$

The hydride ion is a stronger base because it took an H^+ from water, leaving the OH^- ion. Every time a strong base is added to water, it will react with the water leaving the OH^- as the product, so a strong base cannot exist in water.

16.99 a. $2HF(l) \rightleftharpoons H_2F^+ + F^-$

b. NaF will be a base because F^- is a conjugate base of HF.

c. $HClO_4 + HF \rightarrow H_2F^+ + ClO_4^-$

The conjugate acid is H_2F^+.

16.101 The reaction of ammonia with water is given by

$$NH_3(aq) + H_2O(l) \rightleftharpoons NH_4^+(aq) + OH^-(aq)$$

The initial concentration of NH_3 (molar mass 17.03 g/mol) is

$$\text{Molarity} = \frac{4.25 \text{ g } NH_3 \times \dfrac{1 \text{ mol } NH_3}{17.03 \text{ g } NH_3}}{0.2500 \text{ L}} = 0.99\underline{8}2 \text{ M}$$

Since the NH_3 is 0.42 percent reacted, the concentration of OH^- is

$[OH^-] = 0.9982 \text{ M} \times 0.0042 = 0.004\underline{1}9 \text{ M}$

$pOH = -\log [OH^-] = -\log (0.00419) = 2.3\underline{7}8$

$pH = 14 - pOH = 14 - 2.3\underline{7}8 = 11.6\underline{2}2 = 11.62$

■ Solutions to Cumulative-Skills Problems

16.103 For $(HO)_mYO_n$ acids, acid strength increases with n regardless of the number of OH's. The structure of H_3PO_4 is $(HO)_3PO$; because H_3PO_3 and H_3PO_4 have about the same acidity, H_3PO_3 must also have n = 1; thus, m = 2. This leaves one H, which must bond to phosphorus, giving a structure of $(HO)_2(O)PH$. Assuming that only two H's react with NaOH, the mass of NaOH that reacts with 1.00 g of H_3PO_3 (PA) is calculated as follows:

$$1.00 \text{ g PA} \times \frac{1 \text{ mol PA}}{81.994 \text{ g PA}} \times \frac{2 \text{ mol NaOH}}{1 \text{ mol PA}} \times \frac{40.00 \text{ g NaOH}}{1 \text{ mol NaOH}}$$

$$= 0.97\underline{5}6 = 0.976 \text{ g NaOH}$$

16.105 BF_3 acts as a Lewis acid, accepting an electron pair from NH_3:

$$BF_3 + :NH_3 \rightarrow F_3B:NH_3$$

The NH_3 acts as a Lewis base in donating an electron pair to BF_3. When 10.0 g of each are mixed, the BF_3 is the limiting reagent because it has the higher formula weight. The mass of $BF_3:NH_3$ formed is

$$10.0 \text{ g } BF_3 \times \frac{1 \text{ mol } BF_3}{67.81 \text{ g } BF_3} \times \frac{1 \text{ mol } BF_3:NH_3}{1 \text{ mol } BF_3} \times \frac{84.84 \text{ g } BF_3:NH_3}{1 \text{ mol } BF_3:NH_3}$$

$$= 12.\underline{5}1 = 12.5 \text{ g } BF_3:NH_3$$

17. ACID-BASE EQUILIBRIA

■ Solutions to Exercises

Note on significant figures: If the final answer to a solution needs to be rounded off, it is given first with one nonsignificant figure, and the last significant figure is underlined. The final answer is then rounded to the correct number of significant figures. In multiple-step problems, intermediate answers are given with at least one nonsignificant figure; however, only the final answer has been rounded off.

17.1   Abbreviate the formula of lactic acid as HL. To solve, assemble a table of starting, change, and equilibrium concentrations.

Conc. (M)	HL + H₂O	⇌	H₃O⁺	+ L⁻
Starting	0.025		0	0
Change	-x		+x	+x
Equilibrium	0.025 - x		x	x

Substituting into the equilibrium-constant equation gives

$$K_a = \frac{[H_3O^+][L^-]}{[HL]} = \frac{(x)^2}{(0.025 - x)}$$

The value of x equals the value of the molarity of the H_3O^+ ion, which can be obtained from the pH:

$$[H_3O^+] = \text{antilog}(-pH) = \text{antilog}(-2.75) = 0.001\underline{7}8 \text{ M}$$

(continued)

Substitute this value for x into the equation to get

$$K_a = \frac{(x)^2}{(0.025 - x)} = \frac{(0.00178)^2}{(0.025 - 0.00178)} = 1.\underline{3}6 \times 10^{-4} = 1.4 \times 10^{-4}$$

The degree of ionization is

$$\text{Degree of ionization} = \frac{0.00178}{0.025} = 0.071$$

17.2   To solve, assemble a table of starting, change, and equilibrium concentrations. Use HAc as the symbol for acetic acid.

Conc. (M)	HAc + H$_2$O ⇌	H$_3$O$^+$ +	Ac$^-$
Starting	0.10	0	0
Change	-x	+x	+x
Equilibrium	0.10 - x	x	x

Now, substitute these concentrations and the value of K$_a$ into the equilibrium-constant equation for acid ionization:

$$K_a = \frac{[H_3O^+][Ac^-]}{[HAc]} = \frac{(x)^2}{(0.10 - x)} = 1.7 \times 10^{-5}$$

Solve the equation for x, assuming x is much smaller than 0.10, so (0.10 - x) ≅ 0.10.

$$\frac{(x)^2}{(0.10)} \cong 1.7 \times 10^{-5}$$

$$x^2 = 1.7 \times 10^{-5} \times 0.10 = 1.7 \times 10^{-6}$$

$$x = 0.001\underline{3}0 \text{ M}$$

Check to make sure the assumption that (0.10 - x) ≅ 0.10 is valid:

0.10 - 0.001\underline{3}0 = 0.0\underline{9}87, = 0.10   (to two significant figures)

The concentrations of hydronium ion and acetate ion are

$$[H_3O^+] = [Ac^-] = x = 0.0013 = 1.3 \times 10^{-3} \text{ M}$$

(continued)

The pH of the solution is

$$pH = -\log [H_3O^+] = -\log (0.001\underline{3}0) = 2.8\underline{8}4 = 2.88$$

The degree of ionization is

$$\text{Degree of ionization} = \frac{0.00130}{0.10} = 0.01\underline{3}0 = 0.013$$

17.3   Abbreviate the formula for pyruvic acid as HPy. To solve, assemble a table of starting, change, and equilibrium concentrations:

Conc. (M)	HPy + H₂O ⇌	H₃O⁺ +	Py⁻
Starting	0.0030	0	0
Change	-x	+x	+x
Equilibrium	0.0030 - x	x	x

Substitute the equilibrium concentrations and the value of K_a into the equilibrium-constant equation to get

$$K_a = \frac{[H_3O^+][Py^-]}{[HPy]} = \frac{(x)^2}{(0.0030 - x)} = 1.4 \times 10^{-4}$$

Note that the concentration of acid divided by K_a is $0.0030/1.4 \times 10^{-4} = 21$, which is considerably smaller than 100. Thus, you can expect that x cannot be ignored compared with 0.0030. The quadratic formula must be used. Rearrange the preceding equation to put it into the form $ax^2 + bx + c = 0$.

$$x^2 + 1.4 \times 10^{-4} x - 4.20 \times 10^{-7} = 0$$

Substitute into the quadratic formula to get

$$x = \frac{-1.4 \times 10^{-4} \pm \sqrt{(1.4 \times 10^{-4})^2 + 4(4.20 \times 10^{-7})}}{2}$$

$$= \frac{-1.4 \times 10^{-4} \pm 1.303 \times 10^{-3}}{2}$$

Using the positive root: $x = [H_3O^+] = 5.\underline{8}1 \times 10^{-4}$ M. Now you can calculate the pH.

$$pH = -\log [H_3O^+] = -\log (5.82 \times 10^{-4}) = 3.2\underline{3}50 = 3.24$$

17.4 To solve, note that $K_{a1} = 1.3 \times 10^{-2} > K_{a2} = 6.3 \times 10^{-8}$, and hence the second ionization and K_{a2} can be ignored. Assemble a table of starting, change, and equilibrium concentrations:

Conc. (M)	$H_2SO_3 + H_2O \rightleftharpoons$	H_3O^+	HSO_3^-
Starting	0.25	0	0
Change	-x	+x	+x
Equilibrium	0.25 - x	x	x

Substitute into the equilibrium-constant equation for the first ionization.

$$K_{a1} = \frac{[H_3O^+][HSO_3^-]}{[H_2SO_3]} = \frac{(x)^2}{(0.25 - x)} = 1.3 \times 10^{-2} = 0.013$$

This gives $x^2 + 0.013x - 0.00325 = 0$

Note that the concentration of acid divided by K_a is 0.25/0.013 = 19, which is considerably smaller than 100. Thus, you can expect that x cannot be ignored compared with 0.25. Reorganize the above equilibrium-constant expression into the form $ax^2 + bx + c = 0$, and substitute for a, b, and c in the quadratic formula.

$$x = \frac{-0.013 \pm \sqrt{(0.013)^2 + 4(0.00325)}}{2}$$

$$= \frac{-0.013 \pm 0.1147}{2}$$

Using the positive root, $x = [H_3O^+] = 0.05087$ M.

$$pH = -\log(0.05087) = 1.293$$

To calculate $[SO_3^{2-}]$, which will be represented by y, use the second ionization. Assume the starting concentrations of H_3O^+ and HSO_3^- are those from the first equilibrium.

Conc. (M)	$HSO_3^- + H_2O \rightleftharpoons$	H_3O^+	SO_3^{2-}
Starting	0.05087	0	0
Change	-y	+y	+y
Equilibrium	0.0508 - y	0.0508 + y	y

(continued)

Now, substituting into the K_{a2} expression for the second ionization.

$$K_{a2} = \frac{[H_3O^+][SO_3^{2-}]}{[HSO_3^-]} = \frac{(0.0508 + y)(y)}{(0.0508 - y)} = 6.3 \times 10^{-8}$$

Assuming y is much smaller than 0.0508, note that the (0.0508 + y) cancels the (0.0508 - y) term leaving $y \cong K_{a2}$, or

$$y = [SO_3^{2-}] \cong 6.3 \times 10^{-8} \text{ M}$$

17.5   First, convert the pH to pOH, and then to [OH⁻]:

$$pOH = 14.00 - pH = 14.00 - 9.84 = 4.16$$

$$[OH^-] = \text{antilog } (-4.16) = 6.\underline{9}2 \times 10^{-5} \text{ M}$$

Using the symbol Qu for quinine, assemble a table of starting, change, and equilibrium concentrations.

Conc. (M)	Qu + H₂O ⇌	HQu⁺ +	OH⁻
Starting	0.0015	0	0
Change	-x	+x	+x
Equilibrium	0.0015 - x	x	x

Note $x = 6.\underline{9}2 \times 10^{-5}$. Substitute into the equilibrium-constant equation to get

$$K_b = \frac{[HQ^+][OH^-]}{[Qu]} = \frac{(x)^2}{(0.0015 - x)} = \frac{(6.92 \times 10^{-5})^2}{(0.0015 - 6.92 \times 10^{-5})}$$

$$= 3.\underline{3}46 \times 10^{-6} = 3.3 \times 10^{-6}$$

17.6   Assemble a table of starting, change, and equilibrium concentrations:

Conc. (M)	NH₃ + H₂O ⇌	NH₄⁺ +	OH⁻
Starting	0.20	0	0
Change	-x	+x	+x
Equilibrium	0.20 - x	x	x

(continued)

Assume x is small enough to ignore compared with 0.20. Substitute into the equilibrium-constant equation to get :

$$K_b = \frac{[NH_4^+][OH^-]}{[NH_3]} = \frac{(x)^2}{(0.20 - x)} \cong \frac{(x)^2}{(0.20)} = 1.8 \times 10^{-5}$$

Solving for x gives

$$x^2 = (0.20) \times 1.8 \times 10^{-5} = 3.6 \times 10^{-6}$$

$$x = [OH^-] \cong 1.\underline{8}9 \times 10^{-3} \, M \quad \text{(Note x is negligible compared to 0.20.)}$$

Now calculate the hydronium ion concentration.

$$[H_3O^+] = \frac{K_w}{[OH^-]} = \frac{1.0 \times 10^{-14}}{1.89 \times 10^{-3}} = 5.\underline{2}9 \times 10^{-12} = 5.3 \times 10^{-12} \, M$$

17.7 a. Acidic. NH_4NO_3 is the salt of a weak base (NH_4OH) and a strong acid (HNO_3), so a solution of NH_4NO_3 is acidic because of the hydrolysis of NH_4^+.

   b. Neutral. KNO_3 is the salt of a strong base (KOH) and a strong acid (HNO_3), so a solution of NH_4NO_3 is neutral because none of the ions hydrolyze.

   c. Acidic. $Al(NO_3)_3$ is the salt of a weak base [$Al(OH)_3$] and a strong acid (HNO_3), so a solution of $Al(NO_3)_3$ is acidic because of the hydrolysis of Al^{3+}.

17.8 a. Calculate K_b of the F^- ion from the K_a of its conjugate acid, HF:

$$K_b = \frac{K_w}{K_a} = \frac{1.0 \times 10^{-14}}{6.8 \times 10^{-4}} = 1.\underline{4}7 \times 10^{-11} = 1.5 \times 10^{-11}$$

   b. Calculate K_a of $C_6H_5NH_3^+$ from K_b of its conjugate base, $C_6H_5NH_2$:

$$K_a = \frac{K_w}{K_b} = \frac{1.0 \times 10^{-14}}{4.2 \times 10^{-10}} = 2.\underline{3}8 \times 10^{-5} = 2.4 \times 10^{-5}$$

17.9   Assemble the usual table, writing HBen for benzoic acid and Ben⁻ for the benzoate ion. Let 0.015 - x equal the equilibrium concentration of the benzoate anion.

Conc. (M)	Ben⁻ + H₂O	⇌	HBen	+ OH⁻
Starting	0.015		0	0
Change	-x		+x	+x
Equilibrium	0.015 - x		x	x

Calculate K_b for the Benzoate ion from K_a for HBen.

$$K_b = \frac{K_w}{K_a} = \frac{1.0 \times 10^{-14}}{6.3 \times 10^{-5}} = 1.\underline{5}8 \times 10^{-10}$$

Substitute into the equilibrium-constant equation. Assume x is much smaller than 0.015.

$$K_b = \frac{[HBen][OH^-]}{[Ben^-]} = \frac{(x)^2}{(0.015 - x)} \cong \frac{(x)^2}{(0.015)} = 1.\underline{5}8 \times 10^{-10}$$

$$x = [HBen] = [OH^-] \cong 1.\underline{5}39 \times 10^{-6} \text{ M} \quad \text{(x is negligible compared to 0.015.)}$$

Thus, the concentration of benzoic acid in the solution is 1.5×10^{-6} M. The pH is

$$pOH = -\log[OH^-] = -\log(1.539 \times 10^{-6}) = 5.8\underline{1}2$$

$$pH = 14.00 - 5.812 = 8.1\underline{8}8 = 8.19$$

17.10   Assemble the usual table, using starting $[H_3O^+] = 0.20$ M from 0.20 M HCl and letting HFo symbolize $HCHO_2$.

Conc. (M)	HFo + H₂O	⇌	H₃O⁺	+ Fo⁻
Starting	0.10		0.20	0
Change	-x		+x	+x
Equilibrium	0.10 - x		0.20 + x	x

(continued)

Assume x is negligible compared to 0.10 M and 0.20 M, and substitute into the equilibrium-constant equation to get

$$K_a = \frac{[H_3O^+][Fo^-]}{[HFo]} = \frac{(0.20 + x)(x)}{(0.10 - x)} \cong \frac{(0.20)(x)}{(0.10)} = 1.7 \times 10^{-4}$$

$$x = [Fo^-] = 8.\underline{5}0 \times 10^{-5} = 8.5 \times 10^{-5} M$$

The degree of ionization is

$$\text{Degree of ionization} = \frac{8.50 \times 10^{-5}}{0.10} = 8.\underline{5}0 \times 10^{-4} = 8.5 \times 10^{-4}$$

17.11   Assemble the usual table, using a starting $[CHO_2^-]$ of 0.018 M from 0.018 M $NaCHO_2$ and symbolizing $HCHO_2$ as HFo and the CHO_2^- anion as Fo^-.

Conc. (M)	HFo + H₂O	⇌	H₃O⁺	+ Fo⁻
Starting	0.025		0	0.018
Change	-x		+x	+x
Equilibrium	0.025 - x		x	0.018 + x

Substitute into the equilibrium-constant equation. Assume x is negligible compared to 0.025 M and 0.018 M.

$$K_a = \frac{[H_3O^+][Fo^-]}{[HFo]} = \frac{(0.018 + x)(x)}{(0.025 - x)} \cong \frac{(0.018)(x)}{(0.025)} = 1.7 \times 10^{-4}$$

$$x = [H_3O^+] \cong 2.\underline{3}6 \times 10^{-4} M$$

The pH can now be calculated.

$$pH = -\log [H_3O^+] = -\log (2.36 \times 10^{-4}) = 3.6\underline{2}7 = 3.63$$

17.12   Let HOAc represent $HC_2H_3O_2$ and OAc^- represent $C_2H_3O_2^-$. The total volume of the buffer is

$$\text{Total volume} = 30.0 \text{ mL} + 70.0 \text{ mL} = 100.0 \text{ mL} = 0.1000 \text{ L}$$

The moles of HOAc and OAc^- in the buffer are

$$\text{mol HOAc} = 0.15 \text{ M} \times 0.0300 \text{ L} = 0.00450 \text{ mol}$$

$$\text{mol OAc}^- = 0.0700 \text{ M} \times 0.0700 \text{ L} = 0.0140 \text{ mol}$$

(continued)

The concentrations of HOAc and OAc⁻ in the buffer are

$$[HOAc] = \frac{0.00450 \text{ mol}}{0.1000 \text{ L}} = 0.0450 \text{ M}$$

$$[OAc^-] = \frac{0.0140 \text{ mol}}{0.1000 \text{ L}} = 0.140 \text{ M}$$

Now, substitute these starting concentrations into a table.

Conc. (M)	HOAc + H₂O	⇌	H₃O⁺ +	OAc⁻
Starting	0.0450		0	0.140
Change	-x		+x	+x
Equilibrium	0.0450 - x		x	0.140 + x

Substitute the equilibrium concentrations into the equilibrium-constant expression; then assume x is negligible compared to the starting concentrations of both HOAc and OAc⁻.

$$K_a = \frac{[H_3O^+][OAc^-]}{[HOAc]} = \frac{(0.140 + x)(x)}{(0.0450 - x)} \cong \frac{(0.140)(x)}{(0.0450)} = 1.7 \times 10^{-5}$$

$$x = [H_3O^+] \cong 5.46 \times 10^{-6} \text{ M}$$

You can now calculate the pH.

$$pH = -\log [H_3O^+] = -\log (5.46 \times 10^{-6}) = 5.262 = 5.26$$

17.13   First, do the stoichiometric calculation. From Exercise 17.11, [HFo] = 0.025 M and [Fo⁻] = 0.018 M. In one liter of buffer, there are 0.025 mol HFo and 0.018 mol Fo⁻. The moles of OH⁻ (equal to moles NaOH) added are

$$(0.10 \text{ M}) \times 0.0500 \text{ L} = 0.00500 \text{ mol OH}^-$$

The total volume of solution is

$$\text{Total volume} = 1 \text{ L} + 0.0500 \text{ L} = 1.0500 \text{ L}$$

After reaction with the OH⁻, the moles of HFo and Fo⁻ remaining in the solution are:

$$\text{Mol HFo} = (0.025 - 0.00500) \text{ mol} = 0.0200 \text{ mol}$$

$$\text{Mol Fo}^- = (0.018 + 0.00500) \text{ mol} = 0.0230 \text{ mol}$$

(continued)

The concentrations are

$$[HFo] = \frac{0.0200 \text{ mol}}{1.0500 \text{ L}} = 0.0190 \text{ M}$$

$$[Fo^-] = \frac{0.0230 \text{ mol}}{1.0500 \text{ L}} = 0.0219 \text{ M}$$

Now account for the ionization of HFo to Fo⁻ at equilibrium by assembling the usual table.

Conc. (M)	HFo + H₂O ⇌	H₃O⁺ + Fo⁻	
Starting	0.0190	0	0.0219
Change	-x	+x	+x
Equilibrium	0.0190 - x	x	0.0219 + x

Assume x is negligible compared to 0.0190 M and 0.0219 M, and substitute into the equilibrium-constant equation to get

$$K_a = \frac{[H_3O^+][Fo^-]}{[HFo]} = \frac{(0.0219 + x)(x)}{(0.0190 - x)} \cong \frac{(0.0219)(x)}{(0.0190)} = 1.7 \times 10^{-4}$$

$$x = [H_3O^+] = 1.47 \times 10^{-4} \text{ M}$$

Now, calculate the pH.

$$pH = -\log[H_3O^+] = -\log(1.47 \times 10^{-4}) = 3.831 = 3.83$$

17.14 All the OH⁻ reacts with the H₃O⁺ from HCl. Calculate the stoichiometric amounts of OH⁻ and H₃O⁺.

Mol H₃O⁺ = (0.10 mol/L) x 0.025 L = 0.0025 mol

Mol OH⁻ = (0.10 mol NaOH/L) x 0.015 L = 0.0015 mol

The total volume of solution is

Total volume = 0.025 L + 0.015 L = 0.040 L

(continued)

Subtract the mol of OH^- from the mol of H_3O^+, and divide by the total volume to get the concentration of H_3O^+.

$$\text{Mol } H_3O^+ \text{ left} = (0.0025 - 0.0015) \text{ mol} = 0.0010 \text{ mol}$$

$$[H_3O^+] = \frac{0.0010 \text{ mol}}{0.040 \text{ L}} = 0.0250 \text{ M}$$

Now calculate the pH.

$$pH = -\log [H_3O^+] = -\log (0.0250) = 1.602 = 1.60$$

17.15   At the equivalence point, the solution will contain NaF. The molar amount of F^- is equal to the molar amount of HF and is calculated as follows.

$$(0.10 \text{ mol HF/L}) \times 0.025 \text{ L} = 0.0025 \text{ mol } F^-$$

The volume of 0.15 M NaOH added and the total volume of solution are calculated next.

$$\text{Volume NaOH} = \frac{M_{acid}V_{acid}}{M_{base}} = \frac{(0.10 \text{ M})(25 \text{ mL})}{0.15 \text{ M}} = 16.6 \text{ mL}$$

$$\text{Total volume} = 25 \text{ mL} + 16.6 \text{ mL} = 41.6 \text{ mL} = 0.0416 \text{ L}$$

The concentration of F^- at the equivalence point can now be calculated.

$$[F^-] = \frac{0.0025 \text{ mol}}{0.0416 \text{ L}} = 0.0600 \text{ M}$$

Next, consider the hydrolysis of F^-. Start by calculating the hydrolysis constant of F^- from the K_a of its conjugate acid, HF.

$$K_b = \frac{K_w}{K_a} = \frac{1.0 \times 10^{-14}}{6.8 \times 10^{-4}} = 1.47 \times 10^{-11}$$

Now, assemble the usual table of concentrations, assume x is negligible compared to 0.0600, and calculate $[OH^-]$.

Conc. (M)	F^- + H_2O $\rightleftharpoons$	HF	+ OH^-
Starting	0.0600	0	0
Change	-x	+x	+x
Equilibrium	0.0600 - x	x	x

(continued)

Substitute into the equilibrium-constant equation to get

$$K_b = \frac{[HF][OH^-]}{[F^-]} = \frac{(x)^2}{(0.0600 - x)} \cong \frac{(x)^2}{(0.0600)} = 1.\underline{4}7 \times 10^{-11}$$

$$x = [OH^-] = 9.\underline{3}9 \times 10^{-7} \, M$$

Finally, calculate the pOH and then the pH.

$$pOH = -\log [OH^-] = -\log (9.\underline{3}9 \times 10^{-7}) = 6.0\underline{2}6$$

$$pH = 14.00 - 6.0\underline{2}6 = 7.9\underline{7}4 = 7.97$$

17.16 At the equivalence point, the solution will contain NH_4Cl. The molar amount of NH_4^+ is equal to the molar amount of NH_3 and is calculated as follows.

$$(0.20 \text{ mol } NH_3 /L) \times 0.035 \, L = 0.007\underline{0}0 \text{ mol } NH_4^+$$

The volume of 0.12 M HCl added and the total volume of solution are calculated next.

$$\text{Volume HCl} = \frac{M_{base}V_{base}}{M_{acid}} = \frac{(0.20 \, M)(35 \, mL)}{0.12 \, M} = 5\underline{8}.3 \, mL$$

$$\text{Total volume} = 35 \, mL + 58.3 \, mL = 9\underline{3}.3 \, mL = 0.09\underline{3}3 \, L$$

The concentration of NH_4^+ at the equivalence point can now be calculated.

$$[NH_4^+] = \frac{0.00700 \text{ mol}}{0.0933 \, L} = 0.07\underline{5}0 \, M$$

Next, consider the hydrolysis of NH_4^+. Start by calculating the hydrolysis constant of NH_4^+ from the K_b of its conjugate base, NH_3.

$$K_a = \frac{K_w}{K_b} = \frac{1.0 \times 10^{-14}}{1.8 \times 10^{-5}} = 5.\underline{5}6 \times 10^{-10}$$

Now, assemble the usual table of concentrations, assume x is negligible compared to 0.0750, and calculate $[H_3O^+]$.

Conc. (M)	NH_4^+ + H_2O	$\rightleftharpoons$	NH_3 + H_3O^+	
Starting	0.0750		0	0
Change	-x		+x	+x
Equilibrium	0.0750 - x		x	x

(continued)

Substitute into the equilibrium-constant equation to get

$$K_a = \frac{[NH_3][H_3O^+]}{[NH_4^+]} = \frac{(x)^2}{(0.0750 - x)} \cong \frac{(x)^2}{(0.0750)} = 5.\underline{5}6 \times 10^{-10}$$

$$x = [H_3O^+] = 6.\underline{4}6 \times 10^{-6} \text{ M}$$

Finally, calculate the pH.

$$pH = -\log [H_3O^+] = -\log (6.46 \times 10^{-6}) = 5.1\underline{8}9 = 5.19$$

■ Answers to Review Questions

17.1 The equation is

$$HCN(aq) + H_2O(l) \rightleftharpoons H_3O^+(aq) + CN^-(aq)$$

The equilibrium-constant expression is

$$K_a = \frac{[H_3O^+][CN^-]}{[HCN]}$$

17.2 HCN is the weakest acid. Its K_a of 4.9×10^{-10} is < K_a of 1.7×10^{-5} of $HC_2H_3O_2$; $HClO_4$ is a strong acid, of course.

17.3 Both methods involve direct measurement of the concentrations of the hydronium ion and the anion of the weak acid and calculation of the concentration of the un-ionized acid. All concentrations are substituted into the K_a expression to obtain a value for K_a. In the first method, the electrical conductivity of a solution of the weak acid is measured. The conductivity is proportional to the concentration of the hydronium ion and anion. In the second method, the pH of a known starting concentration of weak acid is measured. The pH is converted to $[H_3O^+]$, which will be equal to the [anion].

17.4 The degree of ionization of a weak acid decreases as the concentration of the acid added to the solution increases. Compared to low concentrations, at high concentrations, there is less water for each weak acid molecule to react with as the weak acid ionizes:

$$HA(aq) + H_2O(l) \rightleftharpoons H_3O^+(aq) + A^-(aq)$$

17.5 You can neglect x if C_a/K is $\geq$ 100. In this case, $C_a/K = [(0.0010\ M \div 6.8 \times 10^{-4}) = 1.47]$, which is significantly less than 100, and x cannot be neglected in the $(0.0010 - x)$ term. This says the degree of ionization is significant.

17.6 The ionization of the first H_3O^+ is

$$H_2PHO_3(aq) + H_2O(l) \rightleftharpoons H_3O^+(aq) + HPHO_3^-(aq)$$

The equilibrium-constant expression is

$$K_{a1} = \frac{[H_3O^+][HPHO_3^-]}{[H_2PHO_3]}$$

The ionization of the second H_3O^+ is

$$HPHO_3^-(aq) + H_2O(l) \rightleftharpoons H_3O^+(aq) + PHO_3^{2-}(aq)$$

The equilibrium-constant expression is

$$K_{a2} = \frac{[H_3O^+][PHO_3^{2-}]}{[HPHO_3^-]}$$

17.7 As shown in Example 17.4, the concentration of a -2 anion of a polyprotic acid in a solution of the diprotic acid alone is approximately equal to the value of K_{a2}. For oxalic acid, begin by noting that the $[H_3O^+] \cong [HC_2O_4^-]$. Then, substitute $[H_3O^+]$ for the $[HC_2O_4^-]$ term in the equilibrium-constant equation for K_{a2}:

$$K_{a2} = \frac{[H_3O^+][C_2O_4^{2-}]}{[HC_2O_4^-]} \cong \frac{[H_3O^+][C_2O_4^{2-}]}{[H_3O^+]} = [C_2O_4^{2-}]$$

17.8 The balanced chemical equation for the ionization of aniline is

$$C_6H_5NH_2(aq) + H_2O(l) \rightleftharpoons C_6H_5NH_3^+(aq) + OH^-(aq)$$

The equilibrium-constant equation, or expression for K_b, is defined without an $[H_2O]$ term; this term is included in the value for K_b as discussed in Section 17.3. The expression is

$$K_b = \frac{[C_6H_5NH_3^+][OH^-]}{[C_6H_5NH_2]}$$

17.9 First, decide whether any of the three is a strong base or not. Because all the molecules are among the nitrogen-containing weak bases listed in Table 17.2, none is a strong base. Next, recognize the greater the $[OH^-]$, the stronger the weak base. Because $[OH^-]$ can be calculated from the square root of the product of K_b and concentration, the larger the K_b, the greater the $[OH^-]$ and the stronger the weak base. Thus, CH_3NH_2 is the strongest of these three weak bases because its K_b is the largest.

17.10 Anilinium chloride is not a weak base nor a weak acid but a salt that contains the anilinium ion, $C_6H_5NH_3^+$, and the Cl^- ion. The chloride ion does not hydrolyze because it could only form HCl, a strong acid. The anilinium ion does hydrolyze as follows:

$$C_6H_5NH_3^+ + H_2O(l) \rightleftharpoons C_6H_5NH_2 + H_3O^+$$

The equilibrium-constant expression for this reaction is

$$K_a = \frac{[C_6H_5NH_2][H_3O^+]}{[C_6H_5NH_3^+]}$$

Obtain the value for K_a by calculating the value of K_w/K_b, where K_b is the ionization constant for aniline, $C_6H_5NH_2$.

17.11 The common-ion effect is the shift in an ionic equilibrium caused by the addition of a solute that furnishes an ion that is common to, or takes part in, the equilibrium. If the equilibrium involves the ionization of a weak acid, then the common ion is usually the anion formed by the ionization of the weak acid. If the equilibrium involves the ionization of a weak base, then the common ion is usually the cation formed by the ionization of the weak base. An example is the addition of F^- ion (as NaF) to a solution of the weak acid HF, which ionizes as shown below:

$$HF(aq) + H_2O(l) \rightleftharpoons H_3O^+(aq) + F^-(aq)$$

The effect of adding F^- to this equilibrium is that it causes a shift in the equilibrium composition to the left. The additional F^- reacts with H_3O^+, lowering its concentration and raising the concentration of HF.

17.12 The addition of CH_3NH_3Cl to 0.10 M CH_3NH_2 exerts a common-ion effect that causes the equilibrium below to exhibit a shift in composition to the left:

$$CH_3NH_2(aq) + H_2O(l) \rightleftharpoons CH_3NH_3^+(aq) + OH^-(aq)$$

This shift lowers the equilibrium concentration of the OH^- ion, which increases the $[H_3O^+]$. An increase in $[H_3O^+]$ lowers the pH below 11.8. The shift in composition to the left occurs according to Le Chatelier's principle, which states that a system shifts to counteract any change in composition.

17.13   A buffer is most often a solution of a mixture of two substances that is able to resist pH changes when limited amounts of acid or base are added to it. A buffer must contain a weak acid and its conjugate (weak) base. Strong acids and/or bases cannot form effective buffers because a buffer acts by converting H_3O^+ (strong acid) to the un-ionized (weak) buffer acid and by converting OH^- (strong base) to the un-ionized (weak) buffer base. An example of a buffer pair is a mixture of H_2CO_3 and HCO_3^-, the principal buffer in the blood.

17.14   The capacity of a buffer is the amount of acid or base with which the buffer can react before exhibiting a significant pH change. (A significant change in blood pH might mean 0.01-0.02 pH units; for other systems a significant change might mean 0.5 pH units.) A high-capacity buffer might be of the type discussed for Figure 17.10: one mol of buffer acid and one mol of buffer base. A low-capacity buffer might involve quite a bit less than these amounts: 0.01-0.05 mol of buffer acid and buffer base.

17.15   The pH of a weak base before titration is relatively high, around pH 10 for a 0.1 M solution of a typical weak base. As a strong acid titrant is added, the $[OH^-]$ decreases, and the pH decreases. At 50 percent neutralization, a buffer of equal amounts of buffer acid and base is formed. The $[OH^-]$ equals the K_b, or the pOH equals the pK_b. At the equivalence point, the pH is governed by the hydrolysis of the salt of the weak base formed and is usually in the pH 4-6 region. After the equivalence point, the pH decreases to a level just greater than the pH of the strong acid titrant.

17.16   If the pH is 8.0, an indicator that changes color in the basic region would be needed. Of the indicators mentioned in the text, phenolphthalein (pH 8.2-10.0) and thymol blue would work. In actual practice, cresol red (pH 7.2-8.8) would be the best choice because the pH should be closer to the middle of the range than to one end.

■ Solutions to Practice Problems

Note on significant figures: If the final answer to a solution needs to be rounded off, it is given first with one nonsignificant figure, and the last significant figure is underlined. The final answer is then rounded to the correct number of significant figures. In multiple-step problems, intermediate answers are given with at least one nonsignificant figure; however, only the final answer has been rounded off.

17.27 a. $HBrO(aq) + H_2O(l) \rightleftharpoons H_3O^+(aq) + BrO^-(aq)$

b. $HClO_2(aq) + H_2O(l) \rightleftharpoons H_3O^+(aq) + ClO_2^-(aq)$

c. $HNO_2(aq) + H_2O(l) \rightleftharpoons H_3O^+(aq) + NO_2^-(aq)$

d. $HCN(aq) + H_2O(l) \rightleftharpoons H_3O^+(aq) + CN^-(aq)$

17.29 HAc will be used throughout as an abbreviation for acrylic acid and Ac⁻ for the acrylate ion. At the start, the H_3O^+ from the self-ionization of water is so small it is approximately zero. Once the acrylic acid solution is prepared, some of the 0.10 M HAc ionizes to H_3O^+ and Ac⁻. Then, let x equal the mol/L of HAc that ionize, forming x mol/L of H_3O^+ and x mol/L of Ac⁻ and leaving (0.10 - x) M HAc in solution. We can summarize the situation in tabular form:

Conc. (M)	HAc + H₂O ⇌	H₃O⁺ +	Ac⁻
Starting	0.10	0	0
Change	-x	+x	+x
Equilibrium	0.10 - x	x	x

The equilibrium-constant equation is:

$$K_a = \frac{[H_3O^+][Ac^-]}{[HAc]} = \frac{x^2}{(0.10 - x)}$$

The value of x can be obtained from the pH of the solution:

$$x = [H_3O^+] = \text{antilog}(-pH) = \text{antilog}(-2.63) = 2.\underline{3}4 \times 10^{-3} = 0.00234\ M$$

Note that (0.10 - x) = (0.10 - 0.00234) = 0.09766, which is significantly different from 0.10, so x cannot be ignored in the calculation. Thus, we substitute for x in both the numerator and denominator to obtain the value of K_a:

$$K_a = \frac{x^2}{(0.10 - x)} = \frac{(0.00234)^2}{(0.10 - 0.00234)} = 5.\underline{6}06 \times 10^{-5} = 5.6 \times 10^{-5}$$

17.31 To solve, assemble a table of starting, change, and equilibrium concentrations. Use HBo as the symbol for boric acid and Bo⁻ as the symbol for $B(OH)_4^-$.

Conc. (M)	HBo + H₂O ⇌	H₃O⁺ +	Bo⁻
Starting	0.021	0	0
Change	-x	+x	+x
Equilibrium	0.021 - x	x	x

The value of x equals the value of the molarity of the H_3O^+ ion, which can be obtained from the equilibrium-constant expression.

(continued)

Substitute into the equilibrium-constant expression, and solve for x.

$$K_a = \frac{[H_3O^+][Bo^-]}{[HBo]} = \frac{(x)^2}{(0.021 - x)} = 5.9 \times 10^{-10}$$

Solve the equation for x, assuming x is much smaller than 0.021.

$$\frac{(x)^2}{(0.021 - x)} \cong \frac{(x)^2}{(0.021)} = 5.9 \times 10^{-10}$$

$$x^2 = 5.9 \times 10^{-10} \times (0.021) = 1.\underline{2}39 \times 10^{-11}$$

$$x = [H_3O^+] = 3.\underline{5}19 \times 10^{-6}\,M$$

Check to make sure the assumption that $(0.021 - x) \cong 0.021$ is valid:

$$0.021 - (3.519 \times 10^{-6}) = 0.02099, \text{ or} \cong 0.021 \text{ to two sig. figs.}$$

$$pH = -\log[H_3O^+] = -\log(3.519 \times 10^{-6}) = 5.4\underline{5}3 = 5.45$$

The degree of ionization is

$$\text{degree of ionization} = \frac{3.519 \times 10^{-6}}{(0.021)} = 0.000\underline{16}7 = 0.00017 = 1.7 \times 10^{-4}$$

17.33 To solve, assemble a table of starting, change, and equilibrium concentrations. Use HPaba as a symbol for p-aminobenzoic acid (PABA), and use Paba⁻ as the symbol for the -1 anion.

Conc. (M)	HPaba + H₂O ⇌	H₃O⁺	Paba⁻
Starting	0.055	0	0
Change	-x	+x	+x
Equilibrium	0.055 - x	x	x

Write the equilibrium-constant expression in terms of chemical symbols, and then substitute the terms x and (0.055 - x):

$$K_a = \frac{[H_3O^+][Paba^-]}{[HPaba]} = \frac{(x)^2}{(0.055 - x)} = 2.2 \times 10^{-5}$$

(continued)

Solve the equation for x, assuming x is smaller than 0.055.

$$x^2 = (2.2 \times 10^{-5}) \times 0.055 = 1.\underline{21} \times 10^{-6}$$

$$x = [H_3O^+] = [Paba^-] \cong 1.\underline{10} \times 10^{-3} = 0.0011 = 1.1 \times 10^{-3} \text{ M}$$

Check to make sure the assumption that $(0.055 - x) \cong 0.055$ is valid.

$$0.055 - 0.00110 = 0.0539, \text{ or } 0.054$$

17.35   To solve, first convert the pH to $[H_3O^+]$, which also equals $[C_2H_3O_2^-]$, here symbolized as $[Ac^-]$. Then, assemble the usual table, and substitute into the equilibrium-constant expression to solve for $[HC_2H_3O_2]$, here symbolized as $[HAc]$.

$$[H_3O^+] = \text{antilog } (-2.68) = 2.\underline{089} \times 10^{-3} \text{ M}$$

Conc. (M)	HAc + H₂O ⇌	H₃O⁺ +	Ac⁻
Starting	x	0	0
Change	-2.089×10^{-3}	$+2.089 \times 10^{-3}$	$+2.089 \times 10^{-3}$
Equilibrium	$x - (2.089 \times 10^{-3})$	2.089×10^{-3}	2.089×10^{-3}

Write the equilibrium-constant expression in terms of chemical symbols, and then substitute the x and the $x - (2.089 \times 10^{-3})$ terms into the expression:

$$K_a = \frac{[H_3O^+][Ac^-]}{[HAc]} = \frac{(2.089 \times 10^{-3})^2}{(x - 2.089 \times 10^{-3})} = 1.7 \times 10^{-5}$$

Solve the equation for x, assuming 0.002089 is much smaller than x.

$$x = [HAc] \cong \frac{(2.089 \times 10^{-3})^2}{(1.7 \times 10^{-5})} = 0.2\underline{56} = 0.26 \text{ M}$$

17.37 To solve, assemble the usual table of starting, change, and equilibrium concentrations of HF and F⁻ ions.

Conc. (M)	HF	+	H₂O	⇌	H₃O⁺	+	F⁻
Starting	0.040				0		0
Change	-x				+x		+x
Equilibrium	0.040 - x				x		x

Write the equilibrium-constant expression in terms of chemical symbols, and then substitute the terms x and (0.040 - x):

$$K_a = \frac{[H_3O^+][F^-]}{[HF]} = \frac{(x)^2}{(0.040 - x)} = 6.8 \times 10^{-4}$$

In this case, x cannot be ignored compared to 0.040 M. (If it is ignored, subtracting the calculated $[H_3O^+]$ from 0.040 yields a significant change.) The quadratic formula must be used. Reorganize the equilibrium-constant expression into the form $ax^2 + bx + c = 0$, and substitute for a, b, and c in the quadratic formula.

$$x^2 + (6.8 \times 10^{-4})x - (2.72 \times 10^{-5}) = 0$$

$$x = \frac{-6.8 \times 10^{-4} \pm \sqrt{(6.8 \times 10^{-4})^2 + 4(2.72 \times 10^{-5})}}{2}$$

$$x = \frac{-6.8 \times 10^{-4} \pm 0.01045}{2}$$

Use the positive root.

$$x = [H_3O^+] = 4.88 \times 10^{-3} = 0.0049 = 4.9 \times 10^{-3} \text{ M}$$

$$pH = -\log(4.88 \times 10^{-3}) = 2.311 = 2.31$$

17.39 To solve, assemble the usual table of starting, change, and equilibrium concentrations of $(NO_2)_2C_6H_3CO_2H$, symbolized as HDin, and the $(NO_2)_2C_6H_3CO_2^-$ ion, symbolized as Din⁻.

Conc. (M)	HDin	+	H₂O	⇌	H₃O⁺	+	Din⁻
Starting	2.00				0		0
Change	-x				+x		+x
Equilibrium	2.00 - x				x		x

(continued)

Write the equilibrium-constant expression in terms of chemical symbols, and then substitute the terms x and (2.00 - x):

$$K_a = \frac{[H_3O^+][Din^-]}{[HDin]} = \frac{(x)^2}{(2.00 - x)} = 7.94 \times 10^{-2} = 0.0794$$

In this case, x cannot be ignored compared to 2.00 M. (If it is ignored, subtracting the calculated $[H_3O^+]$ from 2.00 yields a significant change.) The quadratic formula must be used. Reorganize the equilibrium-constant expression into the form $ax^2 + bx + c = 0$, and substitute for a, b, and c in the quadratic formula.

$$x^2 + (0.0794)\,x - 0.1588 = 0$$

$$x = \frac{-0.0794 \pm \sqrt{(0.0794)^2 + 4(0.1588)}}{2}$$

$$x = \frac{-0.0794 \pm 0.8009}{2}$$

Use the positive root.

$$x = [H_3O^+] = 0.360\underline{7} = 0.361 \text{ M}$$

17.41 a. To solve, note that $K_{a1} = 1.2 \times 10^{-3} > K_{a2} = 3.9 \times 10^{-6}$, and hence the second ionization and K_{a2} can be neglected. Assemble a table of starting, change, and equilibrium concentrations. Let $H_2Ph = H_2C_8H_4O_4$ and $HPh^- = H\,C_8H_4O_4^-$.

Conc. (M)	H_2Ph + H_2O	$\rightleftharpoons$	H_3O^+ + HPh^-	
Starting	0.015		0	0
Change	-x		+x	+x
Equilibrium	0.015 - x		x	x

Write the equilibrium-constant expression in terms of chemical symbols, and then substitute x and (0.015 - x):

$$K_{a1} = \frac{[H_3O^+][HPh^-]}{[H_2Ph]} = \frac{(x)^2}{(0.015 - x)} = 1.2 \times 10^{-3} = 0.0012$$

In this case, x cannot be ignored in the (0.015 M - x) term. (If it is ignored, the calculated $[H_3O^+]$ when subtracted from 0.0015 M yields a significant change.) The quadratic formula must be used.

(continued)

Reorganize the equilibrium-constant expression into the form $ax^2 + bx + c = 0$, and substitute for a, b, and c in the quadratic formula.

$$x^2 + (0.0012)x - 1.80 \times 10^{-5} = 0$$

$$x = \frac{-0.0012 \pm \sqrt{(0.0012)^2 - 4(-1.80 \times 10^{-5})}}{2}$$

$$x = \frac{-0.0012 \pm 0.008569}{2}$$

Use the positive root.

$$x = [H_3O^+] \cong 3.\underline{6}84 \times 10^{-3} = 0.0037 = 3.7 \times 10^{-3} \, M$$

b. Because $[HPh^-] \cong [H_3O^+]$, these terms cancel in the K_{a2} expression. This reduces to

$$[Ph^{2-}] = K_{a2} = 3.9 \times 10^{-6} \, M.$$

17.43   The equation is

$$CH_3NH_2(aq) + H_2O(l) \rightleftharpoons CH_3NH_3^+(aq) + OH^-(aq)$$

The K_b expression is

$$K_b = \frac{[CH_3NH_3^+][OH^-]}{[CH_3NH_2]}$$

17.45   To solve, convert the pH to $[OH^-]$:

$$pOH = 14.00 - pH = 14.00 - 11.34 = 2.66$$

$$[OH^-] = \text{antilog}\,(-2.66) = 2.\underline{1}88 \times 10^{-3} \, M$$

Using the symbol EtN for ethanolamine, assemble a table of starting, change, and equilibrium concentrations.

Conc. (M)	EtN + H₂O ⇌	HEtN⁺ +	OH⁻
Starting	0.15	0	0
Change	-x	+x	+x
Equilibrium	$0.15 - (2.188 \times 10^{-3})$	2.188×10^{-3}	2.188×10^{-3}

(continued)

Write the equilibrium-constant expression in terms of chemical symbols, and then substitute the terms, and solve for K_b:

$$K_b = \frac{[HEtN^+][OH^-]}{[EtN]} = \frac{(2.188 \times 10^{-3})^2}{(0.15 - 2.188 \times 10^{-3})} = 3.\underline{2}3 \times 10^{-5} = 3.2 \times 10^{-5}$$

17.47 To solve, assemble a table of starting, change, and equilibrium concentrations:

Conc. (M)	CH_3NH_2 + H_2O $\rightleftharpoons$	$CH_3NH_3^+$ +	OH^-
Starting	0.060	0	0
Change	-x	+x	+x
Equilibrium	0.060 - x	x	x

Write the equilibrium-constant expression in terms of chemical symbols, and then substitute the terms and the value of K_b:

$$K_b = \frac{[CH_3NH_3^+][OH^-]}{[CH_3NH_2]} = \frac{(x)^2}{(0.060 - x)} = 4.4 \times 10^{-4}$$

In this case, x cannot be ignored compared to 0.060 M. (If it is ignored, subtracting the calculated $[H_3O^+]$ from 0.060 yields a significant change.) The quadratic formula must be used. Reorganize the equilibrium-constant expression into the form $ax^2 + bx + c = 0$, and substitute for a, b, and c in the quadratic formula.

$$x^2 + (4.4 \times 10^{-4})x - 2.64 \times 10^{-5} = 0$$

$$x = \frac{-4.4 \times 10^{-4} \pm \sqrt{(4.4 \times 10^{-4})^2 + 4(2.64 \times 10^{-5})}}{2}$$

$$x = \frac{-4.4 \times 10^{-4} \pm 0.01028}{2}$$

Use the positive root.

$$x = [OH^-] = 4.\underline{9}2 \times 10^{-3} = 0.0049 = 4.9 \times 10^{-3} \text{ M}$$

$$pOH = -\log(4.92 \times 10^{-3}) = 2.3\underline{0}7$$

$$pH = 14.00 - 2.307 = 11.6\underline{9}2 = 11.69$$

17.49 a. No hydrolysis occurs because the nitrate ion (NO_3^-) is the anion of a strong acid.

b. Hydrolysis occurs. Equation:

$$OCl^- + H_2O \rightleftharpoons HOCl + OH^-$$

Equilibrium-constant expression:

$$K_b = \frac{K_w}{K_a} = \frac{[HOCl][OH^-]}{[OCl^-]}$$

c. Hydrolysis occurs. Equation:

$$NH_2NH_3^+ + H_2O \rightleftharpoons H_3O^+ + NH_2NH_2$$

Equilibrium-constant expression:

$$K_a = \frac{K_w}{K_b} = \frac{[H_3O^+][NH_2NH_2]}{[NH_2NH_3^+]}$$

d. No hydrolysis occurs because the bromide ion (Br^-) is the anion of a strong acid.

17.51 Acid ionization is

$$Zn(H_2O)_6^{2+}(aq) + H_2O(l) \rightleftharpoons Zn(H_2O)_5(OH)^+(aq) + H_3O^+(aq)$$

17.53 a. $Fe(NO_3)_3$ is a salt of a weak base, $Fe(OH)_3$, and a strong acid, HNO_3, so it would be expected to be acidic. Fe^{3+} is not in Group IA or IIA, so it would be expected to form a metal hydrate ion that would hydrolyze to form an acidic solution.

b. Na_2CO_3 is a salt of a strong base, NaOH, and the anion of a weak acid, HCO_3^-, so it would be expected to be basic.

c. $Ca(CN)_2$ is a salt of a strong base, $Ca(OH)_2$, and a weak acid, HCN, so it would be expected to be basic.

d. NH_4ClO_4 is a salt of a weak base, NH_3, and a strong acid, $HClO_4$, so it would be expected to be acidic.

17.55 a. Both ions hydrolyze:

$$NH_4^+ + H_2O \rightleftharpoons NH_3 + H_3O^+$$

$$C_2H_3O_2^- + H_2O \rightleftharpoons HC_2H_3O_2 + OH^-$$

Calculate the K_a and K_b constants of each to compare them:

$$NH_4^+ \text{ as an acid: } K_a = \frac{K_w}{K_b} = \frac{1.0 \times 10^{-14}}{1.8 \times 10^{-5}} = 5.\underline{5}5 \times 10^{-10}$$

$$C_2H_3O_2^- \text{ as a base: } K_b = \frac{K_w}{K_a} = \frac{1.0 \times 10^{-14}}{1.7 \times 10^{-5}} = 5.\underline{8}8 \times 10^{-10}$$

Because the K_b for the hydrolysis of $C_2H_3O_2^-$ is slightly larger than the constant, K_a, for the hydrolysis of NH_4^+, the solution will be slightly basic but close to pH 7.0.

b. Both ions hydrolyze:

$$C_6H_5NH_3^+ + H_2O \rightleftharpoons C_6H_5NH_2 + H_3O^+$$

$$C_2H_3O_2^- + H_2O \rightleftharpoons HC_2H_3O_2 + OH^-$$

Calculate the K_a and K_b constants of each to compare them:

$$C_6H_5NH_3^+ \text{ as an acid: } K_a = \frac{K_w}{K_b} = \frac{1.0 \times 10^{-14}}{4.2 \times 10^{-10}} = 2.\underline{3}8 \times 10^{-5}$$

$$C_2H_3O_2^- \text{ as a base: } K_b = \frac{K_w}{K_a} = \frac{1.0 \times 10^{-14}}{1.7 \times 10^{-5}} = 5.\underline{8}8 \times 10^{-10}$$

Because the constant, K_a, for the hydrolysis of $C_6H_5NH_3^+$ is larger than the hydrolysis constant, K_b, for the hydrolysis of $C_2H_3O_2^-$, the solution will be acidic and significantly less than pH 7.0.

17.57 a. The reaction is

$$NO_2^- + H_2O \rightleftharpoons HNO_2 + OH^-$$

The constant, K_b, is obtained by dividing K_w by the K_a of the conjugate acid, HNO_2:

$$K_b = \frac{K_w}{K_a} = \frac{1.0 \times 10^{-14}}{4.5 \times 10^{-4}} = 2.\underline{2}2 \times 10^{-11} = 2.2 \times 10^{-11}$$

(continued)

b. The reaction is

$$C_5H_5NH^+ + H_2O \rightleftharpoons C_5H_5N + H_3O^+$$

The constant, K_a, is obtained by dividing K_w by the K_b of the conjugate base, C_5H_5N:

$$K_a = \frac{K_w}{K_b} = \frac{1.0 \times 10^{-14}}{1.4 \times 10^{-9}} = 7.\underline{14} \times 10^{-6} = 7.1 \times 10^{-6}$$

17.59 Assemble the usual table, letting [Pr⁻] equal the equilibrium concentration of the propionate anion (the only ion that hydrolyzes). Then, calculate the K_b of the Pr⁻ ion from the K_a of its conjugate acid, HPr. Assume x is much smaller than the 0.025 M concentration in the denominator, and solve for x in the numerator of the equilibrium-constant expression. Finally, calculate pOH from the [OH⁻] and pH from the pOH.

Conc. (M)	Pr⁻ + H₂O $\rightleftharpoons$ HPr + OH⁻		
Starting	0.025	0	0
Change	-x	+x	+x
Equilibrium	0.025 - x	x	x

$$K_b = \frac{K_w}{K_a} = \frac{1.0 \times 10^{-14}}{1.3 \times 10^{-5}} = 7.\underline{69} \times 10^{-10}$$

Substitute into the equilibrium-constant equation.

$$K_b = \frac{[HPr][OH^-]}{[Pr^-]} = \frac{(x)^2}{(0.025 - x)} \cong \frac{(x)^2}{(0.025)} = 7.\underline{69} \times 10^{-10}$$

$$x = [OH^-] = [HPr] \cong 4.\underline{38} \times 10^{-6} = 4.4 \times 10^{-6} \, M$$

$$pOH = -\log [OH^-] = -\log (4.38 \times 10^{-6}) = 5.3\underline{58}$$

$$pH = 14.00 - 5.358 = 8.6\underline{42} = 8.64$$

17.61 Assemble the usual table, letting [PyNH$^+$] equal the equilibrium concentration of the pryridinium cation (the only ion that hydrolyzes). Then, calculate the K_a of the PyNH$^+$ ion from the K_b of its conjugate base, PyN. Assume x is much smaller than the 0.15 M concentration in the denominator, and solve for x in the numerator of the equilibrium-constant expression. Finally, calculate pH from the [H$_3$O$^+$].

Conc. (M)	PyNH$^+$ + H$_2$O $\rightleftharpoons$ H$_3$O$^+$ + PyN		
Starting	0.15	0	0
Change	-x	+x	+x
Equilibrium	0.15 - x	x	x

$$K_a = \frac{K_w}{K_b} = \frac{1.0 \times 10^{-14}}{1.4 \times 10^{-9}} = 7.\underline{1}4 \times 10^{-6}$$

Write the equilibrium-constant expression in terms of chemical symbols and then substitute the terms and solve for K_a:

$$K_a = \frac{[PyN][H_3O^+]}{[PyNH^+]} = \frac{(x)^2}{(0.15 - x)} \cong \frac{(x)^2}{(0.15)} = 7.\underline{1}4 \times 10^{-6}$$

$$x = [H_3O^+] = [PyN] \cong 1.\underline{0}3 \times 10^{-3} = 0.0010 = 1.0 \times 10^{-3} \text{ M}$$

$$pH = -\log [H^+] = -\log (1.03 \times 10^{-3}) = 2.9\underline{8}7 = 2.99$$

17.63 To solve, assemble a table of starting, change, and equilibrium concentrations for each part. For each part, assume x is much smaller than the 0.75 M starting concentration of HF. Then, solve for x in the numerator of each equilibrium-constant expression by using the product of 6.8×10^{-4} and other terms.

a. 0.75 M Hydrofluoric acid, HF:

Conc. (M)	HF + H$_2$O $\rightleftharpoons$ H$_3$O$^+$ + F$^-$		
Starting	0.75	0	0
Change	-x	+x	+x
Equilibrium	0.75 - x	x	x

$$K_a = \frac{[H_3O^+][F^-]}{[HF]} = \frac{(x)^2}{(0.75 - x)} \cong \frac{(x)^2}{(0.75)} = 6.8 \times 10^{-4}$$

$$x^2 = 6.8 \times 10^{-4} \times (0.75)$$

(continued)

$$x = [H_3O^+] = 0.02258 \text{ M}$$

Check to see if the assumption is valid.

$$0.75 - (0.02258) = 0.727 = 0.73$$

This is a borderline case; the quadratic equation gives $[H_3O^+] = 0.02224$ M, not much different. The degree of ionization is

$$\text{Degree of ionization} = \frac{0.02258}{0.75} = 0.0301 = 0.030$$

b. 0.75 M HF with 0.12 M HCl:

Conc. (M)	HF + H₂O ⇌	H₃O⁺ +	F⁻
Starting	0.75	0.12	0
Change	-x	+x	+x
Equilibrium	0.75 - x	0.12 + x	x

Assuming x is negligible compared to 0.12 and to 0.75, substitute into the equilibrium-constant expression 0.12 for $[H_3O^+]$ from 0.12 M HCl and 0.75 from the HF:

$$K_a = \frac{[H_3O^+][F^-]}{[HF]} = \frac{(0.12 + x)(x)}{(0.75 - x)} \cong \frac{(0.12)(x)}{(0.75)} = 6.8 \times 10^{-4}$$

$$x = [F^-] = \frac{(0.75)(6.8 \times 10^{-4})}{(0.12)} \cong 4.250 \times 10^{-3} \text{ M}$$

Check to see if the assumptions are valid:

$$0.75 - (4.250 \times 10^{-3}) = 0.7457 = 0.75$$

$$0.12 + (4.250 \times 10^{-3}) = 0.1242 = 0.12$$

$$\text{Degree of ionization} = \frac{4.25 \times 10^{-3}}{0.75} = 0.00566 = 0.0057$$

17.65 Assemble the usual table, using a starting NO_2^- of 0.10 M, from 0.10 M KNO_2, and a starting HNO_2 of 0.15 M. Assume x is negligible compared to 0.10 M and 0.15 M, and solve for the x in the numerator.

Conc. (M)	HNO_2 + H_2O $\rightleftharpoons$	H_3O^+	+ NO_2^-
Starting	0.15	0	0.10
Change	-x	+x	+x
Equilibrium	0.15 - x	x	0.10 + x

$$K_a = \frac{[H_3O^+][NO_2^-]}{[HNO_2]} = \frac{(0.10+x)(x)}{(0.15-x)} \cong \frac{(0.10)(x)}{(0.15)} = 4.5 \times 10^{-4}$$

$$x = [H_3O^+] = 6.\underline{7}5 \times 10^{-4} \text{ M}$$

$$pH = -\log[H_3O^+] = -\log(6.75 \times 10^{-4}) = 3.1\underline{7}0 = 3.17$$

17.67 Assemble the usual table, using a starting $CH_3NH_3^+$ of 0.15 M, from 0.15 M CH_3NH_3Cl, and a starting CH_3NH_2 of 0.10 M. Assume x is negligible compared to 0.15 M and 0.10 M, and solve for the x in the numerator.

Conc. (M)	CH_3NH_2 + H_2O $\rightleftharpoons$	OH^-	+ $CH_3NH_3^+$
Starting	0.10	0	0.15
Change	-x	+x	+x
Equilibrium	0.10 - x	x	0.15 + x

$$K_b = \frac{[CH_3NH_3^+][OH^-]}{[CH_3NH_2]} = \frac{(0.15+x)(x)}{(0.10-x)} \cong \frac{(0.15)(x)}{(0.10)} = 4.4 \times 10^{-4}$$

$$x = [OH^-] = 2.\underline{9}3 \times 10^{-4} \text{ M}$$

$$pOH = -\log[OH^-] = -\log(2.93 \times 10^{-4}) = 3.5\underline{3}2$$

$$pH = 14.00 - pOH = 14.00 - 3.532 = 10.4\underline{6}8 = 10.47$$

17.69 Find the mol/L of HF and the mol/L of F⁻, and assemble the usual table. Substitute the equilibrium concentrations into the equilibrium-constant expression; then, assume x is negligible compared to the starting concentrations of both HF and F⁻. Solve for x in the numerator of the equilibrium-constant expression, and calculate the pH from this value.

Total volume = 0.045 L + 0.035 L = 0.080 L

(0.10 mol HF/L) x 0.035 L = 0.0035 mol HF (÷ 0.080 L total volume = 0.04375 M)

(0.15 mol F⁻/L) x 0.045 L = 0.00675 mol F⁻ (÷ 0.080 L total volume = 0.084375 M)

Now, substitute these starting concentrations into the usual table:

Conc. (M)	HF + H_2O $\rightleftharpoons$	H_3O^+ +	F⁻
Starting	0.04375	0	0.084375
Change	-x	+x	+x
Equilibrium	0.04375 - x	x	0.084375 + x

$$K_a = \frac{[H_3O^+][F^-]}{[HF]} = \frac{(x)(0.084375 + x)}{(0.04375 - x)} \cong \frac{(x)(0.084375)}{(0.04375)} = 6.8 \times 10^{-4}$$

$$x = [H_3O^+] = 3.\underline{5}2 \times 10^{-4} \text{ M}$$

$$pH = -\log[H_3O^+] = -\log(3.52 \times 10^{-4}) = 3.4\underline{5}3 = 3.45$$

17.71 First, use the 0.10 M NH_3 and 0.10 M NH_4^+ to calculate the [OH⁻] and pH before HCl is added. Assemble a table of starting, change, and equilibrium concentrations. Assume x is negligible compared to 0.10 M, and substitute the approximate concentrations into the equilibrium-constant expression.

Conc. (M)	NH_3 + H_2O $\rightleftharpoons$	NH_4^+ +	OH⁻
Starting	0.10	0.10	0
Change	-x	+x	+x
Equilibrium	0.10 - x	0.10 + x	x

$$K_b = \frac{[NH_4^+][OH^-]}{[NH_3]} = \frac{(0.10 + x)(x)}{(0.10 - x)} \cong \frac{(0.10)(x)}{(0.10)} = 1.8 \times 10^{-5}$$

$$x = [OH^-] = 18 \times 10^{-5} \text{ M}$$

(continued)

$$pOH = -\log[OH^-] = -\log(1.8 \times 10^{-5}) = 4.7\underline{4}4$$

$$pH = 14.00 - pOH = 14.00 - 4.744 = 9.2\underline{5}52 = 9.26 \text{ (before HCl added)}$$

Now, calculate the pH after the 0.012 L (12 mL) of 0.20 M HCl is added by noting that the H_3O^+ ion reacts with the NH_3 to form additional NH_4^+. Calculate the stoichiometric amount of HCl; then, subtract the moles of HCl from the moles of NH_3. Add the resulting moles of NH_4^+ to the 0.0125 starting moles of NH_4^+ in the 0.125 L of buffer.

$$(0.20 \text{ mol HCl/L}) \times 0.012 \text{ L} = 0.0024 \text{ mol HCl (reacts with } 0.0024 \text{ mol } NH_3)$$

$$\text{Mol } NH_3 \text{ left} = (0.0125 - 0.0024) \text{ mol} = 0.01\underline{0}1 \text{ mol}$$

$$\text{Mol } NH_4^+ \text{ present} = (0.0125 + 0.0024) \text{ mol} = 0.0149 \text{ mol}$$

The concentrations of NH_3 and NH_4^+ are

$$[NH_3] = \frac{0.0101 \text{ mol } NH_3}{0.137 \text{ L}} = 0.07\underline{3}7 \text{ M}$$

$$[NH_4^+] = \frac{0.0149 \text{ mol } NH_3}{0.137 \text{ L}} = 0.1\underline{0}8 \text{ M}$$

Now, account for the ionization of NH_3 to NH_4^+ and OH^- at equilibrium by assembling the usual table. Assume x is negligible compared to 0.0737 M and 0.108 M, and solve the equilibrium-constant expression for x in the numerator. Calculate the pH from x, the $[OH^-]$.

Conc. (M)	NH_3 + H_2O $\rightleftharpoons$	NH_4^+ +	OH^-
Starting	0.0737	0.108	0
Change	-x	+x	+x
Equilibrium	0.0737 - x	0.108 + x	x

$$K_b = \frac{[NH_4^+][OH^-]}{[NH_3]} = \frac{(0.108 + x)(x)}{(0.0737 - x)} \cong \frac{(0.108)(x)}{(0.0737)} = 1.8 \times 10^{-5}$$

$$x = [OH^-] \cong 1.\underline{2}2 \times 10^{-5} \text{ M}$$

$$pOH = -\log[OH^-] = -\log(1.22 \times 10^{-5}) = 4.9\underline{1}3$$

$$pH = 14.00 - pOH = 14.00 - 4.913 = 9.0\underline{8}6 = 9.09 \text{ (after HCl added)}$$

17.73 Use the Henderson-Hasselbalch equation, where $pK_a = -\log K_a$.

$$pH = -\log K_a + \log \frac{[\text{buff. base}]}{[\text{buff. acid}]}$$

$$= -\log (1.4 \times 10^{-3}) + \log \frac{(0.10\,M)}{(0.15\,M)} = 2.6\underline{7}7 = 2.68$$

17.75 Calculate the K_a of the pyridinium ion from the K_b of pyridine, and then use the Henderson-Hasselbalch equation where $pK_a = -\log K_a$.

$$K_a = \frac{K_w}{K_b} = \frac{1.0 \times 10^{-14}}{1.4 \times 10^{-9}} = 7.\underline{1}4 \times 10^{-6}$$

$$pH = -\log K_a + \log \frac{[\text{buff. base}]}{[\text{buff. acid}]}$$

$$= -\log (7.14 \times 10^{-6}) + \log \frac{(0.15\,M)}{(0.10\,M)} = 5.3\underline{2}2 = 5.32$$

17.77 Symbolize acetic acid as HOAc and sodium acetate as Na^+OAc^-. Use the Henderson-Hasselbalch equation to find the log of $[OAc^-]/[HOAc]$. Then, solve for $[OAc^-]$ and for moles of NaOAc in the 2.0 L of solution.

$$pH = -\log K_a + \log \frac{[OAc^-]}{[HOAc]} = 5.00$$

$$5.00 = -\log (1.7 \times 10^{-5}) + \log \frac{[OAc^-]}{(0.10\,M)}$$

$$5.00 = 4.7\underline{7}0 + \log [OAc^-] - \log (0.10)$$

$$\log[OAc^-] = 5.00 - 4.770 - 1.00 = -0.770$$

$$[OAc^-] = 0.1\underline{6}98\,M$$

$$\text{Mol NaOAc} = 0.1698\,\text{mol/L} \times 2.0\,L = 0.3\underline{3}96 = 0.34\,\text{mol}$$

17.79 All the OH⁻ (from the NaOH) reacts with the H_3O^+ from HCl. Calculate the stoichiometric amounts of OH⁻ and H_3O^+, and subtract the mol of OH⁻ from the mol of H_3O^+. Then, divide the remaining H_3O^+ by the total volume of 0.015 L + 0.025 L, or 0.040 L, to find the $[H_3O^+]$. Then calculate the pH.

$$Mol\ H_3O^+ = (0.10\ mol\ HCl/L) \times 0.025\ L\ HCl = 0.0025\ mol\ H_3O^+$$

$$Mol\ OH^- = (0.10\ mol\ NaOH/L) \times 0.015\ L\ NaOH = 0.0015\ mol\ OH^-$$

$$Mol\ H_3O^+\ left = (0.0025 - 0.0015)\ mol\ H_3O^+ = 0.0010\ mol\ H_3O^+$$

$$[H_3O^+] = 0.0010\ mol\ H_3O^+ \div 0.040\ L\ total\ volume = 0.02\underline{5}0\ M$$

$$pH = -\log[H_3O^+] = -\log(0.0250) = 1.6\underline{0}2 = 1.60$$

17.81 Use HBen to symbolize benzoic acid and Ben⁻ to symbolize the benzoate anion. At the equivalence point, equal molar amounts of HBen and NaOH react to form a solution of NaBen. Start by calculating the moles of HBen. Use this to calculate the volume of NaOH needed to neutralize all of the HBen (and use the moles of HBen as the moles of Ben⁻ formed at the equivalence point). Add the volume of NaOH to the original 0.050 L to find the total volume of solution.

$$Mol\ HBen = 1.24\ g\ HBen \div (122.1\ g\ HBen/mol\ HBen) = 0.01016\ mol\ HBen$$

$$Volume\ NaOH = 0.01016\ mol\ NaOH \div (0.180\ mol\ NaOH/L) = 0.056\underline{4}4\ L$$

$$Total\ volume = 0.05644\ L + 0.050\ L\ HBen\ soln = 0.10644\ L$$

$$[Ben^-] = (0.01016\ mol\ Ben^-\ from\ HBen) \div 0.10644\ L = 0.09545\ M$$

Because the Ben⁻ hydrolyzes to OH⁻ and HBen, use this to calculate the [OH⁻]. Start by calculating the K_b constant of Ben⁻ from the K_a of its conjugate acid, HBen. Then, assemble the usual table of concentrations, assume x is negligible, and calculate [OH⁻] and pH.

$$K_b = \frac{K_w}{K_a} = \frac{1.0 \times 10^{-14}}{6.3 \times 10^{-5}} = 1.\underline{5}9 \times 10^{-10}$$

Conc. (M)	Ben⁻ + H₂O ⇌	HBen	+ OH⁻
Starting	0.09545	0	0
Change	-x	+x	+x
Equilibrium	0.09545 - x	x	x

(continued)

$$K_b = \frac{[HBen][OH^-]}{[Ben^-]} = \frac{(x)^2}{(0.09545 - x)} \cong \frac{(x)^2}{(0.09545)} = 1.59 \times 10^{-10}$$

$$x = [OH^-] = 3.\underline{8}95 \times 10^{-6} \, M$$

$$pOH = -\log[OH^-] = -\log(3.895 \times 10^{-6}) = 5.4\underline{0}94$$

$$pH = 14.00 - 5.4094 = 8.5\underline{9}05 = 8.59$$

17.83 Use EtN to symbolize ethylamine and $EtNH^+$ to symbolize the ethylammonium cation. At the equivalence point, equal molar amounts of EtN and HCl react to form a solution of EtNHCl. Start by calculating the moles of EtN. Use this to calculate the volume of HCl needed to neutralize all of the EtN (and use the moles of EtN as the moles of $EtNH^+$ formed at the equivalence point). Add the volume of HCl to the original 0.032 L to find the total volume of solution.

$$(0.087 \text{ mol EtN/L}) \times 0.032 \text{ L} = 0.00278 \text{ mol EtN}$$

$$\text{Volume HCl} = 0.00278 \text{ mol HCl} \div (0.15 \text{ mol HCl/L}) = 0.0185 \text{ L}$$

$$\text{Total volume} = 0.0185 \text{ L} + 0.032 \text{ L EtN soln} = 0.0505 \text{ L}$$

$$[EtNH^+] = (0.00278 \text{ mol EtNH}^+ \text{ from EtN}) \div 0.0505 \text{ L} = 0.0550 \text{ M}$$

Because the $EtNH^+$ hydrolyzes to H_3O^+ and EtN, use this to calculate the $[H_3O^+]$. Start by calculating the K_a constant of $EtNH^+$ from the K_b of its conjugate base, EtN. Then, assemble the usual table of concentrations, assume x is negligible, and calculate $[H_3O^+]$ and pH.

$$K_a = \frac{K_w}{K_b} = \frac{1.0 \times 10^{-14}}{4.7 \times 10^{-4}} = 2.\underline{1}3 \times 10^{-11}$$

Conc. (M)	$EtNH^+ + H_2O \rightleftharpoons EtN + H_3O^+$		
Starting	0.0550	0	0
Change	-x	+x	+x
Equilibrium	0.0550 - x	x	x

$$K_a = \frac{[EtN][H_3O^+]}{[EtNH^+]} = \frac{(x)^2}{(0.0550 - x)} \cong \frac{(x)^2}{(0.0550)} = 2.13 \times 10^{-11}$$

$$x = [H_3O^+] = 1.\underline{0}8 \times 10^{-6} \, M$$

$$pH = -\log[H_3O^+] = -\log(1.08 \times 10^{-6}) = 5.9\underline{6}56 = 5.97$$

17.85 Calculate the stoichiometric amounts of NH_3 and HCl, which forms NH_4^+. Then, divide NH_3 and NH_4^+ by the total volume of 0.500 L + 0.200 L = 0.700 L to find the starting concentrations. Calculate the $[OH^-]$, the pOH, and the pH.

$$Mol\ NH_3 = (0.10\ mol\ NH_3/L) \times 0.500\ L = 0.0500\ mol\ NH_3$$

$$Mol\ HCl = (0.15\ mol\ HCl/L) \times 0.200\ L = 0.0300\ mol\ HCl\ (0.0300\ mol\ NH_4^+)$$

$$Mol\ NH_3\ left = 0.0500\ mol - 0.0300\ mol\ HCl = 0.0200\ mol\ NH_3$$

$$0.0200\ mol\ NH_3 \div 0.700\ L = 0.0286\ M\ NH_3$$

$$0.0300\ mol\ NH_4^+ \div 0.700\ L = 0.0429\ M\ NH_4^+$$

Conc. (M)	NH_3 + H_2O	$\rightleftharpoons$	NH_4^+	+	OH^-
Starting	0.0286		0.0429		0
Change	-x		+x		+x
Equilibrium	0.0286 - x		0.0429 + x		x

$$K_b = \frac{[NH_4^+][OH^-]}{[NH_3]} = \frac{(0.0429 + x)(x)}{(0.0286 - x)} \cong \frac{(0.0429)(x)}{(0.0286)} = 1.8 \times 10^{-5}$$

$$x = [OH^-] = 1.20 \times 10^{-5}\ M$$

$$pOH = -\log [OH^-] = -\log (1.20 \times 10^{-5}) = 4.920$$

$$pH = 14.00 - pOH = 14.00 - 4.920 = 9.079 = 9.08$$

■ Solutions to General Problems

17.87   To solve, assemble a table of starting, change, and equilibrium concentrations. Use HSal to symbolize salicylic acid, and use Sal⁻ for the anion. Start by converting pH to $[H_3O^+]$:

$$[H_3O^+] = \text{antilog} (\text{-pH}) = \text{antilog} (\text{-2.43}) = 3.72 \times 10^{-3} \text{ M}$$

Starting M of HSal = $(2.2 \text{ g} \div 138 \text{ g/mol}) \div 1.00 \text{ L} = 0.0159 \text{ M}$

Conc. (M)	HSal + H₂O ⇌ H₃O⁺ + Sal⁻		
Starting	0.0159	0	0
Change	-x	+x	+x
Equilibrium	0.0159 - x	x	x

The value of x equals the value of the molarity of the H_3O^+ ion, which is 3.72×10^{-3} M. Substitute into the equilibrium-constant expression to find K_a:

$$K_a = \frac{[H_3O^+][Sal^-]}{[HSal]} = \frac{(x)^2}{(0.0159 - x)} = \frac{(3.72 \times 10^{-3})^2}{(0.0159 - 3.72 \times 10^{-3})}$$

$$= 1.\underline{13} \times 10^{-3} = 1.1 \times 10^{-3}$$

17.89   To solve, assemble a table of starting, change, and equilibrium concentrations. Start by converting pH to $[H_3O^+]$:

$$[H_3O^+] = \text{antilog} (\text{-pH}) = \text{antilog} (\text{-1.73}) = 1.86 \times 10^{-2} \text{ M}$$

Starting M of HSO₄⁻ = 0.050 M

Conc. (M)	HSO₄⁻ + H₂O ⇌ H₃O⁺ + SO₄²⁻		
Starting	0.050	0	0
Change	-x	+x	+x
Equilibrium	0.050 - x	x	x

The value of x equals the value of the molarity of the H_3O^+ ion, which is 1.86×10^{-2} M. Substitute into the equilibrium-constant expression to find K_a:

$$K_{a2} = \frac{[H_3O^+][SO_4^{2-}]}{[HSO_4^-]} = \frac{(x)^2}{(0.050 - x)} = \frac{(1.86 \times 10^{-2})^2}{(0.050 - 1.86 \times 10^{-2})}$$

$$= 1.\underline{10} \times 10^{-2} = 1.1 \times 10^{-2}$$

17.91 For the base ionization (hydrolysis) of CN^- to $HCN + OH^-$, the base-ionization constant is

$$K_b = \frac{K_w}{K_a} = \frac{1.0 \times 10^{-14}}{4.9 \times 10^{-10}} = 2.\underline{0}4 \times 10^{-5} = 2.0 \times 10^{-5}$$

For the base ionization (hydrolysis) of CO_3^{2-} to $HCO_3^- + OH^-$, the base-ionization constant is calculated from the ionization constant (K_{a2}) of HCO_3^-, the conjugate acid of CO_3^{2-}.

$$K_b = \frac{K_w}{K_a} = \frac{1.0 \times 10^{-14}}{4.8 \times 10^{-11}} = 2.\underline{0}8 \times 10^{-4} = 2.1 \times 10^{-4}$$

Because the constant of CO_3^{2-} is larger, it is the stronger base.

17.93 Assume Al^{3+} is $Al(H_2O)_6^{3+}$. Assemble the usual table to calculate the $[H_3O^+]$ and pH. Use the usual equilibrium-constant expression.

Conc. (M)	$Al(H_2O)_6^{3+}$ +H_2O $\rightleftharpoons$	H_3O^+ +	$Al(H_2O)_5(OH)^{2+}$
Starting	0.15	0	0
Change	-x	+x	+x
Equilibrium	0.15 - x	x	x

$$K_a = \frac{[Al(H_2O)_5(OH)^{2+}][H_3O^+]}{[Al(H_2O)_6^{3+}]} = \frac{(x)^2}{(0.15 - x)} \cong \frac{(x)^2}{(0.15)} = 1.4 \times 10^{-5}$$

$$x = [H_3O^+] = 1.\underline{4}49 \times 10^{-3} \text{ M}$$

$$pH = -\log [H_3O^+] = -\log (1.449 \times 10^{-3}) = 2.8\underline{3}8 = 2.84$$

17.95   Calculate the concentrations of the tartaric acid (H_2Tar) and the hydrogen tartrate ion ($HTar^-$).

$$(11.0 \text{ g } H_2Tar \div 150.1 \text{ g/mol}) \div 1.00 \text{ L} = 0.07328 \text{ M } H_2Tar$$

$$(20.0 \text{ g } KHTar^- \div 188.2 \text{ g/mol}) \div 1.00 \text{ L} = 0.1063 \text{ M } [HTar^-]$$

Conc. (M)	$H_2Tar + H_2O \rightleftharpoons$	H_3O^+	$+ HTar^-$
Starting	0.07328	0	0.1063
Change	-x	+x	+x
Equilibrium	0.07328 - x	x	0.1063 + x

Ignoring x compared to 0.07328 and 0.1063 and substituting into the K_{a1} expression gives

$$K_{a1} = \frac{[H_3O^+][HTar^-]}{[H_2Tar]} = \frac{(0.1063 + x)(x)}{(0.07328 - x)} \cong \frac{(0.1063)(x)}{(0.07328)} = 1.0 \times 10^{-3}$$

$$x = [H_3O^+] = 6.\underline{8}9 \times 10^{-4} \text{ M}$$

$$pH = -\log [H_3O^+] = -\log (6.89 \times 10^{-4}) = 3.1\underline{6}1 = 3.16$$

17.97   Use the Henderson-Hasselbalch equation where $[H_2CO_3]$ = the buffer acid, $[HCO_3^-]$ = the buffer base, and K_{a1} of carbonic acid is the ionization constant.

$$pH = -\log K_a + \log \frac{[HCO_3^-]}{[H_2CO_3]} = 7.40$$

$$7.40 = -\log (4.3 \times 10^{-7}) + \log \frac{[HCO_3^-]}{[H_2CO_3]}$$

$$\log \frac{[HCO_3^-]}{[H_2CO_3]} = 7.40 - 6.366 = 1.0\underline{3}4$$

$$\frac{[HCO_3^-]}{[H_2CO_3]} = \frac{10.81}{1} = \frac{11}{1}$$

17.99 All the OH⁻ (from the NaOH) reacts with the H_3O^+ from HCl. Calculate the stoichiometric amounts of OH⁻ and H_3O^+, and subtract the mol of OH⁻ from the mol of H_3O^+. Then, divide the remaining H_3O^+ by the total volume of 0.456 L + 0.285 L, or 0.741 L, to find the $[H_3O^+]$. Then, calculate the pH.

$$Mol\ H_3O^+ = (0.10\ mol\ HCl/L) \times 0.456\ L\ HCl = 0.0456\ mol\ H_3O^+$$

$$Mol\ OH^- = (0.15\ mol\ NaOH/L) \times 0.285\ L\ NaOH = 0.0428\ mol\ OH^-$$

$$Mol\ H_3O^+\ left = (0.0456 - 0.0428)\ mol\ H_3O^+ = 0.0028\ mol\ H_3O^+$$

$$[H_3O^+] = 0.0028\ mol\ H_3O^+ \div 0.741\ L\ total\ volume = 0.0038\ M$$

$$pH = -\log[H_3O^+] = -\log(0.0038) = 2.42 = 2.4$$

17.101 Use BzN to symbolize benzylamine and BzNH⁺ to symbolize the benzylammonium cation. At the equivalence point, equal molar amounts of BzN and HCl react to form a solution of BzNHCl. Start by calculating the moles of BzN. Use this number to calculate the volume of HCl needed to neutralize all the BzN (and use the moles of BzN as the moles of BzNH⁺ formed at the equivalence point). Add the volume of HCl to the original 0.025 L to find the total volume of solution.

$$(0.025\ mol\ BzN/L) \times 0.065\ L = 0.00162\ mol\ BzN$$

$$Volume\ HCl = 0.00162\ mol\ HCl \div (0.050\ mol\ HCl/L) = 0.0324\ L$$

$$Total\ volume = 0.025\ L + 0.0324\ L = 0.0574\ L$$

$$[BzNH^+] = (0.00162\ mol\ BzNH^+\ from\ BzN) \div 0.0574\ L = 0.0282\ M$$

Because the BzNH⁺ hydrolyzes to H_3O^+ and BzN, use this to calculate the $[H_3O^+]$. Start by calculating the K_a constant of BzNH⁺ from the K_b of its conjugate base, BzN. Then, assemble the usual table of concentrations, assume x is negligible, and calculate $[H_3O^+]$ and pH.

$$K_a = \frac{K_w}{K_b} = \frac{1.0 \times 10^{-14}}{4.7 \times 10^{-10}} = 2.13 \times 10^{-5}$$

Conc. (M)	BzNH⁺ + H_2O	⇌	H_3O^+ +	BzN
Starting	0.0282		0	0
Change	-x		+x	+x
Equilibrium	0.0282 - x		x	x

(continued)

Substitute into the equilibrium-constant equation.

$$K_a = \frac{[H_3O^+][BzN]}{[BzNH^+]} = \frac{(x)^2}{(0.0282 - x)} \cong \frac{(x)^2}{(0.0282)} = 2.13 \times 10^{-5}$$

$$x = [H_3O^+] = 7.\underline{7}502 \times 10^{-4} \text{ M}$$

$$pH = -\log [H_3O^+] = -\log (7.7502 \times 10^{-4}) = 3.1\underline{1}06 = 3.11$$

17.103 a. $[H_3O^+] \cong 0.100$ M

b. The 0.100-M H_2SO_4 ionizes to 0.100-M H_3O^+ and 0.100-M HSO_4^-. Assemble the usual table, and substitute into the K_{a2} equilibrium-constant expression for H_2SO_4. Solve the resulting quadratic equation.

Conc. (M)	HSO_4^- + H_2O	$\rightleftharpoons$	H_3O^+ +	SO_4^{2-}
Starting	0.100		0.100	0
Change	-x		+x	+x
Equilibrium	0.100 - x		0.100 + x	x

$$x^2 + 0.111x - (1.10 \times 10^{-3}) = 0$$

$$x = \frac{-0.111 \pm \sqrt{(0.111)^2 + 4(1.10 \times 10^{-3})}}{2}$$

$$= \frac{-0.111 \pm 0.1293}{2}$$

$$x = 9.\underline{1}54 \times 10^{-3} \text{ M}$$

$$[H_3O^+] = 0.100 + x = 0.10\underline{0} + 0.009154 = 0.10\underline{9} = 0.11 \text{ M}$$

17.105a. From the pH, calculate the H_3O^+ ion concentration:

$$[H_3O^+] = 10^{-pH} = 10^{-5.82} = 1.\underline{5}1 \times 10^{-6} \text{ M}$$

Use the table approach, giving the starting, change, and equilibrium concentrations.

Conc. (M)	$CH_3NH_3^+ + H_2O$	$\rightleftharpoons$	CH_3NH_2	+	H_3O^+
Starting	0.10		0		0
Change	-x		+x		+x
Equilibrium	0.10 - x		x		x

From the hydronium ion concentration, $x = 1.51 \times 10^{-6}$. Substituting into the equilibrium-constant expression gives

$$K_a = \frac{[CH_3NH_2][H_3O^+]}{[CH_3NH_3^+]} = \frac{(x)^2}{(0.10 - x)} \cong \frac{(1.51 \times 10^{-6})^2}{(0.10)} = 2.\underline{2}8 \times 10^{-11}$$

$$= 2.3 \times 10^{-11}$$

b. Now, use K_w to calculate the value of K_b:

$$K_b = \frac{K_w}{K_a} = \frac{1.0 \times 10^{-14}}{2.28 \times 10^{-11}} = 4.\underline{3}8 \times 10^{-4} = 4.4 \times 10^{-4}$$

c. The equilibrium concentration of $CH_3NH_3^+$ is approximately 0.450 mol/1.00 L = 0.450 M. Use $[CH_3NH_2] \cong 0.250$ M. For K_a, we get

$$K_a = \frac{[CH_3NH_2][H_3O^+]}{[CH_3NH_3^+]} \cong \frac{(0.250)[H_3O^+]}{(0.450)} = 2.\underline{2}8 \times 10^{-11}$$

Solving for $[H_3O^+]$ gives

$$[H_3O^+] \cong \frac{(2.28 \times 10^{-11})(0.450)}{(0.250)} = 4.\underline{1}0 \times 10^{-11} \text{ M}$$

Thus, the pH is

$$pH = -\log(4.10 \times 10^{-11}) = 10.3\underline{8}7 = 10.39$$

17.107 a. True. Weak acids have small K_a values, so most of the solute is present as undissociated molecules.

b. True. Weak acids have small K_a values, so most of the solute is present as the molecule.

c. False. The hydroxide concentration equals the hydronium concentration only in neutral solutions.

d. False. If HA were a strong acid, the pH would be equal to two.

e. False. The H_3O^+ would be 0.010 if HA were a strong acid.

f. True. For every HA molecule that dissociates, one H_3O^+ is generated along with one A^-.

17.109 a. At the equivalence point, moles of acid equal moles of base. Thus,

moles acid $= 0.115$ mol/L x 0.03383 L $= 3.8\underline{9}0 \times 10^{-3}$

The molar mass is

$$\text{Molar mass} = \frac{0.288 \text{ g}}{3.890 \times 10^{-3} \text{ mol}} = 74.\underline{0}4 = 74.0 \text{ g/mol}$$

b. At the 50 percent titration point, pH = pK_a. Since the pH measurement wasn't made, it is necessary to obtain an $[A^-]/[HA]$ ratio. The ratio can be in terms of moles, percent, mL, or M because the units end up canceling out. First, use the pH to obtain the H_3O^+ ion concentration. Express the other concentrations in mL for convenience.

$[H_3O^+] = 10^{-4.92} = 1.\underline{2}0 \times 10^{-5}$ M

$HA_o \cong 33.83$ mL

$[HA] \cong 33.83 - 17.54 = 16.29$ mL

$[A^-] \cong [OH^-] = 17.54$ mL

$$K_a = [H_3O^+] \times \frac{[A^-]}{[HA]} = 1.20 \times 10^{-5} \text{ M} \times \frac{17.54 \text{ mL}}{16.29 \text{ mL}} = 1.\underline{2}9 \times 10^{-5}$$

$$= 1.3 \times 10^{-5}$$

17.111 a. Initial pH: Use the table approach, and give the starting, change, and equilibrium concentrations.

Conc. (M)	$NH_3 + H_2O \rightleftharpoons NH_4^+ + OH^-$		
Starting	0.10	0	0
Change	-x	+x	+x
Equilibrium	0.10 - x	x	x

Substituting into the equilibrium-constant expression gives

$$K_b = \frac{[NH_4^+][OH^-]}{[OH^-]} = \frac{(x)^2}{(0.10 - x)} \cong \frac{(x)^2}{(0.10)} = 1.8 \times 10^{-5}$$

Rearranging and solving for x gives

$$x = [OH^-] = \sqrt{(0.10)(1.8 \times 10^{-5})} = 1.34 \times 10^{-3} \text{ M}$$

$$pOH = -\log(1.34 \times 10^{-3}) = 2.873$$

$$pH = 14 - pOH = 14 - 2.873 = 11.127 = 11.13$$

30 percent titration point: Express the concentrations as percents for convenience.

$[NH_3] \approx 70\%$ and $[NH_4^+] \approx 30\%$

Plug into the equilibrium-constant expression.

$$K_b = \frac{[NH_4^+][OH^-]}{[NH_3]} = \frac{(30\%)(x)}{(70\%)} = 1.8 \times 10^{-5}$$

Solving for x gives

$$x = [OH^-] = 1.8 \times 10^{-5} \times \frac{70\%}{30\%} = 4.20 \times 10^{-5} \text{ M}$$

$$pOH = -\log(4.20 \times 10^{-5}) = 4.377$$

$$pH = 14 - pOH = 14 - 4.377 = 9.623 = 9.62$$

(continued)

50 percent titration point:

$[NH_4^+] = [NH_3]$

$K_b = [OH^-] = 1.8 \times 10^{-5}$ M

$pOH = -\log(1.8 \times 10^{-5}) = 4.7\underline{4}4$

$pH = 14 - pOH = 14 - 4.744 = 9.2\underline{5}6 = 9.26$

100 percent titration point:

The NH_4Cl that is produced has undergone a two-fold dilution.

$[NH_4^+] = 0.05\underline{0}0$ M

Use the table approach, and give the starting, change, and equilibrium concentrations.

Conc. (M)	$NH_4^+ + H_2O \rightleftharpoons$	$NH_3 +$	H_3O^+
Starting	0.0500	0	0
Change	-x	+x	+x
Equilibrium	0.0500 - x	x	x

Now, use K_w to calculate the value of K_a:

$$K_a = \frac{K_w}{K_b} = \frac{1.0 \times 10^{-14}}{1.8 \times 10^{-5}} = 5.\underline{5}5 \times 10^{-10}$$

Plug into the equilibrium-constant expression.

$$K_a = \frac{[NH_3][H_3O^+]}{[NH_4^+]} = \frac{(x)^2}{(0.0500 - x)} \cong \frac{(x)^2}{(0.0500)} = 5.55 \times 10^{-10}$$

Solving for x gives

$$x = [H_3O^+] = \sqrt{0.0500 \times \frac{1.0 \times 10^{-14}}{1.8 \times 10^{-5}}} = 5.\underline{2}7 \times 10^{-6}$$ M

$pH = -\log(5.27 \times 10^{-6}) = 5.2\underline{7}8 = 5.28$

b. The solution is acidic because the NH_4^+ ion reacts with water to produce acid. Ammonium chloride is the salt of a weak base and a strong acid.

17.113a. Select the conjugate pair that has a pK_a value closest to a pH of 2.88.

$$pK_a\ (H_2C_2O_4)\ =\ 1.25$$

$$pK_a\ (H_3PO_4)\ =\ 2.16$$

$$pK_a\ (HCOOH)\ =\ 3.77$$

Therefore, the best pair is H_3PO_4 and $H_2PO_4^-$.

b. Using the Henderson-Hasselbalch equation,

$$pH\ =\ pK_a\ +\ \log \frac{[A^-]}{[HA]}$$

$$2.88\ =\ 2.16\ +\ \log \frac{[A^-]}{[HA]}$$

$$\log \frac{[A^-]}{[HA]}\ =\ 0.72$$

$$\frac{[A^-]}{[HA]}\ =\ 10^{0.72}\ =\ 5.\underline{2}4$$

Therefore, 5.24 times more conjugate base is needed than acid. Since the starting concentrations of H_3PO_4 (HA) and $H_2PO_4^-$ (A^-) are the same, the volume of A^- needed is 5.24 times the volume of HA needed. For 50 mL of buffer, this is

$$50\ mL\ =\ vol\ HA\ +\ vol\ A^-\ =\ vol\ HA\ +\ 5.24\ vol\ HA\ =\ 6.24\ vol\ HA$$

Therefore, the volume of 0.10 M H_3PO_4 required is

$$vol\ H_3PO_4\ =\ \frac{50\ mL}{6.24}\ =\ 8.\underline{0}1\ mL\ =\ 8\ mL$$

The volume of 0.10 M $H_2PO_4^-$ required is

$$50\ mL\ -\ 8.01\ mL\ =\ 4\underline{2}.0\ mL\ =\ 42\ mL$$

17.115a. At the equivalence point, moles of base equal moles of acid. Therefore,

$$M_{NH_2OH} \times 25.0 \text{ mL} = 0.150 \text{ M} \times 35.8 \text{ mL}$$

$$M_{NH_2OH} = \frac{0.150 \text{ M} \times 35.8 \text{ mL}}{25.0 \text{ mL}} = 0.21\underline{4}8 = 0.215$$

b. Concentration of NH_3OH^+ at the equivalence point is

$$[NH_3OH^+] = \frac{(25.0 \text{ mL})(0.215 \text{ M})}{25.0 \text{ mL} + 35.8 \text{ mL}} = 0.088\underline{4}0 \text{ M}$$

$$K_a = \frac{K_w}{K_b} = \frac{1.0 \times 10^{-14}}{1.1 \times 10^{-8}} = 9.\underline{0}9 \times 10^{-7}$$

Use the table approach, and give the starting, change, and equilibrium concentrations.

Conc. (M)	NH_3OH^+ + H₂O	⇌	NH_2OH +	H_3O^+
Starting	0.088\underline{4}0		0	0
Change	-x		+x	+x
Equilibrium	0.088\underline{4}0 - x		x	x

Substituting into the equilibrium-constant expression gives

$$K_a = \frac{[NH_2OH][H_3O^+]}{[NH_3OH^+]} = \frac{(x)^2}{(0.08840 - x)} \cong \frac{(x)^2}{(0.08840)} = 9.\underline{0}9 \times 10^{-7}$$

Rearranging and solving for x gives

$$x = [H_3O^+] = \sqrt{(0.08840)(9.\underline{0}9 \times 10^{-7})} = 2.\underline{8}3 \times 10^{-4} \text{ M}$$

$$pH = -\log(2.\underline{8}4 \times 10^{-4}) = 3.5\underline{4}7 = 3.55$$

c. You need an indicator to change pH around pH of 3 to 4. Therefore, the appropriate indicator is bromophenol blue. Select an indicator that changes color around the equivalence point.

17.117 a. $H_3O^+(aq) + NH_3(aq) \rightarrow NH_4^+(aq) + H_2O(l)$

b. $NH_4^+(aq) + OH^-(aq) \rightarrow NH_3(aq) + H_2O(l)$

c. Initial:

mol NH_3 = 1.0 mol/L x 0.100 L = 0.1$\underline{0}$0 = 0.10 mol

mol NH_4^+ = 0.50 mol/L x 0.100 L x $\dfrac{2 \text{ mol } NH_4^+}{1 \text{ mol } (NH_4)_2SO_4}$ = 0.1$\underline{0}$0 = 0.10 mol

After addition of HCl: The HCl is completely consumed by producing 0.01$\underline{0}$0 mol of NH_4^+. The amount of NH_3 is decreased by 0.01$\underline{0}$0 mol.

mol HCl added = 1.00 mol/L x 0.0100 L = 0.010$\underline{0}$0 mol

mol NH_3 = 0.1$\underline{0}$0 - 0.010$\underline{0}$0 = 0.09$\underline{0}$0 = 0.090 mol

mol NH_4^+ = 0.1$\underline{0}$0 + 0.010$\underline{0}$0 = 0.1$\underline{1}$0 = 0.11 mol

The reaction is:

$NH_3 + H_2O \rightleftharpoons NH_4^+ + OH^-$

Substituting into the equilibrium-constant expression gives

$K_b = \dfrac{[NH_4^+][OH^-]}{[NH_3]} = \dfrac{(0.110)[OH^-]}{(0.0\underline{9}0)} = 1.8 \times 10^{-5}$

Rearranging and solving for [OH⁻] gives

$[OH^-] = 1.8 \times 10^{-5} \times \dfrac{0.0\underline{9}0}{0.110} = \underline{1}.47 \times 10^{-5} M$

pOH = - log ($\underline{1}$.47 x 10⁻⁵) = 4.$\underline{8}$3

pH = 14 - pOH = 14 - 4.$\underline{8}$3 = 9.$\underline{1}$7 = 9.2

d. This is a buffer system, so the ratio of NH_4^+/NH_3 has not changed very much.

17.119a. From the pH $[H_3O^+] = 10^{-7.44} = 3.\underline{6}3 \times 10^{-8}$ M. The reaction is

$$H_2PO_4^-(aq) + H_2O \rightleftharpoons H_3O^+(aq) + HPO_4^{2-}(aq) \qquad K_{a_2} = 6.2 \times 10^{-8}$$

$$K_{a_2} = \frac{[HPO_4^{2-}][H_3O^+]}{[H_2PO_4^-]}$$

Rearranging gives

$$\frac{[H_2PO_4^-]}{[HPO_4^{2-}]} = \frac{[H_3O^+]}{K_{a_2}} = \frac{3.63 \times 10^{-8}}{6.2 \times 10^{-8}} = \frac{0.5\underline{8}55}{1} = 0.59 \text{ to } 1$$

Or, using the Henderson-Hasselbalch equation

$$pH = pK_a + \log \frac{[base]}{[acid]}$$

$$7.44 = 7.21 + \log \frac{[base]}{[acid]}$$

$$\log \frac{[base]}{[acid]} = 0.23$$

$$\frac{[base]}{[acid]} = 10^{0.23} = 1.\underline{7}1$$

You must invert each side. The result is

$$\frac{[acid]}{[base]} = \frac{1}{1.71} = \frac{0.5\underline{8}57}{1} = 0.59 \text{ to } 1$$

b. $[H_3O^+] = K_{a_2} = \dfrac{[HPO_4^{2-}]}{[H_2PO_4^-]}$

If $HPO_4^{2-} = 1.00$ mol, then when 25 percent of it is converted into $H_2PO_4^-$, 0.75 mol remain, and $H_2PO_4^-$ becomes $0.5\underline{8}6 + 0.25 = 0.8\underline{3}6$ mol. Therefore,

$$[H_3O^+] = 6.2 \times 10^{-8} \times \frac{0.836}{0.75} = 6.\underline{9}1 \times 10^{-8}$$

$$pH = -\log(6.\underline{9}1 \times 10^{-8}) = 7.1\underline{6}1 = 7.16$$

(continued)

c. Assume $H_2PO_4^- = 0.586$ mol. Then, $H_2PO_4^-$ remaining $= 0.586 - (0.15 \times 0.586)$

$$= 0.85 \times 0.586 = 0.498 \text{ mol left}$$

The moles of $HPO_4^{2-} = 1.00 + (0.15 \times 0.586) = 1.088$ mol.

$$[H_3O^+] = 6.2 \times 10^{-8} \times \frac{0.498}{1.088} = 2.84 \times 10^{-8} \text{ M}$$

$$pH = -\log(2.84 \times 10^{-8}) = 7.547 = 7.55$$

17.121 a.

$$H_2A + H_2O \rightleftharpoons H_3O^+ + HA^- \qquad K_{a_1}$$

$$HA^- + H_2O \rightleftharpoons H_3O^+ + A^{2-} \qquad K_{a_2}$$

$$H_2A + 2 H_2O \rightleftharpoons 2 H_3O^+ + A^{2-}, \qquad K = K_{a_1} \times K_{a_2}$$

b. $H_2A \gg H_3O^+ = HA^- \gg A^{2-}$

c.
$$H_2A + H_2O \rightleftharpoons H_3O^+ + HA^- \qquad K_{a_1} = 1.0 \times 10^{-3}$$

$$0.0250 - x \qquad\qquad x \qquad x$$

$$K_{a_1} = \frac{[HA^-][H_3O^+]}{[H_2A]} = \frac{(x)^2}{(0.0250 - x)} = 1.0 \times 10^{-3}$$

Rearranging into a quadratic equation and solving for x gives

$$x^2 + (1.0 \times 10^{-3})x + (-2.50 \times 10^{-5}) = 0$$

$$x = \frac{-(1.0 \times 10^{-3}) \pm \sqrt{(1.0 \times 10^{-3})^2 - (4)(1)(-2.50 \times 10^{-5})}}{2}$$

$$x = [H_3O^+] = [HA^-] = 4.52 \times 10^{-3} = 4.5 \times 10^{-3} \text{ M}$$

$$pH = -\log(4.52 \times 10^{-3}) = 2.344 = 2.34$$

$$[H_2A] = 0.0250 - x = 0.0250 - 4.52 \times 10^{-3} = 0.02048 = 0.0205 \text{ M}$$

(continued)

d.
$$HA^- + H_2O \rightleftharpoons H_3O^+ + A^{2-} \qquad K_{a_2} = 4.6 \times 10^{-5}$$

$4.52 \times 10^{-3} - y \qquad\qquad 4.52 \times 10^{-3} + y \quad y$

Assume y is small compared to 4.52×10^{-3} M. Substitute into the rearranged equilibrium-constant equation to get

$$[A^{2-}] = \frac{[HA^-]K_{a_2}}{[H_3O^+]} = \frac{(4.52 \times 10^{-3})(4.6 \times 10^{-5})}{(4.52 \times 10^{-3})} = 4.6 \times 10^{-5} \text{ M}$$

17.123 a. From the pH, $[H_3O^+] = 10^{-4.45} = 3.\underline{55} \times 10^{-5}$ M. The $NaCH_3COO$ is generated by the acid-base reaction, resulting in the formation of a buffer solution. The reaction is

$$CH_3COOH + H_2O \rightleftharpoons H_3O^+ + CH_3COO^- \qquad K_a = 1.7 \times 10^{-5}$$

$0.15 - x \qquad\qquad\qquad x \qquad x$

$$K_a = \frac{[H_3O^+][CH_3COO^-]}{[CH_3COOH]}$$

Rearranging gives

$$\frac{[CH_3COO^-]}{[CH_3COOH]} = \frac{K_a}{[H_3O^+]}$$

$$\frac{x}{0.15 - x} = \frac{1.7 \times 10^{-5}}{3.55 \times 10^{-5}}$$

Solving for x gives x = 0.04\underline{86} mol. Therefore, the molar concentration is

$$[CH_3COO^-] = 0.0486 \text{ mol}/0.375 \text{ L} = 0.1\underline{296} = 0.13 \text{ M}$$

b. moles $OH^- = $ moles CH_3COO^-

$0.04\underline{86}$ mol = 0.25 M x V_{NaOH}

V_{NaOH} = 0.0486 mol/0.25 M = 0.1\underline{94} = 0.19 L (1.9×10^2 mL)

c. The volume of the original acid is V = 0.375 L - 0.1\underline{94} L = 0.1\underline{81} L. Therefore, the concentration of the original acid is

$$[CH_3COOH] = 0.15 \text{ mol}/0.181 \text{ L} = 0.8\underline{29} = 0.83 \text{ M}$$

■ Solutions to Cumulative-Skills Problems

17.125 Use the pH to calculate $[H_3O^+]$, and then use the K_a of 1.7×10^{-5} and the K_a expression to calculate the molarity of acetic acid (HAc), assuming ionization is negligible. Convert molarity to mass percentage using the formula weight of 60.05 g/mol of HAc.

$$[H_3O^+] = \text{antilog } (-2.45) = 3.\underline{5}48 \times 10^{-3} \text{ M}$$

Write the equilibrium-constant expression in terms of chemical symbols, and then substitute the x and the (0.003548 M) terms into the expression:

$$K_a = \frac{[H_3O^+][Ac^-]}{[HAc]} \cong \frac{(0.003548)^2}{(x)}$$

Solve the equation for x, assuming 0.003548 is much smaller than x.

$$x = (0.003548)^2 \div (1.7 \times 10^{-5}) \cong [HAc] \cong 0.7\underline{4}04 \text{ M}$$

$$(0.7404 \text{ mol HAc/L}) \times (60.05 \text{ g/mol}) \times (1 \text{ L}/1090 \text{ g})(100\%)$$

$$= 4.\underline{0}78 = 4.1 \text{ percent HAc}$$

17.127 Find the $[H_3O^+]$ and $[C_2H_3O_2^-]$ by solving for the approximate $[H_3O^+]$, noting x is much smaller than the starting 0.92 M of acetic acid. The usual table is used but is not shown; only the final setup for $[H_3O^+]$ is shown. Use $K_f = 1.858$ °C/m for the constant for water.

$$[H_3O^+] = [C_2H_3O_2^-] = (1.7 \times 10^{-5} \times 0.92)^{1/2} = 0.003\underline{9}54 \text{ M}$$

Total molarity of acid + ions $= 0.92 + 0.003954 = 0.9\underline{2}39$

Molality $= m = 0.9239 \text{ mol} \div (1.000 \text{ L} \times 0.953 \text{ kg H}_2\text{O/L}) = 0.9695 \text{ m}$

Freezing point $= -\Delta T_f = -K_f c_m = (-1.858 \text{ °C/m}) \times 0.9695 \text{ m} = -1.801 = -1.8 \text{ °C}$

17.129 The $[H_3O^+] = $ -antilog $(-4.35) = 4.\underline{4}6 \times 10^{-5}$ M. Note that 0.465 L of 0.0941 M NaOH will produce 0.043756 mol of acetate, Ac^-, ion. Rearranging the K_a expression for acetic acid (HAc) and Ac^- and canceling the volume in the mol/L of each, you obtain:

$$\frac{[HAc]}{[Ac^-]} = \frac{[H^+]}{K_a} \; = \; \frac{4.46 \times 10^{-5}}{1.7 \times 10^{-5}} = \frac{2.63}{1.00} \cong \frac{x \text{ mol HAc}}{0.043756 \text{ mol } Ac^-}$$

$$x = 0.1\underline{1}49 \text{ mol HAc}$$

Total mol HAc added $= 0.1149 + 0.043756 = 0.1587$ mol HAc

Mol/L of pure HAc $= 1049$ g HAc/L $\times$ (1 mol HAc/60.05 g) $= 17.4\underline{6}7$ mol/L

L of pure HAc needed $= 0.1587$ mol HAc $\times$ (L/17.467 mol) $= 0.009\underline{0}8$ L (9.1 mL)

$$HC_2H_3O_2 + H_2O \rightleftharpoons H_3O + C_2H_3O_2^-$$

18. SOLUBILITY AND COMPLEX-ION EQUILIBRIA

■ Solutions to Exercises

Note on significant figures: If the final answer to a solution needs to be rounded off, it is given first with one nonsignificant figure, and the last significant figure is underlined. The final answer is then rounded to the correct number of significant figures. In multiple-step problems, intermediate answers are given with at least one nonsignificant figure; however, only the final answer has been rounded off.

18.1 a. $BaSO_4(s) \rightleftharpoons Ba^{2+}(aq) + SO_4^{2-}(aq)$ $K_{sp} = [Ba^{2+}][SO_4^{2-}]$

b. $Fe(OH)_3(s) \rightleftharpoons Fe^{3+}(aq) + 3OH^-(aq)$ $K_{sp} = [Fe^{3+}][OH^-]^3$

c. $Ca_3(PO_4)_2(s) \rightleftharpoons 3Ca^{2+}(aq) + 2PO_4^{3-}(aq);$ $K_{sp} = [Ca^{2+}]^3[PO_4^{3-}]^2$

18.2   Calculate the molar solubility. Then, assemble the usual concentration table, and substitute the equilibrium concentrations from it into the equilibrium-constant expression. (Because no concentrations can be given for solid AgCl, dashes are written; in later problems, similar spaces will be left blank.)

$$\frac{1.9 \times 10^{-3} \text{ g}}{1 \text{ L}} \times \frac{1 \text{ mol}}{143 \text{ g}} = 1.3\underline{3} \times 10^{-5} \text{ M}$$

(continued)

Set up the table as usual.

Conc. (M)	$AgCl(s)$ ⇌	Ag^+	+	Cl^-
Starting		0		0
Change		$+1.33 \times 10^{-5}$		$+1.33 \times 10^{-5}$
Equilibrium		1.33×10^{-5}		1.33×10^{-5}

$$K_{sp} = [Ag^+][Cl^-] = (1.33 \times 10^{-5})(1.33 \times 10^{-5}) = 1.\underline{7}68 \times 10^{-10} = 1.8 \times 10^{-10}$$

18.3 Calculate the molar solubility. Then, assemble the usual concentration table and substitute from it into the equilibrium-constant expression. (Because no concentrations can be given for solid $Pb_3(AsO_4)_2$, spaces are left blank.)

$$\frac{3.0 \times 10^{-5} \text{ g}}{1 \text{ L}} \times \frac{1 \text{ mol}}{899 \text{ g}} = 3.\underline{3}4 \times 10^{-8} \text{ M}$$

Conc. (M)	$Pb_3(AsO_4)_2(s)$ ⇌	$3Pb^{2+}$	+	$2AsO_4^{3-}$
Starting		0		0
Change		$+3(3.34 \times 10^{-8})$		$+2(3.34 \times 10^{-8})$
Equilibrium		$3(3.34 \times 10^{-8})$		$2(3.34 \times 10^{-8})$

$$K_{sp} = [Pb^{2+}]^3[AsO_4^{3-}]^2 = [3 \times (3.34 \times 10^{-8})]^3(2 \times 3.34 \times 10^{-8})^2$$

$$= 4.\underline{4}89 \times 10^{-36} = 4.5 \times 10^{-36}$$

18.4 Assemble the usual concentration table. Let x equal the molar solubility of $CaSO_4$. When x mol $CaSO_4$ dissolves in one L of solution, x mol Ca^{2+} and x mol SO_4^{2-} form.

Conc. (M)	$CaSO_4(s)$ ⇌	Ca^{2+}	+	SO_4^{2-}
Starting		0		0
Change		$+x$		$+x$
Equilibrium		x		x

Substitute the equilibrium concentrations into the equilibrium-constant expression and solve for x. Then, convert to g $CaSO_4$ per L.

(continued)

$$[Ca^{2+}][SO_4^{2-}] = K_{sp}$$

$$(x)(x) = x^2 = 2.4 \times 10^{-5}$$

$$x = \sqrt{(2.4 \times 10^{-5})} = 4.\underline{8}9 \times 10^{-3} \text{ M}$$

$$\frac{4.89 \times 10^{-3} \text{ mol}}{L} \times \frac{136 \text{ g}}{1 \text{ mol}} = 0.6\underline{6}64 = 0.67 \text{ g/L}$$

18.5 a. Let x equal the molar solubility of BaF_2. Assemble the usual concentration table, and substitute from the table into the equilibrium-constant expression.

Conc. (M)	$BaF_2(s)$	$\rightleftharpoons$	Ba^{2+}	+	$2F^-$
Starting			0		0
Change			+x		+2x
Equilibrium			x		2x

$$[Ba^{2+}][F^-]^2 = K_{sp}$$

$$(x)(2x)^2 = 4x^3 = 1.0 \times 10^{-6}$$

$$x = \sqrt[3]{\frac{1.0 \times 10^{-6}}{4}} = 6.\underline{2}99 \times 10^{-3} = 6.3 \times 10^{-3} \text{ M}$$

b. At the start, before any BaF_2 dissolves, the solution contains 0.15 M F^-. At equilibrium, x mol of solid BaF_2 dissolves to yield x mol Ba^{2+} and 2x mol F^-. Assemble the usual concentration table, and substitute the equilibrium concentrations into the equilibrium-constant expression. As an approximation, assume x is negligible compared to 0.15 M F^-.

Conc. (M)	$BaF_2(s)$	$\rightleftharpoons$	Ba^{2+}	+	$2F^-$
Starting			0		0.15
Change			+x		+2x
Equilibrium			x		0.15 + 2x

$$[Ba^{2+}][F^-]^2 = K_{sp}$$

(continued)

$$(x)(0.15 + 2x)^2 \cong (x)(0.15)^2 \cong 1.0 \times 10^{-6}$$

$$x \cong \frac{1.0 \times 10^{-6}}{(0.15)^2} = 4.\underline{4}44 \times 10^{-5} = 4.4 \times 10^{-5} \text{ M}$$

Note that adding 2x to 0.15 M will not change it (to two significant figures), so 2x is negligible compared to 0.15 M. The solubility of 4.4×10^{-5} M in 0.15 M NaF is lower than the solubility of 6.3×10^{-3} M in pure water.

18.6   Calculate the ion product, Q_c, after evaporation, assuming no precipitation has occurred. Compare it with the K_{sp}.

$$Q_c = [Ca^{2+}][SO_4^{2-}]$$

$$Q_c = (2 \times 0.0052)(2 \times 0.0041) = 8.528 \times 10^{-5}$$

Since $Q_c > K_{sp}$ (2.4×10^{-5}), precipitation occurs.

18.7   Calculate the concentrations of Pb^{2+} and SO_4^{2-}, assuming no precipitation. Use a total volume of 0.456 L + 0.255 L, or 0.711 L.

$$[Pb^{2+}] = \frac{\dfrac{0.00016 \text{ mol}}{L} \times 0.255 \text{ L}}{0.711 \text{ L}} = 5.\underline{7}4 \times 10^{-5} \text{ M}$$

$$[SO_4^{2-}] = \frac{\dfrac{0.00023 \text{ mol}}{L} \times 0.456 \text{ L}}{0.711 \text{ L}} = 1.\underline{4}8 \times 10^{-4} \text{ M}$$

Calculate the ion product, and compare it to K_{sp}.

$$Q_c = [Pb^{2+}][SO_4^{2-}] = (5.74 \times 10^{-5})(1.48 \times 10^{-4}) = 8.\underline{4}9 \times 10^{-9}$$

Because Q_c is less than the K_{sp} of 1.7×10^{-8}, no precipitation occurs, and the solution is unsaturated.

18.8 The solubility of AgCN would increase as the pH decreases because the increasing concentration of H_3O^+ would react with the CN^- to form the weakly ionized acid HCN. As CN^- is removed, more AgCN dissolves to replace the cyanide:

$$AgCN(s) \rightleftharpoons Ag^+(aq) + CN^-(aq) \quad [+ H_3O^+ \rightarrow HCN + H_2O]$$

In the case of AgCl, the chloride ion is the conjugate base of a strong acid and would, therefore, not be affected by any amount of hydrogen ion.

18.9 Because $K_f = 4.8 \times 10^{12}$ and because the starting concentration of NH_3 is much larger than that of the Cu^{2+} ion, you can make a rough assumption that most of the copper(II) is converted to $Cu(NH_3)_4^{2+}$ ion. This ion then dissociates slightly to give a small concentration of Cu^{2+} and additional NH_3. The amount of NH_3 remaining at the start after reacting with 0.015 M Cu^{2+} is

$$[0.100 \text{ M} - (4 \times 0.015 \text{ M})] = 0.040 \text{ M starting } NH_3$$

Assemble the usual concentration table using this starting concentration for NH_3 and assuming the starting concentration of Cu^{2+} is zero.

Conc. (M)	$Cu(NH_3)_4^{2+}$	$\rightleftharpoons$	Cu^{2+}	+	$4NH_3$
Starting	0.015		0		0.040
Change	-x		+x		+4x
Equilibrium	0.015 - x		x		0.040 + 4x

Even though this reaction is the opposite of the equation for the formation constant, the formation-constant expression can be used. Simply substitute all exact equilibrium concentrations into the formation-constant expression; then, simplify the exact equation by assuming x is negligible compared to 0.015 and 4x is negligible compared to 0.040.

$$K_f = \frac{[Cu(NH_3)_4^{2+}]}{[Cu^{2+}][NH_3]^4} = \frac{(0.015 - x)}{(x)(0.040 + 4x)^4} \cong \frac{(0.015)}{(x)(0.040)^4} \cong 4.8 \times 10^{-12}$$

Rearrange and solve for x:

$$x = [Cu^{2+}] \cong (0.015) \div [(4.8 \times 10^{12})(0.040)^4] \cong 1.\underline{22} \times 10^{-9} = 1.2 \times 10^{-9} \text{ M}$$

18.10 Start by calculating the $[Ag^+]$ in equilibrium with the $Ag(CN)_2^-$ formed from Ag^+ and CN^-. Then, use the $[Ag^+]$ to decide whether or not AgI will precipitate by calculating the ion product and comparing it with the K_{sp} of 8.3×10^{-17} for AgI. Assume all the 0.0045 M Ag^+ reacts with CN^- to form 0.0045 M $Ag(CN)_2^-$, and calculate the remaining CN^-. Use these as starting concentrations for the usual concentration table.

$[0.20 \text{ M KCN} - (2 \times 0.0045 \text{ M})] = 0.191$ M starting CN^-

Conc. (M)	$Ag(CN)_2^-$	$\rightleftharpoons$	Ag^+	+	$2CN^-$
Starting	0.0045		0		0.191
Change	-x		+x		+2x
Equilibrium	0.0045 - x		x		0.191 + 2x

Even though this reaction is the opposite of the equation for the formation constant, the formation-constant expression can be used. Simply substitute all exact equilibrium concentrations into the formation-constant expression; then, simplify the exact equation by assuming x is negligible compared to 0.0045 and 2x is negligible compared to 0.191.

$$K_f = \frac{[Ag(CN)_2^-]}{[Ag^+][CN^-]^2} = \frac{(0.0045 - x)}{(x)(0.191 + 2x)^2} \cong \frac{(0.0045)}{(x)(0.191)^2} \cong 5.6 \times 10^{18}$$

Rearrange and solve for x:

$$x = [Ag^+] \cong (0.0045) \div [(5.6 \times 10^{18})(0.191)^2] \cong 2.\underline{2}02 \times 10^{-20} \text{ M}$$

Now, calculate the ion product for AgI:

$$Q_c = [Ag^+][I^-] = (2.20 \times 10^{-20})(0.15) = 3.\underline{3}0 \times 10^{-21} = 3.3 \times 10^{-21}$$

Because Q_c is less than the K_{sp} of 8.3×10^{-17}, no precipitate will form, and the solution is unsaturated.

18.11 Obtain the overall equilibrium constant for this reaction from the product of the individual equilibrium constants of the two individual equations whose sum gives this equation:

$$AgBr(s) \rightleftharpoons Ag^+(aq) + Br^-(aq) \qquad K_{sp} = 5.0 \times 10^{-13}$$

$$Ag^+(aq) + 2S_2O_3^{2-}(aq) \rightleftharpoons Ag(S_2O_3)_2^{3-}(aq) \qquad K_f = 2.9 \times 10^{13}$$

$$AgBr(s) + 2S_2O_3^{2-}(aq) \rightleftharpoons Ag(S_2O_3)_2^{3-}(aq) + Br^-(aq)$$

$$K_c = K_{sp} \times K_f = 14.5$$

Assemble the usual table using 1.0 M as the starting concentration of $S_2O_3^{2-}$ and x as the unknown concentration of $Ag(S_2O_3)_2^{3-}$ formed.

Conc. (M)	AgBr(s)	+ $2S_2O_3^{2-}$	$\rightleftharpoons$	$Ag(S_2O_3)_2^{3-}$	+	Br^-
Starting		1.0		0		0
Change		-2x		+x		+x
Equilibrium		1.0 - 2x		x		x

The equilibrium-constant expression can now be used. Simply substitute all exact equilibrium concentrations into the equilibrium-constant expression. The solution can be obtained without using the quadratic equation.

$$K_c = \frac{[Ag(S_2O_3)_2^{3-}][Br^-]}{[S_2O_3^{2-}]^2} = \frac{(x)^2}{(1.0 - 2x)^2} = 14.5$$

Take the square root of both sides of the two right-hand terms, and solve for x:

$$\frac{x}{(1.0 - 2x)} = 3.\underline{8}08$$

$$x = 3.808(1.0 - 2x)$$

$$7.62x + x = 3.808$$

$$x = 0.4\underline{4}17 = 0.44 \text{ M} \quad \text{(Molar solubility of AgBr in 1.0 M Na}_2S_2O_3)$$

■ Answers to Review Questions

18.1 The solubility equation is $Ni(OH)_2(s) \rightleftharpoons Ni^{2+}(aq) + 2OH^-(aq)$. If the molar solubility of $Ni(OH)_2$ = x molar, the concentrations of the ions in the solution must be x M Ni^{2+} and 2x M OH^-. Substituting into the equilibrium-constant expression gives

$$K_{sp} = [Ni^{2+}][OH^-]^2 = (x)(2x)^2 = 4x^3$$

18.2 Calcium sulfate is less soluble in a solution containing sodium sulfate because the increase in sulfate from the sodium sulfate causes the equilibrium composition in the equation below to shift to the left:

$$CaSO_4(s) \rightleftharpoons Ca^{2+}(aq) + SO_4^{2-}(aq)$$

The result is a decrease in both the calcium ion and the calcium sulfate concentrations.

18.3 Substitute the 0.10 M concentration of chloride into the solubility product expression and solve for $[A^+]$:

$$[Ag^+] = \frac{K_{sp}}{[Cl^-]} = \frac{1.8 \times 10^{-10}}{0.10} = 1.\underline{8}0 \times 10^{-9} = 1.8 \times 10^{-9} \text{ M}$$

18.4 In order to predict whether or not PbI_2 will precipitate when lead nitrate and potassium iodide are mixed, the concentrations of Pb^{2+} and I^- after mixing would have to be calculated first (if the concentrations are not known or are not given). Then, the value of Q_c, the ion product, would have to be calculated for PbI_2. Finally, Q_c would have to be compared with the value of K_{sp}. If $Q_c > K_{sp}$, then a precipitate will form at equilibrium. If Q_c is ≤ than K_{sp}, no precipitate will form.

18.5 Barium fluoride, normally insoluble in water, dissolves in dilute hydrochloric acid because the fluoride ion, once it forms, reacts with the hydronium ion to form weakly ionized HF:

$$BaF_2(s) \rightleftharpoons Ba^{2+}(aq) + 2F^-(aq) \quad [+ 2H_3O^+ \rightarrow 2HF + 2H_2O]$$

18.6 Metal ions such as Pb^{2+} and Zn^{2+} are separated by controlling the $[S^{2-}]$ in a solution of saturated H_2S by means of adjusting the pH correctly. Because the K_{sp} of 2.5×10^{-27} for PbS is smaller than the K_{sp} of 1.1×10^{-21} for ZnS, the pH can be adjusted to make the $[S^{2-}]$ just high enough to precipitate PbS without precipitating ZnS.

18.7   When NaCl is first added to a solution of $Pb(NO_3)_2$, a precipitate of $PbCl_2$ forms. As more NaCl is added, the excess chloride reacts further with the insoluble $PbCl_2$, forming soluble complex ions of $PbCl_3^-$ and $PbCl_4^{2-}$:

$$Pb^{2+}(aq) \ + \ 2Cl^-(aq) \ \rightleftharpoons \ PbCl_2(s)$$

$$PbCl_2(s) \ + \ Cl^-(aq) \ \rightleftharpoons \ PbCl_3^-(aq)$$

$$PbCl_3^-(aq) \ + \ Cl^-(aq) \ \rightleftharpoons \ PbCl_4^{2-}(aq)$$

18.8   When a small amount of NaOH is added to a solution of $Al_2(SO_4)_3$, a precipitate of $Al(OH)_3$ forms at first. As more NaOH is added, the excess hydroxide ion reacts further with the insoluble $Al(OH)_3$, forming a soluble complex ion of $Al(OH)_4^-$.

18.9   The Ag^+, Cu^{2+}, and Ni^{2+} ions can be separated in two steps: (1) Add HCl to precipitate just the Ag^+ as AgCl, leaving the others in solution. (2) After pouring the solution away from the precipitate, add 0.3 M HCl and H_2S to precipitate only the CuS away from the Ni^{2+} ion, whose sulfide is soluble under these conditions.

18.10   By controlling the pH through the appropriate buffer, one can control the $[CO_3^{2-}]$ using the equilibrium reaction:

$$H_3O^+(aq) + CO_3^{2-}(aq) \ \rightleftharpoons \ HCO_3^-(aq) + H_2O(l)$$

Calcium carbonate is much more insoluble than magnesium carbonate and thus will precipitate in weakly basic solution, whereas magnesium carbonate will not. Magnesium carbonate will precipitate only in highly basic solution. (There is the possibility that $Mg(OH)_2$ might precipitate, but the $[OH^-]$ is too low for this to occur.)

■ Solutions to Practice Problems

Note on significant figures: If the final answer to a solution needs to be rounded off, it is given first with one nonsignificant figure, and the last significant figure is underlined. The final answer is then rounded to the correct number of significant figures. In multiple-step problems, intermediate answers are given with at least one nonsignificant figure; however, only the final answer has been rounded off.

18.19 a. NaBr is soluble (Group IA salts are soluble).

b. PbI_2 is insoluble (PbI_2 is an insoluble iodide).

c. $BaCO_3$ is insoluble (Carbonates are generally insoluble).

d. $(NH_4)_2SO_4$ is soluble (All ammonium salts are soluble).

18.21 a. $K_{sp} = [Mg^{2+}][OH^-]^2$ b. $K_{sp} = [Sr^{2+}][CO_3^{2-}]$

      c. $K_{sp} = [Ca^{2+}]^3[AsO_4^{3-}]^2$ d. $K_{sp} = [Fe^{3+}][OH^-]^3$

18.23   Calculate molar solubility, assemble the usual table, and substitute the equilibrium concentrations from it into the equilibrium-constant expression. (Dashes are given for $AgBrO_3(s)$; in later problems, the dashes are omitted.)

$$\frac{7.2 \times 10^{-3} \text{ g}}{L} \times \frac{1 \text{ mol}}{236 \text{ g}} = 3.\underline{0}5 \times 10^{-5} \text{ M}$$

Conc. (M)	$AgBrO_3(s)$ ⇌	Ag^+	+	BrO_3^-
Starting	—	0		0
Change	—	$+3.05 \times 10^{-5}$		$+3.05 \times 10^{-5}$
Equilibrium	—	3.05×10^{-5}		3.05×10^{-5}

$K_{sp} = [Ag^+][BrO_3^-] = (3.05 \times 10^{-5})(3.05 \times 10^{-5}) = 9.\underline{3}02 \times 10^{-10} = 9.3 \times 10^{-10}$

18.25   Calculate the molar solubility. Then, assemble the usual concentration table, and substitute the equilibrium concentrations from it into the equilibrium-constant expression.

$$\frac{0.13 \text{ g}}{0.100 \text{ L}} \times \frac{1 \text{ mol}}{413 \text{ g}} = 3.\underline{1}47 \times 10^{-3} \text{ M}$$

Conc. (M)	$Cu(IO_3)_2(s)$ ⇌	Cu^{2+}	+	$2IO_3^-$
Starting		0		0
Change		$+3.147 \times 10^{-3}$		$+2 \times 3.147 \times 10^{-3}$
Equilibrium		3.147×10^{-3}		$2 \times 3.147 \times 10^{-3}$

$K_{sp} = [Cu^{2+}][IO_3^-]^2 = (3.147 \times 10^{-3})[2(3.147 \times 10^{-3})]^2 = 1.\underline{2}47 \times 10^{-7} = 1.2 \times 10^{-7}$

18.27   Calculate the pOH from the pH, and then convert pOH to $[OH^-]$. Then, assemble the usual concentration table, and substitute the equilibrium concentrations from it into the equilibrium-constant expression.

pOH $= 14.00 - 10.52 = 3.48$

$[OH^-] =$ antilog $(-pOH) =$ antilog $(-3.48) = 3.\underline{3}11 \times 10^{-4}$ M

$[Mg^{2+}] = [OH^-] \div 2 = (3.311 \times 10^{-4}) \div 2 = 1.655 \times 10^{-4}$ M

(continued)

Conc. (M)	Mg(OH)$_2$(s)	$\rightleftharpoons$	Mg^{2+}	+	2OH$^-$
Starting			0		0
Change			+1.655 x 10^{-4}		+3.311 x 10^{-4}
Equilibrium			1.655 x 10^{-4}		3.311 x 10^{-4}

K_{sp} = [Mg^{2+}][OH$^-$]^2 = (1.655 x 10^{-4})(3.311 x 10^{-4})^2 = 1.$\underline{8}$1 x 10^{-11} = 1.8 x 10^{-11}

18.29 Assemble the usual concentration table. Let x equal the molar solubility of SrCO$_3$.
When x mol SrCO$_3$ dissolves in one L of solution, x mol Sr^{2+} and x mol CO$_3^{2-}$ form.

Conc. (M)	SrCO$_3$(s)	$\rightleftharpoons$	Sr^{2+}	+	CO$_3^{2-}$
Starting			0		0
Change			+x		+x
Equilibrium			x		x

Substitute the equilibrium concentrations into the equilibrium-constant expression and
solve for x. Then, convert to grams SrCO$_3$ per liter.

[Sr^{2+}][CO$_3^{2-}$] = K_{sp}

(x)(x) = x^2 = 9.3 x 10^{-10}

x = $\sqrt{(9.3 \times 10^{-10})}$ = 3.$\underline{0}$4 x 10^{-5} M

$$\frac{3.04 \times 10^{-5} \text{ mol}}{L} \times \frac{147.63 \text{ g}}{1 \text{mol SrCO}_3} = 4.\underline{5}0 \times 10^{-3} = 0.0045 \text{ g/L}$$

18.31 Let x equal the molar solubility of PbF$_2$. Assemble the usual concentration table, and
substitute from the table into the equilibrium-constant expression.

Conc. (M)	PbF$_2$(s)	$\rightleftharpoons$	Pb^{2+}	+	2F$^-$
Starting			0		0
Change			+x		+2x
Equilibrium			x		2x

(continued)

$$[Pb^{2+}][F^-]^2 = K_{sp}$$

$$(x)(2x)^2 = 4x^3 = 2.7 \times 10^{-8}$$

$$x = \sqrt[3]{\frac{2.7 \times 10^{-8}}{4}} = 1.\underline{8}8 \times 10^{-3} = 1.9 \times 10^{-3} \, M$$

18.33   At the start, before any $SrSO_4$ dissolves, the solution contains 0.23 M SO_4^{2-}. At equilibrium, x mol of solid $SrSO_4$ dissolves to yield x mol Sr^{2+} and x mol SO_4^{2-}. Assemble the usual concentration table, and substitute the equilibrium concentrations into the equilibrium-constant expression. As an approximation, assume x is negligible compared to 0.23 M SO_4^{2-}.

Conc. (M)	$SrSO_4(s)$	$\rightleftharpoons$	Sr^{2+}	+	SO_4^{2-}
Starting			0		0.23
Change			+x		+x
Equilibrium			x		0.23 + x

$$[Sr^{2+}][SO_4^{2-}] = K_{sp}$$

$$(x)(0.23 + x) \cong (x)(0.23) \cong 2.5 \times 10^{-7}$$

$$x \cong \frac{2.5 \times 10^{-7}}{0.23} = 1.\underline{0}86 \times 10^{-6} \, M$$

$$\frac{1.086 \times 10^{-6} \, mol}{L} \times \frac{183.69 \, g}{1 \, mol \, SrSO_4} = 1.\underline{9}9 \times 10^{-4} = 2.0 \times 10^{-4} \, g/L$$

Note that adding x to 0.23 M will not change it (to two significant figures), so x is negligible compared to 0.23 M.

18.35   Calculate the value of K_{sp} from the solubility using the concentration table. Then, using the common-ion calculation, assemble another concentration table. Use 0.020 M NaF as the starting concentration of F^- ion. Substitute the equilibrium concentrations from the table into the equilibrium-constant expression. As an approximation, assume 2x is negligible compared to 0.020 M F^- ion.

$$\frac{0.016 \, g}{L} \times \frac{1 \, mol \, MgF_2}{62.31 \, g} = 2.\underline{5}67 \times 10^{-4} \, M$$

(continued)

Conc. (M)	MgF$_2$(s)	⇌	Mg^{2+}	+	2F$^-$
Starting			0		0
Change			+2.567 x 10^{-4}		+2 x 2.567 x 10^{-4}
Equilibrium			2.567 x 10^{-4}		2 x 2.567 x 10^{-4}

$$K_{sp} = [Mg^{2+}][F^-]^2 = (2.567 \times 10^{-4})[2(2.567 \times 10^{-4})]^2 = 6.772 \times 10^{-11}$$

Now, use K_{sp} to calculate the molar solubility of MgF$_2$.

Conc. (M)	MgF$_2$(s)	⇌	Mg^{2+}	+	2F$^-$
Starting			0		0.020
Change			+x		+2x
Equilibrium			x		0.020 + 2x

$$[Mg^{2+}][F^-]^2 = K_{sp}$$

$$(x)(0.020 + 2x)^2 \cong (x)(0.020)^2 = 6.772 \times 10^{-11}$$

$$x \cong \frac{6.772 \times 10^{-11}}{(0.020)^2} = 1.693 \times 10^{-7} \text{ M}$$

$$\frac{1.693 \times 10^{-7} \text{ mol}}{L} \times \frac{62.31 g}{1 \text{ mol MgF}_2} = 1.054 \times 10^{-5} = 1.1 \times 10^{-5} \text{ g/L}$$

18.37 The concentration table follows.

Conc. (M)	MgC$_2$O$_4$(s)	⇌	Mg^{2+}	+	C$_2$O$_4^{2-}$
Starting			0		0.020
Change			+x		+x
Equilibrium			x		0.020 + x

The equilibrium-constant equation is

$$K_{sp} = [Mg^{2+}][C_2O_4^{2-}]$$

$$8.5 \times 10^{-5} = x(0.020 + x)$$

$$x^2 + 0.020x - (8.5 \times 10^{-5}) = 0$$

(continued)

Solving the quadratic equation gives

$$x = \frac{-0.020 \pm \sqrt{(0.020)^2 + 4(8.5 \times 10^{-5})}}{2} = 0.0036 \text{ M}$$

The solubility in grams per liter is

$$\frac{0.0036 \text{ mol}}{L} \times \frac{112 \text{ g}}{1 \text{ mol}} = 0.403 = 0.4 \text{ g/L}$$

18.39 a. Calculate Q_c, the ion product of the solution, using the concentrations in the problem as the concentrations present after mixing and assuming no precipitation. Then, compare Q_c with K_{sp} to determine whether precipitation has occurred. Start by defining the ion product with brackets as used for the definition of K_{sp}; then, use parentheses for the concentrations.

$$Q_c = [Ba^{2+}][F^-]^2$$

$$Q_c = (0.020)(0.015)^2 = 4.50 \times 10^{-6}$$

Since $Q_c > K_{sp}$ (1.0×10^{-6}), precipitation will occur.

b. Calculate Q_c, the ion product of the solution, using the concentrations in the problem as the concentrations present after mixing and assuming no precipitation. Then, compare Q_c with K_{sp} to determine whether precipitation has occurred. Start by defining the ion product with brackets as used for the definition of K_{sp}; then, use parentheses for the concentrations.

$$Q_c = [Pb^{2+}][Cl^-]^2$$

$$Q_c = (0.0035)(0.15)^2 = 7.87 \times 10^{-5}$$

Since $Q_c > K_{sp}$ (1.6×10^{-5}), precipitation will occur.

18.41 Calculate the ion product, Q_c, after preparation of the solution, assuming no precipitation has occurred. Compare it with the K_{sp}.

$$Q_c = [Pb^{2+}][CrO_4^{2-}]$$

$$Q_c = (5.0 \times 10^{-4})(5.0 \times 10^{-5}) = 2.50 \times 10^{-8} \text{ M}^2 \ (> K_{sp} \text{ of } 1.8 \times 10^{-14})$$

The solution is supersaturated before equilibrium is reached. At equilibrium, precipitation occurs, and the solution is saturated.

18.43 Calculate the concentrations of Mg^{2+} and OH^-, assuming no precipitation. Use a total volume of 1.0 L + 1.0 L, or 2.0 L. (Note the concentrations are halved when the volume is doubled.)

$$[Mg^{2+}] = \frac{\dfrac{0.0020\ mol}{L} \times 1.0\ L}{2.0\ L} = 1.\underline{0}0 \times 10^{-3}\ M$$

$$[OH^-] = \frac{\dfrac{0.00010\ mol}{L} \times 1.0\ L}{2.0\ L} = 5.\underline{0}0 \times 10^{-5}\ M$$

Calculate the ion product, and compare it to K_{sp}.

$$Q_c = (Mg^{2+})(OH^-)^2 = (1.00 \times 10^{-3})(5.00 \times 10^{-5})^2 = 2.\underline{5}0 \times 10^{-12}$$

Because Q_c is less than the K_{sp} of 1.8×10^{-11}, no precipitation occurs, and the solution is unsaturated.

18.45 Calculate the concentrations of Ba^{2+} and F^-, assuming no precipitation. Use a total volume of 0.045 L + 0.075 L, or 0.120 L.

$$[Ba^{2+}] = \frac{\dfrac{0.0015\ mol}{L} \times 0.045\ L}{0.120\ L} = 5.\underline{6}25 \times 10^{-4}\ M$$

$$[F^-] = \frac{\dfrac{0.0025\ mol}{L} \times 0.075\ L}{0.120\ L} = 1.\underline{5}6 \times 10^{-3}\ M$$

Calculate the ion product and compare it to K_{sp}.

$$Q_c = [Ba^{2+}][F^-]^2 = (5.625 \times 10^{-4})(1.56 \times 10^{-3})^2 = 1.\underline{3}6 \times 10^{-9}$$

Because Q_c is less than the K_{sp} of 1.0×10^{-6}, no precipitation occurs, and the solution is unsaturated.

18.47 A mixture of $CaCl_2$ and K_2SO_4 can only precipitate $CaSO_4$ because KCl is soluble. Use the K_{sp} expression to calculate the $[Ca^{2+}]$ needed to just begin precipitating the 0.020 M SO_4^{2-} (in essentially a saturated solution). Then, convert to moles.

$$[Ca^{2+}][SO_4^{2-}] = K_{sp} = 2.4 \times 10^{-5}$$

$$[Ca^{2+}] = \frac{K_{sp}}{[SO_4^{2-}]} = \frac{2.4 \times 10^{-5}}{2.0 \times 10^{-2}} = 1.\underline{2}0 \times 10^{-3}\ M$$

(continued)

The number of moles in 1.5 L of this calcium-containing solution is (mol $CaCl_2$ = mol Ca^{2+}):

$$1.5 \text{ L} \times (1.20 \times 10^{-3}) \text{ mol/L} = 1.\underline{8}0 \times 10^{-3} = 0.0018 \text{ mol } CaCl_2$$

18.49 Because the $AgNO_3$ solution is relatively concentrated, ignore the dilution of the solution of Cl^- and I^- from the addition of $AgNO_3$. The $[Ag^+]$ just as the AgCl begins to precipitate can be calculated from the K_{sp} expression for AgCl using $[Cl^-]$ = 0.015 M. Therefore, for AgCl

$$[Ag^+][Cl^-] = K_{sp}$$

$$[Ag^+][0.015] = 1.8 \times 10^{-10}$$

$$[Ag^+] = \frac{1.8 \times 10^{-10}}{0.015} = 1.\underline{2}0 \times 10^{-8} \text{ M}$$

The $[I^-]$ at this point can be obtained by substituting the $[Ag^+]$ into the K_{sp} expression for AgI. Therefore, for AgI

$$[Ag^+][I^-] = K_{sp}$$

$$[1.20 \times 10^{-8}][I^-] = 8.3 \times 10^{-17}$$

$$[I^-] = \frac{8.3 \times 10^{-17}}{1.2 \times 10^{-8}} = 6.\underline{9}1 \times 10^{-9} = 6.9 \times 10^{-9} \text{ M}$$

18.51 The net ionic equation is

$$BaF_2(s) + 2 H_3O^+(aq) \rightleftharpoons Ba^{2+}(aq) + 2 HF(aq) + 2 H_2O(l)$$

18.53 Calculate the value of K for the reaction of H_3O^+ with both the SO_4^{2-} and F^- anions as they form by the slight dissolving of the insoluble salts. These constants are the reciprocals of the K_a values of the conjugate acids of these anions.

$$F^-(aq) + H_3O^+(aq) \rightleftharpoons HF(aq) + H_2O(l)$$

$$K = \frac{1}{K_a} = \frac{1}{6.8 \times 10^{-4}} = 1.\underline{4}7 \times 10^3$$

(continued)

$$SO_4^{2-}(aq) + H_3O^+(aq) \rightleftharpoons HSO_4^-(aq) + H_2O(l)$$

$$K = \frac{1}{K_{a2}} = \frac{1}{1.1 \times 10^{-2}} = 90.9$$

Because K for the fluoride ion is relatively larger, more BaF_2 will dissolve in acid than $BaSO_4$.

18.55 The equation is

$$Cu^+(aq) + 2\,CN^-(aq) \rightleftharpoons Cu(CN)_2^-(aq)$$

The K_f expression is

$$K_f = \frac{[Cu(CN)_2^-]}{[Cu^+][CN^-]^2} = 1.0 \times 10^{16}$$

18.57 Assume the only $[Ag^+]$ is that in equilibrium with the $Ag(CN)_2^-$ formed from Ag^+ and CN^-. (In other words, assume all the 0.015 M Ag^+ reacts with CN^- to form 0.015 M $Ag(CN)_2^-$.) Subtract the CN^- that forms the 0.015 M $Ag(CN)_2^-$ from the initial 0.100 M CN^-. Use this as the starting concentration of CN^- for the usual concentration table.

[0.100 M NaCN - (2 x 0.015 M)] = 0.070 M starting CN^-

Conc. (M)	$Ag(CN)_2^-$ $\rightleftharpoons$	Ag^+ +	$2CN^-$
Starting	0.015	0	0.070
Change	-x	+x	+2x
Equilibrium	0.015 - x	x	0.070 + 2x

Even though this reaction is the opposite of the equation for the formation constant, the formation-constant expression can be used. Simply substitute all the exact equilibrium concentrations into the formation-constant expression; then, simplify the exact equation by assuming x is negligible compared to 0.015 and 2x is negligible compared to 0.070.

$$K_f = \frac{[Ag(CN)_2^-]}{[Ag^+][CN^-]^2} = \frac{(0.015 - x)}{(x)(0.070 + 2x)^2} \cong \frac{(0.015)}{(x)(0.070)^2} \cong 5.6 \times 10^{18}$$

Rearrange and solve for x = $[Ag^+]$:

$$x \cong (0.015) \div [(5.6 \times 10^{18})(0.070)^2] \cong 5.46 \times 10^{-19} = 5.5 \times 10^{-19}\ M$$

18.59 Start by calculating the $[Cd^{2+}]$ in equilibrium with the $Cd(NH_3)_4^{2+}$ formed from Cd^{2+} and NH_3. Then, use the $[Cd^{2+}]$ to decide whether or not CdC_2O_4 will precipitate by calculating the ion product and comparing it with the K_{sp} of 1.5×10^{-8} for CdC_2O_4. Assume all the 0.0020 M Cd^{2+} reacts with NH_3 to form 0.0020 M $Cd(NH_3)_4^{2+}$, and calculate the remaining NH_3. Use these as the starting concentrations for the usual concentration table.

$$[0.10 \text{ M NH}_3 - (4 \times 0.0020 \text{ M})] = 0.092 \text{ M starting NH}_3$$

Conc. (M)	$Cd(NH_3)_4^{2+} \rightleftharpoons$	Cd^{2+} +	$4NH_3$
Starting	0.0020	0	0.092
Change	-x	+x	+4x
Equilibrium	0.0020 - x	x	0.092 + 4x

Even though this reaction is the opposite of the equation for the formation constant, the formation-constant expression can be used. Simply substitute all the exact equilibrium concentrations into the formation-constant expression; then, simplify the exact equation by assuming x is negligible compared to 0.0020 and 4x is negligible compared to 0.092.

$$K_f = \frac{[Cd(NH_3)_4^{2+}]}{[Cd^{2+}][NH_3]^4} = \frac{(0.0020 - x)}{(x)(0.092 + 4x)^4} \cong \frac{(0.0020)}{(x)(0.092)^4} \cong 1.0 \times 10^7$$

Rearrange and solve for x:

$$x \cong (0.0020) \div [(1.0 \times 10^7)(0.092)^4] \cong 2.79 \times 10^{-6} \text{ M} \cong [Cd^{2+}]$$

Now, calculate the ion product for CdC_2O_4:

$$Q_c = [Cd^{2+}][C_2O_4^{2-}] = (2.79 \times 10^{-6})(0.010) = 2.79 \times 10^{-8}$$

Because Q_c is greater than the K_{sp} of 1.5×10^{-8}, the solution is supersaturated before equilibrium. At equilibrium, a precipitate will form, and the solution will be saturated.

18.61 Using the rule from Chapter 15, obtain the overall equilibrium constant for this reaction from the product of the individual equilibrium constants of the two individual equations whose sum gives this equation.

$$CdC_2O_4(s) \rightleftharpoons Cd^{2+}(aq) + C_2O_4^{2-}(aq) \qquad K_{sp} = 1.5 \times 10^{-8}$$

$$Cd^{2+}(aq) + 4NH_3(aq) \rightleftharpoons Cd(NH_3)_4^{2+}(aq) \qquad K_f = 1.0 \times 10^7$$

$$CdC_2O_4(s) + 4NH_3(aq) \rightleftharpoons Cd(NH_3)_4^{2+}(aq) + C_2O_4^{2-}(aq)$$

$$K_c = K_{sp} \times K_f = 0.15$$

Assemble the usual table using 0.10 M as the starting concentration of NH_3 and x as the unknown concentration of $Cd(NH_3)_4^{2+}$ formed.

Conc. (M)	$CdC_2O_4(s)$	+	$4NH_3$	$\rightleftharpoons$	$Cd(NH_3)_4^{2+}$	+	$C_2O_4^{2-}$
Starting			0.10		0		0
Change			-4x		+x		+x
Equilibrium			0.10 - 4x		x		x

The equilibrium-constant expression can now be used. Simply substitute all the exact equilibrium concentrations into the equilibrium-constant expression.

$$K_c = \frac{[Cd(NH_3)_4^{2+}][C_2O_4^{2-}]}{[NH_3]^4} = \frac{(x)^2}{(0.10 - 4x)^4} = 0.15$$

Take the square root of both sides of the two right-hand terms, rearrange into a quadratic equation, and solve for x:

$$\frac{x}{(0.10 - 4x)^2} = 0.387$$

$$16x^2 - 3.38x + 0.010 = 0$$

$$x = \frac{3.38 \pm \sqrt{(-3.38)^2 - 4(16)(0.010)}}{2(16)} = 0.208 \text{ (too large) and } 3.001 \times 10^{-3}$$

Using the smaller root, $x = 3.001 \times 10^{-3} = 3.0 \times 10^{-3}$ M (molar solubility of CdC_2O_4)

18.63 The Pb^{2+}, Cd^{2+}, and Sr^{2+} ions can be separated in two steps: (1) Add HCl to precipitate only the Pb^{2+} as $PbCl_2$, leaving the others in solution. (2) After pouring the solution away from the precipitate, add 0.3 M HCl and H_2S to precipitate only the CdS away from the Sr^{2+} ion, whose sulfide is soluble under these conditions.

18.65 a. Ag^+ is not possible because no precipitate formed with HCl.

b. Ca^{2+} is not possible if the compound contains only one cation because Mn^{2+} is indicated by the evidence. However, if the compound consists of two or more cations, Ca^{2+} is possible because no reactions were described involving Ca^{2+}.

c. Mn^{2+} is possible because a precipitate was obtained with basic sulfide ion.

d. Cd^{2+} is not possible because no precipitate was obtained with acidic sulfide solution.

■ Solutions to General Problems

18.67　Assemble the usual concentration table. Let x equal the molar solubility of $PbSO_4$. When x mol $PbSO_4$ dissolves in one L of solution, x mol Pb^{2+} and x mol SO_4^{2-} form.

Conc. (M)	$PbSO_4(s)$ ⇌	Pb^{2+}	+	SO_4^{2-}
Starting		0		0
Change		+x		+x
Equilibrium		x		x

Substitute the equilibrium concentrations into the equilibrium-constant expression and solve for x.

$$[Pb^{2+}][SO_4^{2-}] = K_{sp}$$

$$(x)(x) = x^2 = 1.7 \times 10^{-8}$$

$$x = \sqrt{(1.7 \times 10^{-8})} = 1.\underline{3}03 \times 10^{-4} = 1.3 \times 10^{-4} \text{ M}$$

18.69　Let x equal the molar solubility of Hg_2Cl_2. Assemble the usual concentration table, and substitute from the table into the equilibrium-constant expression.

Conc. (M)	$Hg_2Cl_2(s)$ ⇌	Hg_2^{2+}	+	$2Cl^-$
Starting		0		0
Change		+x		+2x
Equilibrium		x		2x

(continued)

$$[Hg_2^{2+}][Cl^-]^2 = K_{sp}$$

$$(x)(2x)^2 = 4x^3 = 1.3 \times 10^{-18}$$

$$x = \sqrt[3]{\frac{1.3 \times 10^{-18}}{4}} = 6.\underline{8}7 \times 10^{-7} \text{ M}$$

a. Molar solubility = $6.\underline{8}7 \times 10^{-7}$ = 6.9×10^{-7} M

b. $\dfrac{6.87 \times 10^{-7} \text{ mol}}{L} \times \dfrac{472.1 \text{ g Hg}_2\text{Cl}_2}{1 \text{ mol}} = 3.\underline{2}4 \times 10^{-4} = 3.2 \times 10^{-4}$ g/L

18.71 Let x equal the molar solubility of $Ce(OH)_3$. Assemble the usual concentration table, and substitute from the table into the equilibrium-constant expression.

Conc. (M)	$Ce(OH)_3(s)$	$\rightleftharpoons$	Ce^{3+}	+	$3OH^-$
Starting			0		0
Change			+x		+3x
Equilibrium			x		3x

$$[Ce^{3+}][OH^-]^3 = K_{sp}$$

$$(x)(3x)^3 = 27x^4 = 2.0 \times 10^{-20}$$

$$x = \sqrt[4]{\frac{2.0 \times 10^{-20}}{27}} = 5.\underline{2}16 \times 10^{-6} \text{ M}$$

a. Molar solubility = $5.\underline{2}16 \times 10^{-6}$ = 5.2×10^{-6} M

b. $[OH^-]$ = 3x = $1.\underline{5}6 \times 10^{-5}$ M

pOH = $4.8\underline{0}54$ = 4.81

18.73 Calculate the pOH from the pH, and then convert pOH to [OH$^-$]. Then, assemble the usual concentration table, and substitute the equilibrium concentrations from it into the equilibrium-constant expression.

$$pOH = 14.00 - 8.80 = 5.20$$

$$[OH^-] = antilog\ (-pOH) = antilog\ (-5.20) = 6.\underline{3}09 \times 10^{-6}\ M$$

Conc. (M)	Mg(OH)$_2$(s)	$\rightleftharpoons$	Mg^{2+}	+	2OH$^-$
Starting			0		0
Change			+x		—
Equilibrium			x		6.309 × 10^{-6}

$$K_{sp} = [Mg^{2+}][OH^-]^2 = (x)(6.309 \times 10^{-6})^2 = 1.8 \times 10^{-11}$$

$$x = \frac{1.8 \times 10^{-11}}{(6.309 \times 10^{-6})^2} = 0.4\underline{5}2 = 0.45\ M$$

$$\frac{0.452\ mol}{L} \times \frac{58.3\ g\ Mg(OH)_2}{1\ mol} = 2\underline{6}.3 = 26\ g/L$$

18.75 Let x equal the change in M of Mg^{2+} and 0.10 M equal the starting OH$^-$ concentration. Then, assemble the usual concentration table, and substitute the equilibrium concentrations from it into the equilibrium-constant expression. Assume 2x is negligible compared to 0.10 M, and perform an approximate calculation.

Conc. (M)	Mg(OH)$_2$(s)	$\rightleftharpoons$	Mg^{2+}	+	2OH$^-$
Starting			0		0.10
Change			+x		+2x
Equilibrium			x		0.10 + 2x

$$K_{sp} = [Mg^{2+}][OH^-]^2 = (x)(0.10 + 2x)^2 \cong (x)(0.10)^2 \cong 1.8 \times 10^{-11}$$

$$x = \frac{1.8 \times 10^{-11}}{(0.10)^2} = 1.\underline{8}0 \times 10^{-9} = 1.8 \times 10^{-9}\ M$$

18.77 To begin precipitation, you must add just slightly more sulfate ion than that required to give a saturated solution. Use the K_{sp} expression to calculate the $[SO_4^{2-}]$ needed to just begin precipitating the 0.0030 M Ca^{2+}.

$$[Ca^{2+}][SO_4^{2-}] = K_{sp} = 2.4 \times 10^{-5}$$

$$[SO_4^{2-}] = \frac{K_{sp}}{[Ca^{2+}]} = \frac{2.4 \times 10^{-5}}{3.0 \times 10^{-3}} = 8.\underline{0}0 \times 10^{-3} \text{ M}$$

When the sulfate ion concentration slightly exceeds 8.0×10^{-3} M, precipitation begins.

18.79 Calculate the concentrations of Pb^{2+} and Cl^-. Use a total volume of 3.20 L + 0.80 L, or 4.00 L.

$$[Pb^{2+}] = \frac{\dfrac{1.25 \times 10^{-3} \text{ mol}}{L} \times 3.20 \text{ L}}{4.00 \text{ L}} = 1.0\underline{0} \times 10^{-3} \text{ M}$$

$$[Cl^-] = \frac{\dfrac{5.0 \times 10^{-1} \text{ mol}}{L} \times 0.80 \text{ L}}{4.00 \text{ L}} = 1.0\underline{0} \times 10^{-1} \text{ M}$$

Calculate the ion product, and compare it to K_{sp}.

$$Q_c = [Pb^{2+}][Cl^-]^2 = (1.00 \times 10^{-3})(1.00 \times 10^{-1})^2 = 1.\underline{0}0 \times 10^{-5} \text{ M}^3$$

Because the Q_c is less than the K_{sp} of 1.6×10^{-5}, no precipitation occurs, and the solution is not saturated.

18.81 A mixture of $AgNO_3$ and $NaCl$ can precipitate only $AgCl$ because $NaNO_3$ is soluble. Use the K_{sp} expression to calculate the $[Cl^-]$ needed to prepare a saturated solution (just before precipitating the 0.0015 M Ag^+). Then, convert to moles and finally to grams.

$$[Ag^+][Cl^-] = K_{sp} = 1.8 \times 10^{-10}$$

$$[Cl^-] = \frac{K_{sp}}{[Ag^+]} = \frac{1.8 \times 10^{-10}}{1.5 \times 10^{-3}} = 1.\underline{2}0 \times 10^{-7} \text{ M}$$

The number of moles and grams in 0.785 L (785 mL) of this chloride-containing solution is

0.785 L x $(1.20 \times 10^{-7} \text{ mol/L})$ = $9.\underline{4}2 \times 10^{-8}$ mol Cl^- = $9.\underline{4}2 \times 10^{-8}$ mol NaCl

$(9.42 \times 10^{-7} \text{ mol NaCl})$ x $(58.5 \text{ g NaCl/1mol})$ = $5.\underline{5}1 \times 10^{-6}$ = 5.5×10^{-6} g NaCl

This amount of NaCl is too small to weigh on a balance.

18.83   From the magnitude of K_f, assume Fe^{3+} and SCN^- react essentially completely to form 2.00 M $Fe(SCN)^{2+}$ at equilibrium. Use 2.00 M as the starting concentration of $Fe(SCN)^{2+}$ for the usual concentration table.

Conc. (M)	Fe^{3+}	+	SCN^-	⇌	$FeSCN^{2+}$
Starting	0		0		2.00
Change	+x		+x		-x
Equilibrium	x		x		2.00 - x

Substitute all the exact equilibrium concentrations into the formation-constant expression; then, simplify the exact equation by assuming x is negligible compared to 2.00 M.

$$K_f = \frac{[FeSCN^{2+}]}{[Fe^{3+}][SCN^-]} = \frac{(2.00 - x)}{(x)(x)} \cong \frac{(2.00)}{x^2} \cong 9.0 \times 10^2$$

Rearrange and solve for x:

$$x = \sqrt{\frac{2.00}{9.0 \times 10^2}} = 4.\underline{7}1 \times 10^{-2} = 4.7 \times 10^{-2} \text{ M}$$

Fraction dissociated $= [(4.\underline{7}1 \times 10^{-2}) \div (2.00)] \times 100\%$

$= 0.02\underline{3}5 \ (< 0.03, \text{ so acceptable})$

18.85   Obtain the overall equilibrium constant for this reaction from the product of the individual equilibrium constants of the two individual equations whose sum gives this equation:

$AgBr(s) \rightleftharpoons Ag^+(aq) + Br^-(aq) \qquad K_{sp} = 5.0 \times 10^{-13}$

$Ag^+(aq) + 2NH_3(aq) \rightleftharpoons Ag(NH_3)_2^+(aq) \qquad K_f = 1.7 \times 10^7$

$AgBr(s) + 2NH_3(aq) \rightleftharpoons Ag(NH_3)_2^{2+}(aq) + Br^-(aq)$

$K_c = K_{sp} \times K_f = 8.\underline{50} \times 10^{-6}$

Assemble the usual table using 5.0 M as the starting concentration of NH_3 and x as the unknown concentration of $Ag(NH_3)_2^+$ formed.

(continued)

Conc. (M)	AgBr(s)	+	$2NH_3$	$\rightleftharpoons$	$Ag(NH_3)_2{}^+$	+	Br^-
Starting			5.0		0		0
Change			-2x		+x		+x
Equilibrium			5.0 - 2x		x		x

The equilibrium-constant expression can now be used. Simply substitute all the exact equilibrium concentrations into the equilibrium-constant expression; it will not be necessary to simplify the equation because taking the square root of both sides removes the x^2 term.

$$K_c = \frac{[Ag(NH_3)_2{}^+][Br^-]}{[NH_3]^2} = \frac{x^2}{(5.0 - 2x)^2} = 8.50 \times 10^{-6}$$

Take the square root of both sides of the two right-hand terms and solve for x.

$$\frac{x}{(5.0 - 2x)} = 2.915 \times 10^{-3}$$

$$x + (5.8 \times 10^{-3})x = 1.4575 \times 10^{-2}$$

$$x = 1.449 \times 10^{-2} = 1.4 \times 10^{-2} \text{ M} \text{ (the molar solubility of AgBr in 5.0 M } NH_3)$$

18.87 Start by recognizing that, because each zinc oxalate produces one oxalate ion, the solubility of zinc oxalate equals the oxalate concentration:

$$ZnC_2O_4(s) + 4NH_3 \rightleftharpoons Zn(NH_3)_4{}^{2+} + C_2O_4{}^{2-}$$

Thus, the $[C_2O_4{}^{2-}] = 3.6 \times 10^{-4}$ M. Now, calculate the $[Zn^{2+}]$ in equilibrium with the oxalate ion using the K_{sp} expression for zinc oxalate.

$$K_{sp} = [Zn^{2+}][C_2O_4{}^{2-}] = 1.5 \times 10^{-9}$$

$$[Zn^{2+}] = \frac{1.5 \times 10^{-9}}{3.6 \times 10^{-4}} = 4.16 \times 10^{-6} = 4.2 \times 10^{-6} \text{ M}$$

(continued)

To calculate K_f, the $[Zn(NH_3)_4^{2+}]$ term must be calculated. This can be done by recognizing that the molar solubility of ZnC_2O_4 is the sum of the concentration of Zn^{2+} and $Zn(NH_3)_4^{2+}$ ions:

Molar solubility of ZnC_2O_4 = $[Zn^{2+}]$ + $[Zn(NH_3)_4^{2+}]$

3.6×10^{-4} = (4.16×10^{-6}) + $[Zn(NH_3)_4^{2+}]$

$[Zn(NH_3)_4^{2+}]$ = (3.6×10^{-4}) - (4.16×10^{-6}) = $3.\underline{5}58 \times 10^{-4}$ M

Now, the $[NH_3]$ term must be calculated by subtracting the ammonia in $[Zn(NH_3)_4^{2+}]$ from the starting NH_3 of 0.0150 M:

$[NH_3]$ = 0.0150 - 4 $[Zn(NH_3)_4^{2+}]$ = 0.0150 - $4(3.558 \times 10^{-4})$ = $0.013\underline{5}7$ M

Solve for K_f by substituting the known concentrations into the K_f expression:

$$K_f = \frac{[Zn(NH_3)_4^{2+}]}{[Zn^{2+}][NH_3]^4} = \frac{3.558 \times 10^{-4}}{(4.16 \times 10^{-6})(0.01357)^4} = 2.\underline{5}2 \times 10^9 = 2.5 \times 10^9$$

18.89 The OH^- formed by ionization of NH_3 (to NH_4^+ and OH^-) is a common ion that will precipitate Mg^{2+} as the slightly soluble $Mg(OH)_2$ salt. The simplest way to treat the problem is to calculate the $[OH^-]$ of 0.10 M NH_3 before the soluble Mg^{2+} salt is added. As the soluble Mg^{2+} salt is added, the $[Mg^{2+}]$ will increase until the solution is saturated with respect to $Mg(OH)_2$ (the next Mg^{2+} ions to be added will precipitate). Calculate the $[Mg^{2+}]$ at the point at which precipitation begins.

To calculate the $[OH^-]$ of 0.10 M NH_3, let x equal the mol/L of NH_3 that ionize, forming x mol/L of NH_4^+ and x mol/L of OH^- and leaving (0.10 - x) M NH_3 in solution. We can summarize the situation in tabular form:

Conc. (M)	NH_3	+	H_2O	$\rightleftharpoons$	NH_4^+	+	OH^-
Starting	0.10				~0		0
Change	-x				+x		+x
Equilibrium	0.10 - x				x		x

The equilibrium-constant equation is:

$$K_c = \frac{[NH_4^+][OH^-]}{[NH_3]} = \frac{x^2}{(0.10 - x)} \cong \frac{x^2}{(0.10)}$$

(continued)

The value of x can be obtained by rearranging and taking the square root:

$$[OH^-] \cong \sqrt{1.8 \times 10^{-5} \times 0.10} = 1.\underline{3}4 \times 10^{-3} \text{ M}$$

Note that (0.10 - x) is not significantly different from 0.10, so x can be ignored in the (0.10 - x) term.

Now, use the K_{sp} of $Mg(OH)_2$ to calculate the $[Mg^{2+}]$ of a saturated solution of $Mg(OH)_2$, which essentially will be the Mg^{2+} ion concentration when $Mg(OH)_2$ begins to precipitate.

$$[Mg^{2+}] = \frac{K_{sp}}{[OH^-]^2} = \frac{1.8 \times 10^{-11}}{(1.34 \times 10^{-3})^2} = 1.\underline{0}0 \times 10^{-5} = 1.0 \times 10^{-5} \text{ M}$$

18.91 a. Use the solubility information to calculate K_{sp}. The reaction is

$$Cu(IO_3)_2(s) \rightleftharpoons Cu^{2+}(aq) + 2IO_3^-(aq)$$
$$2.7 \times 10^{-3} \qquad 2 \times (2.7 \times 10^{-3})$$

$$K_{sp} = [Cu^{2+}][IO_3^-]^2 = [2.7 \times 10^{-3}][2 \times (2.7 \times 10^{-3})]^2 = 7.\underline{8}7 \times 10^{-8}$$

Set up an equilibrium. The reaction is

$$Cu(IO_3)_2(s) \rightleftharpoons Cu^{2+}(aq) + 2IO_3^-(aq)$$
$$y \qquad 0.35 + 2y \approx 0.35$$

$$K_{sp} = [y][0.35]^2 = 7.87 \times 10^{-8}$$

Molar solubility = $y = 6.\underline{4}2 \times 10^{-7} = 6.4 \times 10^{-7}$ M

b. Set up an equilibrium. The reaction is

$$Cu(IO_3)_2(s) \rightleftharpoons Cu^{2+}(aq) + 2IO_3^-(aq)$$
$$y + 0.35 \approx 0.35 \qquad 2y$$

$$K_{sp} = [0.35][2y]^2 = 7.87 \times 10^{-8}$$

Molar solubility = $y = 2.\underline{3}7 \times 10^{-4} = 2.4 \times 10^{-4}$ M

c. Yes, $Cu(IO_3)_2$ is a 1:2 electrolyte. It takes two IO_3^- ions to combine with one Cu^{2+} ion. The IO_3^- ion is involved as a square term in the K_{sp} expression.

18.93 a. Set up an equilibrium. The reaction and equilibrium-constant expression are

$$PbI_2(s) \rightleftharpoons Pb^{2+}(aq) + 2I^-(aq) \qquad K_{sp} = [Pb^{2+}][I^-]^2 = 6.5 \times 10^{-9}$$

The Pb^{2+} ion concentration is 0.0150 M. Plug this in, and solve for the iodine ion concentration.

$$[I^-] = \sqrt{\frac{6.5 \times 10^{-9}}{0.0150}} = 6.\underline{5}8 \times 10^{-4} = 6.6 \times 10^{-4} M$$

b. Solve the equilibrium-constant expression for the lead ion concentration.

$$[Pb^{2+}] = \frac{6.5 \times 10^{-9}}{(2.0 \times 10^{-3})^2} = 1.\underline{6}3 \times 10^{-3} M$$

The percent of the lead(II) ion remaining in solution is

$$\text{Percent } Pb^{2+} \text{ remaining} = \frac{1.63 \times 10^{-3}}{0.0150} \times 100\% = 1\underline{0}.9 = 11 \text{ percent}$$

18.95 a. The reaction and equilibrium-constant expression are

$$Co(OH)_2(s) \rightleftharpoons Co^{2+}(aq) + 2OH^-(aq) \qquad K_{sp} = [Co^{2+}][OH^-]^2$$

From the molar solubility, $[Co^{2+}] = 5.4 \times 10^{-6}$ M and $[OH^-] = 2 \times (5.4 \times 10^{-6})$ M. Therefore,

$$K_{sp} = [5.4 \times 10^{-6}][2 \times (5.4 \times 10^{-6})]^2 = 6.\underline{3}0 \times 10^{-16} = 6.3 \times 10^{-16}$$

b. From the pOH (14 - pH), the $[OH^-] = 10^{-3.57} = 2.\underline{6}9 \times 10^{-4}$ M. The molar solubility is equal to the cobalt ion concentration at equilibrium.

$$[Co^{2+}] = \frac{K_{sp}}{[OH^-]^2} = \frac{6.3 \times 10^{-16}}{(2.69 \times 10^{-4})^2} = 8.\underline{7}1 \times 10^{-9} = 8.7 \times 10^{-9} M$$

c. The common ion effect (OH^-) in part (b) decreases the solubility of Co^{2+}.

18.97 a.

$$AgCl(s) \rightleftharpoons Ag^+(aq) + Cl^-(aq) \qquad K_{sp} = 1.8 \times 10^{-10}$$

$$\underline{Ag^+(aq) + 2NH_3(aq) \rightleftharpoons Ag(NH_3)_2^+(aq) \qquad K_f = 1.7 \times 10^7}$$

$$AgCl(s) + 2NH_3(aq) \rightleftharpoons Ag(NH_3)_2^+(aq) + Cl^-(aq)$$

$$K = K_{sp} \times K_f = (1.8 \times 10^{-10})(1.7 \times 10^7) = 3.\underline{0}6 \times 10^{-3} = 3.1 \times 10^{-3}$$

(continued)

b. The equilibrium-constant expression is

$$\frac{[Ag(NH_3)_2^+][Cl^-]}{[NH_3]^2} = \frac{y^2}{(0.80)^2} = 3.06 \times 10^{-3}$$

$$y = [Ag(NH_3)_2^+] = 0.04\underline{4}3 = 0.044 \text{ M}$$

mol AgCl dissolved = mol $Ag(NH_3)_2^+$ = 0.044 mol/L x 1.00 L = 0.044 mol

$$\text{mol NH}_3 \text{ reacted} = 0.0443 \text{ mol } Ag(NH_3)_2^+ \times \frac{2 \text{ mol NH}_3}{1 \text{ mol } Ag(NH_3)_2^+} = 0.088\underline{6} \text{ mol}$$

The total moles of NH_3 added = 0.80 M x 1.00 L + 0.0886 = 0.8\underline{8}9 = 0.89 mol

18.99 a. The moles of NH_4Cl (molar mass 53.49 g/mol) are given by

$$\text{mol NH}_4Cl = 26.7 \text{ g} \times \frac{1 \text{ mol NH}_4Cl}{53.49 \text{ g NH}_4Cl} = 0.49\underline{9}2 \text{ mol}$$

The reaction and equilibrium-constant expression are

$$NH_3 + H_2O \rightleftharpoons NH_4^+ + OH^- \qquad K_b = \frac{[NH_4^+][OH^-]}{[NH_3]} = 1.8 \times 10^{-5}$$

Solve the equilibrium-constant expression for [OH⁻]. The molarity of the NH_4Cl is 0.49\underline{9}2 mol/1.0 L = 0.49\underline{9}2 M.

$$[OH^-] = \frac{(1.8 \times 10^{-5})(4.2)}{(0.4992)} = 1.\underline{5}1 \times 10^{-4} = 1.5 \times 10^{-4} \text{ M}$$

b. The reaction and equilibrium-constant expression are

$$Mg(OH)_2 \rightleftharpoons Mg^{2+} + 2OH^- \qquad K_{sp} = [Mg^{2+}][OH^-]^2 = 1.8 \times 10^{-11}$$

Solving for $[Mg^{2+}]$ gives

$$[Mg^{2+}] = \frac{(1.8 \times 10^{-11})}{(1.51 \times 10^{-4})^2} = 7.\underline{8}9 \times 10^{-4} = 7.9 \times 10^{-4} \text{ M}$$

The percent Mg^{2+} that has been removed is given by

$$\text{Percent Mg}^{2+} \text{ removed} = \frac{0.075 - 7.89 \times 10^{-4}}{0.075} \times 100\% = 98.9 = 99 \text{ percent}$$

■ Solutions to Cumulative-Skills Problems

18.101 Using the K_{sp} of 1.1×10^{-21} for ZnS, calculate the $[S^{2-}]$ needed to maintain a saturated solution (without precipitation):

$$[S^{2-}] = \frac{1.1 \times 10^{-21}}{1.5 \times 10^{-4} \, M \, Zn^{2+}} = 7.33 \times 10^{-18} \, M$$

Next, use the overall H_2S ionization expression to calculate the $[H_3O^+]$ needed to achieve this $[S^{2-}]$ level:

$$[H_3O^+] = \sqrt{\frac{1.1 \times 10^{-20} \, (0.10 \, M)}{7.33 \times 10^{-18} \, M}} = 1.\underline{2}25 \times 10^{-2} \, M$$

Finally, calculate the buffer ratio of $[SO_4^{2-}]/[HSO_4^-]$ from the H_2SO_4 K_{a2} expression where K_{a2} has the value 1.1×10^{-2}:

$$\frac{[SO_4^{2-}]}{[HSO_4^-]} = \frac{K_{a2}}{[H_3O^+]} = \frac{1.1 \times 10^{-2}}{1.225 \times 10^{-2}} = \frac{0.8979}{1.000}$$

If $[HSO_4^-] = 0.20 \, M$, then

$$[SO_4^{2-}] = 0.8979 \times 0.20 \, M = 0.1\underline{7}95 = 0.18 \, M$$

18.103 Begin by solving for $[H_3O^+]$ in the buffer. Ignoring changes in $[HCHO_2]$ as a result of ionization in the buffer, you obtain

$$[H_3O^+] \cong 1.7 \times 10^{-4} \times \frac{0.45 \, M}{0.20 \, M} = 3.\underline{8}25 \times 10^{-4} \, M$$

You should verify that this approximation is valid (you obtain the same result from the Henderson-Hasselbalch equation). The equilibrium for the dissolution of CaF_2 in acidic solution is obtained by subtracting twice the acid ionization of HF from the solubility equilibrium of CaF_2:

$CaF_2(s) \rightleftharpoons Ca^{2+}(aq) + 2F^-(aq)$		K_{sp}
$2H_3O^+(aq) + 2F^-(aq) \rightleftharpoons 2HF(aq) + 2H_2O(l)$		$1/(K_a)^2$
$2H_3O^+(aq) + CaF_2(s) \rightleftharpoons Ca^{2+}(aq) + 2HF(aq) + 2H_2O(l)$		$K_c = K_{sp}/(K_a)^2$

(continued)

Therefore, $K_c = (3.4 \times 10^{-11}) \div (6.8 \times 10^{-4})^2 = 7.\underline{35} \times 10^{-5}$. In order to solve the equilibrium-constant equation, you require the concentration of HF, which you obtain from the acid-ionization constant for HF.

$$K_a = \frac{[H_3O^+][F^-]}{[HF]}$$

$$6.8 \times 10^{-4} = 3.825 \times 10^{-4} \times \frac{[F^-]}{[HF]}$$

$$\frac{[F^-]}{[HF]} = 1.778, \text{ or } [F^-] = 1.\underline{7}78 \, [HF]$$

Let x be the solubility of CaF_2 in the buffer. Then, $[Ca^{2+}] = x$ and $[F^-] + [HF] = 2x$. Substituting from the previous equation, you obtain

$$2x = 1.778[HF] + [HF] = 2.778[HF], \text{ or } [HF] = 2x/2.\underline{7}78$$

You can now substitute for $[H_3O^+]$ and $[HF]$ into the equation for K_c.

$$K_c = \frac{[Ca^{2+}][HF]^2}{[H_3O^+]^2} = \frac{x(2x/2.778)^2}{(3.825 \times 10^{-4})^2} = 7.35 \times 10^{-5}$$

$$7.35 \times 10^{-5} = (3.543 \times 10^6)x^3$$

$$x^3 = 2.075 \times 10^{-11}$$

$$x = 2.\underline{7}48 \times 10^{-4} = 2.7 \times 10^{-4} \, M$$

18.105 The net ionic equation is

$$Ba^{2+}(aq) + 2OH^-(aq) + Mg^{2+}(aq) + SO_4^{2-}(aq) \rightleftharpoons BaSO_4(s) + Mg(OH)_2(s)$$

Start by calculating the mol/L of each ion after mixing and before precipitation. Use a total volume of $0.0450 + 0.0670 \, L = 0.112 \, L$.

M of SO_4^{2-} and Mg^{2+} = (0.350 mol/L x 0.0670 L) $\div$ 0.112 L = 0.20$\underline{9}$4 M

M of Ba^{2+} = (0.250 mol/L x 0.0450 L) $\div$ 0.112 L = 0.10$\underline{0}$4 M

M of OH^- = (2 x 0.250 mol/L x 0.0450 L) $\div$ 0.112 L = 0.20$\underline{0}$89 M

(continued)

Assemble a table showing the precipitation of $BaSO_4$ and $Mg(OH)_2$.

Conc. (M)	Ba^{2+}	+	SO_4^{2-}	+	Mg^{2+}	+	$2OH^-$
							$\rightleftharpoons$ $BaSO_4(s)$ + $Mg(OH)_2(s)$
Starting	0.1004		0.2094		0.2094		0.20089
Change	-0.1004		-0.1004		-0.1004		-0.20089
Equilibrium	0.0000		0.1090		0.1090		0.0000

To calculate $[Ba^{2+}]$, use the K_{sp} expression for $BaSO_4$.

$$[Ba^{2+}] = \frac{1.1 \times 10^{-10}}{0.1090\,M\,SO_4^{2-}} = 1.\underline{0}09 \times 10^{-9} = 1.0 \times 10^{-9}\,M$$

$$[SO_4^{2-}] = 0.10\underline{9}0 = 0.109\,M$$

Calculate the $[OH^-]$ using the K_{sp} expression for $Mg(OH)_2$:

$$[OH^-] = \sqrt{\frac{1.8 \times 10^{-11}}{0.1090\,M\,Mg^{2+}}} = 1.\underline{2}8 \times 10^{-5} = 1.3 \times 10^{-5}\,M$$

$$[Mg^{2+}] = 0.10\underline{9}0 = 0.109\,M$$

19. THERMODYNAMICS AND EQUILIBRIUM

■ Solutions to Exercises

Note on units and significant figures: The mol unit is omitted from all thermodynamic parameters such as $S°$, $\Delta S°$, etc. If the final answer to a solution needs to be rounded off, it is given first with one nonsignificant figure, and the last significant figure is underlined. The final answer is then rounded to the correct number of significant figures. In multiple-step problems, intermediate answers are given with at least one nonsignificant figure; however, only the final answer has been rounded off.

19.1   Calculate the work, w, done using $w = F \times d = (mg) \times d$. Then, use w to calculate ΔE.

$w = (mg) \times d = (2.20 \text{ kg} \times 9.80 \text{ m/s}^2) \times 0.250 \text{ m} = 5.3\underline{9}0 \text{ kg·m}^2\text{/s}^2 = 5.39 \text{ J}$

$\Delta U = q + w = (-1.50 \text{ J}) + (5.3\underline{9}0 \text{ J}) = 3.8\underline{9}0 = 3.89 \text{ J}$

19.2   At 1.00 atm and 25 °C, the volume occupied by 1.00 mol of any of the gases in the equation is 22.41 L × (298/273) = 24.46 L. Find the change in volume:

CH_4(24.46 L) + $2O_2$(2 × 24.46 L) → CO_2(24.46 L) + $2H_2O$(l)

$\Delta V = 24.46 \text{ L} - (3 \times 24.46 \text{ L}) = -48.\underline{9}2 \text{ L}$

Next, calculate the work, w, done on the system (its value decreases) using 1.00 atm = 1.013×10^5 Pa and 1.00 L = 1.00×10^{-3} m³. Add this to the heat, q_p, at constant P:

$w = -P\Delta V = -(1.013 \times 10^5 \text{ Pa}) \times (-48.92 \times 10^{-3} \text{ m}^3) = 4.9\underline{5}56 \times 10^3 \text{ J} = 4.96 \text{ kJ}$

$\Delta U = q_p + w = (-890.2 \text{ kJ}) + (+4.9\underline{5}56 \text{ kJ}) = -885.\underline{2}4 = -885.2 \text{ kJ}$

19.3   When the liquid evaporates, it absorbs heat: ΔH_{vap} = 42.6 kJ/mol (42.6 x 10^3 J/mol) at 25 °C, or 298 K. The entropy change, ΔS, is

$$\Delta S = \frac{\Delta H_{vap}}{T} = \frac{42.6 \times 10^3 \text{ J/mol}}{298 \text{ K}} = 142.9 \text{ J/(mol·K)}$$

The entropy of one mol of the vapor equals the entropy of one mol of liquid (161 J/K) plus 142.9 J/K.

$$S° = (161 + 142.9) \text{ J/(mol·K)} = 303.9 = 304 \text{ J/(mol·K)}$$

19.4   a. $\Delta S°$ is positive because there is an increase in moles of gas (Δn_{gas} = +1) from a solid reactant forming a mole of gas. (Entropy increases.)

       b. $\Delta S°$ is positive because there is an increase in moles of gas (Δn_{gas} = +1) from a liquid reactant forming a mole of gas. (Entropy increases.)

       c. $\Delta S°$ is negative because there is a decrease in moles of gas (Δn_{gas} = -1) from liquid and gaseous reactants forming two moles of solid. (Entropy decreases.)

       d. $\Delta S°$ is positive because there is an increase in moles of gas (Δn_{gas} = +1) from solid and liquid reactants forming a mole of gas and four moles of an ionic compound. (Entropy increases.)

19.5   The reaction and standard entropies are given below. Multiply the S° values by their stoichiometric coefficients, and subtract the entropy of the reactant from the sum of the product entropies.

	$C_6H_{12}O_6$	$\rightarrow$	$2C_2H_5OH(l)$	+	$2CO_2(g)$
S°:	212.1		2 x 160.7		2 x 213.7 J/K

$$\Delta S° = \Sigma nS°(\text{products}) - \Sigma mS°(\text{reactants}) =$$

$$[(2 \times 160.7 + 2 \times 213.7) - 212.1] \text{ J/K} = 536.7 \text{ J/K}$$

19.6   The reaction, standard enthalpy changes, and standard entropies are as follows:

	$CH_4(g)$	+	$2O_2(g)$	$\rightarrow$	$CO_2(g)$	+	$2H_2O(g)$
$\Delta H_f°$:	-74.87		0		-393.5		2 x (-285.8) kJ
S°:	186.1		2 x 205.0		213.7		2 x 69.9 J/K

(continued)

Calculate $\Delta H°$ and $\Delta S°$ for the reaction by taking the values for products and subtracting the values for reactants.

$$\Delta H° = \Sigma n \Delta H_f°(\text{products}) - \Sigma m \Delta H_f°(\text{reactants}) =$$

$$[(-393.5 + 2 \times -285.8) - (-74.87)] \text{ kJ} = -890.\underline{2}3 \text{ kJ}$$

$$\Delta S° = \Sigma n S°(\text{products}) - \Sigma m S°(\text{reactants}) =$$

$$[(213.7 + 2 \times 69.9) - (186.1 + 2 \times 205.0)] \text{ J/K} = -242.6 \text{ J/K}$$

Now, substitute into the equation for $\Delta G°$ in terms of $\Delta H°$ and $\Delta S°$ ($= -242.6 \times 10^{-3}$ kJ/K):

$$\Delta G° = \Delta H° - T\Delta S° = -890.23 \text{ kJ} - (298 \text{ K})(-242.6 \times 10^{-3} \text{ kJ/K})$$

$$= -817.\underline{9}3 = -817.9 \text{ kJ}$$

19.7   Write the values of $\Delta G_f°$ multiplied by their stoichiometric coefficients below each formula:

$$\text{CaCO}_3(s) \rightarrow \text{CaO}(s) + \text{CO}_2(g)$$

$\Delta G_f°$:   -1128.8   -603.5   -394.4 kJ

The calculation is

$$\Delta G° \; \Sigma n G_f°(\text{products}) - \Sigma m G_f°(\text{reactants}) =$$

$$[(-603.5) + (-394.4) - (-1128.8)] \text{ kJ} = 130.\underline{9} \text{ kJ}$$

19.8 a.   C(graphite) + 2H$_2$(g) $\rightarrow$ CH$_4$(g)

$\Delta G_f°$:   0   0   -50.80 kJ

$$\Delta G° = [(-50.80) - (0)] \text{ kJ} = -50.80 \text{ kJ (spontaneous reaction)}$$

    b.   2H$_2$(g) + O$_2$(g) $\rightarrow$ 2H$_2$O(l)

$\Delta G_f°$:   0   0   2 x (-237.1) kJ

$$\Delta G° = [(2 \times -237.1) - (0)] \text{ kJ} = -474.\underline{2} \text{ kJ (spontaneous reaction)}$$

(continued)

c.
$$4HCN(g) + 5O_2(g) \rightarrow 2H_2O(l) + 4CO_2(g) + 2N_2(g)$$

$\Delta G_f°$:   4 x 124.7   0   2 x (-237.1)   4 x (-394.4)   0 kJ

$\Delta G° = [(2 \times -237.1) + 4 \times (-394.4) - (4 \times 124.7)]$ kJ

$= -2550.6$ kJ (spontaneous reaction)

d.
$$Ag^+(aq) + I^-(aq) \rightarrow AgI(s)$$

$\Delta G_f°$:   77.12   -51.59   -66.19 kJ

$\Delta G° = [(-66.19) - (77.12 - 51.59)]$ kJ $= -91.72$ kJ (spontaneous reaction)

19.9 a. $K = K_p = P_{CO_2}$

b. $K = K_{sp} = [Pb^{2+}][I^-]^2$

c. $K = \dfrac{P_{CO_2}}{[H^+][HCO_3^-]}$

19.10   First, calculate $\Delta G°$ using the $\Delta G_f°$ values from Table 19.2.

$$CaCO_3(s) \rightleftharpoons CaO(s) + CO_2(g)$$

$\Delta G_f°$:   -1128.8   -603.5   -394.4 kJ

Subtract $\Delta G_f°$ of reactant from that of the products:

$\Delta G_f° = \Sigma n G_f°(\text{products}) - \Sigma m G_f°(\text{reactants}) =$

$[(-603.5) + (-394.4) - (-1128.8)]$ kJ $= 130.9$ kJ

Use the rearranged form of the equation, $\Delta G° = -RT \ln K$, to solve for ln K. To get compatible units, express $\Delta G°$ in joules, and set R equal to 8.31 J/(mol•K). Substituting the numerical values into the expression gives

$$\ln K = \frac{\Delta G°}{-RT} = \frac{130.9 \times 10^3}{-8.31 \times 298} = -52.859$$

$$K = K_p = e^{-52.859} = 1.10 \times 10^{-23} = 1 \times 10^{-23}$$

19.11   First, calculate $\Delta G°$ using the $\Delta G_f°$ values in the exercise.

$$Mg(OH)_2(s) \rightarrow Mg^{2+}(aq) + 2OH^-(aq)$$

$\Delta G_f°$:   -833.7   -454.8   2 x (-157.3) kJ

Hence, $\Delta G°$ for the reaction is

$$\Delta G° = [2 \times (-157.3) + (-454.8) - (-833.7)] \text{ kJ} = 64.\underline{3} \text{ kJ}$$

Now, substitute numerical values into the equation relating ln K and $\Delta G°$.

$$\ln K = \frac{\Delta G°}{-RT} = \frac{64.3 \times 10^3}{-8.31 \times 298} = -25.\underline{9}65$$

$$K = K_{sp} = e^{-25.965} = \underline{5}.289 \times 10^{-12} = 5 \times 10^{-12}$$

19.12   From Appendix C, you have

$$H_2O(l) \rightleftharpoons H_2O(g)$$

$\Delta H_f°$:   -285.8   -241.8 kJ

S°:   69.95   188.7 J/K

Calculate $\Delta H°$ and $\Delta S°$ from these values.

$$\Delta H° = [-241.8 - (-285.8)] \text{ kJ} = 44.\underline{0} \text{ kJ}$$

$$\Delta S° = [188.7 - 69.95] \text{ J/K} = 118.\underline{7}5 \text{ J/K}$$

Substitute $\Delta H°$, $\Delta S°$ (= 0.11875 kJ/K) and T (= 318 K) into the equation for $\Delta G_T°$:

$$\Delta G_T° = \Delta H° - T\Delta S° = 44.0 \text{ kJ} - (318 \text{ K})(0.11875 \text{ kJ/K}) = 6.\underline{2}3 \text{ kJ}$$

Substitute the value of $\Delta G°$ (6.23 x 10^3 J) at 318 K into the equation relating ln K and $\Delta G°$.

$$\ln K = \frac{\Delta G°}{-RT} = \frac{6.23 \times 10^3}{-8.31 \times 318} = -2.\underline{3}60$$

$$K = K_p = e^{-2.360} = 0.0\underline{9}43 = 0.09$$

$K_p = P_{H_2O}$, so the vapor pressure of H_2O is 0.09 atm (71.7 mmHg).
      The value is 0.0946 atm in Table 5.6.

19.13   First, calculate $\Delta H°$ and $\Delta S°$ using the given $\Delta H_f°$ and $S°$ values.

	$MgCO_3(s)$	$\rightleftharpoons$	$MgO(s)$	+	$CO_2(g)$
$\Delta H_f°$:	-1111.7		-601.2		-393.5 kJ
$\Delta S°$:	65.85		26.92		213.7 J/K

$\Delta H° = [-601.2 + (-393.5) - (-1111.7)]$ kJ $= 117.\underline{0}$ kJ

$\Delta S° = [(26.92 + 213.7) - 65.85]$ J/K $= 174.\underline{7}7$ J/K

Substitute these values into the expression relating T, $\Delta H°$, and $\Delta S°$ (= 0.17477 kJ/K).

$$T = \frac{\Delta H°}{\Delta S°} = \frac{117.0 \text{ kJ}}{0.17477 \text{ kJ/K}} = 66\underline{9}.45 \text{ K (lower than that for } CaCO_3)$$

■ Answers to Review Questions

19.1   A spontaneous process is a chemical and/or a physical change that occurs by itself without the continuing intervention of an outside agency. Three examples are (1) a rock on a hilltop rolls down, (2) heat flows from a hot object to a cold one, and (3) iron rusts in moist air. Three examples of nonspontaneous processes are (1) a rock rolls uphill by itself, (2) heat flows from a cold object to a hot one, and (3) rust is converted to iron and oxygen.

19.2   Because the energy is more dispersed in liquids than in solids, liquid benzene contains more entropy than does the same quantity of frozen benzene.

19.3   The second law of thermodynamics states that, for a spontaneous process, the total entropy of a system and its surroundings always increases. As stated in Section 19.2, a spontaneous process actually creates energy dispersal, or entropy.

19.4   The relationship between entropy and enthalpy can be expressed in terms of the following equation

$$\Delta S = \frac{\Delta H - \Delta G}{T}$$

At equilibrium, ΔG equals 0, so the equation reduces to $\Delta H/T$ whereas, when not at equilibrium, $\Delta G \neq 0$, so this is not the case. In contrast to a phase change at equilibrium, the entropy change for a spontaneous chemical reaction (at constant pressure) does not equal $\Delta H/T$ because entropy is created by the spontaneous reaction. This can be an increase in the entropy of the surroundings or of the system. An example of the latter is the reaction $N_2O_4(g) \rightarrow 2NO_2(g)$, where one reactant molecule forms two product molecules, thus increasing the randomness.

19.5   The standard entropy of hydrogen gas at 25 °C can be obtained by starting near 0.0 K as a reference point, where the entropy of perfect crystals of hydrogen is almost zero. Then, warm to room temperature in small increments, and calculate $\Delta S°$ for each incremental temperature change (say, 2 K) by dividing the heat absorbed by the average temperature (1 K is used as the average for 0 K to 2 K), and also take into account the entropy increases that accompany a phase change.

19.6   To predict the sign of $\Delta S°$, look for a change, Δn_{gas}, in the number of moles of gas. If there is an increase in moles of gas in the products (Δn_{gas} is positive), then $\Delta S°$ should be positive. A decrease in moles of gas in the products suggests $\Delta S°$ should be negative.

19.7   Free energy, G, equals H - TS; that is, it is the difference between the enthalpy of a system and the product of temperature and entropy. The free-energy change, ΔG, equals ΔH - $T\Delta S$.

19.8   The standard free-energy change, $\Delta G°$, equals $\Delta H°$ - $T\Delta S°$; that is, it is the difference between the standard enthalpy change of a system and the product of temperature and the standard entropy change of a system. The standard free-energy change of formation is the free-energy change when one mole of a substance is formed from its elements in their stable states at one atm and at a standard temperature, usually 25 °C.

19.9   If $\Delta G°$ for a reaction is negative, the equation for the reaction is spontaneous in the direction written; that is, the reactants form the products as written. If it is positive, then the equation as written is nonspontaneous.

19.10   In principle, if a reaction is carried out so that no entropy is produced, the useful work obtained is the maximum useful work, w_{max}, and is equal to ΔG of the reaction.

19.11   When gasoline burns in an automobile engine, the change in free energy shows up as useful work. Gasoline, a mixture of hydrocarbons such as C_8H_{18} or octane, burns to yield energy, gaseous CO_2, and gaseous H_2O.

19.12   A nonspontaneous reaction can be made to occur by coupling it with a spontaneous reaction having a sufficiently negative $\Delta G°$ to furnish the required energy. (The net $\Delta G°$ of the coupled reactions must be negative.)

19.13   As a spontaneous reaction proceeds, the free energy decreases until equilibrium is
reached at a minimum ΔG. See the diagram below.

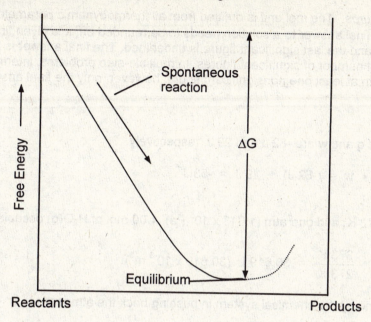

19.14   Because the equilibrium constant is related to $\Delta H°$ and $\Delta S°$ by $-RT \ln K = \Delta H° - T\Delta S°$,
heat measurements alone can be used to obtain it. The standard enthalpy, $\Delta H°$, is the
heat of reaction measured at constant pressure. The standard entropy change, $\Delta S°$,
can be calculated from standard entropies, which are obtained from heat-capacity
data.

19.15   The four combinations are as follows: (1) A negative $\Delta H°$ and a positive $\Delta S°$ always
give a negative $\Delta G°$ and a spontaneous reaction. (2) A positive $\Delta H°$ and a negative
$\Delta S°$ always give a positive $\Delta G°$ and a nonspontaneous reaction. (3) A negative $\Delta H°$
and a negative $\Delta S°$ may give a negative or a positive $\Delta G°$. At low temperatures, $\Delta G°$
will usually be negative and the reaction spontaneous; at high temperatures, $\Delta G°$ will
usually be positive and the reaction nonspontaneous. (4) A positive $\Delta H°$ and a positive
$\Delta S°$ may give a negative or a positive $\Delta G°$. At low temperatures, $\Delta G°$ will usually be
positive and the reaction nonspontaneous; at high temperatures, $\Delta G°$ will usually be
negative and the reaction spontaneous.

19.16   You can estimate the temperature at which a nonspontaneous reaction becomes
spontaneous by substituting zero for $\Delta G°$ into the equation $\Delta G° = \Delta H° - T\Delta S°$ and
then solving for T using the form $T = \Delta H°/\Delta S°$.

■ Solutions to Practice Problems

Note on significant figures: The mol unit is omitted from all thermodynamic parameters such as S°, ΔS°, etc. If the final answer to a solution needs to be rounded off, it is given first with one nonsignificant figure, and the last significant figure is underlined. The final answer is then rounded to the correct number of significant figures. In multiple-step problems, intermediate answers are given with at least one nonsignificant figure; however, only the final answer has been rounded off.

19.25 The values of q and w are -82 J and 29 J, respectively.

$$\Delta U = q + w = (-82\ \text{J}) + 29\ \text{J} = -53\ \text{J}$$

19.27 At 100 °C (373 K) and one atm (1.013×10^5 Pa), 1.00 mol of $H_2O(g)$ occupies

$$22.41\ \text{L} \times \frac{373\ \text{K}}{273\ \text{K}} = 30.\underline{6}19\ \text{L}\ (30.\underline{6}19 \times 10^{-3}\ \text{m}^3)$$

The work done by the chemical system in pushing back the atmosphere is

$$w = -P\Delta V = -(1.013 \times 10^5\ \text{Pa}) \times (30.619 \times 10^{-3}\ \text{m}^3) = -3.10\underline{1}7 \times 10^3\ \text{J}$$

$$= -3.10\underline{1}7\ \text{kJ}$$

$$\Delta U = q_p + w = (40.6\underline{6}\ \text{kJ}) + (-3.1017\ \text{kJ}) = 37.5\underline{8}3 = 37.58\ \text{kJ}$$

19.29 First, determine the enthalpy change for the reaction of 1.20 mol of $CHCl_3$.

$$\Delta H = 1.20\ \text{mol} \times \frac{29.6\ \text{kJ}}{1\ \text{mol}} = 35.\underline{5}2\ \text{kJ} = 3.5\underline{5}2 \times 10^4\ \text{J}$$

Use the equilibrium relation between ΔS and ΔH_{vap} at the boiling point (61.2 °C = 334.4 K):

$$\Delta S = \frac{\Delta H}{T} = \frac{3.552 \times 10^4\ \text{J}}{334.4\ \text{K}} = 10\underline{6}.2 = 106\ \text{J/K}$$

19.31   First, determine the enthalpy change for the condensation of 1.00 mol of $CH_3OH(l)$.
$\Delta H_{cond} = -\Delta H_{vap} = -38.0$ kJ/mol.

The entropy change for this condensation at 25 °C (298 K) is

$$\Delta S = \frac{\Delta H_{cond}}{T} = \frac{-3.80 \times 10^4 J}{298 \text{ K}} = -127.51 \text{ J/K}$$

The entropy of one mole of liquid is calculated using the entropy of one mole of vapor, 255 J/(mol•K).

$$S_{liq} = S_{vap} + \Delta S_{cond} = 255 \text{ J/K} + (-127.51 \text{ J/K}) = 127.48 = 127 \text{ J/K}$$

19.33 a. $\Delta S°$ is negative because there is a decrease in moles of gas ($\Delta n_{gas} = -2$) from three moles of gaseous reactants forming one mole of gaseous product. (Entropy decreases.)

b. $\Delta S°$ is not predictable from the rules given. The molecules N_2, O_2 and NO are of similar size and present in equal number. There is no change in moles of gas ($\Delta n_{gas} = 0$), since two moles of gaseous reactants form two moles of gaseous products. Also, there is no phase change occurring.

c. $\Delta S°$ is positive because there is an increase in moles of gas ($\Delta n_{gas} = +1$) from five moles of gaseous reactants forming six moles of gaseous products. (Entropy increases.)

d. $\Delta S°$ is positive because there is an increase in moles of gas ($\Delta n_{gas} = +1$) from a solid reactant and one mole of gaseous reactant forming two moles of gaseous products. (Entropy increases.)

19.35   The reaction and standard entropies are given below. Multiply the S° values by their stoichiometric coefficients, and subtract the entropy of the reactant from the sum of the product entropies.

a. $2Na(s) + Cl_2(g) \rightarrow 2NaCl(s)$
S°:   2 x 51.46   223.0   2 x 72.12 J/K

$\Delta S° = \Sigma nS°(products) - \Sigma mS°(reactants) =$
[(2 x 72.12) - (2 x 51.46 + 223.0)] J/K = -181.68 = -181.7 J/K

(continued)

b.
$$Ag(s) + \frac{1}{2} Cl_2(g) \rightarrow AgCl(s)$$

S°:   172.9   1/2 x 223.0   96.2   J/K

$$\Delta S° = \Sigma nS°(products) - \Sigma mS°(reactants) =$$

$$[(96.2) - (172.9 + 1/2 \times 223.0)] \text{ J/K} = -188.\underline{1}3 = -188.1 \text{ J/K}$$

c.
$$CS_2(l) + 3O_2(g) \rightarrow CO_2(g) + 2SO_2(g)$$

S°:   151.3   3 x 205.0   213.7   2 x 248.1 J/K

$$\Delta S° = \Sigma nS°(products) - \Sigma mS°(reactants) =$$

$$[(213.7 + 2 \times 248.1) - (151.3 + 3 \times 205.0)] \text{ J/K} = -56.4 \text{ J/K}$$

d.
$$2CH_3OH(l) + 3O_2(g) \rightarrow 2CO_2(g) + 4H_2O(g)$$

S°:   2 x 126.8   3 x 205.0   2 x 213.7   4 x 188.7 J/K

$$\Delta S° = \Sigma nS°(products) - \Sigma mS°(reactants) =$$

$$[(2 \times 213.7 + 4 \times 188.7) - (2 \times 126.8 + 3 \times 205.0)] \text{ J/K} = 313.\underline{6} = 314 \text{ J/K}$$

19.37
$$CH_4(g) + 2O_2(g) \rightarrow CO_2(g) + 2H_2O(l)$$

S°:   186.1   2 x 205.0   213.7   2 x 69.95 J/K

$$\Delta S° = \Sigma nS°(products) - \Sigma mS°(reactants) =$$

$$[(213.7 + 2 \times 69.95) - (186.1 + 2 \times 205.0)] \text{ J/K} = -242.\underline{5}0 \text{ J/K}$$

S decreases as expected from the decrease in moles of gas.

19.39   The reaction, with standard enthalpies of formation and standard entropies written
        underneath, is

$$2CH_3OH(l) + 3O_2(g) \rightarrow 2CO_2(g) + 4H_2O(l)$$

$\Delta H_f°$:   2 x (-238.7)   0   2 x (-393.5)   4 x (-285.8) kJ

S°:   2 x 126.8   3 x 205.0   2 x 213.7   4 x 69.95 J/K

(continued)

Calculate $\Delta H°$ and $\Delta S°$ for the reaction.

$$\Delta H° = \Sigma n\Delta H_f°(\text{products}) - \Sigma m\Delta H_f°(\text{reactants}) =$$

$$[2 \times (-393.5) + 4 \times (-285.8) - 2 \times (-238.7)] \text{ kJ} = -1452.8 \text{ kJ}$$

$$\Delta S° = \Sigma nS°(\text{products}) - \Sigma mS°(\text{reactants}) =$$

$$[(2 \times 213.7 + 4 \times 69.95) - (2 \times 126.8 + 3 \times 205.0)] \text{ J/K}$$

$$= -161.\underline{4}0 = -161.4 \text{ J/K}$$

Now, substitute into the equation for $\Delta G°$ in terms of $\Delta H°$ and $\Delta S°$ ($= -0.161\underline{4}0$ kJ/K).

$$\Delta G° = \Delta H° - T\Delta S° = -1452.8 \text{ kJ} - (298 \text{ K}) \times (-0.16140 \text{ kJ/K})$$

$$= -1404.\underline{7}0 = -1404.7 \text{ kJ}$$

19.41 a. $K(s) + 1/2Br_2(g) \rightarrow KBr(s)$

b. $3/2H_2(g) + C(\text{graphite}) + 1/2Cl_2(g) \rightarrow CH_3Cl(l)$

c. $1/8S_8(\text{rhombic}) + H_2(g) \rightarrow H_2S(g)$

d. $As(s) + 3/2H_2(g) \rightarrow AsH_3(g)$

19.43 Write the values of $\Delta G_f°$ multiplied by their stoichiometric coefficients below each formula; then subtract $\Delta G_f°$ of the reactant from that of the products.

a.

	$CH_4(g)$	+	$2O_2(g)$	$\rightarrow$	$CO_2(g)$	+	$2H_2O(g)$
$\Delta G_f°$:	-50.80		0		-394.4		2 x (-228.6) kJ

$$\Delta G° = \Sigma n\Delta G_f°(\text{products}) - \Sigma m\Delta G_f°(\text{reactants}) =$$

$$[(-394.4) + 2(-228.6) - (-50.80)] \text{ kJ} = -800.8 \text{ kJ}$$

b.

	$CaCO_3(s)$	+	$2H^+(aq)$	$\rightarrow$	$Ca^{2+}(aq)$	+	$H_2O(l)$	+	$CO_2(g)$
$\Delta G_f°$:	-1128.8		0		-553.5		-237.1		-394.4 kJ

$$\Delta G° = \Sigma n\Delta G_f°(\text{products}) - \Sigma m\Delta G_f°(\text{reactants}) =$$

$$[(-553.5) + (-237.1) + (-394.4) - (-1128.8)] \text{ kJ} = -56.2 \text{ kJ}$$

19.45 a. Spontaneous reaction

b. Spontaneous reaction

c. Nonspontaneous reaction

d. Equilibrium mixture; significant amounts of both

e. Nonspontaneous reaction

19.47 Calculate $\Delta H°$ and $\Delta G°$ using the given $\Delta H_f°$ and $\Delta G_f°$ values.

a.
$$Al_2O_3(s) \; + \; 2Fe(s) \; \rightarrow \; Fe_2O_3(s) \; + \; 2Al(s)$$

$\Delta H_f°$:	-1675.7	0	-825.5	0 kJ
$\Delta G_f°$:	-1582.3	0	-743.5	0 kJ

$\Delta H° = [(-825.5) - (-1675.7)] \text{ kJ} = 850.2 \text{ kJ}$

$\Delta G° = [(-743.5) - (-1582.3)] \text{ kJ} = 838.8 \text{ kJ}$

The reaction is endothermic, absorbing 850.2 kJ of heat. The large positive value for $\Delta G°$ indicates the equilibrium composition is mainly reactants.

b.
$$COCl_2(g) \; + \; H_2O(l) \; \rightarrow \; CO_2(g) \; + \; 2HCl(g)$$

$\Delta H_f°$:	-220.1	-285.8	-393.5	2 x (-92.31) kJ
$\Delta G_f°$:	-205.9	-237.1	-394.4	2 x (-95.30) kJ

$\Delta H° = [-393.5 + (2)(-92.31) - (-220.1) - (-285.8)] \text{ kJ} = -72.22 = -72.2 \text{ kJ}$

$\Delta G° = [-394.4 + (2)(-95.30) - (-205.9) - (-237.1)] \text{ kJ} = -142.00 = -142.0 \text{ kJ}$

The reaction is exothermic; the $\Delta G°$ value indicates mainly products at equilibrium.

19.49 Calculate $\Delta G°$ using the given $\Delta G_f°$ values.

$$2H_2(g) \; + \; O_2(g) \; \rightarrow \; 2H_2O(l)$$

$\Delta G_f°$:	0	0	2 x (-237.1) kJ

$\Delta G° = [2(-237.1) - 0] \text{ kJ} = -474.2 \text{ kJ}$

Maximum work equals $\Delta G°$ equals -474.2 kJ. Because maximum work is stipulated, no entropy is produced.

19.51   Calculate $\Delta G°$ per one mol Zn(s) using the given $\Delta G_f°$ values.

$$Zn(s) \quad + \quad Cu^{2+}(aq) \quad \rightarrow \quad Zn^{2+}(aq) \quad + \quad Cu(s)$$

$\Delta G_f°$:   0   65.52   -147.0   0 kJ

$$\Delta G° = [(-147.0) - (65.52)] \text{ kJ} = -212.52 \text{ kJ/mol Zn}$$

$$-212.52 \text{ kJ/mol Zn} \times (4.85 \text{ g} \div 65.39 \text{ g/mol Zn}) = -15.\underline{7}6 = -15.8 \text{ kJ}$$

Maximum work equals $\Delta G°$ equals -15.8 kJ. Because maximum work is stipulated, no entropy is produced.

19.53 a. $K = K_p = \dfrac{P_{CO_2} P_{H_2}}{P_{CO} P_{H_2O}}$

b. $K = K_{sp} = [Mg^{2+}][OH^-]^2$

c. $K = [Li^+]^2[OH^-]^2 P_{H_2}$

19.55   First, calculate $\Delta G°$ using the $\Delta G_f°$ values from Appendix C.

$$H_2(g) \quad + \quad Br_2(g) \quad \rightarrow \quad 2HBr(g)$$

$\Delta G_f°$:   0   0   2 x (-53.50) kJ

$$\Delta G° = [2(-53.50) - 0] \text{ kJ} = -107.00 \text{ kJ}$$

Use the rearranged form of the equation, $\Delta G° = -RT \ln K$, to solve for $\ln K$. To get compatible units, express $\Delta G°$ in joules, and set R equal to 8.31 J/(mol•K).

Substituting the numerical values into the expression gives

$$\ln K = \frac{\Delta G°}{-RT} = \frac{-107.00 \times 10^3}{-8.31 \times 298} = 43.\underline{2}08$$

$$K = e^{43.208} = \underline{5}.82 \times 10^{18} = 6 \times 10^{18}$$

19.57   First, calculate $\Delta G°$ using the $\Delta G_f°$ values from Appendix C.

$$CO(g) \quad + \quad 3H_2(g) \; \rightleftharpoons \; CH_4(g) \quad + \quad H_2O(g)$$

$\Delta G_f°$:   -137.2   0   -50.80   -228.6 kJ

Subtract $\Delta G_f°$ of the reactants from that of the products:

$$\Delta G° = [(-50.80) + (-228.6) - (-137.2)] \; kJ = -142.20 \; kJ$$

Use the rearranged form of the equation, $\Delta G° = -RT \ln K$, to solve for $\ln K$. To get compatible units, express $\Delta G°$ in joules, and set R equal to 8.31 J/(mol•K). Substituting the numerical values into the expression gives

$$\ln K = \frac{\Delta G°}{-RT} = \frac{-142.20 \times 10^3}{-8.31 \times 298} = 57.\underline{4}22$$

$$K = K_p = e^{57.422} = \underline{8}.67 \times 10^{24} = 9 \times 10^{24}$$

19.59   First, calculate $\Delta G°$ using the $\Delta G_f°$ values from Appendix C.

$$Fe(s) \quad + \quad Cu^{2+}(aq) \; \rightleftharpoons \; Fe^{2+}(aq) \quad + \quad Cu(s)$$

$\Delta G_f°$:   0   65.52   -78.87   0 kJ

Hence

$$\Delta G° = [(-78.87) - 65.52] \; kJ = -144.38 \; kJ$$

Now, substitute numerical values into the equation relating $\ln K$ and $\Delta G°$.

$$\ln K = \frac{\Delta G°}{-RT} = \frac{-144.38 \times 10^3}{-8.31 \times 298} = 58.\underline{3}06$$

Therefore,

$$K = K_c = e^{58.306} = \underline{2}.10 \times 10^{25} = 2 \times 10^{25}$$

19.61   From Appendix C, you have

$$C(graphite) \quad + \quad CO_2(g) \quad \rightleftharpoons \quad 2CO(g)$$

$\Delta H_f°$:	0	-393.5	2 x (-110.5) kJ
S°:	5.740	213.7	2 x (197.5) J/K

Calculate $\Delta H°$ and $\Delta S°$ from these values.

$$\Delta H° = [2(-110.5) - (-393.5)] \, kJ = 172.\underline{5} \, kJ$$

$$\Delta S° = [2(197.5) - (5.740 + 213.7)] \, J/K = 175.\underline{56} \, J/K$$

Substitute $\Delta H°$, $\Delta S°$ (= 0.17556 kJ/K), and T (= 1273 K) into the equation for $\Delta G_T°$.

$$\Delta G_T° = \Delta H° - T\Delta S° = 172.5 \, kJ - (1273 \, K)(0.17556 \, kJ/K) = -50.\underline{987} \, kJ$$

Substitute the value of $\Delta G°$ (= -50.987 x 10^3 J) into the equation relating ln K and $\Delta G°$.

$$\ln K = \frac{\Delta G°}{-RT} = \frac{-50.987 \times 10^3}{-8.31 \times 1273} = 4.8\underline{198}$$

$$K = K_p = e^{4.8198} = 12\underline{3}.9 = 1.2 \times 10^2$$

Because K_p is greater than one, the data predict combustion of carbon should form significant amounts of CO product at equilibrium.

19.63   First, calculate $\Delta H°$ and $\Delta S°$ using the given $\Delta H_f°$ and S° values.

$$2NaHCO_3(s) \quad \rightarrow \quad Na_2CO_3(s) \quad + \quad H_2O(g) \quad + \quad CO_2(g)$$

$\Delta H_f°$:	2 x (-950.8)	-1130.8	-241.8	-393.5 kJ
$\Delta S°$:	2 x 101.7	138.8	188.7	213.7 J/K

$$\Delta H° = [(-1130.8) + (-241.8) + (-393.5) - 2(-950.8)] \, kJ = 135.5 \, kJ$$

$$\Delta S° = [(138.8 + 188.7 + 213.7) - 2(101.7)] \, J/K = 337.8 \, J/K \, (0.3378 \, kJ/K)$$

Substitute these values into $\Delta G° = \Delta H° - T\Delta S°$; let $\Delta G° = 0$, and rearrange to solve for T.

$$T = \frac{\Delta H°}{\Delta S°} = \frac{135.5 \, kJ}{0.3378 \, kJ/K} = 40\underline{1}.1 = 401 \, K$$

■ Solutions to General Problems

19.65 The sign of $\Delta S°$ should be positive because there is an increase in moles of gas (Δn_{gas} = +5) as the solid reactant forms five moles of gas. The reaction is endothermic, denoting a positive $\Delta H°$. The fact that the reaction is spontaneous implies the product, $T\Delta S°$, is larger than $\Delta H°$, so $\Delta G°$ is negative, as required for a spontaneous reaction.

19.67 The ΔH value $\sim$ BE(H-H) + BE(Cl-Cl) - BE(H-Cl) $\sim$ [432 + 240 - 2(428)] kJ $\sim$ -184 kJ, and thus the reaction is exothermic. $\Delta S°$ should be positive because there is a increase in energy dispersal with the formation of unsymmetrical molecules from symmetrical H_2 and Cl_2. The reaction should be spontaneous because the contributions of both the ΔH term and the $-T\Delta S$ term are negative.

19.69 When the liquid freezes, it releases heat: ΔH_{fus} = -69.0 J/g at 16.6 °C (289.6 K). The entropy change is

$$\Delta S = \frac{\Delta H_{fus}}{T} = \frac{-69.0 \text{ J/g}}{289.8 \text{ K}} \times \frac{60.05 \text{ g}}{1 \text{ mol}} = -14.\underline{30} = -14.3 \text{ J/(K•mol)}$$

19.71 a. $\Delta S°$ is negative because there is a decrease in moles of gas (Δn_{gas} = -1) from one mole of gaseous reactant forming aqueous and liquid products. (Entropy decreases.)

  b. $\Delta S°$ is positive because there is an increase in moles of gas (Δn_{gas} = +5) from a solid reactant forming five moles of gas. (Entropy increases.)

  c. $\Delta S°$ is positive because there is an increase in moles of gas (Δn_{gas} = +3) from two moles of gaseous reactant forming five moles of gaseous products. (Entropy increases.)

  d. $\Delta S°$ is negative because there is a decrease in moles of gas (Δn_{gas} = -1) from three moles of gaseous reactants forming two moles of gaseous products. (Entropy decreases.)

19.73 $\Delta S°$ is negative because there is a decrease in the moles of gas (Δn_{gas} = -2) from three moles of gaseous reactant forming one mole of gaseous product plus liquid product.

19.75 Calculate $\Delta S°$ from the individual $S°$ values:

$$C_2H_5OH(l) \quad + \quad O_2(g) \quad \rightarrow \quad CH_3COOH(l) \quad + \quad H_2O(l)$$

$S°$:   160.7   205.0   159.8   69.95 J/K

$\Delta S° = \Sigma n S°(\text{products}) - \Sigma m S°(\text{reactants}) =$

$[(159.8 + 69.95) - (160.7 + 205.0)]$ J/K $= -135.\underline{95} = -136.0$ J/K

19.77 Calculate $\Delta G°$ using the $\Delta G_f°$ values from Appendix C.

$$H_2(g) \quad + \quad SO_2(g) \quad \rightarrow \quad H_2S(g) \quad + \quad O_2(g)$$

$\Delta G_f°$:   0   -300.1   -33.33   0 kJ

$\Delta G° = \Sigma n \Delta G_f°(\text{products}) - \Sigma m \Delta G_f°(\text{reactants}) =$

$[(-33.33) - (-300.1)]$ kJ $= 266.\underline{77} = 266.8$ kJ

Because $\Delta G°$ is positive, the reaction is nonspontaneous, as written, at 25 °C.

19.79 At low (room) temperature, $\Delta G°$ or $(\Delta H° - T\Delta S°)$ must be positive, but at higher temperatures, $\Delta G°$ or $(\Delta H° - T\Delta S°)$ must be negative. Thus, at the higher temperatures, the $-T\Delta S°$ term must become more negative than $\Delta H°$. Thus, $\Delta S°$ must be positive and so must $\Delta H°$. If either were negative, $\Delta G°$ would not become negative at higher temperatures.

19.81 First, calculate $\Delta G°$ using the values in Appendix C.

$$CaF_2(s) \quad \rightleftharpoons \quad Ca^{2+}(aq) \quad + \quad 2F^-(aq)$$

$\Delta G_f°$:   -1173.5   -553.5   2 x (-262.0) kJ

Hence, $\Delta G°$ for the reaction is

$\Delta G_f° = [-553.5 + 2(-262.0) - (-1173.5)]$ kJ $= 96.0$ kJ

Now, substitute numerical values into the equation relating ln K and $\Delta G°$.

$$\ln K = \frac{\Delta G°}{-RT} = \frac{96.0 \times 10^3}{-8.31 \times 298} = -38.\underline{766}$$

$$K = K_{sp} = e^{-38.766} = \underline{1}.45 \times 10^{-17} = 1 \times 10^{-17}$$

19.83   From Appendix C, you have

$$COCl_2(g) \quad \rightarrow \quad CO(g) \quad + \quad Cl_2(g)$$

	$COCl_2(g)$	$CO(g)$	$Cl_2(g)$
$\Delta H_f°$:	-220.1	-110.5	0 kJ
$S°$:	283.9	197.5	223.0 J/K

Calculate $\Delta H°$ and $\Delta S°$ from these values.

$$\Delta H° = [(-110.5) - (-220.1)] \text{ kJ} = 109.6 \text{ kJ}$$

$$\Delta S° = [197.5 + 223.0 - 283.9] \text{ J/K} = 136.6 \text{ J/K} \ (0.1366 \text{ kJ/K})$$

At 25 °C: $\Delta G° = \Delta H° - T\Delta S° = 109.6 \text{ kJ} - (298 \text{ K})(0.1366 \text{ kJ/K})$
$$= 68.\underline{89} = 68.9 \text{ kJ}$$

At 800 °C: $\Delta G°_T = \Delta H° - T\Delta S° = 109.6 \text{ kJ} - (1073 \text{ K})(0.1366 \text{ kJ/K})$
$$= -37.\underline{07} = -37.1 \text{ kJ}$$

Thus, $\Delta G°$ changes from a positive value and a nonspontaneous reaction at 25 °C to a negative value and a spontaneous reaction at 800 °C.

19.85 a.

		$\Delta H°$, kJ
$CO_2(g) + 2H_2(g) \rightarrow HCHO(g) + H_2O(g)$		35
$HCHO(g) + 2H_2(g) \rightarrow CH_4(g) + H_2O(g)$		-201
$C(s) + O_2(g) \rightarrow CO_2(g)$		-393
$2H_2O(g) \rightarrow 2H_2(g) + O_2(g)$		484
$C(s) + 2H_2(g) \rightarrow CH_4(g)$		-75 kJ

b.

$$C(s) + 2H_2(g) \quad \rightarrow \quad CH_4(g)$$

	$C(s)$	$2H_2(g)$	$CH_4(g)$
$S°$:	5.740	2 x 130.6	186.1 J/K

$$\Delta S° = \Sigma n S°(\text{products}) - \Sigma m S°(\text{reactants}) =$$

$$[186.1 - 5.740 - 2(130.6)] \text{ J/K} = -80.\underline{84} \text{ J/K} \ (-0.08084 \text{ kJ/K})$$

(continued)

c. $\Delta G°$ for the reaction, which involves one mole of methane, is equal to $\Delta G_f°$ for methane, and is obtained as follows.

$$\Delta G_f° = \Delta G° = \Delta H° - T\Delta S°$$

$$= -75 \text{ kJ} - (298 \text{ K})(-0.08084 \text{ kJ/K})$$

$$= -50.90 = -51 \text{ kJ/mol}$$

19.87 $\Delta H° = [-393.5 + 2(-285.8) - (-238.7)] \text{ kJ} = -726.4 \text{ kJ}$

$$\Delta G° = \Delta H° - T\Delta S°$$

$$-702.2 \text{ kJ} = -726.4 \text{ kJ} - (298 \text{ K})(\Delta S°)$$

$$\Delta S° = -\frac{(-702.2 \text{ kJ}) - (-726.4 \text{ kJ})}{298 \text{ K}} = -0.08120 \text{ kJ/K} = -81.20 \text{ J/K}$$

$$\Delta S° = -81.20 \text{ J/K} = [2(70.0) + 213.7 - (126.8) - (3/2 \text{ mol}) \times S°(O_2)] \text{ J/K}$$

$$S°(O_2) = 205.40 = 205.4 \text{ J/mol•K}$$

19.89 a. The first reaction is

	$SnO_2(s)$	+	$2H_2(g)$	$\rightarrow$	$Sn(s)$	+	$2H_2O(g)$
$\Delta H_f°$	-580.7		0		0		-241.8 kJ
$S°$	52.3		130.6		51.55		188.7 J/K

$$\Delta H° = [2(-241.8) - (-580.7)] \text{ kJ} = 97.1 \text{ kJ} = 97.1 \times 10^3 \text{ J}$$

$$\Delta S° = [2(188.7) + 51.55 - 2(130.6) - 52.3] \text{ J/K} = 115.45 \text{ J/K}$$

The second reaction is

	$SnO_2(s)$	+	$C(s)$	$\rightarrow$	$Sn(s)$	+	$CO_2(g)$
$\Delta H_f°$	-580.7		0		0		-393.5 kJ
$S°$	52.3		5.740		51.5		213.7 J/K

$$\Delta H° = [-393.5 - (-580.7)] \text{ kJ} = 187.2 \text{ kJ} = 187.2 \times 10^3 \text{ J}$$

$$\Delta S° = [213.7 + 51.5 - 5.740 - 52.3] \text{ J/K} = 207.16 \text{ J/K}$$

(continued)

b. For H_2, at what temperature does $\Delta G = 0$?

$$0 = \Delta H° - T\Delta S°$$

$$T = \frac{\Delta H}{\Delta S} = \frac{97.1 \times 10^3 \text{ J}}{115.45 \text{ J/K}} = 84\underline{1}.0 = 841 \text{ K}$$

At temperatures greater than 841 K, the reaction will be spontaneous.

For C, at what temperature does $\Delta G = 0$?

$$T = \frac{\Delta H}{\Delta S} = \frac{187.2 \times 10^3 \text{ J}}{207.16 \text{ J/K}} = 90\underline{3}.6 = 904 \text{ K}$$

At temperatures greater than 904 K, the reaction will be spontaneous.

c. From a consideration of temperature, the process with hydrogen would be preferred. But hydrogen is very expensive and carbon is cheap. On this basis, carbon would be preferred. Tin is produced commercially using carbon as the reducing agent.

19.91 a. Formic acid is favored as it is of lower energy than CO and H_2O.

b. The change in entropy for the decomposition of formic acid is positive as a mole of gas is produced. So, the change in entropy would be the driving force for this reaction, favoring the formation of products.

19.93 a. If $\Delta G°$ is negative, then K must be greater than one. Consequently, the products will predominate.

b. The molecules must have enough energy to react when they collide with each other. So, it depends upon the activation energy for the reaction. Usually it is necessary to heat solids for a reaction to occur as it is difficult to have effective collisions.

19.95 a. $C_4H_{10}(g) + 13/2 O_2(g) \rightarrow 4CO_2(g) + 5H_2O(l)$

$$\Delta H° = \frac{-49.50 \text{ kJ}}{1.000 \text{ g}} \times \frac{58.12 \text{ g}}{1 \text{ mol}} = -2876.94 \text{ kJ/mol}$$

$\Delta H° = \Sigma n \Delta H_f°(\text{products}) - \Sigma m \Delta H_f°(\text{reactants})$

$-2876.94 \text{ kJ} = [4(-393.5) + 5(-285.8) - (1 \text{ mol}) \times \Delta H_f°(C_4H_{10})] \text{ kJ}$

$\Delta H_f°(C_4H_{10}) = -126.06 = -126 \text{ kJ/mol}$

b. $\Delta G° = [4(-394.4) + 5(-237.1) - (-17.2)] \text{ kJ} = -2745.9 \text{ kJ}$

c. $\Delta G° = \Delta H° - T\Delta S°$

$-2745.9 \text{ kJ} = -2876.94 \text{ kJ} - (298 \text{ K}) \Delta S°$

$$\Delta S° = -\frac{(-2745.9 \text{ kJ}) - (-2876.94 \text{ kJ})}{298 \text{ K}} = -0.4397 \text{ kJ/K} = -440. \text{ J/K (for one mole)}$$

19.97 a. $\Delta H° = [-1285 - (-1288.3)] \text{ kJ} = 3.3 \text{ kJ} = 3.3 \times 10^3 \text{ J}$

$\Delta S° = [89 - 158.2] \text{ J/K} = -69.2 \text{ J/K}$

$\Delta G° = \Delta H° - T\Delta S°$

$\Delta G° = 3.3 \times 10^3 \text{ J} - (298 \text{ K})(-69.2 \text{ J/K}) = 23.921 \times 10^3 \text{ J}$

Now, substitute numerical values into the equation relating ln K and $\Delta G°$.

$$\ln K = \frac{\Delta G°}{-RT} = \frac{23.921 \times 10^3}{-8.31 \times 298} = -9.6599$$

$K = e^{-9.6599} = 6.37 \times 10^{-5} = 6 \times 10^{-5}$

b. The change in entropy is negative, greater order, so this causes H_3PO_4 to be a weak acid. The enthalpy change hinders the acid strength of H_3PO_4, and the entropy is a very important term.

■ Solutions to Cumulative-Skills Problems

19.99 For the dissociation of HBr, assume ΔH and ΔS are constant over the temperature range from 25 °C to 375 °C, and calculate the value of each to use to calculate K at 375 °C. Start by calculating $\Delta H°$ and $\Delta S°$ at 25 °C, using $\Delta H_f°$ and S° values.

	2HBr(g)	→	H_2(g)	+	Br_2(g)
$\Delta H_f°$:	2 x (-36.44)		0		30.91 kJ
S°:	2 x 198.6		130.6		245.3 J/K

Calculate $\Delta H°$ and $\Delta S°$ from these values.

$$\Delta H° = [30.91 - 2(-36.44)] \text{ kJ} = 103.79 \text{ kJ}$$

$$\Delta S° = [245.3 + 130.6 - 2(198.6)] \text{ J/K} = -21.3 \text{ J/K}$$

Substitute $\Delta H°$, $\Delta S°$ (= -0.02130 kJ/K), and T (648 K) into the equation for $\Delta G_T°$.

$$\Delta G_T° = \Delta H° - T\Delta S° = 103.79 \text{ kJ} - (648 \text{ K})(-0.02130 \text{ kJ/K}) = 117.\underline{5}9 \text{ kJ}$$

$$= 117.\underline{5}9 \times 10^3 \text{ J}$$

Now, substitute numerical values into the equation relating ln K and $\Delta G°$ (= $\Delta G_T°$).

$$\ln K = \frac{\Delta G°}{-RT} = \frac{117.59 \times 10^3}{-8.31 \times 648} = -21.\underline{8}37$$

$$K = e^{-21.837} = 3.\underline{2}8 \times 10^{-10.}$$

Assuming x equals $[H_2]$ equals $[Br_2]$ and assuming [HBr] = (1.00 - 2x) $\cong$ 1.00 atm, substitute into the equilibrium expression:

$$K = \frac{[H_2][Br_2]}{[HBr]^2} = \frac{(x)(x)}{(1.00)^2} = 3.28 \times 10^{-10}$$

Solve for the approximate pressure of x:

$$x = \sqrt{(3.28 \times 10^{-10})(1.00)^2} = 1.\underline{8}1 \times 10^{-5} \text{ atm}$$

(continued)

The percent dissociation at 1.00 atm is

$$\text{Percent dissociation} = \frac{2(1.81 \times 10^{-5} \text{ atm})}{1.00 \text{ atm}} \times 100\% = 0.0036 \text{ percent}$$

Based on Le Chatelier's principle, pressure has no effect on equilibrium. Therefore, the percent dissociation is 0.004 percent at 1.00 atm and at 10.0 atm.

19.101 For the dissociation of NH_3, assume ΔH and ΔS are constant over the temperature range from 25 °C to 345 °C, and calculate values of each to calculate K at 345 °C.

	$2NH_3(g)$	$\rightarrow$	$3H_2(g)$	+	$N_2(g)$
$\Delta H_f°$:	2 x (-45.90)		0		0 kJ
S°:	2 x 192.7		3 x 130.6		191.6 J/K

Calculate $\Delta H°$ and $\Delta S°$ from these values.

$$\Delta H° = [0 - 2(-45.90)] \text{ kJ} = 91.80 \text{ kJ}$$

$$\Delta S° = [3(130.6) + 191.6 - 2(192.7)] \text{ J/K} = 198.0 \text{ J/K}$$

Substitute $\Delta H°$, $\Delta S°$ (= 0.1980 kJ/K), and T (618 K) into the equation for $\Delta G_T°$.

$$\Delta G_T° = \Delta H° - T\Delta S° = 91.8 \text{ kJ} - (618 \text{ K})(0.1980 \text{ kJ/K}) = -30.\underline{5}64 \text{ kJ}$$

$$= -30.564 \times 10^3 \text{ J}$$

Now, substitute numerical values into the equation relating ln K and $\Delta G°$ (= $\Delta G_T°$).

$$\ln K = \frac{\Delta G°}{-RT} = \frac{-30.564 \times 10^3}{-8.31 \times 618} = 5.9\underline{5}1$$

$$K = K_p = e^{5.951} = 3\underline{8}4.2$$

Now obtain K_c.

$$K_c = K_p(RT)^{-2}$$

$$= (384.2)(0.0821 \times 618)^{-2} = 0.1\underline{4}92$$

(continued)

The starting concentration of NH_3 is 1.00 mol/20.0L = 0.0500M. You obtain the following table:

	$2NH_3(g)$	$\rightarrow$	$3H_2(g)$	+	$N_2(g)$
Starting	0.0500		0		0
Change	-2x		+3x		+x
Equilibrium	0.0500 - 2x		3x		x

The equilibrium equation is

$$K_c = \frac{[H_2]^3[N_2]}{[NH_3]^2}$$

or

$$\frac{(3x)^3 x}{(0.0500 - 2x)^2} = 0.1492$$

$$\frac{x^4}{(0.0500 - 2x)^2} = \frac{0.1492}{27} = 5.528 \times 10^{-3}$$

Taking the square root of both sides of this equation gives

$$\frac{x^2}{(0.0500 - 2x)} = 0.07\underline{4}35$$

This can be rearranged into the following quadratic equation.

$$x^2 + (0.1487)x - (3.717 \times 10^{-3}) = 0$$

From the quadratic formula, you obtain

$$x = \frac{-0.1487 \pm \sqrt{(0.1487)^2 + 4(3.717 \times 10^{-3})}}{2}$$

The positive root is

$$x = 0.01\underline{9}42 \text{ M}$$

(continued)

Hence,

$$[NH_3] = 0.0500 - 2(0.01942) = 0.01\underline{1}14 \text{ M}$$

$$\text{Percent } NH_3 \text{ dissociated} = \left(1 - \frac{0.01114 \text{ M}}{0.0500 \text{ M}} \right) \times 100\%$$

$$= 7\underline{7}.7 = 78 \text{ percent}$$

19.103 First, calculate $\Delta G°$ at each temperature, using $\Delta G° = - RT \ln K$:

25.0 °C: $\Delta G° = - (0.008314 \text{ kJ/K})(298.2 \text{ K})(\ln 1.754 \times 10^{-5}) = 27.15\underline{5}1 \text{ kJ}$

50.0 °C: $\Delta G°_T = - (0.008314 \text{ kJ/K})(323.2 \text{ K})(\ln 1.633 \times 10^{-5}) = 29.62\underline{3}7 \text{ kJ}$

Next, solve two equations in two unknowns assuming $\Delta H°$ and $\Delta S°$ are constant over the range of 25.0 °C to 50.0 °C. Use 0.2982 K(kJ/J) and 0.3232 K(kJ/J) to convert $\Delta S°$ in J to $T\Delta S°$ in kJ.

1. $27.1551 \text{ kJ} = \Delta H° - [0.2982 \text{ K(kJ/J)} \Delta S°]$

2. $29.6237 \text{ kJ} = \Delta H° - [0.3232 \text{ K(kJ/J)} \Delta S°]$

Then, rearrange equation 2, and substitute for $\Delta H°$ into equation 2:

3a. $\Delta H° = 0.3232 \text{ K(kJ/J)} \Delta S° + 29.6237 \text{ kJ}$

3b. $27.1551 \text{ kJ} = [0.3232 \text{ K(kJ/J)} \Delta S° + 29.6237 \text{ kJ}] - [0.2982 \text{ K(kJ/J)} \Delta S°]$

Solve for $\Delta S°$:

$$\Delta S° = \frac{(29.6237 - 27.1551) \text{ kJ}}{(0.2982 - 0.3232) \text{ K/(kJ/J)}} = -98.\underline{7}4 = -98.7 \text{ J/K}$$

Substitute this value into equation 3a and solve for $\Delta H°$:

$$\Delta H° = [(0.3232) \text{ K(kJ/J)} \times (-98.74 \text{ J/K})] + 29.6237 \text{ kJ} = -2.2\underline{8}9 = -2.29 \text{ kJ}$$

20. ELECTROCHEMISTRY

■ Solutions to Exercises

Note on significant figures: If the final answer to a solution needs to be rounded off, it is given first with one nonsignificant figure, and the last significant figure is underlined. The final answer is then rounded to the correct number of significant figures. In multiple-step problems, intermediate answers are given with at least one nonsignificant figure; however, only the final answer has been rounded off.

20.1 Assign oxidation numbers to the skeleton equation (Step 1).

$$\overset{0}{I_2} + \overset{+5}{NO_3^-} \rightarrow \overset{+5}{IO_3^-} + \overset{+4}{NO_2}$$

Separate into two incomplete half-reactions (Step 2). Note that iodine is oxidized (increases in oxidation number), and nitrogen is reduced (decreases in oxidation number).

$$I_2 \rightarrow IO_3^-$$

$$NO_3^- \rightarrow NO_2$$

Balance each half-reaction separately. The oxidation half-reaction is not balanced in I, so place a two in front of IO_3^- (Step 3a). Then add six H_2O's to the left side to balance O atoms (Step 3b), and add twelve H^+ ions to the right side to balance H atoms (Step 3c). Finally, add ten electrons to the right side to balance the charge (Step 3d). The balanced oxidation half-reaction is

$$I_2 + 6H_2O \rightarrow 2IO_3^- + 12H^+ + 10e^-$$

The reduction half-reaction is balanced in N (Step 3a). Add one H_2O to the right side to balance O atoms (Step 3b), and add two H^+ ion to the left side to balance H atoms. Finally, add one electron to the left side to balance the charge (Step 3d).

(continued)

The balanced reduction half-reaction is

$$NO_3^- + 2H^+ + e^- \rightarrow NO_2 + H_2O$$

Multiply the reduction half-reaction by five so that, when added, the electrons cancel (Step 4a).

$$I_2 + 6H_2O \rightarrow 2IO_3^- + 12H^+ + 10e^-$$

$$\underline{10NO_3^- + 20H^+ + 10e^- \rightarrow 10NO_2 + 10H_2O}$$

$$I_2 + 10NO_3^- + 20H^+ + 6H_2O + \cancel{10e^-} \rightarrow$$
$$2IO_3^- + 10NO_2 + 12H^+ + 10H_2O + \cancel{10e^-}$$

Simplify the equation by canceling the twelve H^+ and six H_2O that appear on both sides. The coefficients do not need to be reduced (Step 4b). The net ionic equation is

$$I_2(s) + 10NO_3^-(aq) + 8H^+(aq) \rightarrow 2IO_3^-(aq) + 10NO_2(g) + 4H_2O(l)$$

20.2 After balancing the equation as if it were in acid solution, you obtain the following:

$$H_2O_2 + 2ClO_2 \rightarrow 2ClO_2^- + O_2 + 2H^+$$

Add two OH^- to both sides of the equation (Step 5), and replace the two H^+ and two OH^- on the right side with two H_2O. No further cancellation is required. The balanced equation for the reaction in basic solution is

$$H_2O_2 + 2ClO_2 + 2OH^- \rightarrow 2ClO_2^- + O_2 + 2H_2O$$

20.3 Silver ion is reduced at the silver electrode (cathode). The half-reaction is

$$Ag^+(aq) + e^- \rightarrow Ag(s)$$

The nickel electrode (anode) is where oxidation occurs. The half-reaction is

$$Ni(s) \rightarrow Ni^{2+}(aq) + 2e^-$$

(continued)

Electron flow in the external circuit is from the nickel electrode (anode) to the silver electrode (cathode). Positive ions will flow in the solution portion of the circuit opposite to the direction of the electrons. A sketch of the cell is given below:

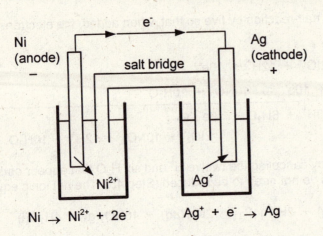

$$Ni \rightarrow Ni^{2+} + 2e^- \qquad Ag^+ + e^- \rightarrow Ag$$

20.4   The notation for the cell is $Zn(s)|Zn^{2+}(aq)||H^+(aq)|H_2(g)|Pt(s)$.

20.5   The half-cell reactions are

$$Cd(s) \rightarrow Cd^{2+}(aq) + 2e^-$$
$$2H^+(aq) + 2e^- \rightarrow H_2(g)$$

Summing the half-cell reactions gives the overall cell reaction.

$$Cd(s) + 2H^+(aq) \rightarrow Cd^{2+}(aq) + H_2(g)$$

20.6   The half-reactions are

$$Zn(s) \rightarrow Zn^{2+}(aq) + 2e^-$$
$$Cu^{2+}(aq) + 2e^- \rightarrow Cu(s)$$

n equals two, and the maximum work for the reaction as written is

$$w_{max} = -nFE_{cell} = -2 \times 9.65 \times 10^4\,C \times 1.10\,V = -2.1\underline{2}3 \times 10^5\,V{\cdot}C = -2.12 \times 10^5\,J$$

For 6.54 g of zinc metal, the maximum work is

$$6.54\,g\,Zn \times \frac{1\,mol\,Zn}{65.39\,g\,Zn} \times \frac{-2.123 \times 10^5\,J}{1\,mol\,Zn} = -2.1\underline{2}3 \times 10^4 = -2.12 \times 10^4\,J$$

20.7 The half-reactions and corresponding electrode potentials are as follows

$$Ag^+(aq) + e^- \rightarrow Ag(s) \qquad\qquad\qquad 0.80\ V$$

$$NO_3^-(aq) + 4H^+(aq) + 4e^- \rightarrow NO(g) + 2H_2O(l) \qquad 0.96\ V$$

The stronger oxidizing agent is the one involved in the half-reaction with the more positive standard electrode potential, so NO_3^- is the stronger oxidizing agent.

20.8 In this reaction, Cu^{2+} is the oxidizing agent on the left side; I_2 is the oxidizing reagent on the right side. The corresponding standard electrode potentials are

$$Cu^{2+}(aq) + 2e^- \rightarrow Cu(s); \qquad E° = 0.34\ V$$

$$I_2(s) + 2e^- \rightarrow 2I^-(aq); \qquad E° = 0.54\ V$$

The stronger oxidizing agent is the one involved in the half-reaction with the more positive standard electrode potential, so I_2 is the stronger oxidizing agent. The reaction is nonspontaneous as written.

20.9 The reduction half-reactions and standard electrode potentials are

$$Zn^{2+}(aq) + 2e^- \rightarrow Zn(s); \qquad E°_{Zn} = -0.76\ V$$

$$Cu^{2+}(aq) + 2e^- \rightarrow Cu(s); \qquad E°_{Cu} = 0.34\ V$$

Reverse the first half-reaction and its half-cell potential to obtain

$$Zn(s) \rightarrow Zn^{2+}(aq) + 2e^-; \qquad -E°_{Zn} = 0.76\ V$$

$$Cu^{2+}(aq) + 2e^- \rightarrow Cu(s); \qquad E°_{Cu} = 0.34\ V$$

Obtain the cell emf by adding the half-cell potentials.

$$E°_{cell} = E°_{Cu} - E°_{Zn} = 0.34\ V + 0.76\ V = 1.10\ V$$

20.10 The half-cell reactions, the corresponding half-cell potentials, and their sums are displayed below:

$$Sn^{2+}(aq) \rightarrow Sn^{4+}(aq) + 2e^- \qquad\qquad -E° = 0.15\ V$$

$$\underline{2Hg^{2+}(aq) + 2e^- \rightarrow Hg_2^{2+}(aq) \qquad\qquad E° = 0.90\ V}$$

$$Sn^{2+}(aq) + 2Hg^{2+}(aq) \rightarrow Sn^{4+}(aq) + Hg_2^{2+}(aq) \qquad E°_{cell} = 0.75\ V$$

(continued)

Note that each half-reaction involves two electrons; hence n equals two. Also, $E°_{cell}$ equals 0.75 V, and the faraday constant, F, is 9.65×10^4 C. Therefore,

$$\Delta G° = -nFE°_{cell} = -2 \times 9.65 \times 10^4 \text{ C} \times 0.75 \text{ V} = -1.\underline{4}475 \times 10^5 \text{ J} = -1.4 \times 10^5 \text{ J}$$

Thus, the standard free-energy change is -1.4×10^2 kJ.

20.11 Write the equation with $\Delta G_f°$'s beneath each substance.

$$Mg(s) \ + \ Cu^{2+}(aq) \ \rightarrow \ Mg^{2+}(aq) \ + \ Cu(s)$$

| $\Delta G_f°$: | 0 | 65.0 | -456.0 | 0 kJ |

Hence,

$$\Delta G° = \Sigma n \Delta G_f°(\text{products}) - \Sigma m \Delta G_f°(\text{reactants})$$

$$= [(-456.0) - 65.0] \text{ kJ} = -521.\underline{0} \text{ kJ} = -5.21\underline{0} \times 10^5 \text{ J}$$

Obtain n by splitting the reaction into half-reactions.

$$Mg(s) \ \rightarrow \ Mg^{2+}(aq) + 2e^-$$
$$Cu^{2+}(aq) + 2e^- \ \rightarrow \ Cu(s)$$

Each half-reaction involves two electrons, so n equals two. Therefore,

$$\Delta G° = -nFE°_{cell}$$

$$-5.210 \times 10^5 \text{ J} = -2 \times 9.65 \times 10^4 \text{ C} \times E°_{cell}$$

Rearrange and solve for $E°_{cell}$. Recall that $J = C \cdot V$.

$$E°_{cell} = \frac{-5.210 \times 10^5 \text{ J}}{-2 \times 9.65 \times 10^4 \text{ C}} = 2.6\underline{9}94 = 2.70 \text{ V}$$

20.12 The half-reactions and standard electrode potentials are

$$Fe(s) \ \rightarrow \ Fe^{2+}(aq) + 2e^-; \qquad -E°_{Fe} = 0.41 \text{ V}$$
$$Sn^{4+}(aq) + 2e^- \ \rightarrow \ Sn^{2+}(s); \qquad E°_{Sn^{2+}} = 0.15 \text{ V}$$

The standard emf for the cell is

$$E°_{cell} = E°_{Fe} - E°_{Sn^{2+}} = 0.41 \text{ V} + 0.15 \text{ V} = 0.56 \text{ V}$$

(continued)

Note that n equals two. Substitute into the equation relating E° and K. Also K = K_c.

$$0.56 \text{ V} = \frac{0.0592}{2} \log K_c$$

Solving for K_c, you get

$$\log K_c = 1\underline{8}.91$$

Take the antilog of both sides:

$$K_c = \text{antilog } (18.91) = 8.1 \times 10^{1\underline{8}} = 10^{19}$$

20.13 The half-cell reactions, the corresponding half-cell potentials, and their sums are displayed below:

$$
\begin{array}{ll}
Zn(s) \rightarrow Zn^{2+}(aq) + 2e^- & -E° = 0.76 \text{ V} \\
\underline{2Ag^+(aq) + 2e^- \rightarrow 2Ag(s)} & \underline{E° = 0.80 \text{ V}} \\
Zn(s) + 2Ag^+(aq) \rightarrow Zn^{2+}(aq) + 2Ag(s) & E°_{cell} = 1.56 \text{ V}
\end{array}
$$

Note that n equals two. The reaction quotient is

$$Q = \frac{[Zn^{2+}]}{[Ag^+]^2} = \frac{0.200}{(0.00200)^2} = 5.0\underline{0} \times 10^4$$

The standard emf is 1.56 V, so the Nernst equation becomes

$$E_{cell} = E°_{cell} - \frac{0.0592}{n} \log Q$$

$$= 1.56 - \frac{0.0592}{2} \log (5.00 \times 10^4)$$

$$= 1.56 - 0.139\underline{0}9 = 1.4\underline{2}09 = 1.42 \text{ V}$$

20.14 The half-cell reactions, the corresponding half-cell potentials, and their sums are displayed below:

$$Zn(s) \rightarrow Zn^{2+}(aq) + 2e^{-} \qquad\qquad -E^{\circ} = 0.76 \text{ V}$$

$$\underline{Ni^{2+}(aq) + 2e^{-} \rightarrow Ni(s) \qquad\qquad\qquad E^{\circ} = -0.23 \text{ V}}$$

$$Zn(s) + Ni^{2+}(aq) \rightarrow Zn^{2+}(aq) + Ni(s) \qquad E^{\circ}_{cell} = 0.53 \text{ V}$$

Note that n equals two. The standard emf is 0.53 V, and the emf is 0.34 V, so the Nernst equation becomes

$$E_{cell} = E^{\circ}_{cell} - \frac{0.0592}{n} \log Q$$

$$0.34 \text{ V} = 0.53 \text{ V} - \frac{0.0592}{2} \log Q$$

Rearrange and solve for log Q

$$\log Q = \frac{2}{0.0592} \times (0.53 - 0.34) = 6.\underline{4}18$$

Take the antilog of both sides

$$Q = \frac{[Zn^{2+}]}{[Ni^{2+}]} = \text{antilog} (6.418) = 2.\underline{6}23 \times 10^{6}$$

Substitute in $[Zn^{2+}] = 1.00$ M and solve for $[Ni^{2+}]$.

$$\frac{1.00 \text{ M}}{[Ni^{2+}]} = 2.624 \times 10^{6}$$

$$[Ni^{2+}] = 3.\underline{8}1 \times 10^{-7} = 4 \times 10^{-7} \text{ M}$$

20.15 a. The cathode reaction is $K^{+}(l) + e^{-} \rightarrow K(l)$.

The anode reaction is $2Cl^{-}(l) \rightarrow Cl_{2}(g) + 2e^{-}$.

b. The cathode reaction is $K^{+}(l) + e^{-} \rightarrow K(l)$.

The anode reaction is $4OH^{-}(aq) \rightarrow O_{2}(g) + 2H_{2}O(g) + 4e^{-}$.

20.16   The species you should consider for half-reactions are Ag^+ and H_2O. Two possible cathode reactions are

$$Ag^+(aq) + e^- \rightarrow Ag(s); \; E° = 0.80 \text{ V}$$

$$2H_2O(l) + 2e^- \rightarrow H_2(g) + 2OH^-(aq); \; E° = -0.83 \text{ V}$$

Because the silver electrode potential is larger than the reduction potential of water, Ag^+ is reduced. The only possible anode reaction is the oxidation of water. The expected half-reactions are

$$Ag^+(aq) + e^- \rightarrow Ag(s)$$

$$2H_2O(l) \rightarrow O_2(g) + 4H^+(aq) + 4e^-$$

20.17   The conversion of grams of silver to coulombs required to deposit this amount of silver is

$$0.365 \text{ g Ag} \times \frac{1 \text{ mol Ag}}{107.9 \text{ g}} \times \frac{1 \text{ mol } e^-}{1 \text{ mol Ag}} \times \frac{9.65 \times 10^4 \text{ C}}{1 \text{ mol } e^-} = 326.4 \text{ C}$$

The time lapse, 216 min, equals 1.296×10^3 s. Thus,

$$\text{Current} = \frac{\text{charge}}{\text{time}} = \frac{326.4 \text{ C}}{1.296 \times 10^4 \text{ s}} = 2.5188 \times 10^{-2} = 2.52 \times 10^{-2} \text{ A}$$

20.18   When the current flows for 1.85×10^4 s, the amount of charge is

$$0.0565 \text{ A} \times 1.85 \times 10^4 \text{ s} = 1.045 \times 10^3 \text{ C}$$

Note that four moles of electrons are equivalent to one mol of O_2. Hence,

$$1.045 \times 10^3 \text{ C} \times \frac{1 \text{ mol } e^-}{9.65 \times 10^4 \text{ C}} \times \frac{1 \text{ mol } O_2}{4 \text{ mol } e^-} \times \frac{32.00 \text{ g } O_2}{1 \text{ mol } O_2}$$

$$= 0.08665 = 0.0866 \text{ g } O_2$$

■ Answers to Review Questions

20.1   A voltaic cell is an electrochemical cell in which a spontaneous reaction generates an electric current (energy). An electrolytic cell is an electrochemical cell that requires electrical current (energy) to drive a nonspontaneous reaction to the right.

20.2 In both the voltaic and electrolytic cells, the cathode is the electrode at which reduction occurs, and the anode is the electrode at which oxidation occurs.

20.3 The SI unit of electrical potential is the volt (V).

20.4 The faraday (F) is the magnitude of charge on one mole of electrons; it equals 9.65×10^4 C, or 9.65×10^4 J/V.

20.5 It is necessary to measure the voltage of a voltaic cell when no current is flowing because the cell voltage exhibits its maximum value only when no current flows. Even if the current flows just for the time of measurement, the voltage drops enough so that what is measured is significantly less than the maximum.

20.6 Standard electrode potentials are defined relative to a standard electrode potential of zero volts (0.00, 0.000 V, etc.) for the $H^+/H_2(g)$ electrode. Because the cell emf is measured using the hydrogen electrode at standard conditions and a second electrode at standard conditions, the cell emf equals the E° of the half-reaction at the second electrode.

20.7 The SI unit of energy equals joules equals coulombs x volts.

20.8 The mathematical relationships are as follows:

$$\Delta G° = -nFE°_{cell}$$

$$\Delta G° = -RT \ln K$$

Combining these two equations gives

$$\ln K = \frac{nFE°_{cell}}{RT}$$

20.9 The first step in the corrosion of iron is

$$2Fe(s) + O_2(g) + 2H_2O(l) \rightarrow 4OH^- + 2Fe^{2+}$$

The Nernst equation for this reaction is

$$E_{cell} = E°_{cell} - \frac{0.0592}{4} \log [OH^-]^4[Fe^{2+}]^2$$

If the pH increases, the [OH⁻] increases, and thus E_{cell} becomes more negative (this predicts the reaction becomes less spontaneous). If the pH decreases, the [OH⁻] decreases, and thus E_{cell} becomes more positive (this predicts the reaction becomes more spontaneous).

20.10   The zinc-carbon cell has a zinc can as the anode; the cathode is a graphite rod surrounded by a paste of manganese dioxide and carbon black. Around this is a second paste of ammonium and zinc chlorides. The electrode reactions involve oxidation of zinc metal to zinc(II) ion and reduction of $MnO_2(s)$ to $Mn_2O_3(s)$ at the cathode. The lead storage battery consists of a spongy lead anode and a lead dioxide cathode, both immersed in aqueous sulfuric acid. At the anode, the lead is oxidized to lead sulfate; at the cathode, lead dioxide is reduced to lead sulfate.

20.11   A fuel cell is essentially a battery that does not use up its electrodes. Instead, it operates with a continuous supply of reactants (fuel). An example is the hydrogen-oxygen fuel cell in which oxygen is reduced at one electrode to the hydroxide ion, and hydrogen is oxidized at the other electrode to water (H in the +1 oxidation state). Such a cell produces electrical energy in a spacecraft for long periods of time.

20.12   During the rusting of iron, one end of a drop of water exposed to air acts as one electrode of a voltaic cell; at this electrode, an oxygen molecule is reduced by four electrons to four hydroxide ions. Oxidation of metallic iron to iron(II) ion at the center of the drop of water supplies the electrons, and the center serves as the other electrode of the voltaic cell. Thus, electrons flow from the center of the drop through the iron to the end of the drop.

20.13   When iron or steel is connected to an active metal such as zinc, a voltaic cell is formed with zinc as the anode and iron as the cathode. Any type of moisture forms the electrolyte solution, and the zinc metal is then oxidized to zinc(II) ion in preference to the oxidation of iron metal. Oxygen is reduced at the cathode to hydroxide ions. If iron or steel is exposed to oxygen while connected to a less active metal such as tin, a voltaic cell is formed with iron as the anode and tin as the cathode, and iron is oxidized to iron(II) ion rather than tin being oxidized to tin(II) ion. Thus, exposed iron corrodes rapidly in a tin can. Fortunately, as long as the iron is covered by the tin, it cannot corrode.

20.14   The addition of an ionic species such as strongly ionized sulfuric acid facilitates the passage of current through the solution.

20.15   Sodium metal can be prepared by electrolysis of molten sodium chloride.

20.16   The anode reaction in the electrolysis of molten potassium hydroxide is

$$4OH^- \rightarrow O_2(g) + 2H_2O(g) + 4e^-$$

20.17   The reason different products are obtained is that water instead of Na^+ is reduced at the cathode during the electrolysis of aqueous NaCl. This is because water has a more positive E° (smaller decomposition voltage). At the anode, water instead of chloride ion is oxidized because water has a less positive E° (smaller decomposition voltage).

20.18   The Nernst equation for the electrode reaction of $2Cl^-(aq) \rightarrow Cl_2(g) + 2e^-$ is

$$E = -1.36 \text{ V} - \frac{0.0592}{2} \log \frac{1 \text{ atm}}{[Cl^-]^2} = -1.36 \text{ V} + 0.0592 \log [Cl^-]$$

This equation implies that E increases as [Cl⁻] increases. For a sufficiently large [Cl⁻], Cl⁻ will be more readily oxidized than the water solvent.

■ Solutions to Practice Problems

Note on significant figures: If the final answer to a solution needs to be rounded off, it is given first with one nonsignificant figure, and the last significant figure is underlined. The final answer is then rounded to the correct number of significant figures. In multiple-step problems, intermediate answers are given with at least one nonsignificant figure; however, only the final answer has been rounded off.

20.29   In balancing oxidation-reduction reactions in acid, the four steps in the text will be followed. For part a, each step is shown. For the other parts, only a summary is shown.

a. Assign oxidation numbers to the skeleton equation (Step 1).

$$\overset{+6}{Cr_2O_7^{2-}} + \overset{+3}{C_2O_4^{2-}} \rightarrow \overset{+3}{Cr^{3+}} + \overset{+4}{CO_2}$$

Separate into two incomplete half-reactions (Step 2). Note that carbon is oxidized (increases in oxidation number), and chromium is reduced (decreases in oxidation number).

$$C_2O_4^{2-} \rightarrow CO_2$$
$$Cr_2O_7^{2-} \rightarrow Cr^{3+}$$

Balance each half-reaction separately. The oxidation half-reaction is not balanced in C, so place a two in front of CO_2 (Step 3a). Finally, add two electrons to the right side to balance the charge (Step 3d). The balanced oxidation half-reaction is

$$C_2O_4^{2-} \rightarrow 2CO_2 + 2e^-$$

(continued)

The reduction half-reaction is not balanced in Cr, so place a two in front of Cr^{3+} (Step 3a). Add seven H_2O to the right side to balance O atoms (Step 3b), and add fourteen H^+ ion to the left side to balance H atoms (step 3c). Finally, add six electrons to the left side to balance the charge (Step 3d). The balanced reduction half-reaction is

$$Cr_2O_7^{2-} + 14H^+ + 6e^- \rightarrow 2Cr^{3+} + 7H_2O$$

Multiply the oxidation half-reaction by three so that, when added, the electrons cancel (Step 4a).

$$3C_2O_4^{2-} \rightarrow 6CO_2 + 6e^-$$
$$Cr_2O_7^{2-} + 14H^+ + 6e^- \rightarrow 2Cr^{3+} + 7H_2O$$
$$\overline{Cr_2O_7^{2-} + 3C_2O_4^{2-} + 14H^+ + \cancel{6e^-} \rightarrow 2Cr^{3+} + 6CO_2 + 7H_2O + \cancel{6e^-}}$$

The equation does not need to be simplified any further (Step 4b). The net ionic equation is

$$Cr_2O_7^{2-} + 3C_2O_4^{2-} + 14H^+ \rightarrow 2Cr^{3+} + 6CO_2 + 7H_2O$$

b. The two balanced half-reactions are

$$Cu \rightarrow Cu^{2+} + 2e^- \qquad \text{(oxidation)}$$
$$NO_3^- + 4H^+ + 3e^- \rightarrow NO + 2H_2O \qquad \text{(reduction)}$$

Multiply the oxidation half-reaction by three and the reduction half-reaction by four, and then add together. Cancel the six electrons from each side. No further simplification is needed. The balanced equation is

$$3Cu + 2NO_3^- + 8H^+ \rightarrow 3Cu^{2+} + 2NO + 4H_2O$$

c. The two balanced half-reactions are

$$HNO_2 + H_2O \rightarrow NO_3^- + 3H^+ + 2e^- \qquad \text{(oxidation)}$$
$$MnO_2 + 4H^+ + 2e^- \rightarrow Mn^{2+} + 2H_2O \qquad \text{(reduction)}$$

Add the two half-reactions together and cancel the two electrons from each side. Also, cancel three H^+ and one H_2O from each side. The balanced equation is

$$MnO_2 + HNO_2 + H^+ \rightarrow Mn^{2+} + NO_3^- + H_2O$$

(continued)

d. The two balanced half-reactions are

$$Mn^{2+} + 4H_2O \rightarrow MnO_4^- + 8H^+ + 5e^- \qquad \text{(oxidation)}$$
$$PbO_2 + SO_4^{2-} + 4H^+ + 2e^- \rightarrow PbSO_4 + 2H_2O \qquad \text{(reduction)}$$

Multiply the oxidation half-reaction by two and the reduction half-reaction by five, and then add together. Cancel the ten electrons from each side. Also, cancel sixteen H^+ and eight H_2O from each side. The balanced equation is

$$5PbO_2 + 2Mn^{2+} + 5SO_4^{2-} + 4H^+ \rightarrow 5PbSO_4 + 2MnO_4^- + 2H_2O$$

e. The two balanced half-reactions are

$$HNO_2 + H_2O \rightarrow NO_3^- + 3H^+ + 2e^- \qquad \text{(oxidation)}$$
$$Cr_2O_7^{2-} + 14H^+ + 6e^- \rightarrow 2Cr^{3+} + 7H_2O \qquad \text{(reduction)}$$

Multiply the oxidation half-reaction by three, and then add together. Cancel the six electrons from each side. Also, cancel nine H^+ and three H_2O from each side. The balanced equation is

$$3HNO_2 + Cr_2O_7^{2-} + 5H^+ \rightarrow 2Cr^{3+} + 3NO_3^- + 4H_2O$$

20.31 In balancing oxidation-reduction reactions in basic solution, it will first be balanced as if it were in acidic solution; then the extra two steps in the text will be followed.

a. The two balanced half-reactions are

$$Mn^{2+} + 2H_2O \rightarrow MnO_2 + 4H^+ + 2e^- \qquad \text{(oxidation)}$$
$$H_2O_2 + 2H^+ + 2e^- \rightarrow 2H_2O \qquad \text{(reduction)}$$

Add the two half-reactions together, and cancel the two electrons from each side. Also, cancel two H^+ and two H_2O from each side. The balanced equation in acidic solution is

$$Mn^{2+} + H_2O_2 \rightarrow MnO_2 + 2H^+$$

Now add two OH^- to each side (Step 5). Simplify by combining the H^+ and OH^- to give H_2O. No further simplification is required (Step 6). The balanced equation in basic solution is

$$Mn^{2+} + H_2O_2 + 2OH^- \rightarrow MnO_2 + 2H_2O$$

(continued)

b. The two balanced half-reactions are

$$NO_2^- + H_2O \rightarrow NO_3^- + 2H^+ + 2e^- \qquad \text{(oxidation)}$$

$$MnO_4^- + 4H^+ + 3e^- \rightarrow MnO_2 + 2H_2O \qquad \text{(reduction)}$$

Multiply the oxidation half-reaction by three and the reduction half-reaction by two, and then add together. Cancel the six electrons from each side. Also, cancel twelve H^+ and six H_2O from each side. The balanced equation in acidic solution is

$$2MnO_4^- + 3NO_2^- + 2H^+ \rightarrow 2MnO_2 + 3NO_3^- + H_2O$$

Now add two OH^- to each side. Simplify by combining the H^+ and OH^- to give H_2O. Then cancel one H_2O from each side. The balanced equation in basic solution is

$$2MnO_4^- + 3NO_2^- + H_2O \rightarrow 2MnO_2 + 3NO_3^- + 2OH^-$$

c. The two balanced half-reactions are

$$Mn^{2+} + 2H_2O \rightarrow MnO_2 + 4H^+ + 2e^- \qquad \text{(oxidation)}$$

$$ClO_3^- + 2H^+ + e^- \rightarrow ClO_2 + H_2O \qquad \text{(reduction)}$$

Multiply the reduction half-reaction by two and then add together. Cancel the two electrons from each side. Also, cancel four H^+ and two H_2O from each side. The balanced equation in acidic solution is

$$Mn^{2+} + 2ClO_3^- \rightarrow MnO_2 + 2ClO_2$$

Since there are no H^+ on either side, no further simplification is needed. The balanced equation in basic solution is identical to the balanced equation in acidic solution.

d. The two balanced half-reactions are

$$NO_2 + H_2O \rightarrow NO_3^- + 2H^+ + e^- \qquad \text{(oxidation)}$$

$$MnO_4^- + 4H^+ + 3e^- \rightarrow MnO_2 + 2H_2O \qquad \text{(reduction)}$$

Multiply the oxidation half-reaction by three and then add together. Cancel the three electrons from each side. Also, cancel four H^+ and two H_2O from each side. The balanced equation in acidic solution is

$$MnO_4^- + 3NO_2 + H_2O \rightarrow MnO_2 + 3NO_3^- + 2H^+$$

(continued)

Now add two OH$^-$ to each side. Simplify by combining the H$^+$ and OH$^-$ to give H$_2$O. Then cancel one H$_2$O from each side. The balanced equation in basic solution is

$$MnO_4^- + 3NO_2 + 2OH^- \rightarrow MnO_2 + 3NO_3^- + H_2O$$

e. The two balanced half-reactions are

$$Cl_2 + 6H_2O \rightarrow 2ClO_3^- + 12H^+ + 10e^- \qquad \text{(oxidation)}$$

$$Cl_2 + 2e^- \rightarrow 2Cl^- \qquad \text{(reduction)}$$

Multiply the reduction half-reaction by five and then add together. Cancel the ten electrons from each side. Also, divide all the coefficients by two. The balanced equation in acidic solution is

$$3Cl_2 + 3H_2O \rightarrow 5Cl^- + ClO_3^- + 6H^+$$

Now add six OH$^-$ to each side. Simplify by combining the H$^+$ and OH$^-$ to give H$_2$O. Then cancel three H$_2$O from each side. The balanced equation in basic solution is

$$3Cl_2 + 6OH^- \rightarrow 5Cl^- + ClO_3^- + 3H_2O$$

20.33 a. The two balanced half-reactions are

$$8H_2S \rightarrow S_8 + 16H^+ + 16e^- \qquad \text{(oxidation)}$$

$$NO_3^- + 2H^+ + e^- \rightarrow NO_2 + H_2O \qquad \text{(reduction)}$$

Multiply the reduction half-reaction by sixteen, and then add together. Cancel the sixteen electrons and sixteen H$^+$ from each side. The balanced equation is

$$8H_2S + 16NO_3^- + 16H^+ \rightarrow S_8 + 16NO_2 + 16H_2O$$

b. The two balanced half-reactions are

$$Cu \rightarrow Cu^{2+} + 2e^- \qquad \text{(oxidation)}$$

$$NO_3^- + 4H^+ + 3e^- \rightarrow NO + 2H_2O \qquad \text{(reduction)}$$

Multiply the oxidation half-reaction by three and the reduction half-reaction by two, and then add together. Cancel the six electrons from each side. The balanced equation is

$$2NO_3^- + 3Cu + 8H^+ \rightarrow 2NO + 3Cu^{2+} + 4H_2O$$

(continued)

c. The two balanced half-reactions are

$$SO_2 + 2H_2O \rightarrow SO_4^{2-} + 4H^+ + 2e^- \qquad \text{(oxidation)}$$

$$MnO_4^- + 8H^+ + 5e^- \rightarrow Mn^{2+} + 4H_2O \qquad \text{(reduction)}$$

Multiply the oxidation half-reaction by five and the reduction half-reaction by two, and then add together. Cancel the ten electrons from each side. Also, cancel sixteen H^+ and eight H_2O from each side. The balanced equation is

$$2MnO_4^- + 5SO_2 + 2H_2O \rightarrow 5SO_4^{2-} + 2Mn^{2+} + 4H^+$$

d. The two balanced half-reactions are

$$Sn(OH)_3^- + 3H_2O \rightarrow Sn(OH)_6^{2-} + 3H^+ + 2e^- \qquad \text{(oxidation)}$$

$$Bi(OH)_3 + 3H^+ + 3e^- \rightarrow Bi + 3H_2O \qquad \text{(reduction)}$$

Multiply the oxidation half-reaction by three and the reduction half-reaction by two, and then add together. Cancel the six electrons from each side. Also, cancel six H^+ and six H_2O from each side. The balanced equation in acidic solution is

$$2Bi(OH)_3 + 3Sn(OH)_3^- + 3H_2O \rightarrow 3Sn(OH)_6^{2-} + 2Bi + 3H^+$$

Now add three OH^- to each side. Simplify by combining the H^+ and OH^- to give H_2O. Then cancel three H_2O on each side. The balanced equation in basic solution is

$$2Bi(OH)_3 + 3Sn(OH)_3^- + 3OH^- \rightarrow 3Sn(OH)_6^{2-} + 2Bi$$

20.35 a. The two balanced half-reactions are

$$I^- + 3H_2O \rightarrow IO_3^- + 6H^+ + 6e^- \qquad \text{(oxidation)}$$

$$MnO_4^- + 4H^+ + 3e^- \rightarrow MnO_2 + 2H_2O \qquad \text{(reduction)}$$

Multiply the reduction half-reaction by two, and then add together. Cancel the six electrons from each side. Also, cancel six H^+ and three H_2O from each side. The balanced equation in acidic solution is

$$2MnO_4^- + I^- + 2H^+ \rightarrow 2MnO_2 + IO_3^- + H_2O$$

Now add two OH^- to each side. Simplify by combining the H^+ and OH^- to give H_2O. Then cancel one H_2O on each side. The balanced equation in basic solution is

$$2MnO_4^- + I^- + H_2O \rightarrow 2MnO_2 + IO_3^- + 2OH^-$$

(continued)

b. The two balanced half-reactions are

$$2Cl^- \rightarrow Cl_2 + 2e^- \qquad \text{(oxidation)}$$

$$Cr_2O_7^{2-} + 14H^+ + 6e^- \rightarrow 2Cr^{3+} + 7H_2O \qquad \text{(reduction)}$$

Multiply the oxidation half-reaction by three, and then add together. Cancel the six electrons from each side. The balanced equation is

$$Cr_2O_7^{2-} + 6Cl^- + 14H^+ \rightarrow 2Cr^{3+} + 3Cl_2 + 7H_2O$$

c. The two balanced half-reactions are

$$S_8 + 16H_2O \rightarrow 8SO_2 + 32H^+ + 32e^- \qquad \text{(oxidation)}$$

$$NO_3^- + 4H^+ + 3e^- \rightarrow NO + 2H_2O \qquad \text{(reduction)}$$

Multiply the oxidation half-reaction by three and the reduction half-reaction by thirty two, and then add together. Cancel the ninety six electrons from each side. Also, cancel ninety six H^+ and forty eight H_2O from each side. The balanced equation is

$$3S_8 + 32NO_3^- + 32H^+ \rightarrow 32NO + 24SO_2 + 16H_2O$$

d. The two balanced half-reactions are

$$H_2O_2 \rightarrow O_2 + 2H^+ + 2e^- \qquad \text{(oxidation)}$$

$$MnO_4^- + 4H^+ + 3e^- \rightarrow MnO_2 + 2H_2O \qquad \text{(reduction)}$$

Multiply the oxidation half-reaction by three and the reduction half-reaction by two, and then add together. Cancel the six electrons from each side. Also, cancel six H^+ from each side. The balanced equation in acidic solution is

$$3H_2O_2 + 2MnO_4^- + 2H^+ \rightarrow 3O_2 + 2MnO_2 + 4H_2O$$

Now add two OH^- to each side. Simplify by combining the H^+ and OH^- to give H_2O. Then cancel two H_2O on each side. The balanced equation in basic solution is

$$3H_2O_2 + 2MnO_4^- \rightarrow 3O_2 + 2MnO_2 + 2H_2O + 2OH^-$$

e. The two balanced half-reactions are

$$Zn \rightarrow Zn^{2+} + 2e^- \qquad \text{(oxidation)}$$

$$2NO_3^- + 12H^+ + 10e^- \rightarrow N_2 + 6H_2O \qquad \text{(reduction)}$$

Multiply the oxidation half-reaction by five, and then add together. Cancel the ten electrons from each side. The balanced equation is

$$5Zn + 2NO_3^- + 12H^+ \rightarrow N_2 + 5Zn^{2+} + 6H_2O$$

20.37   Sketch of the cell:

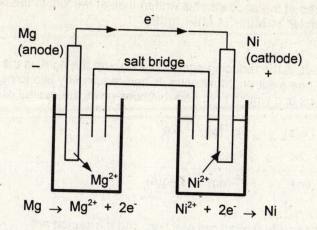

$Mg \rightarrow Mg^{2+} + 2e^-$ $Ni^{2+} + 2e^- \rightarrow Ni$

20.39   Sketch of the cell:

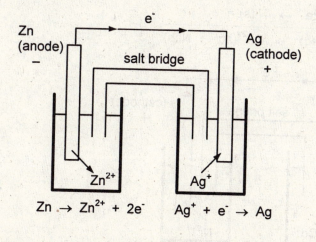

$Zn \rightarrow Zn^{2+} + 2e^-$ $Ag^+ + e^- \rightarrow Ag$

20.41   The electrode half-reactions and the overall cell reaction are

Anode: $Zn(s) + 2OH^-(aq) \rightarrow Zn(OH)_2(s) + 2e^-$

Cathode: $Ag_2O(s) + H_2O(l) + 2e^- \rightarrow 2Ag(s) + 2OH^-(aq)$

Overall: $Zn(s) + Ag_2O(s) + H_2O(l) \rightarrow Zn(OH)_2(s) + 2Ag(s)$

20.43   Because of its less negative $E°$, Pb^{2+} is reduced at the cathode and is written on the right; $Ni(s)$ is oxidized at the anode and is written first, at the left, in the cell notation. The notation is $Ni(s)|Ni^{2+}(aq)||Pb^{2+}(aq)|Pb(s)$.

20.45   Because of its less negative E°, H$^+$ is reduced at the cathode and is written on the right; Ni(s) is oxidized at the anode and is written first, at the left, in the cell notation. The notation is Ni(s)|Ni^{2+}(1 M)||H$^+$(1 M)|H$_2$(g)|Pt.

20.47   The Fe(s), on the left, is the reducing agent. The Ag$^+$, on the right, is the oxidizing agent, gaining just one electron. Multiplying its half-reaction by two to equalize the numbers of electrons and writing both half-reactions gives the overall cell reaction:

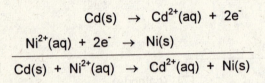

$$Fe(s) \rightarrow Fe^{2+}(aq) + 2e^-$$

$$2Ag^+(aq) + 2e^- \rightarrow 2Ag(s)$$

$$\overline{Fe(s) + 2Ag^+(aq) \rightarrow Fe^{2+}(aq) + 2Ag(s)}$$

20.49   The half-cell reactions, the overall cell reaction, and the sketch are

$$Cd(s) \rightarrow Cd^{2+}(aq) + 2e^-$$

$$Ni^{2+}(aq) + 2e^- \rightarrow Ni(s)$$

$$\overline{Cd(s) + Ni^{2+}(aq) \rightarrow Cd^{2+}(aq) + Ni(s)}$$

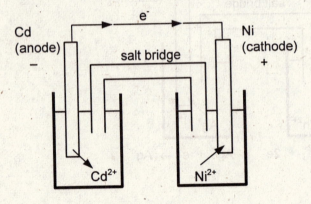

20.51   The half-cell reactions are

$$2Fe^{3+}(aq) + 2e^- \rightarrow 2Fe^{2+}(aq)$$
$$Zn(s) \rightarrow Zn^{2+}(aq) + 2e^-$$

n equals two, and the maximum work for the reaction as written is

$$w_{max} = -nFE_{cell} = -2 \times 9.65 \times 10^4 \text{ C} \times 0.72 \text{ V} = -1.\underline{3}89 \times 10^5 \text{ C} \cdot \text{V}$$

$$= -1.\underline{3}89 \times 10^5 \text{ J}$$

Because this is the work obtained by reduction of two mol of Fe^{3+}, the work for one mol is

$$w_{max} = (-1.389 \times 10^5 \text{ J})/2 \text{ mol} = -6.\underline{9}45 \times 10^4 = -6.9 \times 10^4 \text{ J} = -69 \text{ kJ}$$

20.53   The half-cell reactions are

$$2Ag^+(aq) + 2e^- \rightarrow 2Ag(s)$$
$$Ni(s) \rightarrow Ni^{2+}(aq) + 2e^-$$

n equals two, and the maximum work for the reaction as written is

$$w_{max} = -nFE_{cell} = -2 \times 9.65 \times 10^4 \text{ C} \times 0.97 \text{ V} = -1.\underline{8}7 \times 10^5 \text{ C} \cdot \text{V}$$

$$= -1.\underline{8}7 \times 10^5 \text{ J}$$

For 30.0 g of nickel, the maximum work is

$$30.0 \text{ g Ni} \times \frac{1 \text{ mol Ni}}{58.69 \text{ g Ni}} \times \frac{-1.87 \times 10^5 \text{ J}}{1 \text{ mol Ni}} = -9.\underline{5}6 \times 10^4 \text{ J} = -96 \text{ kJ}$$

20.55   The half-reactions and corresponding electrode potentials are as follows

$NO_3^-(aq) + 4H^+(aq) + 3e^- \rightarrow NO(g) + 2H_2O(l)$	0.96 V
$O_2(g) + 4H^+(aq) + 4e^- \rightarrow 2H_2O(l)$	1.23 V
$MnO_4^-(aq) + 8H^+(aq) + 5e^- \rightarrow Mn^{2+}(aq) + 4H_2O(l)$	1.49 V

The order by increasing oxidizing strength is $NO_3^-(aq)$, $O_2(g)$, $MnO_4^-(aq)$.

20.57 The half-reactions and corresponding electrode potentials are as follows

$$Zn^{2+}(aq) + 2e^- \rightarrow Zn(s) \qquad\qquad -0.76\ V$$
$$Fe^{2+}(aq) + 2e^- \rightarrow Fe(s) \qquad\qquad -0.41\ V$$
$$Cu^{2+}(aq) + e^- \rightarrow Cu^+(aq) \qquad\qquad 0.16\ V$$

Zn(s) is the strongest, and $Cu^+(aq)$ is the weakest.

20.59 a. In this reaction, Sn^{4+} is the oxidizing agent on the left side; Fe^{3+} is the oxidizing reagent on the right side. The corresponding standard electrode potentials are

$$Sn^{4+}(aq) + 2e^- \rightarrow Sn^{2+}(aq) \qquad\qquad E° = 0.15\ V$$
$$Fe^{3+}(aq) + e^- \rightarrow Fe^{2+}(aq) \qquad\qquad E° = 0.77\ V$$

The stronger oxidizing agent is the one involved in the half-reaction with the more positive standard electrode potential, so Fe^{3+} is the stronger oxidizing agent. The reaction is nonspontaneous as written.

b. In this reaction, MnO_4^- is the oxidizing agent on the left side; O_2 is the oxidizing reagent on the right side. The corresponding standard electrode potentials are

$$O_2(g) + 4H^+(aq) + 4e^- \rightarrow 2H_2O(l) \qquad\qquad E° = 1.23\ V$$
$$MnO_4^-(aq) + 8H^+(aq) + 5e^- \rightarrow Mn^{2+}(aq) + 4\ H_2O(l) \qquad E° = 1.49\ V$$

The stronger oxidizing agent is the one involved in the half-reaction with the more positive standard electrode potential, so MnO_4^- is the stronger oxidizing agent. The reaction is spontaneous as written.

20.61 The reduction half-reactions and standard electrode potentials are

$$Br_2(l) + 2e^- \rightarrow 2Br^-(aq) \qquad\qquad E° = 1.07\ V$$
$$Cl_2(g) + 2e^- \rightarrow 2Cl^-(aq) \qquad\qquad E° = 1.36\ V$$
$$F_2(g) + 2e^- \rightarrow 2F^-(aq) \qquad\qquad E° = 2.87\ V$$

From these, you see that the order of increasing oxidizing strength is Br_2, Cl_2, F_2. Therefore, chlorine gas will oxidize Br^- but will not oxidize F^-. The balanced equation for the reaction is

$$Cl_2(g) + 2Br^-(aq) \rightarrow 2Cl^-(aq) + Br_2(l)$$

20.63   The half-reactions and standard electrode potentials are

$$Cr(s) \rightarrow Cr^{3+}(aq) + 3e^- \qquad -E°_{Cr} = 0.74 \text{ V}$$
$$Hg_2^{2+}(aq) + 2e^- \rightarrow 2Hg(l) \qquad E°_{Hg} = 0.80 \text{ V}$$

Obtain the cell emf by adding the half-cell potentials.

$$E°_{cell} = E°_{Hg} - E°_{Cr} = 0.80 \text{ V} + 0.74 \text{ V} = 1.54 \text{ V}$$

20.65   The half-reactions and standard electrode potentials are

$$Cr(s) \rightarrow Cr^{3+}(aq) + 3e^- \qquad -E°_{Cr} = 0.74 \text{ V}$$
$$I_2(s) + 2e^- \rightarrow 2I^-(aq) \qquad E°_{I_2} = 0.54 \text{ V}$$

Obtain the cell emf by adding the half-cell potentials.

$$E°_{cell} = E°_{Cr} - E°_{I_2} = 0.54 \text{ V} + 0.74 \text{ V} = 1.28 \text{ V}$$

20.67   The half-cell reactions, the corresponding half-cell potentials, and their sums are displayed below:

$$3Cu(s) \rightarrow 3Cu^{2+}(aq) + 6e^- \qquad -E° = -0.34 \text{ V}$$
$$2NO_3^-(aq) + 8H^+(aq) + 6e^- \rightarrow 2NO(g) + 4H_2O(l) \qquad E° = 0.96 \text{ V}$$

$$3Cu(s) + 2NO_3^-(aq) + 8H^+(aq) \rightarrow$$
$$3Cu^{2+}(aq) + 2NO(g) + 4H_2O(l) \qquad E°_{cell} = 0.62 \text{ V}$$

Note that each half-reaction involves six electrons, hence n = 6. Therefore,

$$\Delta G° = -nFE°_{cell} = -6 \times 9.65 \times 10^4 \text{ C} \times 0.62 \text{ V} = -3.\underline{5}8 \times 10^5 \text{ J} = -3.6 \times 10^5 \text{ J}$$

Thus, the standard free-energy change is -3.6×10^2 kJ.

20.69   The half-cell reactions, the corresponding half-cell potentials, and their sums are displayed below:

$$2I^-(aq) \rightarrow I_2(s) + 2e^- \qquad -E° = -0.54 \text{ V}$$
$$Cl_2(g) + 2e^- \rightarrow 2Cl^-(aq) \qquad E° = 1.36 \text{ V}$$

$$2I^-(aq) + Cl_2(g) \rightarrow I_2(s) + 2Cl^-(aq) \qquad E°_{cell} = 0.82 \text{ V}$$

(continued)

Note that each half-reaction involves two electrons, hence n = 2. Therefore,

$$\Delta G° = -nFE°_{cell} = -2 \times 9.65 \times 10^4 \, C \times 0.82 \, V = -1.\underline{5}8 \times 10^5 \, J = -1.6 \times 10^5 \, J$$

Thus, the standard free-energy change is -1.6 × 10² kJ.

20.71 Write the equation with $\Delta G_f°$'s beneath each substance.

	Mg(s)	+	2Ag⁺(aq)	→	Mg²⁺(aq)	+	2Ag(s)
$\Delta G_f°$:	0		2 x 77.111		-456.01		0 kJ

Hence,

$$\Delta G° = \Sigma n \Delta G_f°(\text{products}) - \Sigma m \Delta G_f°(\text{reactants})$$

$$= [-456.01 - 2 \times 77.111] \, kJ = -610.\underline{2}3 \, kJ = -6.10\underline{2}3 \times 10^5 \, J$$

Obtain n by splitting the reaction into half-reactions.

$$Mg(s) \rightarrow Mg^{2+}(aq) + 2e^-$$
$$2Ag^+(aq) + 2e^- \rightarrow 2Ag(s)$$

Each half-reaction involves two electrons, so n = 2. Therefore,

$$\Delta G° = -nFE°_{cell}$$

$$-6.1023 \times 10^5 \, J = -2 \times 9.65 \times 10^4 \, C \times E°_{cell}$$

Rearrange and solve for $E°_{cell}$. Recall that J = C•V.

$$E°_{cell} = \frac{-6.1023 \times 10^5 \, J}{-2 \times 9.65 \times 10^4 \, C} = 3.1\underline{6}18 = 3.16 \, V$$

20.73 Write the equation with $\Delta G_f°$'s beneath each substance:

	PbO₂(s)	+	2HSO₄⁻(aq)	+	2H⁺(aq)	+	Pb(s)	→	2PbSO₄(s)	+	2H₂O(l)
$\Delta G_f°$:	-189.2		2(-752.87)		0		0		2(-811.24)		2(-237.192) kJ

Hence,

$$\Delta G° = [2 \times -811.24 + 2 \times -237.192 - -189.2 - 2 \times -752.87] \, kJ$$

$$= -401.\underline{9}24 \, kJ = -4.0192 \times 10^5 \, J$$

(continued)

Obtain n by splitting the reaction into half-reactions.

$$Pb(s) + HSO_4^-(aq) \rightarrow PbSO_4(s) + H^+(aq) + 2e^-$$

$$PbO_2(s) + HSO_4^-(aq) + 3H^+(aq) + 2e^- \rightarrow PbSO_4(s) + 2H_2O(l)$$

Each half-reaction involves two electrons, so n = 2. Therefore,

$$\Delta G^\circ = -nFE^\circ_{cell}$$

$$-4.0192 \times 10^5 \text{ J} = -2 \times 9.65 \times 10^4 \text{ C} \times E^\circ_{cell}$$

Rearrange and solve for E°_{cell}. Recall that J = C•V.

$$E^\circ_{cell} = \frac{-4.0192 \times 10^5 \text{ J}}{-2 \times 9.65 \times 10^4 \text{ C}} = 2.0\underline{8}25 = 2.08 \text{ V}$$

20.75 The half-reactions and standard electrode potentials are

$$Cu(s) \rightarrow Cu^{2+}(aq) + 2e^- \qquad -E^\circ_{Cu} = -0.34 \text{ V}$$

$$2Fe^{3+}(aq) + 2e^- \rightarrow 2Fe^{2+}(aq) \qquad E^\circ_{Fe^{3+}} = 0.77 \text{ V}$$

The standard emf for the cell is

$$E^\circ_{cell} = E^\circ_{Fe^{3+}} - E^\circ_{Cu} = 0.77 \text{ V} - 0.34 \text{ V} = 0.43 \text{ V}$$

Note that n = 2. Substitute into the equation relating E° and K. Note that K = K_c.

$$0.43 \text{ V} = \frac{0.0592}{2} \log K_c$$

Solving for K_c, you get

$$\log K_c = 1\underline{4}.52$$

Take the antilog of both sides:

$$K_c = \text{antilog}(14.52) = 3.4 \times 10^{\underline{14}} = 10^{14}$$

20.77   The half-reactions and standard electrode potentials are

$$Cu^+(aq) \rightarrow Cu^{2+}(aq) + e^- \qquad -E° = -0.16 \text{ V}$$

$$Cu^+(aq) + e^- \rightarrow Cu(s) \qquad E° = 0.52 \text{ V}$$

The standard emf for the cell is

$$E°_{cell} = 0.52 \text{ V} - 0.16 \text{ V} = 0.36 \text{ V}$$

Note that n equals 1. Substitute into the equation relating E° and K. Note that K equals K_c.

$$0.36 \text{ V} = \frac{0.0592}{1} \log K_c$$

Solving for K_c, you get

$$\log K_c = 6.\underline{0}81$$

Take the antilog of both sides:

$$K_c = \text{antilog} (6.081) = \underline{1}.2 \times 10^6 = 1 \times 10^6$$

20.79   The half-cell reactions, the corresponding half-cell potentials, and their sums are displayed below:

$$2Cr(s) \rightarrow 2Cr^{3+}(aq) + 6e^- \qquad -E° = 0.74 \text{ V}$$

$$3Ni^{2+}(aq) + 6e^- \rightarrow 3Ni(s) \qquad E° = -0.23 \text{ V}$$

$$\overline{2Cr(s) + 3Ni^{2+}(aq) \rightarrow 2Cr^{3+}(aq) + 3Ni(s) \qquad\qquad E°_{cell} = 0.51 \text{ V}}$$

Note that n equals six. The reaction quotient is

$$Q = \frac{[Cr^{3+}]^2}{[Ni^{2+}]^3} = \frac{(1.0 \times 10^{-3})^2}{(1.5)^3} = 2.\underline{9}62 \times 10^{-7}$$

The standard emf is 0.51 V, so the Nernst equation becomes

$$E_{cell} = E°_{cell} - \frac{0.0592}{n} \log Q$$

$$= 0.51 - \frac{0.0592}{6} \log (2.962 \times 10^{-7})$$

$$= 0.51 - (-0.06\underline{4}41) = 0.5\underline{7}4 = 0.57 \text{ V}$$

20.81 The half-cell reactions, the corresponding half-cell potentials, and their sums are displayed below:

$$10Br^-(aq) \rightarrow 5Br_2(l) + 10e^- \qquad -E° = 1.07\ V$$

$$2MnO_4^-(aq) + 16H^+(aq) + 10e^- \rightarrow 2Mn^{2+}(aq) + 8H_2O(l) \qquad E° = 1.49\ V$$

$$2MnO_4^-(aq) + 10Br^-(aq) + 16H^+(aq) \rightarrow$$

$$2Mn^{2+}(aq) + 5Br_2(l) + 8H_2O(l) \qquad E°_{cell} = 0.42\ V$$

Note that n equals ten. The reaction quotient is

$$Q = \frac{[Mn^{2+}]^2}{[MnO_4^{2-}]^2[Br^-]^{10}[H^+]^{16}} = \frac{(0.15)^2}{(0.010)^2(0.010)^{10}(1.0)^{16}} = 2.25 \times 10^{22}$$

The standard emf is 0.09 V, so the Nernst equation becomes

$$E_{cell} = E°_{cell} - \frac{0.0592}{n}\log Q = 0.42 - \frac{0.0592}{10}\log(2.25 \times 10^{22})$$

$$= 0.42 - (0.1323) = 0.2876 = 0.29\ V$$

20.83 The overall reaction is

$$Cd(s) + Ni^{2+}(aq) \rightarrow Cd^{2+}(aq) + Ni(s)$$

Note that n equals two. The reaction quotient is

$$Q = \frac{[Cd^{2+}]}{[Ni^{2+}]} = \frac{[Cd^{2+}]}{1.0} = [Cd^+]$$

The standard emf is 0.170 V, and the emf is 0.240 V, so the Nernst equation becomes

$$E_{cell} = E°_{cell} - \frac{0.0592}{n}\log Q$$

$$0.240 = 0.170 - \frac{0.0592}{2}\log Q$$

Rearrange and solve for log Q

$$\log Q = \frac{2}{0.0592} \times (0.240 - 0.170) = -2.3648$$

(continued)

Take the antilog of both sides

$$Q = [Cd^{2+}] = antilog\ (-2.3648) = \underline{4}.31 \times 10^{-3} = 0.004\ M$$

The Cd^{2+} concentration is 0.004 M.

20.85 a. The cathode reaction is $Ca^{2+}(l) + 2e^- \rightarrow Ca(l)$.

The anode reaction is $S^{2-}(l) \rightarrow S(l) + 2e^-$.

b. The cathode reaction is $Cs^+(l) + e^- \rightarrow Cs(l)$.

The anode reaction is $4OH^-(l) \rightarrow O_2(g) + 2H_2O(g) + 4e^-$.

20.87 a. The species you should consider for half-reactions are Na^+, SO_4^{2-}, and H_2O. The possible cathode reactions are

$$Na^+(aq) + e^- \rightarrow Na(s) \qquad\qquad E° = -2.71\ V$$
$$2H_2O(l) + 2e^- \rightarrow H_2(g) + 2OH^-(aq) \qquad\qquad E° = -0.83\ V$$

Because the electrode potential for H_2O is larger (less negative), it is easier to reduce.

The possible anode reactions are

$$2H_2O(l) \rightarrow O_2(g) + 4H^+(aq) + 4e^- \qquad\qquad E° = -1.23\ V$$
$$2SO_4^{2-}(aq) \rightarrow S_2O_4^{2-}(aq) + 2e^- \qquad\qquad E° = -2.01\ V$$

Because the electrode potential for H_2O is less negative, it is easier to oxidize.

The expected half-reactions are

$$2H_2O(l) \rightarrow O_2(g) + 4H^+(aq) + 4e^- \qquad\qquad E° = -1.23\ V$$
$$4H_2O(l) + 4e^- \rightarrow 2H_2(g) + 4OH^-(aq) \qquad\qquad E° = -0.83\ V$$

The overall reaction is

$$2H_2O(l) \rightarrow 2H_2(g) + O_2(g)$$

(continued)

b. The species you should consider for half-reactions are K^+, Br^-, and H_2O. The possible cathode reactions are

$$K^+(aq) + e^- \rightarrow K(s) \qquad\qquad E° = -2.92 \text{ V}$$
$$2H_2O(l) + 2e^- \rightarrow H_2(g) + 2OH^-(aq) \qquad E° = -0.83 \text{ V}$$

Because the electrode potential for H_2O is larger (less negative), it is easier to reduce. The possible anode reactions are

$$2H_2O(l) \rightarrow O_2(g) + 4H^+(aq) + 4e^- \qquad E° = -1.23 \text{ V}$$
$$2Br^-(aq) \rightarrow Br_2(l) + 2e^- \qquad\qquad E° = -1.07 \text{ V}$$

Because the electrode potential for Br^- is less negative, it is easier to oxidize.

The expected half-reactions are

$$2Br^-(aq) \rightarrow Br_2(l) + 2e^- \qquad\qquad E° = -1.07 \text{ V}$$
$$2H_2O(l) + 2e^- \rightarrow H_2(g) + 2OH^-(aq) \qquad E° = -0.83 \text{ V}$$

The overall reaction is

$$2Br^-(aq) + 2H_2O(l) \rightarrow Br_2(l) + H_2(g) + 2OH^-$$

20.89 The conversion of grams of aluminum (3.61 kg = 3.61×10^3 g) to coulombs required to give this amount of aluminum is

$$3.61 \times 10^3 \text{ g} \times \frac{1 \text{ mol Al}}{26.98 \text{ g}} \times \frac{3 \text{ mol } e^-}{1 \text{ mol Al}} \times \frac{9.65 \times 10^4 \text{ C}}{1 \text{ mol } e^-}$$

$$= 3.8\underline{7}3 \times 10^7 = 3.87 \times 10^7 \text{ C}$$

20.91 The conversion of coulombs to grams of lithium is

$$5.00 \times 10^3 \text{ C} \times \frac{1 \text{ mol } e^-}{9.65 \times 10^4 \text{ C}} \times \frac{1 \text{ mol Li}}{1 \text{ mol } e^-} \times \frac{6.941 \text{ g Li}}{1 \text{ mol Li}} = 0.359\underline{6}$$

$$= 0.360 \text{ g Li}$$

■ Solutions to General Problems

20.93 a. The half-reactions and net ionic equation are

$$3Fe(s) \rightarrow 3Fe^{2+}(aq) + 6e^-$$

$$2NO_3^-(aq) + 8H^+(aq) + 6e^- \rightarrow 2NO(g) + 4H_2O(l)$$

$$3Fe(s) + 2NO_3^-(aq) + 8H^+(aq) \rightarrow 3Fe^{2+}(aq) + 2NO(g) + 4H_2O(l)$$

b. The half-reactions and net ionic equation are

$$3Fe^{2+}(aq) \rightarrow 3Fe^{3+}(aq) + 3e^-$$

$$NO_3^-(aq) + 4H^+(aq) + 3e^- \rightarrow NO(g) + 2H_2O(l)$$

$$3Fe^{2+}(aq) + NO_3^-(aq) + 4H^+(aq) \rightarrow 3Fe^{3+}(aq) + NO(g) + 2H_2O(l)$$

c. Add the results from parts (a) and (b)

$$3Fe(s) + 2NO_3^-(aq) + 8H^+(aq) \rightarrow 3Fe^{2+}(aq) + 2NO(g) + 4H_2O(l)$$

$$3Fe^{2+}(aq) + NO_3^-(aq) + 4H^+(aq) \rightarrow 3Fe^{3+}(aq) + NO(g) + 2H_2O(l)$$

$$Fe(s) + NO_3^-(aq) + 4H^+(aq) \rightarrow Fe^{3+}(aq) + NO(g) + 2H_2O(l)$$

20.95 a. The two balanced half-reactions are

$$8S^{2-} \rightarrow S_8 + 16e^- \qquad \text{(oxidation)}$$

$$MnO_4^- + 4H^+ + 3e^- \rightarrow MnO_2 + 2H_2O \qquad \text{(reduction)}$$

Multiply the oxidation half-reaction by three and the reduction half-reaction by sixteen, and then add together. Cancel the forty eight electrons from each side. The balanced equation in acidic solution is

$$16MnO_4^- + 24S^{2-} + 64H^+ \rightarrow 16MnO_2 + 3S_8 + 32H_2O$$

Now add sixty four OH^- to each side. Simplify by combining the H^+ and OH^- to give H_2O. Then cancel thirty two H_2O on each side. The balanced equation in basic solution is

$$16MnO_4^-(aq) + 24S^{2-}(aq) + 32H_2O(l) \rightarrow 16MnO_2(aq) + 3S_8(aq) + 64OH^-(l)$$

(continued)

b. The two balanced half-reactions are

$$HSO_3^- + H_2O \rightarrow SO_4^{2-} + 3H^+ + 2e^- \qquad \text{(oxidation)}$$

$$IO_3^- + 6H^+ + 6e^- \rightarrow I^- + 3H_2O \qquad \text{(reduction)}$$

Multiply the oxidation half-reaction by three, and then add together. Cancel the six electrons from each side. Also, cancel six H^+ and three H_2O from each side. The balanced equation is

$$IO_3^-(aq) + 3HSO_3^-(aq) \rightarrow I^- + 3SO_4^{2-}(aq) + 3H^+(aq)$$

c. The two balanced half-reactions are

$$Fe(OH)_2 + H_2O \rightarrow Fe(OH)_3 + H^+ + e^- \qquad \text{(oxidation)}$$

$$CrO_4^{2-} + 4H^+ + 3e^- \rightarrow Cr(OH)_4^- \qquad \text{(reduction)}$$

Multiply the oxidation half-reaction by three, and then add together. Cancel the three electrons and cancel three H^+ from each side. The balanced equation in acidic solution is

$$CrO_4^{2-}(aq) + 3Fe(OH)_2(s) + H^+(aq) + 3H_2O(l) \rightarrow Cr(OH)_4^-(aq) + 3Fe(OH)_3(s)$$

Now add one OH^- to each side. Simplify by combining the H^+ and OH^- to give H_2O. Then, combine the H_2Os on the left side into a single term. The balanced equation in basic solution is

$$CrO_4^{2-}(aq) + 3Fe(OH)_2(aq) + 4H_2O(l) \rightarrow Cr(OH)_4^-(aq) + 3Fe(OH)_3(s) + OH^-(aq)$$

d. The two balanced half-reactions are

$$Cl_2 + 2H_2O \rightarrow 2ClO^- + 4H^+ + 2e^- \qquad \text{(oxidation)}$$

$$Cl_2 + 2e^- \rightarrow 2Cl^- \qquad \text{(reduction)}$$

Add the two half-reactions together. Cancel the two electrons from each side. Divide all the coefficients by a factor of two. The balanced equation in acidic solution is

$$Cl_2 + H_2O \rightarrow Cl^- + ClO^- + 2H^+$$

Now add two OH^- to each side. Simplify by combining the H^+ and OH^- to give H_2O. Then cancel one H_2O on each side. The balanced equation in basic solution is

$$Cl_2(aq) + 2OH^-(aq) \rightarrow Cl^-(aq) + ClO^-(aq) + H_2O(l)$$

20.97   This reaction takes place in basic solution. The skeleton equation is

$$Fe(OH)_2(s) + O_2(g) \rightarrow Fe(OH)_3(s)$$

The two balanced half-reactions are

$Fe(OH)_2 + H_2O \rightarrow Fe(OH)_3 + H^+ + e^-$ (oxidation)

$O_2 + 4H^+ + 4e^- \rightarrow 2H_2O$ (reduction)

Multiply the oxidation half-reaction by four, and then add together. Cancel the four electrons from each side. Also, cancel four H^+ and two H_2O from each side. The balanced equation in acidic or in basic solution is

$$4Fe(OH)_2(s) + O_2(g) + 2H_2O(l) \rightarrow 4Fe(OH)_3(s)$$

20.99   The cell notation is $Ca(s)|Ca^{2+}(aq)||Cl^-(aq)|Cl_2(g)|Pt(s)$. The reactions are

Anode: $Ca(s) \rightarrow Ca^{2+}(aq) + 2e^-$ $-E° = 2.76$ V

Cathode: $Cl_2(g) + 2e^- \rightarrow 2Cl^-(aq)$ $E° = 1.36$ V

$E°_{cell} = 1.36$ V $+ 2.76$ V $= 4.12$ V

20.101 a. In this reaction, Fe^{3+} is the oxidizing agent on the left side; Ni^{2+} is the oxidizing reagent on the right side. The corresponding standard electrode potentials are

$Ni^{2+}(aq) + 2e^- \rightarrow Ni(s)$ $E° = -0.23$ V

$Fe^{3+}(aq) + e^- \rightarrow Fe^{2+}(aq)$ $E° = 0.77$ V

The stronger oxidizing agent is the one involved in the half-reaction with the more positive standard electrode potential, so Fe^{3+} is the stronger oxidizing agent. The oxidation of nickel by iron(III) is a spontaneous reaction.

b. In this reaction, Fe^{3+} is the oxidizing agent on the left side; Sn^{4+} is the oxidizing reagent on the right side. The corresponding standard electrode potentials are

$Sn^{4+}(aq) + 2e^- \rightarrow Sn^{2+}(aq)$ $E° = 0.15$ V

$Fe^{3+}(aq) + e^- \rightarrow Fe^{2+}(aq)$ $E° = 0.77$ V

The stronger oxidizing agent is the one involved in the half-reaction with the more positive standard electrode potential, so Fe^{3+} is the stronger oxidizing agent. The oxidation of tin(II) by iron(III) is a spontaneous reaction.

20.103 The half-cell reactions, the corresponding half-cell potentials, and their sums are displayed below:

$$Pb(s) \rightarrow Pb^{2+}(aq) + 2e^- \qquad\qquad -E° = 0.13 \text{ V}$$

$$\underline{2H^+(aq) + 2e^- \rightarrow H_2(g) \qquad\qquad E° = 0.00 \text{ V}}$$

$$Pb(s) + 2H^+(aq) \rightarrow Pb^{2+}(aq) + H_2(g) \qquad\qquad E°_{cell} = 0.13 \text{ V}$$

Note that n equals two. Next, use K_{sp} to calculate $[Pb^{2+}]$.

$$[Pb^{2+}] = \frac{K_{sp}}{[SO_4{}^{2-}]} = \frac{1.7 \times 10^{-8}}{1.0} = 1.7 \times 10^{-8} \text{ M}$$

The reaction quotient for the cell reaction is

$$Q = \frac{[Pb^{2+}] \times P_{H_2}}{[H^+]^2} = \frac{(1.7 \times 10^{-8})(1.0)}{(1.0)^2} = 1.7 \times 10^{-8}$$

The standard emf is 0.13 V , so the Nernst equation becomes

$$E_{cell} = E°_{cell} - \frac{0.0592}{n} \log Q$$

$$E_{cell} = 0.13 \text{ V} - \frac{0.0592}{2} \log (1.7 \times 10^{-8})$$

$$= 0.13 \text{ V} - (-0.2299 \text{ V}) = 0.3599 = 0.36 \text{ V}$$

20.105 a. Note that E° = 0.010 V and n equals two. Substitute into the equation relating E° and K.

$$E° = \frac{0.0592}{n} \log K$$

$$0.010 \text{ V} = \frac{0.0592}{2} \log K$$

Solving for K, you get

$$\log K = 0.3378$$

Take the antilog of both sides:

$$K = K_c = \text{antilog} (0.3378) = 2.176 = 2.2$$

(continued)

b. Substitute into the equilibrium expression using $[Sn^{2+}] = x$, and $[Pb^{2+}] = 1.0\ M - x$.

$$K_c = \frac{[Sn^{2+}]}{[Pb^{2+}]} = \frac{x}{1.0 - x} = 2.176$$

$$x = 0.6\underline{8}51\ M$$

$$[Pb^{2+}] = 1.0\ M - x = 0.\underline{3}149 = 0.3\ M$$

20.107 a. The number of faradays is

$$1.0\ \text{mol Na}^+ \times \frac{1\ \text{mol e}^-}{1\ \text{mol Na}^+} \times \frac{1\ F}{1\ \text{mol e}^-} = 1.0\ F$$

The number of coulombs is

$$1.0\ F \times \frac{9.65 \times 10^4\ C}{1\ F} = 9.\underline{6}5 \times 10^4 = 9.7 \times 10^4\ C$$

b. The number of faradays is

$$1.0\ \text{mol Cu}^{2+} \times \frac{2\ \text{mol e}^-}{1\ \text{mol Cu}^{2+}} \times \frac{1\ F}{1\ \text{mol e}^-} = 2.0\ F$$

The number of coulombs is

$$2.0\ F \times \frac{9.65 \times 10^4\ C}{1\ F} = 1.\underline{9}3 \times 10^5 = 1.9 \times 10^5\ C$$

c. The number of faradays is

$$1.0\ \text{g H}_2\text{O} \times \frac{1\ \text{mol H}_2\text{O}}{18.01\ \text{g H}_2\text{O}} \times \frac{2\ \text{mol e}^-}{1\ \text{mol H}_2\text{O}} \times \frac{1\ F}{1\ \text{mol e}^-} = 0.1\underline{1}10 = 0.11\ F$$

The number of coulombs is

$$0.1110\ F \times \frac{9.65 \times 10^4\ C}{1\ F} = 1.\underline{0}7 \times 10^4 = 1.1 \times 10^4\ C$$

(continued)

d. The number of faradays is

$$1.0 \text{ g Cl}^- \times \frac{1 \text{ mol Cl}^-}{35.45 \text{ g Cl}^-} \times \frac{2 \text{ mol e}^-}{2 \text{ mol Cl}^-} \times \frac{1 \text{ F}}{1 \text{ mol e}^-} = 0.02\underline{8}20 = 0.028 \text{ F}$$

The number of coulombs is

$$0.02820 \text{ F} \times \frac{9.65 \times 10^4 \text{ C}}{1 \text{ F}} = 2.\underline{7}22 \times 10^3 = 2.7 \times 10^3 \text{ C}$$

20.109 From the electrolysis information, calculate the total moles of I_2 formed

$$65.4 \text{ s} \times (10.5 \times 10^{-3} \text{ A}) \times \frac{1 \text{C}}{1 \text{A} \cdot \text{s}} \times \frac{1 \text{ mol e}^-}{9.65 \times 10^4 \text{ C}} \times \frac{1 \text{ mol } I_2}{2 \text{ mol e}^-}$$

$$= 3.5\underline{5}8 \times 10^{-6} \text{ mol } I_2$$

From the balanced equation

$$3.558 \times 10^{-6} \text{ mol } I_2 \times \frac{1 \text{ mol } H_3AsO_3}{1 \text{ mol } I_2} \times \frac{1 \text{ mol As}}{1 \text{ mol } H_3AsO_3} \times \frac{79.42 \text{ g As}}{1 \text{ mol As}}$$

$$= 2.8\underline{2}57 \times 10^{-4} = 2.83 \times 10^{-4} \text{ g As}$$

20.111a. The spontaneous chemical reaction and maximum cell potential are

$$
\begin{array}{lll}
\text{Cd(s)} & \rightarrow & \text{Cd}^{2+}\text{(aq)} + 2\text{e}^- & \quad -E^\circ = 0.40 \text{ V} \\
2\text{Ag}^+\text{(aq)} + 2\text{e}^- & \rightarrow & 2\text{Ag(s)} & \quad E^\circ = 0.80 \text{ V} \\
\hline
\text{Cd(s)} + 2\text{Ag}^+\text{(aq)} & \rightarrow & 2\text{Ag(s)} + \text{Cd}^{2+}\text{(aq)} & \quad E^\circ{}_{cell} = 1.20 \text{ V}
\end{array}
$$

b. Addition of S^{2-} would greatly decrease the Cd^{2+}(aq) concentration and help shift the equilibrium to the right, thus forming more Ag.

c. No effect. The size of the electrode makes no difference in the potential.

20.113 Anode: $2H_2O(l) \rightarrow O_2(g) + 4H^+(aq) + 4e^-$

Cathode: $Cu^{2+}(aq) + 2e^- \rightarrow Cu(s)$

20.115 a. First, find the moles of silver.

$$2.48 \text{ g Ag} \times \frac{1 \text{ mol Ag}}{107.9 \text{ g Ag}} = 0.02298 \text{ mol Ag}$$

The number of coulombs is

$$0.02298 \text{ mol Ag} \times \frac{1 \text{ mol e}^-}{1 \text{ mol Ag}} \times \frac{9.65 \times 10^4 \text{ C}}{1 \text{ mol e}^-} = 2218 \text{ C}$$

Since amp x sec = coul, the number of seconds is

$$\text{time} = \frac{2218 \text{ C}}{1.50 \text{ amp}} = 1.478 \times 10^3 = 1.48 \times 10^3 \text{ s}$$

b. For the same amount of current, the moles of Cr would be

$$2217 \text{ C} \times \frac{1 \text{ mol e}^-}{9.65 \times 10^4 \text{ C}} \times \frac{1 \text{ mol Cr}}{3 \text{ mol e}^-} \times \frac{52.00 \text{ g Cr}}{1 \text{ mol Cr}}$$

$$= 0.3984 = 0.398 \text{ g}$$

20.117 a. Note that 3.50 hours = 12600 s, and the efficiency is 90.0 percent.

$$2.75 \text{ amp} \times 12600 \text{ s} \times \frac{1 \text{ F}}{9.65 \times 10^4 \text{ C}} \times 0.900 = 0.3232 = 0.323 \text{ F}$$

b. The moles of Au that were deposited is

$$\text{mol Au} = 21.221 \text{ g Au} \times \frac{1 \text{ mol Au}}{197.0 \text{ g}} = 0.107720 \text{ mol}$$

Since 1 F = 1 mol e⁻, the ratio of moles of e⁻ to moles Au is

$$\frac{\text{mol e}^-}{\text{mol Au}} = \frac{0.3232}{0.10772} = \frac{3.00}{1}$$

Thus, the ion must be Au^{3+}. The reaction is

$$Au^{3+} + 3e^- \rightarrow Au$$

20.119a. The half-cell reactions, the corresponding half-cell potentials, and their sums are displayed below:

$$Zn(s) \rightarrow Zn^{2+}(aq) + 2e^- \qquad\qquad -E° = 0.76 \text{ V}$$

$$\underline{Cu^{2+}(aq) + 2e^- \rightarrow Cu(s) \qquad\qquad\qquad E° = 0.34 \text{ V}}$$

$$Cu^{2+}(aq) + Zn(s) \rightarrow Zn^{2+}(aq) + Cu(s) \qquad E° = 1.10 \text{ V}$$

Use the Nernst equation to calculate the voltage of the cell.

$$E = E° - \frac{0.0592}{n} \log Q = E° - \frac{0.0592}{n} \log \frac{[Zn^{2+}]}{[Cu^{2+}]}$$

Note that n equals two, $[Zn^{2+}] = 0.200$ M, and $[Cu^{2+}] = 0.0100$ M.

$$E° = 1.10 \text{ V} - \frac{0.0592}{2} \log \frac{[0.200]}{[0.0100]}$$

$$= 1.10 \text{ V} - 0.03851 \text{ V} = 1.061 = 1.06 \text{ V}$$

b. First, calculate the moles of electrons passing through the cell.

$$1.00 \text{ amp} \times 225 \text{ s} \times \frac{1 \text{ mol } e^-}{9.65 \times 10^4 \text{ C}} = 0.002332 \text{ mol } e^-$$

The moles of Cu deposited are

$$0.002332 \text{ mol } e^- \times \frac{1 \text{ mol Cu}}{2 \text{ mol } e^-} = 0.001166 \text{ mol Cu}$$

The moles of Cu remaining in the 1.00 L of solution are

$$0.0100 - 0.001166 = 0.008834 = 0.0088 \text{ mol}$$

Since the volume of the solution is 1.00 L, the molarity of Cu^{2+} is 0.0088 M.

20.121 a. Write the cell reaction with the ΔG°_f's beneath

$$\begin{array}{lcccc} & Cd(s) \;+ & Co^{2+}(aq) \;\rightarrow & Cd^{2+}(aq) \;+ & Co(s) \\ \Delta G^\circ_f\text{:} & 0 & -51.5 & -77.6 & 0\ \text{kJ} \end{array}$$

Hence,

$$\Delta G^\circ = \Sigma n\Delta G_f^\circ(\text{products}) - \Sigma m\Delta G_f^\circ(\text{reactants})$$

$$= [-77.6 + 51.5]\ \text{kJ} = -26.\underline{1}\ \text{kJ} = -2.61 \times 10^4\ \text{J}$$

b. Next, determine the standard emf for the cell. Note that n equals two.

$$\Delta G^\circ = -nFE^\circ$$

$$-2.61 \times 10^4\ \text{J} = -2(9.65 \times 10^4\ \text{C}) \times E^\circ$$

Solve for E° to get

$$E^\circ = \frac{-2.61 \times 10^4\ \text{J}}{-2(9.65 \times 10^4\ \text{C})} = 0.13\underline{5}2 = 0.135\ \text{V}$$

The half-reactions and voltages are

$$\begin{array}{lcl} Cd(s) \rightarrow Cd^{2+}(aq) + 2e^- & & -E^\circ_{ox} = 0.40\ \text{V} \\ Co^{2+}(aq) + 2e^- \rightarrow Co(s) & & E^\circ_{red} = ? \\ \hline Cd(s) + Co^{2+}(aq) \rightarrow Cd^{2+}(aq) + Co(s) & & E^\circ = 0.1352\ \text{V} \end{array}$$

The cell emf is

$$E^\circ = E^\circ_{red} + (-E^\circ_{ox})$$

$$0.1352\ \text{V} = E^\circ_{red} + 0.40\ \text{V}$$

$$E^\circ_{red} = 0.1352\ \text{V} - 0.40\ \text{V} = -0.2\underline{6}47 = -0.26\ \text{V}$$

■ Solutions to Cumulative-Skills Problems

20.123 The half-cell reactions, the corresponding half-cell potentials, and their sums are displayed below:

$$3Zn(s) \rightarrow 3Zn^{2+}(aq) + 6e^- \qquad -E° = 0.76 \text{ V}$$
$$2Cr^{3+}(aq) + 6e^- \rightarrow 2Cr(s) \qquad E° = -0.74 \text{ V}$$

$$2Cr^{3+}(aq) + 3Zn(s) \rightarrow 3Zn^{2+}(aq) + 2Cr(s) \qquad E° = 0.02 \text{ V}$$

Note that n equals six. Therefore,

$$\Delta G° = -nFE° = -6(9.65 \times 10^4 \text{ C})(0.02 \text{ V}) = -1.158 \times 10^4 \text{ J} = -11.58 \text{ kJ}$$

Write the cell reaction with the $\Delta H°_f$'s beneath

$$2Cr^{3+}(aq) + 3Zn(s) \rightarrow 3Zn^{2+}(aq) + 2Cr(s)$$
$$\Delta H°_f: \quad 2(-1971) \qquad 0 \qquad 3(-152.4) \qquad 0 \text{ kJ}$$

Hence,

$$\Delta H° = \Sigma n \Delta H_f°(\text{products}) - \Sigma m \Delta H_f°(\text{reactants})$$

$$= [3(-152.4) - 2(-1971)] \text{ kJ} = 3484.8 = 3485 \text{ kJ}$$

Now calculate $\Delta S°$.

$$\Delta G° = \Delta H° - T\Delta S°$$

$$-11.58 \text{ kJ} = 3484.8 \text{ kJ} - 298 \text{ K} \times \Delta S°$$

Solving for $\Delta S°$ gives

$$\Delta S° = \frac{3484.8 \text{ kJ} + 11.58 \text{ kJ}}{298 \text{ K}} = 11.73 = 11.7 \text{ kJ/K}$$

20.125 The half-cell reactions, the corresponding half-cell potentials, and their sums are displayed below:

$$4Br^-(aq) \rightarrow 2Br_2(l) + 4e^- \qquad -E° = -1.07 \text{ V}$$

$$O_2(g) + 4H^+(aq) + 4e^- \rightarrow 2H_2O(l) \qquad E° = 1.23 \text{ V}$$

$$O_2(g) + 4H^+(aq) + 4Br^-(aq) \rightarrow 2H_2O(l) + 2Br_2(l) \qquad E° = 0.16 \text{ V}$$

Now, convert the pH to $[H^+]$

$$[H^+] = \text{antilog } (-3.60) = 2.51 \times 10^{-4} \text{ M}$$

Under standard conditions $[Br^-] = 1$ M, and the pressure of O_2 is one atm. Thus, substitute into the Nernst equation, where n equals four.

$$E = E° - \frac{0.0592}{n} \log Q = E° - \frac{0.0592}{n} \log \frac{1}{[H^+]^4 [Br^-]^4 \times P_{O_2}}$$

$$= 0.16 \text{ V} - \frac{0.0592}{4} \log \frac{1}{(2.51 \times 10^{-4})^4}$$

$$= 0.16 \text{ V} - 0.2131 \text{ V} = -0.0531 = -0.05 \text{ V}$$

Thus, the reaction is nonspontaneous at this $[H^+]$.

20.127 Use the K_a to calculate [H+] for the buffer.

$$K_a = \frac{[H^+][OCN^-]}{[HOCN]} = 3.5 \times 10^{-4}$$

Rearrange, and solve for $[H^+]$, assuming [HOCN] and $[OCN^-]$ remain constant in the buffer. Thus, $[H^+] = K_a = 3.5 \times 10^{-4}$ M.

In Problem 20.123, E° for this cell reaction was found to be 0.16 V. Under standard conditions $[Br^-] = 1$ M, and the pressure of O_2 is one atm. Thus, substitute into the Nernst equation, where n equals four.

$$E = E° - \frac{0.0592}{n} \log Q = E° - \frac{0.0592}{n} \log \frac{1}{[H^+]^4 [Br^-]^4 \times P_{O_2}}$$

$$= 0.16 \text{ V} - \frac{0.0592}{4} \log \frac{1}{(3.5 \times 10^{-4})^4}$$

$$= 0.16 \text{ V} - 0.20459 \text{ V} = -0.04455 = -0.04 \text{ V}$$

Thus, the reaction is nonspontaneous at this $[H^+]$.

20.129 The half-cell reactions, the corresponding half-cell potentials, and their sums are displayed below:

$$H_2(g) \rightarrow 2H^+(aq) + 2e^- \qquad\qquad -E° = -0.00 \text{ V}$$

$$\underline{2Ag^+(aq) + 2e^- \rightarrow 2Ag(s) \qquad\qquad\qquad E° = 0.80 \text{ V}}$$

$$H_2(g) + 2Ag^+(aq) \rightarrow 2H^+(aq) + 2Ag(s) \qquad E° = 0.80 \text{ V}$$

The standard hydrogen electrode has $[H^+]$ = 1.0 M and the pressure of H_2 equals one atm. Substitute into the Nernst equation, where E = 0.45 V and n equals two.

$$E = E° - \frac{0.0592}{n} \log Q = E° - \frac{0.0592}{n} \log \frac{[H^+]^2}{[Ag^+]^2 \times P_{H_2}}$$

$$0.45 \text{ V} = 0.80 \text{ V} - \frac{0.0592}{2} \log \frac{1}{[Ag^+]^2}$$

Using the properties of logs, rearrange to get

$$0.45 \text{ V} = 0.80 \text{ V} + 0.0592 \log[Ag^+]$$

Solve for $[Ag^+]$

$$\log[Ag^+] = \frac{0.45 - 0.80}{0.0592} = -5.\underline{9}12$$

$$[Ag^+] = 10^{-5.912} = \underline{1}.224 \times 10^{-6} \text{ M}$$

Finally, determine the solubility product, using $[SCN^-]$ = 0.10 M

$$K_{sp} = [Ag^+][SCN^-] = (1.224 \times 10^{-6})(0.10) = \underline{1}.224 \times 10^{-7} = 1 \times 10^{-7}$$

21. NUCLEAR CHEMISTRY

■ Solutions to Exercises

Note on significant figures: If the final answer to a solution needs to be rounded off, it is given first with one nonsignificant figure, and the last significant figure is underlined. The final answer is then rounded to the correct number of significant figures. In multiple-step problems, intermediate answers are given with at least one nonsignificant figure; however, only the final answer has been rounded off.

21.1   The nuclide symbol for potassium-40 is $^{40}_{19}K$. Similarly, the nuclide symbol for calcium-40 is $^{40}_{20}Ca$. The equation for beta emission is

$$^{40}_{19}K \; \rightarrow \; ^{40}_{20}Ca \; + \; ^{0}_{-1}e$$

21.2   Plutonium-239 has the nuclide symbol is $^{239}_{94}Pu$. An alpha particle has the symbol $^{4}_{2}He$. The nuclear equation is

$$^{239}_{94}Pu \; \rightarrow \; ^{A}_{Z}X \; + \; ^{4}_{2}He$$

From the superscripts, you can write

239 = A + 4, or A = 235

Similarly, from the subscripts you can write

94 = Z + 2, or Z = 92

Hence, A = 235 and Z = 92, so the product is $^{235}_{92}X$ Because element 92 is uranium, symbol U, you write the product nucleus as $^{235}_{92}U$.

21.3 a. $^{118}_{50}$Sn has atomic number 50. It has 50 protons and 68 neutrons. Because its atomic number is less than 83 and it has an even number of protons and neutrons, it is expected to be stable.

b. $^{76}_{33}$As has atomic number 33. It has 33 protons and 43 neutrons. Because stable odd-odd nuclei are rare, you would expect $^{76}_{33}$As to be one of the radioactive isotopes.

c. $^{227}_{89}$Ac has atomic number 89. Because its atomic number is greater than 83, $^{227}_{89}$Ac is radioactive.

21.4 a. $^{13}_{7}$N has seven protons and six neutrons (fewer neutrons than protons). Its mass number is less than that of N = 14, so it is expected to decay by electron capture or positron emission (more likely, because this is a light isotope).

b. $^{26}_{11}$Na has 11 protons and 15 neutrons. It is expected to decay by beta emission.

21.5 a. The abbreviated notation is: $^{40}_{20}$Ca (d,p) $^{41}_{20}$Ca

b. The nuclear equation is: $^{12}_{6}$C + $^{2}_{1}$H $\rightarrow$ $^{13}_{6}$C + $^{1}_{1}$H

21.6 You can write the nuclear equation as follows:

$$^{A}_{Z}X + ^{1}_{0}n \rightarrow ^{14}_{6}C + ^{1}_{1}H$$

To balance this equation in charge (subscripts) and mass number (superscripts), write the equations

$A + 1 = 14 + 1$ (from superscripts)

$Z + 0 = 6 + 1$ (from subscripts)

Hence, A = 14 and Z = 7. Therefore, the nucleus that produces carbon-14 by this reaction is $^{14}_{7}$N.

21.7   Because an activity of 1.0 Ci is 3.7×10^{10} nuclei/s, the rate of decay in this sample is

$$\text{Rate} = 13 \text{ Ci} \times \frac{3.7 \times 10^{10} \text{ nuclei/s}}{1.0 \text{ Ci}} = 4.\underline{8}1 \times 10^{11} \text{ nuclei/s}$$

The number of nuclei in this 2.5-μg (2.5×10^{-6} g) sample of $^{99m}_{43}\text{Tc}$ is

$$2.5 \times 10^{-6} \text{ g Tc-99m} \times \frac{1 \text{ mol Tc-99m}}{99 \text{ g Tc-99m}} \times \frac{6.02 \times 10^{23} \text{ Tc-99m nuclei}}{1 \text{ mol Tc-99m}}$$

$$= 1.\underline{5}2 \times 10^{16} \text{ Tc-99m nuclei}$$

The decay constant is

$$k = \frac{\text{rate}}{N_t} = \frac{4.81 \times 10^{11} \text{ nuclei/s}}{1.52 \times 10^{16} \text{ nuclei}} = 3.\underline{1}6 \times 10^{-5} = 3.2 \times 10^{-5}\text{/s}$$

21.8   First, substitute the value of k into the equation relating half-life to the decay constant:

$$t_{1/2} = \frac{0.693}{k} = \frac{0.693}{4.18 \times 10^{-9} \text{ /s}} = 1.6\underline{5}8 \times 10^{8} \text{ s}$$

Convert the half-life from seconds to years:

$$1.6\underline{5}8 \times 10^{8} \text{ s} \times \frac{1 \text{ min}}{60 \text{ s}} \times \frac{1 \text{ h}}{60 \text{ min}} \times \frac{1 \text{ d}}{24 \text{ h}} \times \frac{1 \text{ y}}{365 \text{ d}} = 5.2\underline{5}7 = 5.26 \text{ y}$$

21.9   The conversion of the half-life to seconds gives

$$28.1 \text{ y} \times \frac{365 \text{ d}}{1 \text{ y}} \times \frac{24 \text{ h}}{1 \text{ d}} \times \frac{60 \text{ min}}{1 \text{ h}} \times \frac{60 \text{ s}}{1 \text{ min}} = 8.8\underline{6}1 \times 10^{8} \text{ s}$$

Because $t_{1/2} = 0.693/k$, solve this for k, and substitute the half-life in seconds.

$$k = \frac{0.693}{t_{1/2}} = \frac{0.693}{8.861 \times 10^{8} \text{ s}} = 7.8\underline{2}08 \times 10^{-10} \text{ /s}$$

Before substituting into the rate equation, you need to know the number of nuclei in a sample containing 5.2×10^{-9} g of strontium-90.

(continued)

$$5.2 \times 10^{-9} \text{ g Sr-90} \times \frac{1 \text{ mol Sr-90}}{90 \text{ g Sr-90}} \times \frac{6.02 \times 10^{23} \text{ Sr-90 nuclei}}{1 \text{ mol Sr-90}}$$

$$= 3.\underline{4}78 \times 10^{13} \text{ Sr-90 nuclei}$$

Now, substitute into the rate equation:

$$\text{Rate} = kN_t = (7.8208 \times 10^{-10}/\text{s}) \times (3.478 \times 10^{13} \text{ nuclei}) = 2.\underline{7}2 \times 10^{4} \text{ nuclei/s}$$

Calculate the activity by dividing the rate (disintegrations of nuclei per second) by 3.70×10^{10} disintegrations of nuclei per second per curie.

$$\text{Activity} = 2.72 \times 10^{4} \text{ nuclei/s} \times \frac{1.0 \text{ Ci}}{3.7 \times 10^{10} \text{ nuclei/s}}$$

$$= 7.\underline{3}51 \times 10^{-7} = 7.4 \times 10^{-7} \text{ Ci}$$

21.10   The decay constant, k, is $0.693/t_{1/2}$. If you substitute this into the equation

$$\ln \frac{N_t}{N_o} = -kt = -\frac{0.693 \, t}{t_{1/2}}$$

Substitute the values t = 25.0 y and $t_{1/2}$ = 10.76 y to get

$$\ln \frac{N_t}{N_o} = \frac{-0.693 \times 25.0 \text{ y}}{10.76 \text{ y}} = -1.6\underline{1}01$$

The fraction of krypton-85 remaining after 25.0 years is N_t/N_o.

$$\frac{N_t}{N_o} = e^{-1.6101} = 0.1\underline{9}99 = 0.20$$

21.11   The ratio of rates of disintegration is

$$\frac{N_o}{N_t} = \frac{15.3}{4.5} = 3.\underline{4}00$$

Therefore, substituting this value of N_o/N_t and $t_{1/2}$ = 5730 y into the previous equation gives

$$t = \frac{t_{1/2}}{0.693} \ln \frac{N_o}{N_t} = \frac{5730 \text{ y}}{0.693} \ln (3.400) = 1.\underline{0}1 \times 10^{4}$$

$$= 1.0 \times 10^{4} \text{ years}$$

21.12 a. Write the nuclear masses below each nuclide symbol. Then calculate Δm.

$$^{234}_{90}\text{Th} \quad \rightarrow \quad ^{234}_{91}\text{Pa} \quad + \quad ^{0}_{-1}\text{e}$$

234.04359   234.04330   0.000549 amu

Hence,

$$\Delta m = (234.04330 + 0.000549 - 234.04359) \text{ amu} = -0.000259 \text{ amu}$$

The mass change for molar amounts in this reaction is -0.000259 g, or -2.59×10^{-7} kg. The energy change is

$$\Delta E = (\Delta m)c^2 = (-2.59 \times 10^{-7} \text{ kg})(3.00 \times 10^8 \text{ m/s})^2 = -2.33 \times 10^{10} \text{ J/mol}$$

For 1.00 g Th-234, the energy change is

$$1.00 \text{ g Th-234} \times \frac{1 \text{ mol Th-234}}{234 \text{ g Th-234}} \times \frac{-2.33 \times 10^{10} \text{ J}}{1 \text{ mol Th-234}} = -9.96 \times 10^7 = -1.0 \times 10^8 \text{ J}$$

b. Convert the mass change for the reaction from amu to grams.

$$\Delta m = -0.000259 \text{ amu} \times \frac{1 \text{ g}}{1 \text{ amu} \times 6.02 \times 10^{23}} = -4.30 \times 10^{-28} \text{ g}$$

$$= -4.30 \times 10^{-31} \text{ kg}$$

$$\Delta E = (\Delta m)c^2 = (-4.30 \times 10^{-31} \text{ kg})(3.00 \times 10^8 \text{ m/s})^2 = -3.87 \times 10^{-14} \text{ J}$$

Convert this to MeV:

$$\Delta E = -3.87 \times 10^{-14} \text{ J} \times \frac{1 \text{ MeV}}{1.602 \times 10^{-13} \text{ J}} = -0.241 = -0.24 \text{ MeV}$$

■ Answers to Review Questions

21.1   The two types of nuclear reactions and their equations are

Radioactive decay: $^{238}_{92}\text{U} \rightarrow ^{234}_{90}\text{Th} + ^{4}_{2}\text{He}$

Nuclear bombardment reactions: $^{27}_{13}\text{Al} + ^{4}_{2}\text{He} \rightarrow ^{30}_{15}\text{P} + ^{1}_{0}\text{n}$

21.2   Magic numbers are the numbers of nuclear particles in completed shells of protons or neutrons. Examples of nuclei with magic numbers of protons are ^4_2He, $^{16}_8O$, and $^{40}_{20}Ca$.

21.3   To predict whether a nucleus will be stable, look for nuclei that have one of the magic numbers of protons and neutrons. Also look for nuclei that have an even number of protons and an even number of neutrons. Nuclei that fall in the band of stability are also very stable. There are no stable nuclei above atomic number 83.

21.4   The five common types of radioactive decay and the usual condition that leads to each type are listed below (see Table 21.2).

Alpha emission: $Z > 83$.

Beta emission: N/Z is too large.

Positron emission: N/Z is too small.

Electron capture: N/Z is too small.

Gamma emission: The nucleus is in an excited state.

21.5   The isotopes that begin each of the natural radioactive decay series are uranium-238, uranium-235, and thorium-232.

21.6   The equations are as follows:

a. $^{14}_7N + ^4_2He \rightarrow ^{17}_8O + ^1_1H$

b. $^{27}_{13}Al + ^4_2He \rightarrow ^{30}_{15}P + ^1_0n$

21.7   Particle accelerators are devices used to accelerate electrons, protons, alpha particles and ions to very high speeds. They operate by accelerating the charged particle toward a plate with a charge opposite to that of the particle. Particle accelerators are required to accelerate alpha particles to speeds high enough to penetrate nuclei of large positive charge that normally scatter alpha particles.

21.8   Before the discovery of transuranium elements, it was thought that americium (Z = 95) and curium (Z = 96) should be placed after actinium (Z = 89) in the periodic table as d-block transition elements. However, Seaborg and others discovered these elements had properties similar to the lanthanides and placed them in a second series under the lanthanides.

21.9   The Geiger counter measures alpha particles by means of a tube filled with gas. When particles pass through the tube, they ionize the gas, freeing electrons, which creates a pulse of current that is detected by electronic equipment and is counted. A scintillation counter consists of phosphor, a substance that emits photons when struck by radiation; zinc sulfide is used for the detection of alpha particles; and sodium iodide containing thallium(I) iodide is used for gamma rays. The photons travel from the phosphor to a photoelectric detector, such as a photomultiplier, which magnifies the effect and gives a pulse of electric current that is measured.

21.10 A curie (Ci) equals 3.700×10^{10} nuclear disintegrations per second. A rad is the dosage of radiation that deposits 1×10^{-2} J of energy per kilogram of tissue. A rem is a unit of radiation dosage for biological destruction; it equals the rad multiplied by the relative biological effectiveness (RBE).

21.11 It will take cesium-137 three times its half-life of 30.2 y, or 90.6 y, to decay to 1/8 its original mass: 1 to 1/2 to 1/4 to 1/8 = 3 half-lives.

21.12 Because the $^{40}_{18}Ar$ was produced by radioactive decay of $^{40}_{19}K$, half the initial amount of $^{40}_{19}K$ has decomposed. The age equals the half-life of 1.28×10^9 y.

21.13 A radioactive tracer is a radioactive isotope added to a chemical, biological, or physical system to study it. For instance, ^{131}I is used as a tracer in the study of the dissolving of lead(II) iodide and its equilibrium in a saturated solution.

21.14 Isotope dilution is a technique designed to determine the quantity of a substance in a mixture or to determine the total volume of a solution by adding a known amount of an isotope to it. After removing a portion of the mixture, the fraction by which the isotope has been diluted provides a way of determining the quantity of substance or the total volume of solution.

21.15 Neutron activation analysis is an analysis based on the conversion of stable isotopes to radioactive isotopes by bombarding a sample with neutrons. An unstable nucleus results that then emits gamma rays or radioactive particles (such as beta particles). The amount of stable isotope is proportional to the measured emission.

21.16 The reason the deuteron, ^2_1H, has a mass smaller than the sum of the masses of its constituents is that, when nucleons come together to form a nucleus, energy is released. There must, therefore, be an equivalent decrease in mass because mass and energy are equivalent.

21.17 Iron-56 has a binding energy per nucleon near the maximum value. Two light nuclei, such as two C-12 nuclei, will undergo fusion (with the release of energy) as long as the product nuclei are lighter than iron-56 (which is the case with Na-23 and H-1).

21.18 The nuclear fission reactor operates by means of a chain reaction of nuclear fissions controlled to produce energy without explosion.

■ Solutions to Practice Problems

Note on significant figures: If the final answer to a solution needs to be rounded off, it is given first with one nonsignificant figure, and the last significant figure is underlined. The final answer is then rounded to the correct number of significant figures. In multiple-step problems, intermediate answers are given with at least one nonsignificant figure; however, only the final answer has been rounded off.

21.29 $^{87}_{37}Rb \rightarrow \, ^{87}_{38}Sr + \, ^{0}_{-1}e$

21.31 $^{232}_{90}Th \rightarrow \, ^{228}_{88}Ra + \, ^{4}_{2}He$

21.33 Let X be the product nucleus. The nuclear equation is

$$^{18}_{9}F \rightarrow \, ^{A}_{Z}X + \, ^{0}_{1}e$$

From the superscripts: 18 = A + 0, or A = 18; from the subscripts: 9 = Z + 1, or Z = 8. Thus, the product of the reaction is $^{18}_{8}X$, and because element 8 is oxygen, symbol O, the nuclear equation is

$$^{18}_{9}F \rightarrow \, ^{18}_{8}O + \, ^{0}_{1}e$$

21.35 Let X be the product nucleus. The nuclear equation is

$$^{210}_{84}Po \rightarrow \, ^{A}_{Z}X + \, ^{4}_{2}He$$

From the superscripts: 210 = A + 4, or A = 206; from the subscripts: 84 = Z + 2, or Z = 82. Thus, the product nucleus is $^{206}_{82}X$, and because element 82 is lead, symbol Pb, the nuclear equation is

$$^{210}_{84}Po \rightarrow \, ^{206}_{82}Pb + \, ^{4}_{2}He$$

21.37 a. Neither nucleus has an atomic number that is a magic number of protons. Find how many neutrons are in each nucleus.

Sb: No. of neutrons = A - Z = 122 - 51 = 71

Xe: No. of neutrons = A - Z = 136 - 54 = 82

Because 82 is a magic number for neutrons (implying stability of nucleus), you predict $^{136}_{54}$Xe is stable and $^{122}_{51}$Sb is radioactive.

b. $^{204}_{82}$Pb has a magic number of protons (82), so it is expected to be the stable nucleus, and $^{204}_{85}$At is radioactive (atomic number greater than 83).

c. Rb does not have an atomic number that is a magic number of protons. Find the numbers of neutrons in the two isotopes.

$^{87}_{37}$Rb : No. of neutrons = 87 - 37 = 50

$^{80}_{37}$Rb : No. of neutrons = 80 - 37 = 43

Because 50 is a magic number for neutrons, you predict $^{87}_{37}$Rb is stable, and $^{80}_{37}$Rb is radioactive.

21.39 a. α-emission is most likely for nuclei with Z > 83

b. positron emission (more likely; Z < 20) or electron capture because mass no. is less than that of Cu = 63.5

c. β-emission (mass no. > He = 4)

21.41 α-emission decreases the mass number by four; β-emission does not affect the mass number. $^{219}_{86}$Rn belongs to the $^{235}_{92}$U decay series because the difference in mass numbers is sixteen, which is divisible by four. $^{220}_{86}$Rn belongs to the $^{232}_{90}$Th decay series because the difference in mass numbers is twelve, which is divisible by four.

21.43 a. $^{26}_{12}$Mg (d,α) $^{24}_{11}$Na   b. $^{16}_{8}$O (n,p) $^{16}_{7}$N

21.45 a. $^{27}_{13}$Al + $^{2}_{1}$H $\rightarrow$ $^{25}_{12}$Mg + $^{4}_{2}$He   b. $^{10}_{5}$B + $^{4}_{2}$He $\rightarrow$ $^{13}_{6}$C + $^{1}_{1}$H

21.47 $\dfrac{12.6 \text{ MeV}}{1 \text{ proton}} = \dfrac{12.6 \times 10^6 \text{ eV}}{1 \text{ proton}} \times \dfrac{1.602 \times 10^{-19} \text{ J}}{1 \text{ eV}} \times \dfrac{1 \text{ kJ}}{10^3 \text{ J}} \times \dfrac{6.02 \times 10^{23} \text{ protons}}{1 \text{ mol}}$

$$= 1.2\underline{1}5 \times 10^9 = 1.22 \times 10^9 \text{ kJ/mol}$$

21.49 a. $^6_3\text{Li} + ^1_0\text{n} \rightarrow ^A_Z\text{X} + ^3_1\text{H}$

From the superscripts: $6 + 1 = A + 3$, or $A = 6 + 1 - 3 = 4$; from the subscripts: $3 + 0 = Z + 1$, or $Z = 3 - 1 = 2$. The product nucleus is ^4_2X. The element with $Z = 2$ is helium (He), so the missing nuclide is ^4_2He.

b. The reaction may be written

$$^{232}_{90}\text{Th} + ^A_Z\text{X} \rightarrow ^{235}_{92}\text{U} + ^1_0\text{n}$$

From the superscripts: $232 + A = 235 + 1$, or $A = 235 + 1 - 232 = 4$; from the subscripts: $90 + Z = 92 + 0$, or $Z = 92 - 90 = 2$. The projectile nucleus is ^4_2X. The element with $Z = 2$ is helium, so the missing nuclide is ^4_2He, or α. The reaction is then written

$$^{232}_{90}\text{Th}\,(\alpha,\text{n})\,^{235}_{92}\text{U}$$

21.51 The reaction may be written

$$^A_Z\text{X} + ^4_2\text{He} \rightarrow ^{242}_{96}\text{Cm} + ^1_0\text{n}$$

From the superscripts: $A + 4 = 242 + 1$, or $A = 242 + 1 - 4 = 239$; from the subscripts: $Z + 2 = 96 + 0$, or $Z = 96 - 2 = 94$. The element with $Z = 94$ is plutonium (the target nucleus was $^{239}_{94}\text{Pu}$).

21.53 The rate of decay is 8.94×10^{10} nuclei/s. The number of nuclei in the sample is

$$0.250 \times 10^{-3} \text{ g H-3} \times \dfrac{1 \text{ mol H-3}}{3.02 \text{ g H-3}} \times \dfrac{6.02 \times 10^{23} \text{ nuclei}}{1 \text{ mol H-3}} = 4.9\underline{8}3 \times 10^{19} \text{ nuclei}$$

The rate equation is rate = kN_t. Solve for k.

$$k = \dfrac{\text{rate}}{N_t} = \dfrac{8.94 \times 10^{10} \text{ nuclei/s}}{4.98 \times 10^{19} \text{ nuclei}} = 1.7\underline{9}4 \times 10^{-9} = 1.79 \times 10^{-9} \text{ s}^{-1}$$

21.55 Find the rate of decay from the activity.

$$\text{Rate} = 20.4 \text{ Ci} \times \frac{3.700 \times 10^{10} \text{ nuclei/s}}{1 \text{ Ci}} = 7.5\underline{4}8 \times 10^{11} \text{ nuclei/s}$$

Convert the mass of S-35 to the number of nuclei. The molar mass in grams is approximately equal to the mass number.

$$0.48 \times 10^{-3} \text{ g S-35} \times \frac{1 \text{ mol S-35}}{35 \text{ g S-35}} \times \frac{6.02 \times 10^{23} \text{ nuclei}}{1 \text{ mol S-35}} = 8.2\underline{5}6 \times 10^{18} \text{ nuclei}$$

Solve the rate equation for k and substitute.

$$k = \frac{\text{rate}}{N_t} = \frac{7.548 \times 10^{11} \text{ nuclei/s}}{8.256 \times 10^{18} \text{ nuclei}} = 9.\underline{1}4 \times 10^{-8} = 9.1 \times 10^{-8} \text{ s}^{-1}$$

21.57 $t_{1/2} = \dfrac{0.693}{k} = \dfrac{0.693}{1.7 \times 10^{-21} \text{ /s}} \times \dfrac{1 \text{ h}}{3600 \text{ s}} \times \dfrac{1 \text{ d}}{24 \text{ h}} \times \dfrac{1 \text{ y}}{365 \text{ d}} = 1.\underline{2}92 \times 10^{13}$

$$= 1.3 \times 10^{13} \text{ y}$$

21.59 $k = \dfrac{0.693}{t_{1/2}} = \dfrac{0.693}{5.73 \times 10^{3} \text{ y}} \times \dfrac{1 \text{ y}}{365 \text{ d}} \times \dfrac{1 \text{ d}}{24 \text{ h}} \times \dfrac{1 \text{ h}}{3600 \text{ s}} = 3.8\underline{3}505 \times 10^{-12}$

$$3.84 \times 10^{-12} \text{/s}$$

21.61 Find k from the half-life.

$$k = \frac{0.693}{t_{1/2}} = \frac{0.693}{2.69 \text{ d}} \times \frac{1 \text{ d}}{24 \text{ h}} \times \frac{1 \text{ h}}{3600 \text{ s}} = 2.9\underline{8}2 \times 10^{-6} \text{/s}$$

Before substituting into the rate equation, find the number of gold nuclei from the mass.

$$0.86 \times 10^{-3} \text{ g Au-198} \times \frac{1 \text{ mol Au-198}}{198 \text{ g Au-198}} \times \frac{6.02 \times 10^{23} \text{ Au-198 nuclei}}{1 \text{ mol Au-198}}$$

$$= 2.\underline{6}1 \times 10^{18} \text{ Au-198 nuclei}$$

(continued)

Now find the rate.

$$\text{Rate} = kN_t = (2.982 \times 10^{-6}/\text{s})(2.61 \times 10^{18} \text{ nuclei}) = 7.\underline{7}9 \times 10^{12} \text{ nuclei/s}$$

$$\text{Activity} = 7.79 \times 10^{12} \text{ nuclei/s} \times \frac{1 \text{ Ci}}{3.700 \times 10^{10} \text{ nuclei/s}} = 2\underline{1}0.7 = 2.1 \times 10^{2} \text{ Ci}$$

21.63 Find k from the half-life.

$$k = \frac{0.693}{t_{1/2}} = \frac{0.693}{14.3 \text{ d}} \times \frac{1 \text{ d}}{24 \text{ h}} \times \frac{1 \text{ h}}{3600 \text{ s}} = 5.6\underline{0}9 \times 10^{-7}/\text{s}$$

Solve the rate equation for N_t.

$$N_t = \frac{\text{rate}}{k} = \frac{6.0 \times 10^{12} \text{ nuclei/s}}{5.609 \times 10^{-7} \text{ /s}} = 1.\underline{0}7 \times 10^{19} \text{ nuclei}$$

Convert N_t to the mass of P-32.

$$1.07 \times 10^{19} \text{ P-32 nuclei} \times \frac{1 \text{ mol P-32}}{6.02 \times 10^{23} \text{ P-32 nuclei}} \times \frac{32 \text{ g P-32}}{1 \text{ mol P-32}}$$

$$= 5.\underline{6}8 \times 10^{-4} = 5.7 \times 10^{-4} \text{ g P-32}$$

21.65 Substituting for $k = 0.693/t_{1/2}$ gives

$$\ln \frac{N_t}{N_o} = -kt = \frac{-0.693 \text{ t}}{t_{1/2}} = \frac{-0.693 \, (12.0 \text{ h})}{15.0 \text{ h}} = -0.55\underline{4}4$$

Taking the antilog of both sides of this equation gives

$$\frac{N_t}{N_o} = e^{-0.5544} = 0.57\underline{4}41$$

After 12.0 h, 57.4 percent of the Na-24 remains.

$$6.0 \text{ μg} \times 0.57441 = 3.\underline{4}46 \text{ μg} = 3.4 \text{ μg}$$

21.67 After 1.97 sec, the amount of nitrogen-17 remaining is 100% - 28.0% = 72.0 percent, so $N_t/N_o = 0.720$.

$$\ln \frac{N_t}{N_o} = -kt = \frac{-0.693\,t}{t_{1/2}}$$

$$\ln (0.720) = \frac{-0.693(1.97\ s)}{t_{1/2}}$$

Solve for $t_{1/2}$.

$$t_{1/2} = \frac{-0.693(1.97\ s)}{\ln(0.720)} = 4.1\underline{5}58 = 4.16\ s$$

21.69 At t = 0, the rate is 125 nuclei/s, and at time t = 10.0 d, the rate is 107 nuclei/s. Thus,

$$\ln \frac{N_t}{N_o} = -kt = = \frac{-0.693\,t}{t_{1/2}}$$

Substituting

$$\ln \left(\frac{107\ nuclei/s}{125\ nuclei/s} \right) = \frac{-0.693\,(10.0\ d)}{t_{1/2}}$$

Solve for the half-life

$$t_{1/2} = \frac{-0.693\,(10.0\ d)}{\ln \left(\dfrac{107}{125} \right)} = 44.\underline{5}7 = 44.6\ d$$

21.71 Use the equation for the number of nuclei in a sample after a time, t.

$$\ln \left(\frac{N_t}{N_o} \right) = -kt = \frac{-0.693\,t}{t_{1/2}}$$

Rearrange this to give an expression for t.

$$t = \frac{t_{1/2}}{0.693} \ln \frac{N_o}{N_t}$$

(continued)

At t = 0, the rate is 15.3 nuclei/s, and at a later time t, the rate is 8.1 nuclei/s. Since the half-life for carbon-14 is 5730 years,

$$t = \frac{5730 \text{ y}}{0.693} \ln\left(\frac{15.3}{8.1}\right) = 5.\underline{2}58 \times 10^3 = 5.3 \times 10^3 \text{ y}$$

21.73 The half-life of carbon-14 is 5730 y, and the age of the sandals is 9.0×10^3 y. Substitute into the equation

$$\ln \frac{N_t}{N_o} = \frac{-0.693 \, t}{t_{1/2}} = \frac{-0.693 \, (9.0 \times 10^3 \text{ y})}{5730 \text{ y}} = -1.\underline{0}88$$

Taking the antilog of both sides, you obtain

$$\frac{N_t}{N_o} = \frac{(\text{activity})_t}{(\text{activity})_o} = e^{-1.088} = 0.\underline{3}367$$

So, the activity for a gram sample is

$$(\text{activity})_t = 15.3 \text{ disintegrations/(min•g)} \times 0.3367$$
$$= \underline{5}.15 \text{ disintegrations/(min•g)}$$
$$= 5 \text{ disintegrations/(min•g)}$$

21.75 $$\Delta m = \frac{\Delta E}{c^2} = \frac{-393.5 \times 10^3 \text{ J}}{(3.00 \times 10^8 \text{ m/s})^2} = \frac{-3.935 \times 10^5 \text{ kg•m}^2/\text{s}^2}{(3.00 \times 10^8 \text{ m/s})^2}$$

$$= 4.3\underline{7}2 \times 10^{-12} = 4.37 \times 10^{-12} \text{ kg} = 4.37 \times 10^{-9} \text{ g}$$

21.77 Write the atomic masses below each nuclide symbol and calculate Δm.

$$^{2}_{1}\text{H} \quad + \quad ^{3}_{1}\text{H} \quad \rightarrow \quad ^{4}_{2}\text{He} \quad + \quad ^{1}_{0}\text{n}$$

Masses: 2.01400   3.01605   4.00260   1.008665 amu

$$\Delta m = [4.00260 + 1.008665] - [2.01400 + 3.01605] = -0.01878\underline{5}$$

The energy change for one mole is

$$\Delta E = (\Delta m)c^2 = (-0.018785 \times 10^{-3} \text{ kg})(2.998 \times 10^8 \text{ m/s})^2$$

$$= -1.68\underline{8}39 \times 10^{12} \text{ kg•m}^2/\text{s}^2 = -1.688 \times 10^{12} \text{ J}$$

(continued)

Finally, calculate the energy change in MeV for one ^2_1H nucleus.

$$\frac{-1.68839 \times 10^{12} \text{ J}}{1 \text{ mol}} \times \frac{1 \text{ mol}}{6.022 \times 10^{23} \text{ nuclei}} \times \frac{1 \text{ MeV}}{1.602 \times 10^{-13} \text{ J}}$$

$$= -17.5\underline{0}1 = -17.50 \text{ MeV}$$

21.79   Mass of three protons = 3×1.00728 amu = 3.02184 amu

Mass of three neutrons = 3×1.008665 amu = 3.025995 amu

Total mass of nucleons = $(3.02184 + 3.025995)$ amu = $6.0478\underline{3}5$ amu

Mass defect = total nucleon mass - nuclear mass = $(6.047835 - 6.01512)$ amu

$$= 0.0327\underline{1}5 \text{ amu}$$

$$\Delta E = (\Delta m)c^2 = 0.032715 \text{ amu} \times \frac{1 \text{ g}}{6.022 \times 10^{23} \times 1 \text{ amu}} \times \frac{1 \text{ kg}}{10^3 \text{ g}}$$

$$\times (2.998 \times 10^8 \text{ m/s})^2 \times \frac{1 \text{ MeV}}{1.602 \times 10^{-13} \text{ J}} = 30.4\underline{7}9 \text{ MeV}$$

Binding energy per nucleon = $\dfrac{30.479 \text{ MeV}}{6 \text{ nucleons}} = 5.07\underline{9}9 = 5.080$ MeV/nucleon

■ Solutions to General Problems

21.81   Na-20, having fewer neutrons than the stable Na-23, is expected to decay to a nucleus with a lower atomic number (and hence a higher N/Z ratio) by electron capture or positron emission. Na-26, having more neutrons than the stable isotope, is expected to decay by beta emission to give a nucleus with a higher atomic number (and hence a lower N/Z ratio).

21.83 The overall reaction may be written

$$^{235}_{92}U \rightarrow ^{207}_{82}Pb + n\,^4_2He + m\,^0_{-1}e$$

From the superscripts:

$$235 = 207 + (n \times 4) + (m \times 0)$$

$$n = \frac{235 - 207}{4} = 7$$

From the subscripts, and with n equal to seven, you get

$$92 = 82 + (n \times 2) + m \times (-1)$$

$$m = 82 - 92 + (7 \times 2) = 4$$

Therefore, there are seven α emissions and four β emissions.

21.85 $^{209}_{83}Bi + ^4_2He \rightarrow ^A_{85}At + 2\,^1_0n$

From the superscripts: $209 + 4 = A + (2 \times 1)$, or $A = 213 - 2 = 211$.

The reaction is

$$^{209}_{83}Bi + ^4_2He \rightarrow ^{211}_{85}At + 2\,^1_0n$$

21.87 $^{238}_{92}Bi + ^{12}_6C \rightarrow ^A_ZX + 4\,^1_0n$

From the superscripts: $238 + 12 = A + (4 \times 1)$, or $A = 250 - 4 = 246$; from the subscripts: $92 + 6 = Z + (4 \times 0)$, or $Z = 98$.

The element with z = 98 is californium (Cf), so the equation is

$$^{238}_{92}Bi + ^{12}_6C \rightarrow ^{246}_{98}Cf + 4\,^1_0n$$

21.89 Use the equation for the number of nuclei in a sample after a time, t.

$$\ln \frac{N_t}{N_o} = \frac{-0.693 \, t}{t_{1/2}}$$

Rearrange this to give an expression for t.

$$t = \frac{t_{1/2}}{0.693} \ln \frac{N_o}{N_t}$$

Substituting

$$t = \frac{(12.3 \text{ y})}{0.693} \ln \frac{N_o}{0.70 \, N_o} = 6.3\underline{2} = 6.3 \text{ y}$$

21.91 In the annihilation, the final mass is zero; that is, all mass is converted to energy. Therefore,

$$\Delta m = 0 - \text{mass of positron} - \text{mass of electron}$$

$$= -2(0.000549 \text{ amu}) \times \frac{1 \text{ g}}{6.022 \times 10^{23} \text{ amu}} \times \frac{1 \text{ kg}}{10^3 \text{ g}} = -1.8\underline{2}33 \times 10^{-30} \text{ kg}$$

The energy of each photon is $-\Delta E/2$.

$$E_{photon} = \frac{-\Delta E}{2} = \frac{-(\Delta m)c^2}{2} = \frac{-(-1.8233 \times 10^{-30} \text{ kg})(2.998 \times 10^8 \text{ m/s})^2}{2}$$

$$= 8.1\underline{9}39 \times 10^{-14} \text{ J}$$

The energy of a photon is related to its wavelength by the equation

$$E = \frac{hc}{\lambda}$$

where h is Planck's constant. Therefore,

$$\lambda = \frac{hc}{E} = \frac{(6.626 \times 10^{-34} \text{ J} \cdot \text{s})(2.998 \times 10^8 \text{ m/s})}{(8.1939 \times 10^{-14} \text{ J})} = 2.4\underline{2}43 \times 10^{-12}$$

$$= 2.42 \times 10^{-12} \text{ m}$$

The wavelength of the photons is 2.42 pm.

21.93   Write the atomic masses below each nuclide symbol and calculate Δm.

$$_0^1 n \quad + \quad _{92}^{235} U \quad \rightarrow \quad _{53}^{136} I \quad + \quad _{39}^{96} Y \quad + \quad 4 _0^1 n$$

Masses: 1.008665   235.04392   135.8401   95.8629   1.008665 amu

$$\Delta m = [135.8401 + 95.8629 + 4 \times 1.008665 - (1.008665 + 235.04392)] \text{ amu}$$

$$= -0.314925 \text{ amu}$$

When one mol of U-235 decays, the change in mass is -0.314925 g. Therefore, for 5.00 kg of U-235, the change in mass is

$$5.00 \times 10^3 \text{ g U-235} \times \frac{1 \text{ mol U-235}}{235.04392 \text{ g U-235}} \times \frac{-0.314925 \text{ g}}{1 \text{ mol U-235}} = -6.69927 \text{ g}$$

Converting mass to energy gives:

$$E = (\Delta m)c^2 = (-6.69927 \times 10^{-3} \text{ kg})(2.998 \times 10^8 \text{ m/s})^2 = -6.0213 \times 10^{14} \text{ J}$$

For the combustion of carbon:

$$\text{C(graphite)} \quad + \quad O_2(g) \quad \rightarrow \quad CO_2(g)$$

ΔH_f° 0   0   -394 kJ/mol

For the reaction, $\Delta H = \Delta E = -394$ kJ/mol.

$$5.00 \times 10^3 \text{ g} \times \frac{1 \text{ mol C}}{12.01 \text{ g C}} \times \frac{-394 \text{ kJ}}{1 \text{ mol C}} = -1.64029 \times 10^5 = -1.64 \times 10^5 \text{ kJ}$$

The energy released in the fission of 5.00 kg of U-235 (6.02×10^{11} kJ) is larger by several orders of magnitude than the energy released by burning 5.00 kg of C (1.64×10^5 kJ).

21.95 a. The balanced equation is

$$_{20}^{47} Ca \rightarrow _{21}^{47} Sc + _{-1}^{0} \beta$$

b. Use the rate law to find the initial amount. The time is 48 hours (2.0 d), and the half-life is 4.536 d. Substituting

$$\ln \frac{N_t}{N_o} = -kt = \frac{-0.693 \, t}{t_{1/2}} = \frac{-0.693 \, (2.0 \text{ d})}{4.536 \text{ d}} = -0.3055$$

(continued)

Taking the antilog of each side gives

$$\frac{N_t}{N_o} = \frac{A_t}{A_o} = e^{-0.3055} = 0.7367$$

Since A_t is 10.0 μg, you can solve for A_o.

$$A_o = \frac{10.0 \text{ μg}}{0.7367} = 13.57 = 14 \text{ μg Ca-47}$$

Convert this to mass of $CaSO_4$.

$$13.57 \text{ μg Ca-47} \times \frac{136.15 \text{ g CaSO}_4}{40.08 \text{ g Ca}} = 46.10 = 46 \text{ μg } (4.6 \times 10^{-5} \text{ g})$$

21.97 a. The balanced equations are

$$^{82}_{35}Br \rightarrow ^{82}_{36}Kr + ^{0}_{-1}\beta$$

$$2H^{82}_{35}Br \text{ (g)} \rightarrow H_2 + 2 ^{82}_{36}Kr$$

b. Use the rate law to find the fraction remaining. The time is 12.0 h (0.500 d), and the half-life is 1.471 d. Substituting

$$\ln \frac{N_t}{N_o} = -kt = \frac{-0.693 \, t}{t_{1/2}} = \frac{-0.693 \, (0.500 \text{ d})}{1.471 \text{ d}} = -0.23555$$

Taking the antilog of each side gives

$$\frac{N_t}{N_o} = \frac{A_t}{A_o} = e^{-0.23555} = 0.79013$$

Therefore, the moles remaining after 12.0 h is

$$A_t = 0.0150 \text{ mol} \times 0.79013 = 0.011851 \text{ mol}$$

The reaction can be summarized as follows:

2HBr	→	H₂	+	2Kr
0.0150 - 2y		y		2y

(continued)

From this, you can determine y.

0.0150 mol - 2y = 0.011851 mol, or y = 0.001574 mol

To calculate the pressure in the flask, you need the total moles of gas. This is

total mol = mol HBr + mol H_2 + mol Kr = (0.0150 - 2y) + y + 2y

= 0.0150 + y = 0.0150 + 0.001574 = 0.016574 mol

Therefore, the pressure in the flask is

$$P = \frac{nRT}{V} = \frac{0.016574 \text{ mol} \times 0.0821 \text{ L} \cdot \text{atm/K} \cdot \text{mol} \times 295 \text{ K}}{1.00 \text{ L}}$$

= 0.401416 = 0.401 atm

■ Solutions to Cumulative-Skills Problems

21.99   Calculate the number of P-32 nuclei in the sample (N_t). For this, you need the formula weight of N_3PO_4 containing 15.6 percent of P-32 and (100.0 - 15.6)% = 84.4 percent naturally occurring P. The formula weight of Na_3PO_4 containing naturally occurring P is 163.9 amu; the formula weight of Na_3PO_4 with 100 percent P-32 is 165.0. The formula weight of Na_3PO_4 containing 15.6 percent P-32 is obtained from the weighted average of the formula weights:

(163.9 amu x 0.844) + (165.0 amu x 0.156) = 164.1 amu

The moles of P in the sample equal

0.0545 g Na_3PO_4 x $\dfrac{1 \text{ mol } Na_3PO_4}{164.1 \text{ g } Na_3PO_4}$ x $\dfrac{1 \text{ mol P}}{1 \text{ mol } Na_3PO_4}$ = 3.321 x 10^{-4} mol P

and the moles of P-32 equal

(3.321 x 10^{-4} mol P) x 0.156 = 5.181 x 10^{-5} mol P-32

Then, the number of P-32 nuclei is

(5.181 x 10^{-5} mol P-32) x $\dfrac{6.022 \times 10^{23} \text{ P-32 nuclei}}{1 \text{ mol P-32}}$ = 3.120 x 10^{19} P-32 nuclei

(continued)

Now, calculate the rate of disintegrations. Because the rate = kN_t, first find the value of k in reciprocal seconds:

$$k = \frac{0.693}{t_{1/2}} = \frac{0.693}{14.3 \text{ d}} \times \frac{1 \text{ d}}{24 \text{ h}} \times \frac{1 \text{ h}}{3600 \text{ s}} = 5.6\underline{0}9 \times 10^{-7} \text{ /s}$$

Therefore, the rate of disintegrations is

$$\text{Rate} = kN_t = (5.609 \times 10^{-7} \text{/s}) \times (3.120 \times 10^{19} \text{ P-32 nuclei}) = 1.7\underline{5}0 \times 10^{13}$$

$$= 1.75 \times 10^{13} \text{ P-32 nuclei/s}$$

21.101 First, determine the mass of polonium-210 nuclei in 1.0000 g of PoO_2.

$$1.0000 \text{ g PoO}_2 \times \frac{210 \text{ g Po}}{242 \text{ g PoO}_2} = 0.86\underline{7}768 \text{ g Po-210}$$

Use the rate law to find the fraction of polonium-210 that decayed. The time is 48.0 h (2.00 d), and the half-life is 138.4 d. Substituting

$$\ln \frac{N_t}{N_o} = -kt = \frac{-0.693 \text{ t}}{t_{1/2}} = \frac{-0.693 (2.00 \text{ d})}{138.4 \text{ d}} = -0.01\underline{0}014$$

Take the antilog of each side to get the fraction of polonium-210 remaining.

$$\frac{N_t}{N_o} = e^{-0.010014} = 0.9900\underline{3}5$$

Thus, the fraction of polonium-210 that decayed is 1 - 0.990035 = 0.009964. This can now be converted to moles of helium formed.

$$0.867768 \text{ g Po} \times 0.00996\underline{4} \times \frac{1 \text{ mol Po}}{210 \text{ g Po}} \times \frac{1 \text{ mol He}}{1 \text{ mol Po}} = 4.\underline{1}17 \times 10^{-5} \text{ mol He}$$

Now, calculate the volume of He at 25 °C and 735 mmHg.

$$V = \frac{nRT}{P} = \frac{(4.117 \times 10^{-5} \text{ mol})(0.0821 \text{ L} \cdot \text{atm/(K} \cdot \text{mol)})(298 \text{ K})}{(735/760) \text{ atm}}$$

$$= 0.001\underline{0}4 = 0.0010 \text{ L} = 1.0 \text{ mL}$$

21.103 $2p + 2n \rightarrow$ He-4

On a mole basis, the mass difference, Δm, is

$$\Delta m = [4.00260 - 2(1.00728) - 2(1.00867)] \text{ g/mol} = -0.02930 \text{ g/mol}$$

$$= -2.930 \times 10^{-5} \text{ kg/mol}$$

The energy evolved per mole is

$$\Delta E = (\Delta m)c^2 = -2.930 \times 10^{-5} \text{ kg/mol} \times (2.998 \times 10^8 \text{ m/s})^2$$

$$= -2.63258 \times 10^{12} \text{ kg·m}^2/\text{s}^2 = -2.63258 \times 10^{12} \text{ J} = -2.63258 \times 10^9 \text{ kJ/mol}$$

Next calculate $\Delta H°$ for burning of ethane:

$$C_2H_6(g) \quad + \quad 7/2O_2 \quad \rightarrow \quad 2CO_2(g) \quad + \quad 3H_2O(g)$$

$\Delta H_f° = -84.667 \qquad\qquad 0 \qquad\quad 2(-393.5) \qquad 3(-241.826) \text{ kJ}$

$$\Delta H° = [2(-393.5) + 3(-241.826) - (-84.667)] \text{ kJ} = -1427.81 \text{ kJ/mol ethane}$$

Now calculate the mol of ethane needed to obtain 2.63258×10^9 kJ heat:

$$-2.63258 \times 10^9 \text{ kJ} \times \frac{1 \text{ mol ethane}}{-1427.81 \text{ kJ}} = 1.84379 \times 10^6 \text{ mol ethane}$$

Finally, convert moles to liters at 25 °C and 725 mmHg.

$$V = \frac{nRT}{P} = \frac{(1.84379 \times 10^6 \text{ mol}) [0.0821 \text{ L·atm/(K·mol)}] (298 \text{ K})}{(725/760) \text{ atm}}$$

$$= 4.726 \times 10^7 = 4.73 \times 10^7 \text{ L}$$

22. CHEMISTRY OF THE MAIN-GROUP ELEMENTS

■ Answers to Review Questions

22.1   The mineral source is given after the name of the metal: lithium–lithium aluminum silicate; sodium–sodium chloride; magnesium–seawater (Dow process) and dolomite or magnesite; calcium–calcium oxide; aluminum–bauxite; tin–cassiterite; and lead–galena.

22.2   The reactions are

$$2Li(s) + 2H_2O(l) \rightarrow 2LiOH(aq) + H_2(g)$$

$$4Li(s) + O_2(g) \rightarrow 2Li_2O(s)$$

$$2Na(s) + 2H_2O(l) \rightarrow 2NaOH(aq) + H_2(g)$$

$$2Na(s) + O_2(g) \rightarrow Na_2O_2(s)$$

22.3   The reaction is similar to that of lithium carbonate with calcium hydroxide:

$$Li_2CO_3(aq) + Ba(OH)_2(aq) \rightarrow 2LiOH(aq) + BaCO_3(s)$$

22.4   Potassium is expected to be more reactive than lithium because metals become more reactive going down Group IA. This is partly because potassium is much larger, so its $4s$ electron is lost more readily than the $2s$ electron of lithium.

22.5   The reaction is $2Na(s) + 2C_2H_5OH(l) \rightarrow H_2(g) + 2NaOC_2H_5(aq)$.

22.6 Cathode reaction: $Na^+(l) + e^- \rightarrow Na(l)$

Anode reaction: $2Cl^-(l) \rightarrow Cl_2(g) + 2e^-$

b. Cathode reaction: $Na^+(l) + e^- \rightarrow Na(l)$

Anode reaction: $4OH^-(aq) \rightarrow O_2(g) + 2H_2O(g) + 4e^-$

22.7 Sodium hydroxide is manufactured by the electrolysis of aqueous sodium chloride, which also produces chlorine gas as a major product.

22.8 The uses are given after each compound: sodium chloride–used for making sodium hydroxide and in seasoning; sodium hydroxide–used in aluminum production and in producing sodium compounds such as soap; and sodium carbonate–used to make glass and as washing soda with many detergent preparations.

22.9 The main step in the Solvay process involves the reaction of carbon dioxide with ammonia and sodium chloride to form sodium bicarbonate:

$$NH_3(g) + H_2O(l) + CO_2(g) + NaCl(aq) \rightarrow NaHCO_3(s) + NH_4Cl(aq)$$

22.10 $2Mg(s) + O_2(g) \rightarrow 2MgO(s)$

$Mg(s) + H_2O(g) \rightarrow MgO(s) + H_2(g)$

$2Mg(s) + CO_2(g) \rightarrow 2MgO(s) + C(s)$

22.11 a. Calcium oxide is prepared industrially from calcium carbonate:

$$CaCO_3(s) \rightarrow CaO(s) + CO_2(g)$$

b. Calcium hydroxide is prepared from the reaction of calcium oxide and water:

$$CaO(s) + H_2O(l) \rightarrow Ca(OH)_2(aq)$$

22.12 $CaCO_3(s) + 2HCl(aq) \rightarrow CO_2(g) + CaCl_2(aq) + H_2O(l)$

22.13 $Ca(OH)_2 + CO_2(g) \rightarrow CaCO_3(s) + H_2O(l)$

$Ca(OH)_2 + Na_2CO_3(aq) \rightarrow CaCO_3(s) + 2NaOH(aq)$

22.14 $Fe_2O_3(s) + 2Al(s) \rightarrow 2Fe(l) + Al_2O_3(s)$

22.15 Some major uses of aluminum oxide are making abrasives for grinding tools, fusing with small amounts of other metal oxides to make synthetic sapphires and rubies, and making industrial ceramics.

22.16 To purify municipal water, aluminum sulfate and calcium hydroxide are added to waste water, forming a gelatinous precipitate of aluminum hydroxide. Colloidal particles of clay (usually present in the waste water) and other substances adhere to the aluminum hydroxide, whose particles are large enough to be filtered from the water to purify it.

22.17 Lead(IV) oxide, PbO_2, is formed by first packing a paste of PbO into the lead metal grids of the storage battery. When the battery is charged at the factory, the PbO is oxidized by electrolysis to PbO_2. This gives the proper cathode for a new battery.

22.18 Lead pigments are no longer used for house paints because of the possibility of lead poisoning. If chips of lead paint are eaten by children, or if the dust from lead paint that has been removed is breathed by adults or children, lead(II) ion can enter the blood stream. The lead(II) ion can ultimately inhibit the production of red blood cells, causing anemia. (The Pb^{2+} ion can also be absorbed in the brains of children and cause irreversible brain damage.)

22.19 $Pb(NO_3)_2(aq) + CrO_4^{2-}(aq) \rightarrow PbCrO_4(s) + 2NO_3^-(aq)$

22.20 The steam-reforming process is where steam and hydrocarbons from natural gas or petroleum react at high temperature and pressure in the presence of a catalyst to form carbon monoxide and hydrogen. A typical reaction involving propane (C_3H_8) is

$$C_3H_8(g) + 3H_2O(g) \xrightarrow[\Delta]{Ni} 3CO(g) + 7H_2(g)$$

The carbon monoxide is removed from the mixture by reacting with steam in the presence of a catalyst to give carbon dioxide and more hydrogen.

$$CO(g) + H_2O(g) \xrightarrow[\Delta]{Catalyst} CO_2(g) + H_2(g)$$

Finally, the carbon dioxide is removed by dissolving it in a basic aqueous solution.

22.21 The three isotopes of hydrogen and their symbols are as follows: protium, $_1^1H$, or H, which is most abundant; deuterium, $_1^2H$, or D; and tritium, $_1^3H$, or T. Tritium is radioactive, with a half-life of 12.3 years.

22.22 The combustion of hydrogen produces more heat per gram than any other fuel (120 kJ/g). Unlike hydrocarbons, it is a clean fuel since the product is environmentally benign water. These features, in the face of a dwindling supply of hydrocarbons, indicate hydrogen gas may become the favorite fuel of the twenty-first century.

22.23 A binary hydride is a compound that contains hydrogen and one other element. There are three categories of binary hydrides. The first type is an ionic hydride, which contains the hydride ion, H^-, and is formed via reaction with an alkali metal or larger group IIA metal. An example is LiH. A second type is a covalent hydride, which is a molecular compound in which hydrogen is covalently bonded to another element. An example is NH_3. The third type is a metallic hydride formed from a transition metal and hydrogen. These compounds contain hydrogen spread throughout a metal crystal and occupying the holes in the crystal lattice, sometimes in nonstoichiometric amounts. Thus, the composition is variable. An example is $TiH_{1.7}$.

22.24 Catenation is the ability of an atom to bond covalently to like atoms, as in ethylene, $H_2C=CH_2$.

22.25 Carbon monoxide attaches to the iron in the hemoglobin of red blood cells and blocks the combination of hemoglobin with oxygen molecules normally carried by the hemoglobin.

22.26 $CO_2(g) + H_2O(l) \rightleftharpoons H_2CO_3(aq)$

$H_2CO_3(aq) \rightleftharpoons H^+(aq) + HCO_3^-(aq)$

$HCO_3^-(aq) \rightleftharpoons H^+(aq) + CO_3^{2-}(aq)$

22.27 Silicones have a wide variety of applications. Due to the strong Si–O and Si–C bonds, they are generally unreactive and can be synthesized to perform in temperature ranges from -75 to 400 °F. The oils can be found in cosmetics, lipstick, car wax, and hydraulic fluids. The elastomers are used for medical tubing, heart-valve implants, electrical tape, caulk, and gaskets. Silicone resins are used to insulate electrical equipment and in the molding process of electrical circuit boards.

22.28 The formulas could include SiH_4 and its hydrocarbon analog methane, CH_4, and Si_2H_6 with its hydrocarbon analog ethane, C_2H_6.

22.29 The reaction of chlorine with silicon to form silicon tetrachloride is

$$Si(s) + 2Cl_2(g) \longrightarrow SiCl_4(g)$$

22.30 Certain bacteria in the soil and in the roots of plants convert N_2 to ammonium and nitrate compounds. The plants then use these nitrogen compounds to make proteins and other complex nitrogen compounds. Animals eat the plants. Ultimately, the animals die, and bacteria in the decaying organic matter convert the nitrogen compounds back to N_2.

22.31 Rutherford removed oxygen from the air by burning a substance that would combine with the oxygen. He also removed any carbon dioxide formed by the burning by reacting it with KOH. The gas left contained primarily nitrogen, as well as small amounts (< 1%) of noble gases.

22.32 These oxides are nitrous oxide, N_2O; nitric oxide, NO; N_2O_3; nitrogen dioxide, NO_2; dinitrogen tetroxide, N_2O_4; and N_2O_5. The oxidation numbers in each are +1, +2, +3, +4, +4, and +5, respectively.

22.33 Natural gas, or CH_4, plus steam react to form CO and H_2. The CO is reacted with steam to form CO_2 plus additional H_2. The CO_2 is removed from the H_2 by dissolving it in basic aqueous solution; the H_2 is then reacted with N_2 in the Haber process to form NH_3.

22.34 Ammonia is burned in the presence of a platinum catalyst to form NO, which is then reacted with O_2 to form NO_2. The NO_2 is dissolved in water to form HNO_3 and NO. The NO is recycled back to the second step to react with more O_2 to form more NO_2.

22.35 White phosphorus is a molecular solid with the formula, P_4. The phosphorus atoms are arranged at the corners of a regular tetrahedron with a 60° P–P–P bond angle, an angle considerably smaller than the normal tetrahedral bond angle. This accounts for the weakness of the P–P bonds and their high reactivity, as stronger bonds such as P–O can replace them.

22.36 $P_4(s) + 5O_2(g) \rightarrow P_4O_{10}(s)$

$P_4O_{10}(s) + 6H_2O(l) \rightarrow 4H_3PO_4(aq)$

22.37 The first method is the treatment of $Ca_3(PO_4)_2$ with sulfuric acid, giving phosphoric acid and insoluble calcium sulfate. The second method is the treatment of $Ca_3(PO_4)_2$ with HF, giving phosphoric acid and insoluble calcium fluoride.

22.38 $H_3PO_4 + $ HO—P—O—P—OH $\rightarrow$ HO—P—O—P—O—P—OH $+ H_2O$

(with H–O groups above and O below each P)

22.39 Polyphosphates are added to detergents to form complexes with metal ions and thus prevent their precipitation onto clothes.

22.40 Priestley prepared oxygen by heating mercury(II) oxide:

$2HgO(s) \xrightarrow{\Delta} 2Hg(l) + O_2(g)$

22.41 The most important commercial means of producing oxygen is by distillation of liquid air. Air is filtered to remove dust particles, cooled to freeze out water and carbon dioxide, liquefied, and finally warmed until nitrogen and argon distill, leaving liquid oxygen behind.

22.42　Oxides are binary oxygen compounds in which oxygen is in the -2 oxidation state, whereas in peroxides, the oxidation number of oxygen is -1 and the anion is O_2^{2-}. In the superoxides, the oxidation number of oxygen is -1/2 and the anion is O_2^-. An example of each is H_2O (oxide), H_2O_2 (peroxide), and KO_2 (superoxide).

22.43　CrO_3 is an example of an acidic oxide; Cr_2O_3, MgO, and Fe_3O_4 are basic oxides.

22.44　Three natural sources of sulfur or sulfur compounds are sulfate minerals, sulfide minerals, and coal or petroleum products.

22.45　Rhombic sulfur is a yellow crystalline solid with a lattice consisting of crown-shaped S_8 molecules; that is, eight sulfur atoms are arranged in a crown-shaped ring.

22.46　The monoclinic sulfur allotrope can be prepared from rhombic sulfur by first melting the rhombic sulfur and then cooling it to crystals of monoclinic sulfur. A liquid sulfur allotrope of long spiral chains of sulfur atoms can be prepared by heating rhombic sulfur above 160 °C but keeping it below 200 °C. Plastic sulfur allotrope can be prepared by pouring this liquid sulfur allotrope into water. Finally, gaseous allotropes of S_8, S_6, S_4, and S_2 molecules can be formed by boiling sulfur at 445 °C.

22.47　The Frasch process involves melting sulfur deposits with superheated water, using air to force the melted sulfur upward to the surface, and cooling it to form solid sulfur.

22.48　The initial burning of hydrogen sulfide produces some sulfur as well as sulfur dioxide:

$$8H_2S(g) + 4O_2(g) \rightarrow S_8(s) + 8H_2O(g)$$

$$2H_2S(g) + 3O_2(g) \rightarrow 2SO_2(g) + 2H_2O(g)$$

The sulfur dioxide reacts with hydrogen sulfide to form more sulfur:

$$16H_2S(g) + 8SO_2(g) \rightarrow 3S_8(s) + 16H_2O(g)$$

22.49 a.　$2HCl(aq) + ZnS(s) \rightarrow ZnCl_2(aq) + H_2S(g)$

　　b.　$S_8(s) + 8O_2(g) \rightarrow 8SO_2(g)$

22.50　First step:　　$S_8(s) + 8O_2(g) \rightarrow 8SO_2(g)$

　　Second step:　$2SO_2(g) + O_2(g) \rightarrow 2SO_3(g)$

　　Third step:　　$SO_3(g) + H_2O(l) \rightarrow H_2SO_4(aq)$

22.51 a. $16H_2S(g) + 8SO_2(g) \rightarrow 3S_8(s) + 16H_2O(g)$

   b. $Cr_2O_7^{2-}(aq) + 3SO_2(g) + 2H^+(aq) \rightarrow 2Cr^{3+}(aq) + 3SO_4^{2-}(aq) + H_2O(l)$

   c. $Cu(s) + 2H_2SO_4(l) \rightarrow CuSO_4(aq) + 2H_2O(l) + SO_2(g)$

   d. $8Na_2SO_3(aq) + S_8(s) \rightarrow 8Na_2S_2O_3(aq)$

22.52 $4HCl + MnO_2(s) \rightarrow MnCl_2(aq) + Cl_2(g) + 2H_2O(l)$

22.53 a. $I_2(aq) + Cl^-(aq) \rightarrow NR$

   b. $Cl_2(aq) + 2Br^-(aq) \rightarrow Br_2(aq) + 2Cl^-(aq)$

   c. $Br_2(aq) + 2I^-(aq) \rightarrow I_2(s) + 2Br^-(aq)$

   d. $Br_2(aq) + Cl^-(aq) \rightarrow NR$

22.54 Add chlorine water and methylene chloride. For NaCl, there will be no reaction; for NaBr, the lower organic layer will turn orange; and for NaI, the lower organic layer will turn violet.

22.55 Chlorine is used in preparing chlorinated hydrocarbons, as a bleaching agent, and as a disinfectant.

22.56 Sodium hypochlorite is prepared by reacting chlorine with NaOH:

$$Cl_2(g) + 2NaOH(aq) \rightarrow NaClO(aq) + NaCl(aq) + H_2O(l)$$

22.57 Industrially, sodium chloride is electrolyzed to form chlorine gas. The chlorine gas is then heated with sodium hydroxide solution, forming $NaClO_3$ (and NaCl). The ClO_3^- is electrolyzed at the anode to form the ClO_4^- anion. This is mixed with sulfuric acid and the $HClO_4$ is distilled at reduced pressure (below 92 °C) to isolate $HClO_4$.

22.58 Bartlett found that PtF_6 reacted with molecular oxygen. Because the first ionization energy of xenon was slightly less than that of molecular oxygen, he reasoned that PtF_6 ought to react with xenon also.

■ Solutions to Practice Problems

Note on significant figures: If the final answer to a solution needs to be rounded off, it is given first with one nonsignificant figure, and the last significant figure is underlined. The final answer is then rounded to the correct number of significant figures. In multiple-step problems, intermediate answers are given with at least one nonsignificant figure; however, only the final answer has been rounded off.

22.69 $CO_2(g) + NH_3(g) + NaCl(aq) + H_2O(l) \xrightarrow{\Delta} NaHCO_3(s) + NH_4Cl(aq)$

   $2NaHCO_3(s) \rightarrow Na_2CO_3(s) + CO_2(g) + H_2O(g)$

   $Na_2CO_3(s) + Ca(OH)_2(aq) \rightarrow 2NaOH(aq) + CaCO_3(s)$

22.71 a. $2K(s) + Br_2(l) \rightarrow 2KBr(s)$

   b. $2K(s) + 2H_2O(l) \rightarrow 2KOH(aq) + H_2(g)$

   c. $2NaOH(s) + CO_2(g) \rightarrow Na_2CO_3(s) + H_2O(l)$

   d. $Li_2CO_3(aq) + 2HNO_3(aq) \rightarrow H_2O(l) + 2LiNO_3(aq) + CO_2(g)$

   e. $K_2SO_4(aq) + Pb(NO_3)_2(aq) \rightarrow PbSO_4(s) + 2KNO_3(aq)$

22.73 $^{227}_{89}Ac \rightarrow ^{223}_{87}Fr + ^{4}_{2}He$

22.75 $Ca(OH)_2$ can be identified directly by adding an anion that will precipitate the Ca^{2+} (and not the Na^+). For example, adding CO_3^{2-} will precipitate $CaCO_3$ but will not precipitate Na^+ ion.

22.77 $BaCl_2$ can be separated by adding SO_4^{2-} ion, precipitating $BaSO_4$ and leaving $MgCl_2$ in solution because it is soluble (see Table 4.1). Filtering the solution will separate the $BaSO_4$. The filtrate contains soluble $MgCl_2$.

22.79 $^{230}_{90}Th \rightarrow ^{226}_{88}Ra + ^{4}_{2}He$

22.81 a. $BaCO_3(s) \xrightarrow{\Delta} BaO(s) + CO_2(g)$

b. $Ba(s) + 2H_2O(l) \rightarrow Ba(OH)_2(aq) + H_2(g)$

c. $Mg(OH)_2(s) + 2HNO_3(aq) \rightarrow 2H_2O(l) + Mg(NO_3)_2(aq)$

d. $Mg(s) + NiCl_2(aq) \rightarrow Ni(s) + MgCl_2(aq)$

e. $2NaOH(aq) + MgSO_4(aq) \rightarrow Mg(OH)_2(s) + Na_2SO_4(aq)$

22.83 The equation is $Ca(HCO_3)_2(aq) + Ca(OH)_2(aq) \rightarrow 2CaCO_3(s) + 2H_2O(l)$. Using a 1:1 ratio of $Ca(OH)_2$ to Ca^{2+} ion in the reaction, calculate the mass of $Ca(OH)_2$:

$$0.0250 \text{ L } Ca^{2+} \times \frac{0.12 \text{ mol } Ca^{2+}}{\text{L } Ca^{2+}} \times \frac{1 \text{ mol } Ca(OH)_2}{1 \text{ mol } Ca^{2+}} \times \frac{74.09 \text{ g } Ca(OH)_2}{1 \text{ mol } Ca(OH)_2}$$

$$= 0.2\underline{22} = 0.22 \text{ g } Ca(OH)_2$$

22.85 $Al(H_2O)_6^{3+}(aq) + HCO_3^-(aq) \rightarrow Al(H_2O)_5OH^{2+}(aq) + H_2O(l) + CO_2(g)$

22.87 Test portions of solutions of each compound with each other; the results can differentiate the compounds. For example, if one solution is poured into one of the other solutions and gives no precipitate, then that means $BaCl_2$ was mixed with KOH and the third solution is $Al_2(SO_4)_3$. Adding the third solution of $Al_2(SO_4)_3$ to both of the first two solutions will form a precipitate with only $BaCl_2$. [$Al_2(SO_4)_3$ and excess KOH form soluble $Al(OH)_4^-$.] If, instead, one solution is poured into one of the other solutions and a precipitate forms, that means $BaCl_2$ was mixed with $Al_2(SO_4)_3 \rightarrow BaSO_4$, and the third solution is KOH, etc. Thus, all three are identified.

22.89 $Sn(H_2O)_6^{2+}(aq) + H_2O(l) \rightarrow Sn(H_2O)_5(OH)^+(aq) + H_3O^+(aq)$

22.91 The half-reactions and their sum are as follows:

$PbO_2 + 4H^+ + 2e^- \rightarrow Pb^{2+} + 2H_2O$

$2Cl^- \rightarrow Cl_2 + 2e^-$

_____

$PbO_2 + 4H^+ + 2Cl^- \rightarrow Pb^{2+} + Cl_2 + 2H_2O$

Adding 2Cl⁻ to both sides gives

$PbO_2 + 4HCl \rightarrow PbCl_2 + Cl_2 + 2H_2O$

22.93 a. $Al_2O_3(s) + 3H_2SO_4(aq) \rightarrow Al_2(SO_4)_3(aq) + 3H_2O(l)$

b. $Al(s) + 3AgNO_3(aq) \rightarrow 3Ag(s) + Al(NO_3)_3(aq)$

c. $Pb(NO_3)_2(aq) + 2NaI(aq) \rightarrow PbI_2(s) + 2NaNO_3(aq)$

d. $8Al(s) + 3Mn_3O_4(s) \rightarrow 9Mn(s) + 4Al_2O_3(s)$

e. $2Ga(OH)_3(s) \rightarrow Ga_2O_3(s) + 3H_2O(g)$

22.95   For 3.5×10^4 kg (3.5×10^7 g) of hydrogen, you get

$$3.5 \times 10^7 \text{ g } H_2 \times \frac{1 \text{ mol } H_2}{2.016 \text{ g } H_2} \times \frac{-484 \text{ kJ}}{2 \text{ mol } H_2} = -4.201 \times 10^9 = -4.2 \times 10^9 \text{ kJ}$$

Thus, 4.2×10^9 kJ of heat is evolved in the process.

22.97 a. The oxidation state of H in CaH_2 is -1.

b. The oxidation state of H in H_2O is +1.

c. The oxidation state of C in CH_4 is -4.

d. The oxidation state of S in H_2SO_4 is +6.

22.99 a. Carbon has four valence electrons. These are directed tetrahedrally, and the orbitals should be sp^3 hybrid orbitals. Each C–H bond is formed by the overlap of a $1s$ orbital of a hydrogen atom with one of the singly occupied sp^3 hybrid orbitals of the carbon atom.

b. Silicon has four valence electrons. (Two more electrons are added for the -2 charge.) These are directed octahedrally, and the orbitals should be sp^3d^2 hybrid orbitals. Each Si–F bond is formed by the overlap of a $2p$ orbital of a fluorine atom with one of the singly occupied sp^3d^2 hybrid orbitals of the silicon atom.

c. Carbon has four valence electrons. The single C–C bond is a sigma bond formed by the overlap of the sp^2 orbital of the middle carbon and the sp^3 orbital of the CH_3 carbon. The other three orbitals of the CH_3 carbon are used to form the C–H bonds by overlapping with the orbital of the hydrogen atom. The C=C double bond is a sigma bond formed by the overlap of the sp^2 hybrid orbitals of those carbon atoms, and a pi bond formed by the overlap of the unhybridized p orbitals. The C–H bonds on the middle carbon and the CH_2 carbon are formed by the overlap of the sp^2 orbital of carbon and the s orbital of hydrogen.

(continued)

d. Silicon has four valence electrons. These are directed tetrahedrally, and the orbitals should be sp^3 hybrid orbitals. Each Si–H bond is formed by the overlap of a $1s$ orbital of a hydrogen atom with one of the singly occupied sp^3 hybrid orbitals of the silicon atom.

22.101 a. The equation with ΔH°_f's recorded beneath each substance is

$$CH_4(g) \rightarrow C(graphite) + 2H_2(g)$$

| ΔH°_f: | -74.87 | 0 | 0 kJ |

$$\Delta H^\circ = [0 - (-74.87)]\ kJ = 74.87\ kJ$$

b. The equation with ΔH°_f's recorded beneath each substance is

$$C_2H_6(g) \rightarrow C_2H_4(g) + H_2(g)$$

| ΔH°_f: | -84.68 | 52.47 | 0 kJ |

$$\Delta H^\circ = [52.47 - (-84.68)]\ kJ = 137.15\ kJ$$

22.103 a. $CO_2(g) + Ba(OH)_2(aq) \rightarrow BaCO_3(s) + H_2O(l)$

b. $MgCO_3(s) + 2HBr(aq) \rightarrow CO_2(g) + H_2O(l) + MgBr_2(aq)$

22.105 $C(s) + O_2(air) \rightarrow CO_2(g)$

$CO_2(g) + NaOH(aq) \rightarrow NaHCO_3(aq)$

$2NaHCO_3(aq) \xrightarrow{\Delta} Na_2CO_3(s) + H_2O(l) + CO_2(g)$

22.107 The equation is $Mg_3N_2(s) + 6H_2O(l) \rightarrow 3Mg(OH)_2(aq) + 2NH_3(g)$. Using a 1:2 ratio of Mg_3N_2 to NH_3, calculate the mass of NH_3 formed as follows:

$$7.50\ g\ Mg_3N_2 \times \frac{1\ mol\ Mg_3N_2}{100.915\ g\ Mg_3N_2} \times \frac{2\ mol\ NH_3}{1\ mol\ Mg_3N_2} \times \frac{17.03\ g\ NH_3}{1\ mol\ NH_3}$$

$$= 2.5\underline{3}1 = 2.53\ g$$

22.109 Prepare HNO_3 from NH_3:

$$4NH_3(g) + 5O_2(g) \xrightarrow{Pt} 4NO(g) + 6H_2O(g)$$

$$2NO(g) + O_2(g) \rightarrow 2NO_2(g)$$

$$3NO_2(g) + H_2O(l) \rightarrow 2HNO_3(aq) + NO(g)$$

To prepare N_2O, use HNO_3 just prepared:

$$NH_3(g) + HNO_3(aq) \rightarrow NH_4NO_3(aq)$$

$$NH_4NO_3(aq) \rightarrow NH_4NO_3(s)$$

$$NH_4NO_3(s) \xrightarrow{\Delta} N_2O(g) + 2H_2O(g)$$

22.111 The reduction of NO_3^- ion to NH_4^+ ion is an eight-electron reduction, and the oxidation of zinc to zinc(II) ion is a two-electron oxidation. Balancing the equation involves multiplying the zinc half-reaction by four to achieve an eight-electron oxidation. The final equation is

$$4Zn(s) + NO_3^-(aq) + 10H^+(aq) \rightarrow 4Zn^{2+}(aq) + NH_4^+(aq) + 3H_2O(l)$$

22.113 PBr_4^+ has four pairs of electrons around the P atom, arranged in a tetrahedral fashion. The hybridization of P is sp^3. Each bond is formed by the overlap of a $4p$ orbital from Br with an sp^3 hybrid orbital from P.

22.115 The balanced half-reactions are

Oxidation: $H_3PO_3 + H_2O \rightarrow H_3PO_4 + 2H^+ + 2e^-$

Reduction: $H_2SO_4 + 2H^+ + 2e^- \rightarrow SO_2 + 2H_2O$

The overall balanced equation is

$$H_3PO_3 + H_2SO_4 \rightarrow H_3PO_4 + SO_2 + H_2O$$

22.117 In the equation, there is a 2:1 ratio of H_3PO_4 to $Ca_3(PO_4)_2$ that can be used to calculate the mass of H_3PO_4 from the mass of $Ca_3(PO_4)_2$. The mass of $Ca_3(PO_4)_2$ in the 30.0 g of rock is

$$0.746 \times 30.0 \text{ g rock} = 22.\underline{3}8 \text{ g Ca}_3(PO_4)_2 (= \text{CaP})$$

The mass of H_3PO_4 can be calculated from this mass as follows:

$$22.38 \text{ g CaP} \times \frac{1 \text{ mol CaP}}{310.2 \text{ g CaP}} \times \frac{2 \text{ mol H}_3PO_4}{1 \text{ mol CaP}} \times \frac{98.0 \text{ g H}_3PO_4}{1 \text{ mol H}_3PO_4}$$

$$14.\underline{1}4 = 14.1 \text{ g H}_3PO_4$$

22.119a. $4Li(s) + O_2(g) \rightarrow 2Li_2O(s)$

b. Organic materials burn in excess O_2 to give CO_2 and H_2O. The nitrogen becomes N_2:

$$4CH_3NH_2(g) + 9O_2(g) \rightarrow 4CO_2(g) + 2N_2(g) + 10H_2O(g)$$

c. $2(C_2H_5)_2S + 15O_2 \rightarrow 8CO_2(g) + 10H_2O(g) + 2SO_2(g)$

22.121a. $x_S + 6x_F = 0$

The oxidation number of F in compounds is always -1.

$x_S = -6x_F = -6(-1) = +6$

b. $x_S + 3x_O = 0$

The oxidation number of O in most compounds is -2.

$x_S = -3x_O = -3(-2) = +6$

c. $x_S + 2x_H = 0$

The oxidation number of H in most compounds is +1.

$x_S = -2x_H = -2(+1) = -2$

(continued)

d. $x_{Ca} + x_S + 3x_O = 0$

The oxidation number of Ca in compounds is +2; the oxidation number of O in most compounds is -2.

$$x_S = -x_{Ca} - 3x_O = -(+2) - 3(-2) = +4$$

22.123 The reduction of $8H_2SeO_3$ to Se_8 is a 32-electron reduction, and the oxidation of $8H_2S$ to S_8 is a 16-electron oxidation. Balancing the equation requires multiplying the H_2S half-reaction by two to achieve a 32-electron oxidation. The final equation is

$$8H_2SeO_3(aq) + 16H_2S(g) \rightarrow Se_8(s) + 2S_8(s) + 24H_2O(l)$$

22.125 In the equation, there is a 2:1 ratio of $NaHSO_3$ to Na_2CO_3 that can be used to calculate the mass of $NaHSO_3$ from 25.0 g of Na_2CO_3. The mass of $NaHSO_3$ is calculated as follows:

$$25.0 \text{ g } Na_2CO_3 \times \frac{1 \text{ mol } Na_2CO_3}{106.0 \text{ g } Na_2CO_3} \times \frac{2 \text{ mol } NaHSO_3}{1 \text{ mol } Na_2CO_3} \times \frac{104.0 \text{ g } NaHSO_3}{1 \text{ mol } NaHSO_3}$$

$$= 49.\underline{0}56 = 49.1 \text{ g } NaHSO_3$$

22.127 $Ba(ClO_3)_2(aq) + H_2SO_4(aq) \rightarrow 2HClO_3(aq) + BaSO_4(s)$

22.129 The balanced half-reactions are

Oxidation: $2HCl \rightarrow Cl_2 + 2H^+ + 2e^-$

Reduction: $K_2Cr_2O_7 + 14H^+ + 6e^- \rightarrow 2Cr^{3+} + 2K^+ + 7H_2O$

Multiply the oxidation half-reaction by three. After adding and canceling the hydrogen ions and electrons, the resulting overall balanced equation is

$$K_2Cr_2O_7 + 6HCl + 8H^+ \rightarrow 2Cr^{3+} + 3Cl_2 + 2K^+ + 7H_2O$$

If eight Cl^- are added to each side, the equation can also be written as a molecular equation.

$$K_2Cr_2O_7 + 14HCl \rightarrow 2CrCl_3 + 3Cl_2 + 2KCl + 7H_2O$$

22.131a. The electron-dot formula of Cl_2O is

The VSEPR model predicts a bent (angular) molecular geometry. You can describe the four electron pairs on O using sp^3 hybrid orbitals. The diagramming for the bond formation follows:

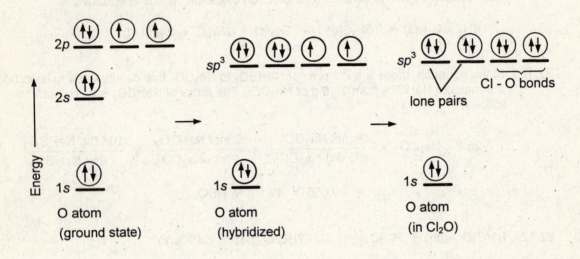

b. An electron-dot formula of BrO_3^- is

$$\left[\begin{array}{c} :\!\ddot{O}\!: \\ :\!\ddot{Br}\!:\!\ddot{O}\!: \\ :\!\ddot{O}\!: \end{array} \right]^{-}$$

(continued)

The VSEPR model predicts a trigonal pyramidal geometry. You can describe the four electron pairs on Br using sp^3 hybrid orbitals. The diagramming for the bond formation follows:

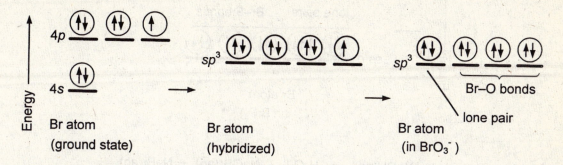

(Note: The additional electron accounts for the -1 charge of the ion; the bonds to O atoms are coordinate covalent.)

c. The electron-dot formula of BrF_3 is

$$:\ddot{F}:$$
$$:\ddot{Br}:\ddot{F}:$$
$$:\ddot{F}:$$

The five electron pairs on Br have a trigonal bipyramidal arrangement. Putting the lone pairs in equatorial positions to reduce repulsions gives a T-shaped molecular geometry for BrF_3. You can describe the five electron pairs on Br in terms of sp^3d hybrid orbitals. The diagramming for the bond formation follows:

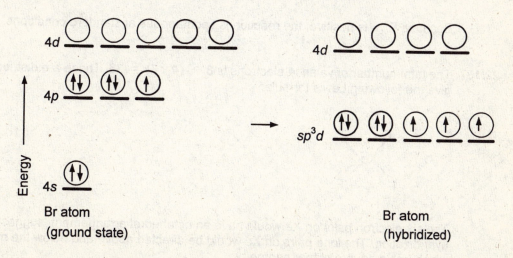

(continued)

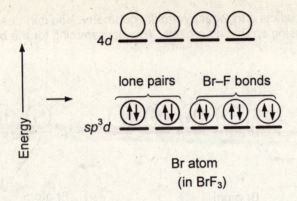

lone pairs   Br–F bonds

Energy

sp^3d

Br atom
(in BrF_3)

22.133 a. $Br_2(aq) + 2NaOH(aq) \rightarrow H_2O(l) + NaOBr(aq) + NaBr(aq)$

b. Assume by analogy with H_2SO_4 and NaCl that the usual heating causes a loss of one H^+ per molecule of acid. Stronger heating would result in a loss of additional H^+ from $H_2PO_4^-$.

$$NaBr(s) + H_3PO_4(aq) \xrightarrow{\Delta} HBr(g) + NaH_2PO_4(aq)$$

22.135 Reverse the iron half-reaction and the sign of its E°, and double the half-reaction to obtain the same number of electrons as the OCl⁻ half-reaction:

$2Fe^{2+}(aq) \rightarrow 2Fe^{3+}(aq) + 2e^-$		$-E° = -0.77 V$
$OCl^-(aq) + H_2O(l) + 2e^- \rightarrow Cl^-(aq) + 2OH^-(aq)$		$E° = 0.90 V$

$$OCl^-(aq) + H_2O(l) + 2Fe^{2+}(aq) \rightarrow 2Fe^{3+}(aq) + Cl^-(aq) + 2OH^-(aq)$$

$$E°_{cell} = 0.13 V$$

Because $E°_{cell}$ is positive, the reaction is spontaneous at standard conditions.

22.137 The total number of valence electrons is $8 + (4 \times 7) = 36$. These are distributed to give the following Lewis formula:

The six electron pairs on Xe would have an octahedral arrangement, suggesting sp^3d^2 hybridization. The lone pairs on Xe would be directed above and below the molecule, which has a square planar geometry.

22.139 From the information given,

$$XeF_2 \rightarrow Xe + O_2 + F^- \text{ (basic solution; not balanced)}$$

The oxygen (and hydrogen) will be balanced by OH^- and H_2O from the basic solution. Balance half-reactions.

Reduction: $XeF_2 \rightarrow Xe + F^-$

Balance F atoms: $XeF_2 \rightarrow Xe + 2F^-$

Balance charge: $2e^- + XeF_2 \rightarrow Xe + 2F^-$

Oxidation: $\qquad\qquad H_2O \rightarrow O_2$

Balance O atoms: $2H_2O \rightarrow O_2$

Balance H atoms: $2H_2O \rightarrow 4H^+ + O_2$

Balance charge: $2H_2O \rightarrow 4H^+ + O_2 + 4e^-$

Convert to base: $2H_2O + 4OH^- \rightarrow 4H_2O + O_2 + 4e^-$

Double the Xe half-reaction, and add half-reactions to get the overall reaction:

$$4e^- + 2XeF_2 \rightarrow 2Xe + 4F^-$$
$$\underline{4OH^- \rightarrow O_2 + 2H_2O + 4e^-}$$
$$2XeF_2 + 4OH^- \rightarrow 2Xe + 4F^- + O_2 + 2H_2O$$

■ Solutions to General Problems

22.141 The equations are

$$2HCl + Mg(OH)_2(s) \rightarrow MgCl_2 + 2H_2O$$

$$HCl + NaOH \rightarrow NaCl + H_2O.$$

Using a 1:1 ratio of HCl to NaOH, calculate the mol of HCl reacting with NaOH:

$$\left[\frac{0.4987 \text{ mol HCl}}{L} \times 0.05000 \text{ L} \right] - \left[\frac{0.2456 \text{ mol NaOH}}{L} \times 0.03942 \text{ L} \right]$$

$$= 0.015253 \text{ mol HCl}$$

(continued)

The mass of $Mg(OH)_2$ is

$$0.015253 \text{ mol HCl } \times \frac{1 \text{ mol Mg(OH)}_2}{2 \text{ mol HCl}} \times \frac{58.33 \text{ g Mg(OH)}_2}{1 \text{ mol Mg(OH)}_2}$$

$$= 0.44486 \text{ g Mg(OH)}_2$$

The mass percentage is

$$\frac{0.44486 \text{ g}}{5.436 \text{ g}} \times 100\% = 8.1837 = 8.184 \text{ percent}$$

22.143 The equation with $\Delta H°_f$'s recorded beneath each substance is

$$Fe_2O_3(s) \;+\; 2Al(s) \;\rightarrow\; 2Fe(s) \;+\; Al_2O_3(s)$$

$\Delta H°_f$:   -825.5   0   0   -1675.5 kJ

$$\Delta H° = [-1675.5 - (-825.5)] \text{ kJ} = -850.0 \text{ kJ}$$

For one mole of iron

$$\frac{-850.0 \text{ kJ}}{2 \text{ mol Fe}} = -425.0 \text{ kJ/mol Fe}$$

22.145 From the ideal gas law, the moles of CO_2 are

$$n = \frac{PV}{RT} = \frac{(745/760 \text{ atm})(0.03456 \text{ L})}{(0.0821 \text{ L} \cdot \text{atm/K} \cdot \text{mol})(294 \text{ K})} = 0.0014035 \text{ mol CO}_2$$

The equation is $CaCO_3(s) + 2HCl \rightarrow CO_2(g) + CaCl_2(aq) + H_2O(l)$. Using a 1:1 mole ratio of CO_2 to $CaCO_3$, the mass of $CaCO_3$ is

$$0.0014035 \text{ mol CO}_2 \times \frac{1 \text{ mol CaCO}_3}{1 \text{ mol CO}_2} \times \frac{100.1 \text{ g CaCO}_3}{1 \text{ mol CaCO}_3} = 0.14049 \text{ g CaCO}_3$$

The mass percentage of $CaCO_3$ is

$$\frac{0.14049 \text{ g}}{0.1662 \text{ g}} \times 100\% = 84.53 = 84.5 \text{ percent CaCO}_3$$

22.147 The equation is $NaCl + NH_3 + H_2O + CO_2 \rightarrow NaHCO_3 + NH_4Cl$. Using a 1:1 mole ratio of NaCl to $NaHCO_3$, the mass of NaCl is

$$10.00 \text{ g NaHCO}_3 \times \frac{1 \text{ mol NaHCO}_3}{84.00 \text{ g NaHCO}_3} \times \frac{1 \text{ mol NaCl}}{1 \text{ mol NaHCO}_3} \times \frac{58.44 \text{ g NaCl}}{1 \text{ mol NaCl}}$$

$$= 6.95\underline{7}1 = 6.957 \text{ g NaCl}$$

22.149 Write the reaction with the $\Delta H°_f$'s and $S°$'s below

$$SrCO_3(s) \rightarrow SrO(s) + CO_2(g)$$

$\Delta H°_f$:	-1220.1	-592.0	-393.5 kJ
$S°$:	97.1	55.52	213.7 J/K

$$\Delta H° = [-592.0 - 393.5 - (-1220.1)] \text{ kJ} = 234.6 \text{ kJ}$$

$$\Delta S° = [55.52 + 213.7 - 97.1] \text{ J/K} = 172.\underline{1}2 \text{ J/K} = 0.17212 \text{ kJ/K}$$

Finally, determine the temperature.

$$\Delta G° = \Delta H° - T\Delta S°$$

$$0 = 234.6 \text{ kJ} - T(0.17212 \text{ kJ/K})$$

$$T = \frac{234.6 \text{ kJ}}{0.17212 \text{ kJ/K}} = 136\underline{3}.0 = 1363 \text{ K}$$

22.151 The overall disproportionation reaction can be considered as the sum of the following reactions:

$$2e^- + 2In^+(aq) \rightarrow 2In(s)$$
$$In^+(aq) \rightarrow In^{3+}(aq) + 2e^-$$
$$\overline{3In^+(aq) \rightarrow 2In(s) + In^{3+}}$$

$$E°_{cell} = E°_{cathode} - E°_{anode} = [(-0.21) - (-0.40)] \text{ V} = 0.19 \text{ V}$$

$$\Delta G° = -nFE°_{cell} = -(2)(9.65 \times 10^4 \text{ C})(0.19 \text{ J/C}) = -3.\underline{6}67 \times 10^4 \text{ J} = -37 \text{ kJ}$$

$\Delta G°$ for the reaction as written is negative, so the disproportionation does occur spontaneously.

22.153 Using the ideal gas law, the amount of CO_2 is

$$n = \frac{PV}{RT} = \frac{(30.0/760 \text{ atm})(1.00 \text{ L})}{(0.0821 \text{ L} \bullet \text{atm/K} \bullet \text{mol})(298 \text{ K})} = 1.6\underline{1}4 \times 10^{-3} \text{ mol } CO_2$$

The balanced equation is $2LiOH(s) + CO_2(g) \rightarrow Li_2CO_3(s) + H_2O(g)$, so use the mole ratio one mol CO_2 to two mol LiOH. Therefore, the mass of LiOH is

$$1.614 \times 10^{-3} \text{ mol } CO_2 \times \frac{2 \text{ mol LiOH}}{1 \text{ mol } CO_2} \times \frac{23.95 \text{ g LiOH}}{1 \text{ mol LiOH}}$$

$$= 7.7\underline{3}1 \times 10^{-2} = 7.73 \times 10^{-2} \text{ g LiOH}$$

22.155 Assume a sample of 100.0 g fertilizer. This contains 17.1 g P. Convert this to the mass of $Ca(H_2PO_4)_2 \bullet H_2O$ in the 100.0 g of fertilizer.

$$17.1 \text{ g P} \times \frac{1 \text{ mol P}}{30.97 \text{ g P}} \times \frac{1 \text{ mol } Ca(H_2PO_4)_2 \bullet H_2O}{2 \text{ mol P}}$$

$$\times \frac{252.1 \text{ g } Ca(H_2PO_4)_2 \bullet H_2O}{1 \text{ mol } Ca(H_2PO_4)_2 \bullet H_2O} = 69.\underline{5}89 \text{ g } Ca(H_2PO_4)_2$$

The mass percent $Ca(H_2PO_4)_2 \bullet H_2O$ in the fertilizer is

$$\text{Mass percent } Ca(H_2PO_4)_2 \bullet H_2O = \frac{69.589 \text{ g } Ca(H_2PO_4)_2 \bullet H_2O}{100 \text{ g fertilizer}} \times 100\%$$

$$= 69.\underline{5}89 = 69.6 \text{ percent}$$

22.157 $2NaCl + 2H_2O \rightarrow 2NaOH + H_2 + Cl_2$ (electrolysis)

$2NaOH + Cl_2 \rightarrow NaClO + NaCl + H_2O$ (spontaneous)

One mol of NaClO is produced for every two mol of NaCl. Thus, the reaction requires two electrons per mol of NaClO produced.

(continued)

Convert the 1.00×10^3 L of NaOCl to time:

$$1.00 \times 10^3 \text{ L} \times \frac{1000 \text{ mL}}{1 \text{ L}} \times \frac{1.00 \text{ g soln}}{1 \text{ mL}} \times \frac{5.25 \text{ g NaOCl}}{100 \text{ g soln}} \times \frac{1 \text{ mol NaOCl}}{74.44 \text{ g NaOCl}}$$

$$\times \frac{2 \text{ mol e}^-}{1 \text{ mol NaOCl}} \times \frac{9.65 \times 10^4 \text{ C}}{1 \text{ mol e}^-} \times \frac{1 \text{ s}}{3.00 \times 10^3 \text{ C}} \times \frac{1 \text{ h}}{3600 \text{ s}}$$

$$= 12.\underline{6}0 = 12.6 \text{ h}$$

22.159   The reactions are

$$H_2O + NaOCl + 2I^- \rightarrow I_2 + NaCl + 2OH^-$$
$$I_2 + 2Na_2S_2O_3 \rightarrow 2NaI + Na_2S_4O_6$$

The mass of NaOCl is

$$0.0346 \text{ L Na}_2\text{S}_2\text{O}_3 \times \frac{0.100 \text{ mol Na}_2\text{S}_2\text{O}_3}{1 \text{ L}} \times \frac{1 \text{ mol I}_2}{2 \text{ mol Na}_2\text{S}_2\text{O}_3} \times \frac{1 \text{ mol NaOCl}}{1 \text{ mol I}_2}$$

$$\times \frac{74.44 \text{ g NaOCl}}{1 \text{ mol NaOCl}} = 0.12\underline{8}8 \text{ g NaOCl}$$

If the density of bleach is taken as 1.00 g/mL, then 5.00 mL = 5.00 g. Thus, the mass percent is

$$\text{Mass percent NaOCl} = \frac{\text{mass NaOCl}}{\text{mass bleach}} \times 100\% = \frac{0.1288 \text{ g}}{5.00 \text{ g}} \times 100\%$$

$$= 2.5\underline{7}56 = 2.58 \text{ percent}$$

23. THE TRANSITION ELEMENTS AND COORDINATION COMPOUNDS

■ Solutions to Exercises

23.1 a. Pentaamminechlorocobalt(III) chloride

b. Potassium aquapentacyanocobaltate(III)

c. Pentaaquahydroxoiron(III) ion

23.2 a. $K_4[Fe(CN)_6]$ b. $[Co(NH_3)_4Cl_2]Cl$ c. $PtCl_4^{2-}$

23.3 a. No geometric isomers.

b.

(continued)

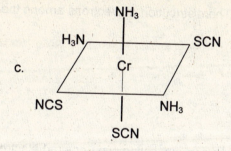

c.

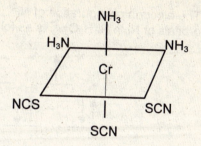

d. No geometric isomers.

23.4 a. No optical isomers.

b.

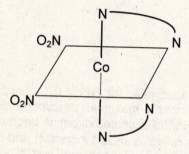

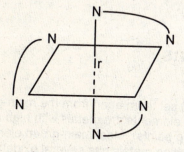

c.

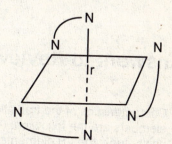

d. No optical isomers.

23.5   The electron configuration of Ni^{2+} is $[Ar]3d^8$. The distribution of electrons among the d orbitals of Ni in $Ni(H_2O)_6^{2+}$ is as follows:

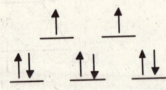

Note there is only one possible distribution of electrons, giving two unpaired electrons.

23.6   The electronic configuration of the Co^{2+} ion is $[Ar]3d^7$. The distribution of the d electrons in $CoCl_4^{2-}$ is as follows:

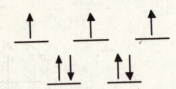

23.7   The approximate wavelength of the maximum absorption for $Fe(H_2O)_6^{3+}$, which is pale purple, is 530 nm. The approximate wavelength of the maximum absorption for $Fe(CN)_6^{3-}$, which is red, is 500 nm. The shift is in the expected direction because CN^- is a more strongly bonding ligand than H_2O. As a result, Δ should increase, and the wavelength of the absorption should decrease when H_2O is replaced by CN^-.

■ Answers to Review Questions

23.1   Characteristics of the transition elements that set them apart from the main-group elements are the following: (1) The transition elements are metals with high melting points (only the IIB elements have low melting points). Most main-group elements have low melting points. (2) Each of the transition metals has several oxidation states (except for the IIIB and IIB elements). Most main-group metals have only one oxidation state in addition to zero. (3) Transition-metal compounds are often colored, and many are paramagnetic. Most main-group compounds are colorless and diamagnetic.

23.2   Technetium has the electron configuration $[Kr] 4d^5 5s^2$.

23.3   Molybdenum has the highest melting point of any element in the fifth period because it has the maximum number of unpaired electrons, which contributes to the strength of the metal bonding.

23.4   One reason iron, cobalt, and nickel are similar in properties is because these elements have similar covalent radii.

23.5   Nickel falls in the fourth period; as such, it has a much smaller covalent radius than the corresponding metals, palladium and platinum in the fifth and sixth periods. However, palladium and platinum have very similar covalent radii.

23.6 $Cr(s) + 2HCl(aq) + 6H_2O(l) \rightarrow Cr(H_2O)_6^{2+} + 2Cl^-(aq) + H_2(g)$

        $Cu(s) + HCl(aq) \rightarrow NR$

23.7 $Cr_2O_3(s) + 6HCl(aq) + 9H_2O(l) \rightarrow 2Cr(H_2O)_6^{3+} + 6Cl^-(aq)$

23.8   Four of the water molecules are associated with the copper(II) ion, and the fifth is hydrogen bonded to the sulfate ion as well as to water molecules on the copper(II) ion. Heating changes the blue color to a white color, the color of anhydrous $CuSO_4$. The blue color is associated with $Cu(H_2O)_4^{2+}$. When the water molecules leave the copper ion, the blue color is lost.

23.9   Two Cu^{2+} ions gain a total of two electrons in forming one Cu_2O. The HCHO loses two electrons in forming $HCOO^-$ ion. Thus, the final equation is

        $$2Cu^{2+}(aq) + HCHO(aq) + 5OH^-(aq) \rightarrow Cu_2O(s) + HCOO^-(aq) + 3H_2O(l)$$

23.10   Werner showed that the electrical conductance of a solution of $[Pt(NH_3)_4Cl_2]Cl_2$ corresponded to that of three ions in solution and that two of the chloride ions could be precipitated as AgCl whereas the other two could not.

23.11   A complex ion is a metal atom or ion with Lewis bases attached to it through coordinate covalent bonds. A ligand is a Lewis base attached to a metal ion in a complex; it may be either a molecule or an anion, rarely a cation. The coordination number of a metal atom in a complex ion is the total number of bonds the metal forms with ligands. An example of a complex ion is $Fe(CN)_6^{4-}$; an example of a ligand is CN^-; and the coordination number of the preceding complex ion is six.

23.12   A bidentate ligand is a ligand that bonds to a metal ion through two atoms. Two examples are ethylenediamine, $H_2N-C_2H_4-NH_2$, and the oxalate ion, $^-O_2C-CO_2^-$.

23.13

23.14 The three properties are isomerism, paramagnetism, and color (or absorption of visible and ultraviolet radiation).

23.15 a. Ionization isomerism involves isomers that are alike in that the same anions are present in the formula, but different anions are coordinated to the metal ion. For example, the sulfate ion is coordinated to cobalt in $[Co(NH_3)_5(SO_4)]Br$, but the bromide ion is coordinated to cobalt in $[Co(NH_3)_5Br]SO_4$.

b. Hydrate isomerism involves differences in the placement of water molecules in the complex ion. For example, $CrCl_3 \cdot 6H_2O$ exists as $[Cr(H_2O)_6]Cl_3$, $[Cr(H_2O)_4Cl_2]Cl \cdot 2H_2O$, and one other isomer.

c. Coordination isomers are those in which both the cation and anion are complex, and the ligands are distributed between the two metal atoms in different ways. For example,

$[Cu(NH_3)_4][PtCl_4]$ and $[Pt(NH_3)_4][CuCl_4]$

d. Linkage isomers are those in which two different donor atoms on the ligand may bond to the metal ion. For example, the SCN^- can bond to a metal ion through the sulfur atom or through the nitrogen atom.

23.16 Geometric isomers are isomers in which the atoms are joined to one another in the same way but differ because some atoms occupy different relative positions in space. In $[Pt(NH_3)_2Cl_2]$, the two NH_3's (or Cl's) can be arranged trans or cis to one another. Optical isomers are isomers that are nonsuperimposable mirror images of one another. See Figure 23.16 for an example of two cobalt optical isomers.

23.17 A *d* optical isomer rotates the plane of polarized light to the right (dextrorotatory), and an *l* ioptical isomer rotates the plane to the left (levorotatory).

23.18 A racemic mixture is a mixture of 50 percent of the *d* isomer and 50 percent of the *l* isomer. One method of resolving a racemic mixture is to prepare a salt with an optically active ion of the opposite charge, and crystallize the salts. They will no longer be optical isomers and will have different solubilities, so one can be precipitated before the other.

23.19 According to valence bond theory, a ligand orbital containing two electrons overlaps an unoccupied orbital on the metal ion.

23.20 a. In the high-spin complex ion, all 3*d* orbitals of Fe^{2+} are occupied (four of the 3*d* orbitals each contain only one electron). Because they are occupied, those orbitals cannot be used for ligand bonding. Instead, d^2sp^3 hybrid orbitals form from the 4*s*, the three 4*p*, and two of the 4*d* orbitals. Each of six ligands donates a pair of electrons to one of these sp^3d^2 hybrid orbitals.

   b. In the low-spin complex ion, the six electrons in the 3*d* orbitals are paired, so they occupy only three of the 3*d* orbitals. Then sp^3d^2 hybrid orbitals form from two of the 3*d*, the 4*s*, and the three 4*p* orbitals. Each of six ligands donates a pair of electrons to one of these d^2sp^3 hybrid orbitals.

23.21 The *d* orbitals of a transition-metal atom may have different energies in the octahedral field of six negative charges because the electron pairs of the ligands point directly at the d_{z^2} and $d_{x^2-y^2}$ orbitals. These orbitals are raised much more in energy than the other three *d* orbitals because they occupy space between the ligands. Thus, there is a crystal field splitting between the first two *d* orbitals mentioned and the d_{xy}, d_{xz}, and d_{yz} orbitals.

23.22 a. A high-spin Fe(II) octahedral complex is

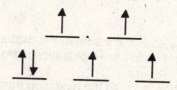

   b. A low-spin Fe(II) octahedral complex is

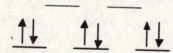

23.23 Crystal field splitting is the difference in energy between the two sets of *d* orbitals for a given structure (such as octahedral) in complex ions. It is determined experimentally by measuring the energy of light absorbed by complex ions.

23.24 The spectrochemical series is the arrangement of ligands in order of the relative size of the crystal field splittings (Δ) they induce in the d orbitals of a given oxidation state of a given metal ion. The order is the same, no matter what metal or oxidation state is involved. For Cl^-, H_2O, NH_3, and CN^-, the order of increasing crystal field splitting is

$$Cl^- < H_2O < NH_3 < CN^-$$

where CN^- always acts as a strong-bonding ligand.

23.25 Pairing energy, P, is the energy required to place two electrons in the same orbital. If the crystal field splitting (Δ) is small because of weak-bonding ligands, then the pairing energy will be larger, and the complex will be high-spin. If the crystal field splitting (Δ) is large because of strong-bonding ligands, then the pairing energy will be smaller, and the complex will be low-spin.

23.26 The complex absorbing red light would appear as a mixture of blue and green (approximately).

■ Solutions to Practice Problems

Note on significant figures: If the final answer to a solution needs to be rounded off, it is given first with one nonsignificant figure, and the last significant figure is underlined. The final answer is then rounded to the correct number of significant figures. In multiple-step problems, intermediate answers are given with at least one nonsignificant figure; however, only the final answer has been rounded off.

23.33 a. The charge on the carbonate ion is -2. For $FeCO_3$ to be neutral, the oxidation number of iron must be +2.

b. The oxidation number of oxygen is -2. For the sum of the oxidation numbers of all atoms to be zero, manganese must be in the +4 oxidation state.

c. The oxidation number of the chlorine is -1, so the oxidation number of copper must be +2.

d. The oxidation number of oxygen is -2. The oxidation number of chlorine is -1. For the sum of the oxidation numbers to be zero, the oxidation number of chromium must be +6.

$$+6 + 2(-2) + 2(-1) = 0$$

23.35   The half-reactions are

$$Fe^{2+} \rightarrow Fe^{3+} + e^-$$

$$4H^+ + NO_3^- + 3e^- \rightarrow NO + 2H_2O$$

The balanced equation is

$$3Fe^{2+} + NO_3^- + 4H^+ \rightarrow 3Fe^{3+} + NO + 2H_2O$$

23.37 a. Four. There are four cyanide groups coordinated to the gold atom.

   b. Six. There are four ammonia molecules and two water molecules coordinated to the cobalt.

   c. Four. Each of the ethylenediamine molecules bonds to the gold atom through two nitrogen atoms.

   d. Six. Each ethylenediamine molecule bonds to the chromium atom through two nitrogen atoms, and the oxalate ion bonds to the chromium through two oxygen atoms.

23.39 a. The charge on the $Ni(CN)_4^{2-}$ ion is -2 to balance the charge of +2 from the two K^+ ions. Each cyanide ion has a charge of -1. The sum of the oxidation number of nickel and the charge on the cyanide ions must equal the charge, so

   $$-2 = [4 \times (-1)] + 1 \times [ox. no. (Ni)]$$

   Ox. no. (Ni) = -2 - (-4) = +2

   b. The charge on ethylenediamine is zero, so the oxidation number of Mo is equal to the charge on the complex ion.

   Ox. no. (Mo) = +3

   c. Oxalate ion has a charge of -2, so

   Ox. no. of Cr = charge of complex ion -3 x (charge of oxalate ion)
   = (-3) - [3 x (-2)] = +3

   d. Chloride ion has a charge of -1, so the charge on the complex ion is +2. The NH_3 ligands are neutral and contribute nothing to the charge of the complex ion.

   Ox. no. of Co = charge of complex ion - charge of nitrite ligand
   = +2 - (-1) = +3

23.41 a. The charge of each chloride ligand is -1; the charge on the oxalate ligand is -2. The ammonia ligands are neutral.

Ox. no. of Cr = charge of complex ion - charge of oxalate

- 2 x (charge of chloride)

$$= -1 - (-2) - 2(-1) = +3$$

b.

Formula	Name
NH_3	Ammine
Cl^-	Chloro
$C_2O_4^{2-}$	Oxalato

c. Six. The chromium atom has one bond to each of the NH_3 ligands and Cl- ligands and two bonds to the $C_2O_4^{2-}$ ligand.

d. If each NH_3 were replaced by one Cl^-, and the $C_2O_4^{2-}$ ligand were replaced by two Cl^- ligands, there would be a total of six Cl^- ligands bonded to a chromium atom in the +3 oxidation state.

Charge on complex ion = ox. no. of Cr + 6 x (charge on chloride ligand)

$$= +3 + [6 \times (-1)] = -3$$

23.43 a. Potassium hexafluoroferrate(III)

b. Diamminediaquacopper(II) ion

c. Ammonium aquapentafluoroferrate(III)

d. Dicyanoargentate(I) ion

23.45 a. Pentacarbonyliron(0)

b. Dicyanobis(ethylenediamine)rhodium(III) ion

c. Tetraamminesulfatochromium(III) chloride

d. Tetraoxomanganate(VII) ion (Permanganate is the usual name.)

23.47 a. The charge on the complex ion equals

Ox. no. of Mn + 6 x (charge on cyanide ion) = +3 + 6(-1) = -3

Hence, the formula is $K_3[Mn(CN)_6]$.

b. The charge on the complex ion equals

Ox. no. of Zn + 4 x (charge on cyanide ion) = 2 + 4(-1) = -2

Hence, the formula is $Na_2[Zn(CN)_4]$.

c. The charge on the complex ion equals

Ox. no. of Co + 2 x (charge on chloride ion)

Note the ammine ligand (NH_3) is neutral. You get +3 + 2(-1) = +1. The formula is $[Co(NH_3)_4Cl_2]NO_3$.

d. The charge on the cation equals the ox. no. of Cr (+3). The charge on the anion equals

Ox. no. of Cu + 4 x (charge on chloride ion) = +2 + 4(-1) = -2

The formula is $[Cr(NH_3)_6]_2[CuCl_4]_3$.

23.49 a.

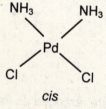

cis *trans*

b. No geometric isomerism.

c. No geometric isomerism.

(continued)

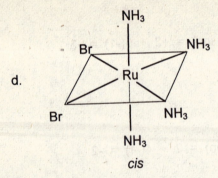

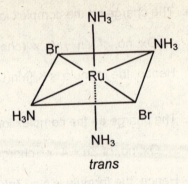

d.

cis

trans

23.51 a.

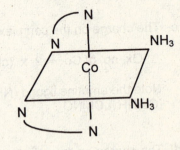

b. No optical isomers.

23.53 a. V^{3+} has two d electrons arranged as shown:

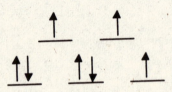

There are two unpaired electrons.

b. Co^{2+} has seven d electrons. In the high-spin case, they are arranged as follows:

There are three unpaired e⁻.

(continued)

c. Mn^{3+} has four d electrons. In the low-spin case, they are arranged as follows:

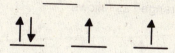

There are two unpaired e^-.

23.55 a. Pt^{2+} has eight electrons in the $5d$ subshell. Because the complex is diamagnetic, the crystal field felt by the d orbitals is most likely square planar (no low-spin tetrahedral complexes are known). The arrangement of the d electrons is

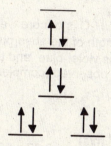

b. Co^{2+} has seven d electrons. One unpaired electron implies a low-spin complex. A square planar field will lead to low-spin complexes, so the geometry is probably square planar with the d electrons arranged as follows:

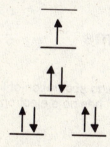

c. Fe^{3+} has five d electrons. If they are all unpaired, that is, a high-spin complex, the field felt by the metal ion is most likely tetrahedral with the d electrons arranged as follows:

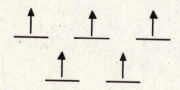

(continued)

d. Co^{2+} has seven d electrons. Three unpaired electrons imply a high-spin complex. A tetrahedral field will lead to high-spin complexes, so the geometry is probably tetrahedral with the d electrons arranged as follows:

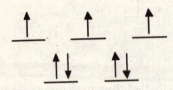

23.57   Purple (from Table 23.7)

23.59   Yes. According to the spectrochemical series, H_2O is a more weakly bonding ligand than NH_3, so Δ should decrease. The wavelength of the absorption should increase ($\lambda = hc/\Delta$). The light absorbed by $Co(NH_3)_6^{3+}$ is violet-blue, and the replacement of one NH_3 by H_2O shifts this toward blue. Thus, the observed (complementary) color should shift toward red, as is observed.

23.61 $\Delta = \dfrac{hc}{\lambda} = \dfrac{(6.626 \times 10^{-34} \text{ J} \cdot \text{s})(2.998 \times 10^8 \text{ m/s})}{(500 \times 10^{-9} \text{ m})} \times 6.02 \times 10^{23} \text{ /mol}$

$= 2.3\underline{9}17 \times 10^5 = 2.39 \times 10^5 \text{ J/mol}$ (239 kJ/mol)

■ Solutions to General Problems

23.63   The color in transition-metal complexes is due to absorption of light when a d electron moves to a higher energy level. Because Sc^{3+} has no d electrons, it is expected to be colorless.

23.65 If $[Co(NH_3)_4Cl_2]^+$ had a regular planar hexagonal geometry, two geometric isomers would be expected.

The known existence of two isomers of the complex does not rule out hexagonal geometry as a possibility.

23.67 The reaction and equilibrium-constant equation are

$$Cu(NH_3)_4^{2+} \rightleftharpoons Cu^{2+} + 4 NH_3$$

$$K_c = 2.1 \times 10^{-13} = \frac{[Cu^{2+}][NH_3]^4}{[Cu(NH_3)_4^{2+}]}$$

Concentration (M)	Cu^{2+}(aq) +	$4NH_3$(aq)	$\rightleftharpoons$	$Cu(NH_3)_4^{2+}$(aq)
Start	0.10	0.40		0
Change	-x	-4x		+x
Equilibrium	0.10 - x	0.40 - 4x		x

Substituting into the equilibrium expression, you obtain

$$2.1 \times 10^{-13} = \frac{(0.10 - x)[4(0.10 - x)]^4}{x}$$

To simplify, let $y = 0.10 - x$. Then $x = 0.10 - y$. The above equation becomes

$$2.1 \times 10^{-13} = \frac{y(4y)^4}{(0.10 - y)}$$

(continued)

Assume y is small compared to 0.10. Then $0.10 - y \cong 0.10$.

$$2.1 \times 10^{-13} = \frac{y(4y)^4}{(0.10)}$$

$$y^5 = \frac{(2.1 \times 10^{-13})(0.10)}{4^4} = 8.\underline{2}03 \times 10^{-17}$$

$$y = \sqrt[5]{8.203 \times 10^{-17}} = 6.\underline{0}64 \times 10^{-4}$$

Going back to the above assumption, you find that, to two significant figures, $0.10 - 0.0006\underline{1} = 0.10$. The assumption was valid.

At equilibrium:

$[Cu^{2+}] = (0.10 - x) M = y = 6.\underline{0}64 \times 10^{-4} = 6.1 \times 10^{-4} M$

$[NH_3] = 4(0.10 - x) M = 4y = 4 \times 6.\underline{0}64 \times 10^{-4} = 2.\underline{4}3 \times 10^{-3} = 2.4 \times 10^{-3} M$

$[Cu(NH_3)_4{}^{2+}] = x = (0.10 - y) = 0.10 M$

24. ORGANIC CHEMISTRY

■ Solutions to Exercises

24.1   The condensed structural formula is

$$CH_3CHCH_2CH_3$$
with CH_3 attached to the second carbon.

24.2 a. Geometric isomers are possible.

cis -2-hexene   trans -2-hexene

   b. No geometric isomers are possible because there are two H atoms attached to the
       second carbon of the double bond.

24.3   According to Markownikoff's rule, when HBr is added across the double bond in
        1-butene, the H will add to carbon one (the C atom with the most bonds to H atoms)
        and the Br will add to carbon two.

$$H_2C = CH-CH_2-CH_3 + HBr \rightarrow H_3C-CH-CH_2-CH_3$$
with Br attached to the second carbon.

        The product is 2-bromobutane.

24.4 a. The longest continuous chain is numbered as follows:

$$\begin{array}{c} CH_3 \\ \underset{1}{} \underset{2}{} \underset{3|}{} \underset{4}{} \\ CH_3CHCHCH_3 \\ | \\ CH_3 \end{array}$$

The name of the compound is 2,3-dimethylbutane.

b. To give the substituents the smaller numbers, the longest continuous chain is numbered as follows:

$$\begin{array}{c} \underset{4}{} \underset{5}{} \underset{6}{} \\ CH_2CH_2CH_3 \\ \underset{1}{} \underset{2}{} \underset{3|}{} \\ CH_3CHCHCH_2CH_3 \\ | \\ CH_3 \end{array}$$

The name of the compound is 3-ethyl-2-methylhexane.

24.5 First, write out the carbon skeleton for octane.

$$\begin{array}{c} \underset{1|}{}\ \underset{2|}{}\ \underset{3|}{}\ \underset{4|}{}\ \underset{5|}{}\ \underset{6|}{}\ \underset{7|}{}\ \underset{8|}{} \\ -C-C-C-C-C-C-C-C- \\ |\ \ |\ \ |\ \ |\ \ |\ \ |\ \ |\ \ | \end{array}$$

Then, attach the alkyl groups.

$$\begin{array}{c} CH_3 \\ \underset{1|}{}\ \underset{2|}{}\ \underset{3|}{}\ \underset{4|}{}\ \underset{5|}{}\ \underset{6|}{}\ \underset{7|}{}\ \underset{8|}{} \\ -C-C-C-C-C-C-C-C- \\ |\ \ |\ \ |\ \ |\ \ |\ \ |\ \ |\ \ | \\ CH_3 \end{array}$$

Finally, fill out the structure with H atoms.

$$\begin{array}{c} CH_3 \\ | \\ CH_3CH_2CCH_2CH_2CH_2CH_2CH_3 \\ | \\ CH_2 \end{array}$$

24.6 a. The numbering of the carbon chain is

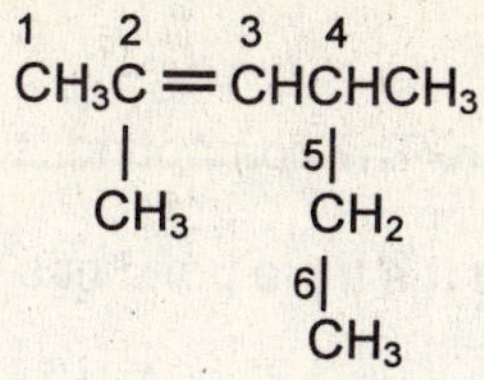

$$
\begin{array}{cccc}
1 & 2 & 3 & 4 \\
\end{array}
$$
$$
CH_3C = CHCHCH_3
$$
$$
\quad\;\; | \qquad\quad 5| \\
\quad\;\; CH_3 \qquad CH_2 \\
\qquad\qquad\quad 6| \\
\qquad\qquad\quad CH_3
$$

Because the longest chain containing a double bond has six carbons, this is a hexene. It is a 2-hexene because the double bond is between carbons two and three. The name of the compound is 2,4-dimethyl-2-hexene.

b. The numbering of the longest chain with a double bond is

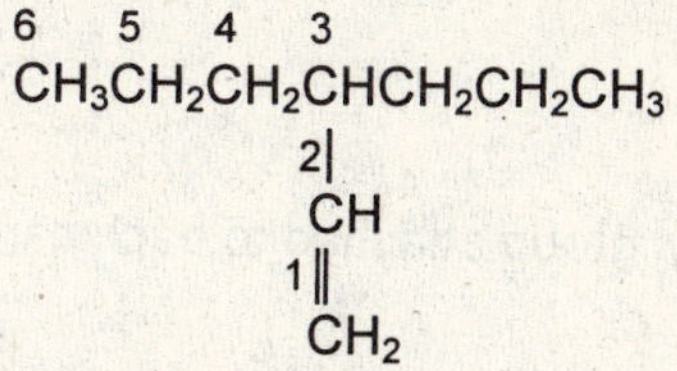

$$
\begin{array}{cccc}
6 & 5 & 4 & 3 \\
\end{array}
$$
$$
CH_3CH_2CH_2CHCH_2CH_2CH_3
$$
$$
\qquad\qquad\quad 2| \\
\qquad\qquad\quad CH \\
\qquad\qquad 1\| \\
\qquad\qquad\quad CH_2
$$

The longest chain containing the double bond has six carbon atoms; therefore, this is a hexene. It is a 1-hexene because the double bond is between carbons one and two. The name of the compound is 3-propyl-1-hexene.

24.7　First, write out the carbon skeleton for 2-heptene.

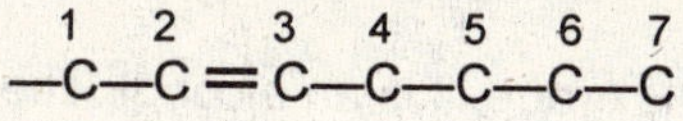

$$
\begin{array}{ccccccc}
1 & 2 & 3 & 4 & 5 & 6 & 7 \\
\end{array}
$$
$$
-C-C = C-C-C-C-C
$$

Then, add the alkyl groups.

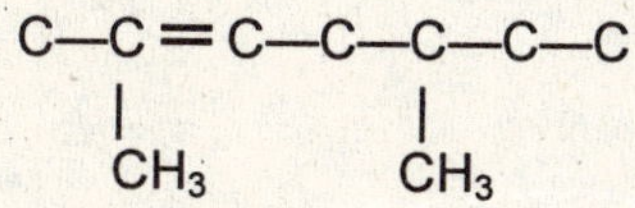

$$
C-C = C-C-C-C-C
$$
$$
\;\; | \qquad\qquad | \\
\;\; CH_3 \qquad\quad CH_3
$$

Finally, add the H atoms.

$$
CH_3CH = CHCH_2CHCH_2CH_3
$$
$$
\quad\; | \qquad\qquad\quad | \\
\quad\; CH_3 \qquad\qquad CH_3
$$

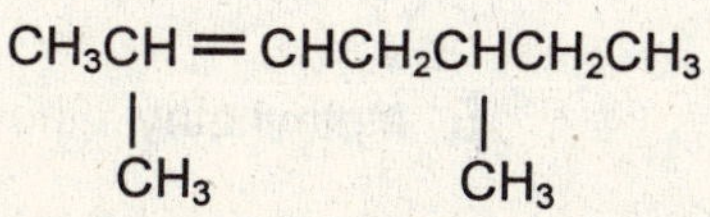

24.8 The two isomers of $CH_3CH_2CH = CHCH_2CH_3$ are *cis*-3-hexene and *trans*-3-hexene.

24.9 a. There are only three carbons in the chain. The compound is propyne.

  b. The longest continuous chain containing the triple bond has five carbon atoms. The compound is 3-methyl-1-pentyne.

24.10 a. This compound has an ethyl group attached to a benzene ring.

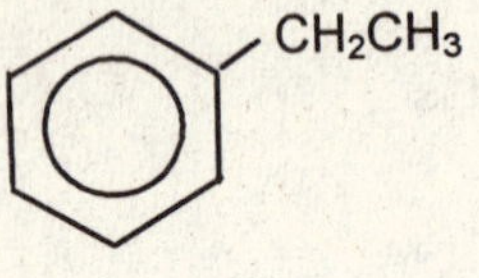

  b. This compound has one phenyl group attached to each carbon atom in the ethane molecule.

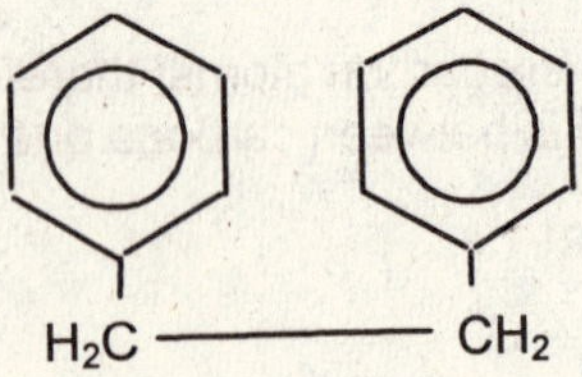

24.11 Because the compound has an –OH group, it is an alcohol. The longest carbon chain in the molecule has six carbons.

$$
\begin{array}{cccccc}
& & & \text{OH} & & \\
6 & 5 & 4 & 3| & 2 & 1 \\
\end{array}
$$

CH₃CH₂CH₂CCH₂CH₃
             |
           CH₂CH₃

The name of the compound is 3-ethyl-3-hexanol.

24.12 a. Dimethyl ether   **b.** Methyl ethyl ether

24.13 a. There are two alkyl groups attached to the carbonyl; therefore, the compound is a ketone. There are five carbon atoms in the chain. The name of the compound is 2-pentanone.

(continued)

b. There is a hydrogen atom attached to the carbonyl, so the compound is an aldehyde. The numbering of the stem carbon chain is

$$\overset{O}{\overset{\|}{\underset{1}{}}}$$

$$\underset{}{H-\overset{1}{C}}-\overset{2}{CH_2}\overset{3}{CH_2}\overset{4}{CH_3}$$

The name of the compound is butanal.

■ Answers to Review Questions

24.1 The formula of an alkane with thirty carbon atoms is $C_{30}H_{62}$.

24.2 $H_3C-CH_2-CH_2-CH_2-CH_3$ $H_3C-CH-CH_2-CH_2-CH_3$
                                                     |
                                                     CH_3

$H_3C-CH_2-CH-CH_2-CH_3$ $H_3C-CH-CH-CH_3$ CH_3
              | | | |
              CH_3 CH_3 CH_3 $CH_3-C-CH_2-CH_3$
                                                                       |
                                                                       CH_3

24.3 The structures of a seven-carbon alkane, cycloalkane, alkene, and aromatic hydrocarbon are

$CH_3-CH_2-CH_2-CH_2-CH_2-CH_2-CH_3$

$CH_3=CH-CH_2-CH_2-CH_2-CH_2-CH_3$

(toluene)

24.4 The two isomers of 2-butene are the *cis* and *trans* geometric isomers:

cis -2-butene   *trans* -2-butene

In the *cis*-2-butene, the two methyl groups are on the same side of the double bond; in the *trans* isomer, they are on opposite sides.

24.5 The structural formulas for the isomers of ethyl-methylbenzene are

24.6 Methane: source–natural gas; use–home fuel.
Octane: source–petroleum; use–auto fuel.
Ethylene: source–petroleum refining; use–chemical industry raw material.
Acetylene: source–methane; use–acetylene torch.

24.7

CH_3CH_2Cl	CH_2ClCH_2Cl	$CHCl_2CHCl_2$	CCl_3CCl_3
CH_3CHCl_2	$CH_2ClCHCl_2$	$CHCl_2CCl_3$	
CH_3CCl_3	CH_2ClCCl_3		

24.8 A substitution reaction is a reaction in which part of the reagent molecule is substituted for a hydrogen atom on a hydrocarbon or hydrocarbon group. For example,

$$CH_4 + Cl_2 \rightarrow CH_3Cl + HCl$$

An addition reaction is a reaction in which parts of the reagent are added to each carbon atom of a carbon-carbon multiple bond, which then becomes a C—C single bond. For example,

$$CH_2 = CH_2 + Br_2 \rightarrow CH_2Br–CH_2Br$$

24.9 The major product of HCl plus acetylene should be $Cl_2HC–CH_3$ because Markownikoff's rule predicts this.

24.10   The structures are

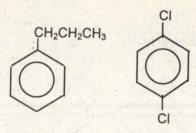

propylbenzene   paradichlorobenzene

24.11. A functional group is a reactive portion of a molecule that undergoes predictable reactions no matter what the rest of the molecule is like. An example is a C=C bond, which always reacts with bromine or other addition reagents to add part of each reagent to each carbon atom.

24.12   An aldehyde is different from a ketone, carboxylic acid, and ester in that it always has a hydrogen atom attached to the carbonyl group in addition to a hydrocarbon group.

24.13   Methanol: source–$CO + H_2$; use–solvent.

Ethanol: source–fermentation of glucose; use–solvent.

Ethylene glycol: source–ethylene; use–antifreeze.

Glycerol: source–from soap making; use–foods.

Formaldehyde: source–oxidation of methanol; use–plastics and resins.

24.14 a. CO is a carbonyl group (ketone).

b. CH_3O—C is an ether group.

c. C=C is a double bond.

d. COOH is a carboxylic acid group.

e. CHO is an aldehyde (carbonyl).

f. CH_2OH, or -OH, is a hydroxyl group (primary alcohol).

■ Solutions to Practice Problems

24.23 The condensed structural formula is $CH_3CH_2CH_2CH_3$.

24.25 a.

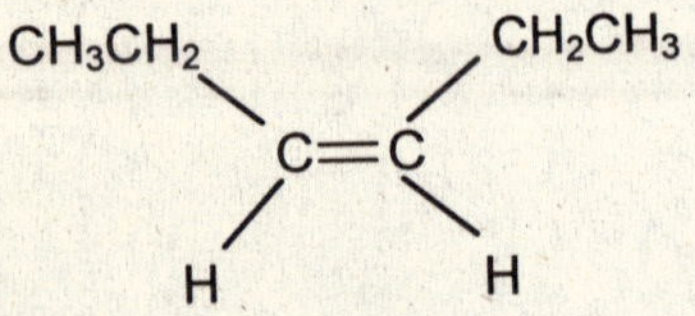

cis -3-hexene

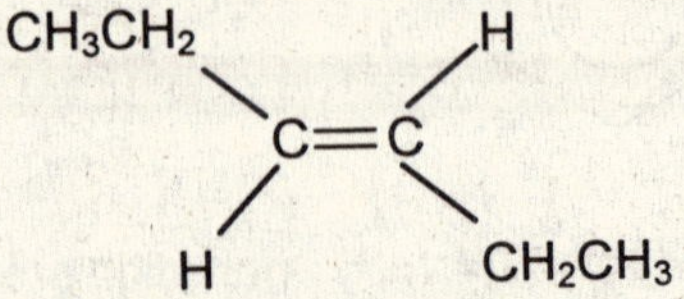

trans -3-hexene

b.

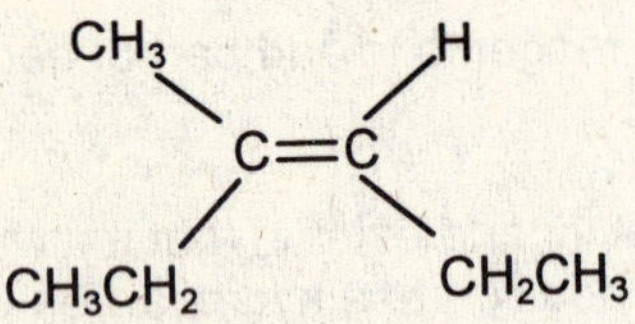

cis -3-methyl-3-hexene

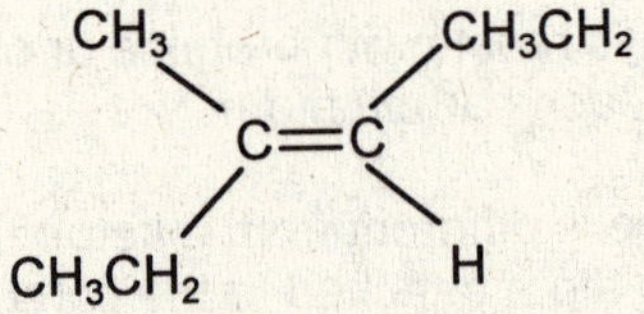

trans -3-methyl-3-hexene

24.27 a. $C_2H_4 + 3O_2 \rightarrow 2CO_2 + 2H_2O$

b.
$$CH_2\!=\!CH_2 + MnO_4^- + H_2O \rightarrow \overset{\displaystyle OH\ OH}{\underset{\displaystyle |\ \ |}{CH_2CH_2}} + MnO_2$$

Oxidation:
$$CH_2\!=\!CH_2 \rightarrow \overset{\displaystyle OH\ OH}{\underset{\displaystyle |\ \ |}{CH_2CH_2}}$$

Balance O:
$$2OH^- + CH_2\!=\!CH_2 \rightarrow \overset{\displaystyle OH\ OH}{\underset{\displaystyle |\ \ |}{CH_2CH_2}}$$

Balance e^-:
$$2OH^- + CH_2\!=\!CH_2 \rightarrow \overset{\displaystyle OH\ OH}{\underset{\displaystyle |\ \ |}{CH_2CH_2}} + 2e^-$$

(continued)

Reduction: $MnO_4^- \rightarrow MnO_2$

Balance O: $MnO_4^- + 2H_2O \rightarrow MnO_2 + 4OH^-$

Balance e^-: $MnO_4^- + 2H_2O + 3e^- \rightarrow MnO_2 + 4OH^-$

Add half-reactions:

$$3\left(2OH^- + CH_2\!=\!CH_2 \rightarrow \underset{\substack{|\ \ \ | \\ CH_2CH_2}}{\overset{OH\ OH}{}} + 2e^-\right)$$

$$2\left(MnO_4^- + 2H_2O + 3e^- \rightarrow MnO_2 + 4OH^-\right)$$

$$\cancel{6}OH^- + 3CH_2\!=\!CH_2 + 2MnO_4^- + 4H_2O \rightarrow 3\ \underset{\substack{|\ \ \ | \\ CH_2CH_2}}{\overset{OH\ OH}{}} + 2MnO_2 + \overset{2}{\cancel{8}}OH^-$$

$$3CH_2\!=\!CH_2 + 2MnO_4^- + 4H_2O \rightarrow 3\ \underset{\substack{|\ \ \ | \\ CH_2CH_2}}{\overset{OH\ OH}{}} + 2MnO_2 + 2OH^-$$

c. $CH_2\!=\!CH_2 + Br_2 \rightarrow \underset{\substack{|\ \ \ | \\ CH_2CH_2}}{\overset{Br\ Br}{}}$

d.

e.

24.29 $C_2H_6 + Cl_2 \rightarrow C_2H_5Cl + HCl$

24.31   According to Markownikoff's rule, the major product is the one obtained when the H atom adds to the carbon atom of the double bond that already has more hydrogen atoms attached to it. Therefore, 2-bromo-2-methylpropane is the major product.

$$CH_3-\underset{\underset{CH_3}{|}}{C}=CH_2 + HBr \longrightarrow CH_3-\underset{\underset{CH_3}{|}}{\overset{\overset{Br}{|}}{C}}-CH_3$$

24.33 a.

$$\underset{\underset{CH_3\ \ CH_3}{|\quad\ \ |}}{\overset{\overset{CH_3}{|}}{\underset{1\ \ \ \ 2\ \ 3\quad\ \ 4\ \ \ 5}{CH_3CHCHCHCH_3}}}$$

longest chain   2,3,4-trimethylpentane

b.

$$\underset{\underset{CH_3\quad\quad\quad\ \ CH_3}{|\quad\quad\quad\quad\ \ |}}{\overset{\overset{CH_3\quad\quad\quad\ \ CH_3}{|\quad\quad\quad\quad\ \ |}}{\underset{1\ \ \ 2\ \ 3\ \ \ 4\ \ \ 5\ \ \ 6\ \ 7}{CH_3CCH_2CH_2CH_2CCH_3}}}$$

longest chain   2,2,6,6-tetramethylheptane

c.

$$\underset{\underset{\underset{3\quad\ \ 2\quad\ \ 1}{CH_2CH_2CH_3}}{|}}{\overset{4\quad\ 5\quad\ 6\quad\ 7\quad 8}{CH_3CH_2CHCH_2CH_2CH_2CH_3}}$$

longest chain   4-ethyloctane

d.

$$\underset{\underset{\underset{2\quad\quad 1}{CH_2CH_3}}{|}}{\overset{\overset{CH_3}{|}}{\underset{3\quad\ 4\ 5\quad 6\quad\ 7}{CH_3CHCHCH_2CH_2CH_3}}}\quad\underset{8}{\overset{|}{CH_3}}$$

longest chain   3,4-dimethyloctane

24.35 a.

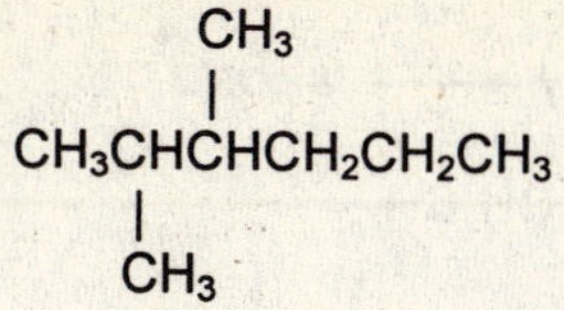

$$\underset{\underset{CH_3}{|}}{CH_3}CH\underset{\underset{CH_3}{|}}{CH}CH_2CH_2CH_3$$

b.

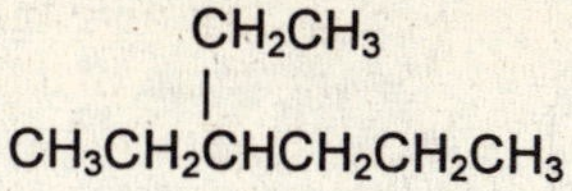

$$CH_3CH_2\underset{\underset{CH_2CH_3}{|}}{CH}CH_2CH_2CH_3$$

c.

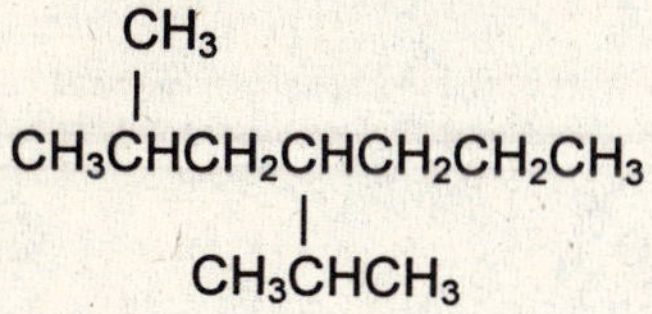

$$CH_3\underset{\underset{CH_3CHCH_3}{|}}{CH}CH_2\underset{\underset{|}{|}}{CH}CH_2CH_2CH_3$$

d.

$$CH_3-\underset{\underset{CH_3}{|}}{\overset{\overset{CH_3}{|}}{C}}-\underset{\underset{CH_3}{|}}{\overset{\overset{CH_3}{|}}{C}}-CH_2CH_3$$

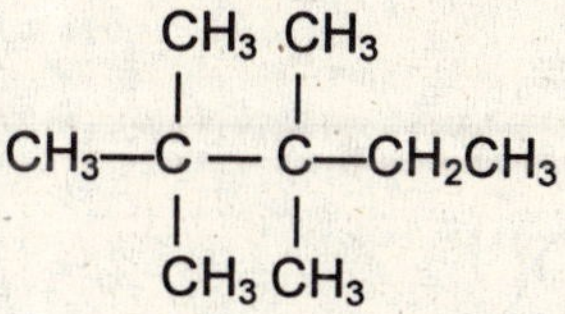

24.37 a. 2-pentene

b. 2,5-dimethyl-2-hexene

24.39 a.

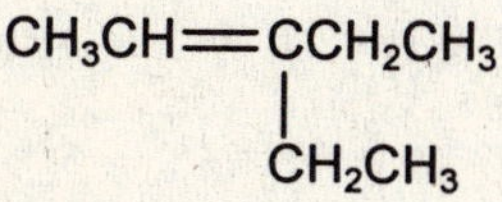

$$CH_3CH=\underset{\underset{CH_2CH_3}{|}}{C}CH_2CH_3$$

b.

$$CH_3\underset{\underset{CH_3}{|}}{C}=CH\underset{\underset{CH_2CH_3}{|}}{CH}CH_2CH_3$$

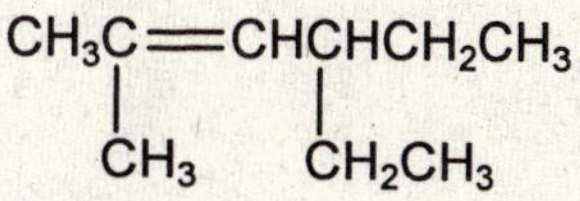

24.41 The two isomers are *cis*-2-pentene and *trans*-2-pentene.

24.43 a. 2-butyne

b. 3-methyl-1-pentyne

24.45 a.

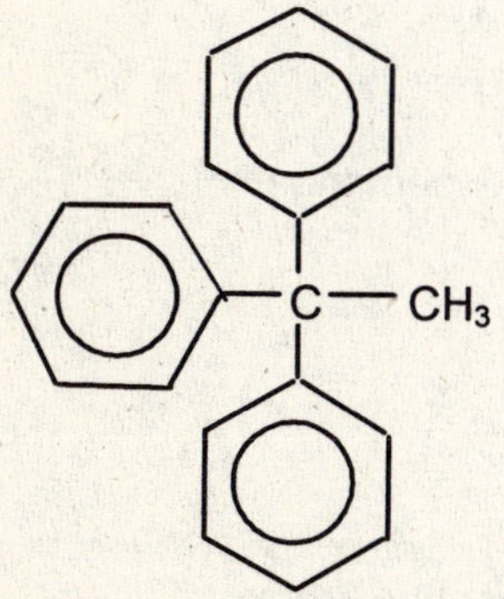

b.

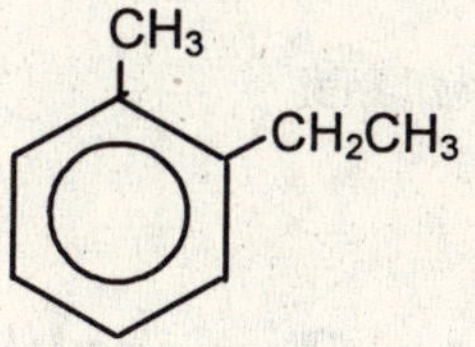

24.47 a. 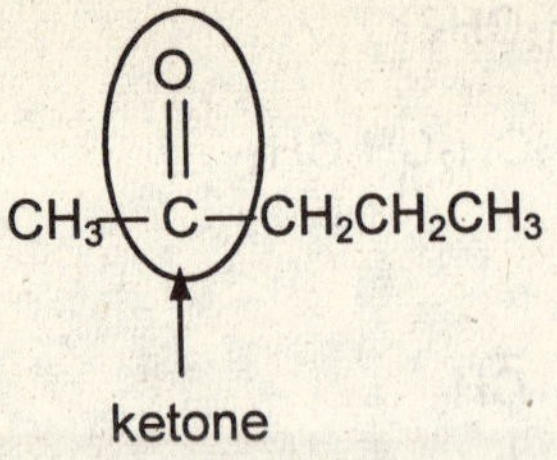 $CH_3 - C - CH_2CH_2CH_3$

(O double bond on C, circled) ← ketone

b. 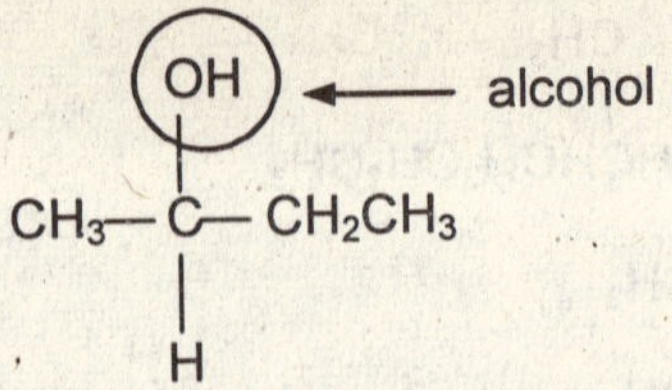 $CH_3 - C - CH_2CH_3$

(OH circled) ← alcohol, H below C

c. $HO - C - CH_2CH_3$

(O double bond on C, circled) → carboxylic acid

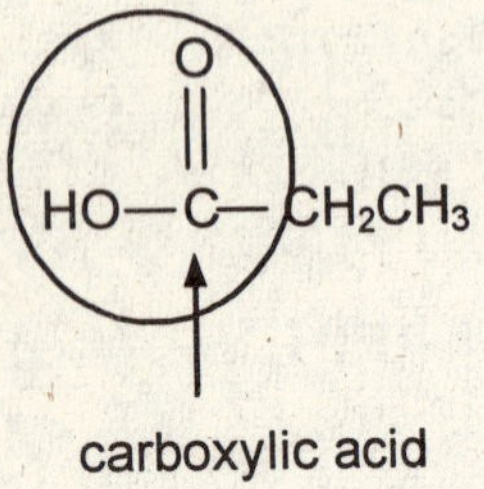

d. 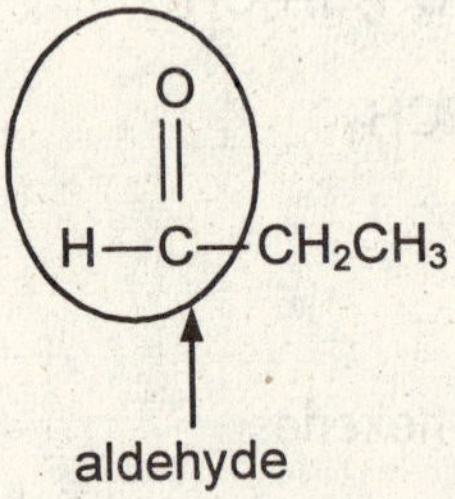 $H - C - CH_2CH_3$

(O double bond on C, circled) → aldehyde

24.49 a. 1-pentanol

b. 2-pentanol

c. 2-propyl-1-pentanol

d. 6-methyl-4-octanol

24.51 a. secondary alcohol

b. secondary alcohol

c. primary alcohol

d. primary alcohol

24.53 a. ethyl propyl ether

b. methyl isopropyl ether

24.55 a. butanone

b. butanal

c. 4,4-dimethylpentanal

d. 3-methyl-2-pentanone

■ Solutions to General Problems

24.57 a. 3-methylbutanoic acid

b. *trans* -5-methyl-2-hexene

c. 2,5-dimethyl-4-heptanone

d. 4-methyl-2-pentyne

24.59 a.

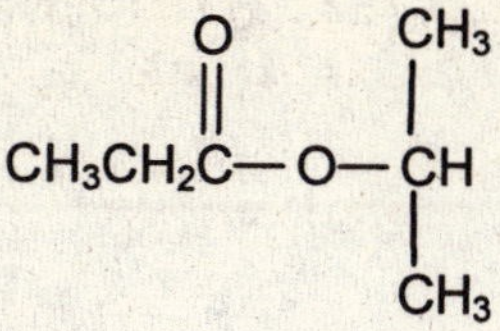

$$CH_3CH_2\overset{\displaystyle O}{\overset{\displaystyle \|}{C}}-O-\underset{\displaystyle CH_3}{\overset{\displaystyle CH_3}{CH}}$$

b.

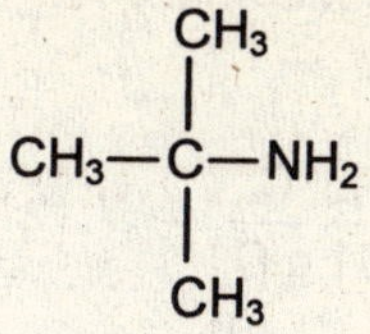

$$CH_3-\underset{\displaystyle CH_3}{\overset{\displaystyle CH_3}{C}}-NH_2$$

c.

$$CH_3CH_2CH_2CH_2\underset{\displaystyle CH_3}{\overset{\displaystyle CH_3}{C}} ? COOH$$

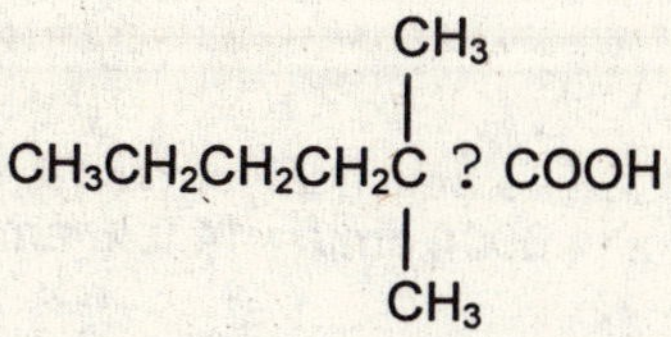

d.

$$\underset{\displaystyle H}{\overset{\displaystyle CH_3CH_2}{}}C=C\underset{\displaystyle H}{\overset{\displaystyle CH_2CH_3}{}}$$

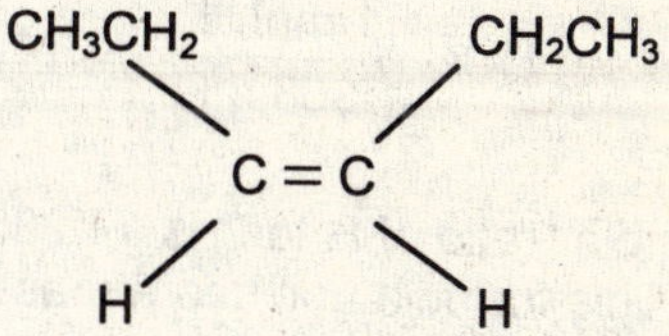

24.61 a. Addition of dichromate ion in acidic solution to propionaldehyde will cause the reagent to change from orange to green as the aldehyde is oxidized. Under similar conditions, acetone (a ketone) would not react.

b. Addition of a solution of Br_2 in CCl_4 to CH_2=CH—$C\equiv C$—CH=CH_2 would cause the bromine color to disappear as the Br_2 was added to the multiple bonds. Addition of benzene to Br_2 in CCl_4 results in no reaction. Aromatic rings are not susceptible to attack by Br_2 in the absence of a catalyst.

24.63 a. ethylene, CH_2=CH_2

b. There must be an aromatic ring in the compound or the double bonds would react with Br_2. The compound is toluene, and the correct formula is

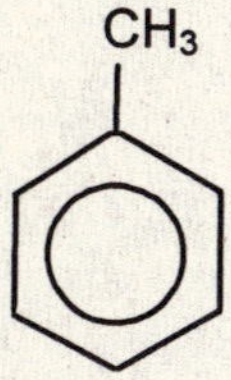

CH₃ (attached to benzene ring)

c. methylamine, CH_3NH_2

d. methanol, CH_3OH

24.65   Assume 100.0 g of the unknown. This contains 85.6 g C and 14.4 g H. Convert these amounts to moles.

$$85.6 \text{ g} \times \frac{1 \text{ mol C}}{12.01 \text{ g C}} = 7.1\underline{2}7 \text{ mol C}$$

$$14.4 \text{ g H} \times \frac{1 \text{ mol H}}{1.008 \text{ g H}} = 14.\underline{2}857 \text{ mol H}$$

The molar ratio of H to C is 14.2857:7.12, or 2.00:1. The empirical formula is therefore CH_2. This formula unit has a mass of $[12.01 + 2 \times (1.008)]$ amu = 14.026 amu.

$$\frac{56.1 \text{ amu}}{1 \text{ molecule}} \times \frac{1 \text{ formula unit}}{14.026 \text{ amu}} = \frac{4.00 \text{ formula units}}{1 \text{ molecule}}$$

The molecular formula is $(CH_2)_4$, or C_4H_8. The formula (C_nH_{2n}) indicates the compound is either an alkene or a cycloalkane. Because it reacts with water and H_2SO_4, it must be an alkene. The product of the addition of H_2O to a double bond is an alcohol. Because the alcohol produced can be oxidized to a ketone, it must be a secondary alcohol. The only secondary alcohol with four carbon atoms is 2-butanol.

$$\underset{\overset{|}{\text{OH}}}{CH_3CH_2CHCH_3}$$

The original hydrocarbon from which it was produced is either 1-butene or 2-butene.

$$CH_3CH = CHCH_3 + H_2O \xrightarrow{H_2SO_4} \underset{\overset{|}{\text{OH}}}{CH_3CH_2CHCH_3}$$

or

$$CH_3CH_2CH = CH_2 + H_2O \xrightarrow{H_2SO_4} \underset{\overset{|}{\text{OH}}}{CH_3CH_2CHCH_3}$$

25. POLYMER MATERIALS: SYNTHETIC AND BIOLOGICAL

■ Solutions to Exercises

25.1   This addition polymer is formed when vinylidene chloride adds to itself across the double bond.

$$\ldots + CH_2{=}CCl_2 + CH_2{=}CCl_2 + CH_2{=}CCl_2 + \ldots \rightarrow$$

$$-CH_2{-}CCl_2{-}CH_2{-}CCl_2{-}CH_2{-}CCl_2{-}$$

■ Answers to Review Questions

25.1   An addition reaction is a polymer formed by linking together many molecules by addition reactions. The monomers have multiple bonds that will undergo addition reactions. An example is the formation of polypropylene from propene.

$$\ldots + \underset{CH_3}{CH}{=}CH_2 + \underset{CH_3}{CH}{=}CH_2 + \underset{CH_3}{CH}{=}CH_2 + \ldots \longrightarrow$$

$$-\underset{CH_3}{CH}{-}CH_2{-}\underset{CH_3}{CH}{-}CH_2{-}\underset{CH_3}{CH}{-}CH_2{-}$$

(continued)

A condensation polymer is formed by linking together many molecules by condensation reactions. An example is the formation of nylon.

$$\ldots + H\underset{\underset{H}{|}}{N}(CH_2)_6\underset{\underset{H}{|}}{N}-H \ + \ HO-\underset{\underset{O}{\|}}{C}(CH_2)_4\underset{\underset{O}{\|}}{C}-OH$$

$$+ \ H\underset{\underset{H}{|}}{N}(CH_2)_6\underset{\underset{H}{|}}{N}-H \ + \ HO-\underset{\underset{O}{\|}}{C}(CH_2)_4\underset{\underset{O}{\|}}{C}-OH \ + \ \ldots \ \xrightarrow{\Delta}$$

$$\sim\underset{\underset{H}{|}}{N}(CH_2)_6\underset{\underset{H}{|}}{N}-\underset{\underset{O}{\|}}{C}(CH_2)_4\underset{\underset{O}{\|}}{C}-\underset{\underset{H}{|}}{N}(CH_2)_6\underset{\underset{H}{|}}{N}-\underset{\underset{O}{\|}}{C}(CH_2)_4\underset{\underset{O}{\|}}{C}\sim \ + \ nH_2O$$

25.2   The addition reaction for one isoprene molecule to another is

$$CH_2{=}\underset{\underset{CH_3}{|}}{C}{-}CH{=}CH_2 \ + \ CH_2{=}\underset{\underset{CH_3}{|}}{C}{-}CH{=}CH_2 \ \longrightarrow \ CH_2{=}\underset{\underset{CH_3}{|}}{C}{-}CH{-}CH_2CH_2{-}\underset{\underset{CH_3}{|}}{C}CH{=}CH_2$$

25.3   The chain polymer obtained by the reaction of ethylene glycol with malonic acid can be represented by

$$\sim OCH_2CH_2O{-}\underset{\underset{O}{\|}}{C}CH_2\underset{\underset{O}{\|}}{C}{-}OCH_2CH_2O{-}\underset{\underset{O}{\|}}{C}CH_2\underset{\underset{O}{\|}}{C}\sim$$

25.4   The *pi* electrons in polyacetylene would have to shift as follows.

25.5   Experiments show that iodine inserts itself into the plastic between polyacetylene molecules, where it forms triiodide ion. Each ion carries a negative charge that was abstracted from a *pi* orbital of the polymer molecule. The polymer material is left with a positive charge, or hole. The triiodide ion can facilitate the movement of an electron from an uncharged polymer molecule to a positively charged molecule (a hole). In effect, the hole moves. The movement of holes through the material constitutes an electric current.

25.6   The primary structure of a protein refers to the order, or sequence, of the amino-acid units in the protein polymer. What makes one protein different from another of the same size is the arrangement of the various possible amino acids in the sequence. For example, there are 120 different sequences possible for a polypeptide with just five different amino acids. The basis of the unique conformation of a protein is the folding and coiling into a three-dimensional conformation in aqueous solution due to the different side-chain amino-acid units. In stable conformations, the nonpolar side chains are buried within the structure away from water, and the polar groups are on the surface of the conformation, where they can hydrogen bond with water.

25.7   The secondary structure of a protein is the simpler coiled or parallel arrangement of the protein chain. The tertiary structure refers to the folded nature of the structure.

25.8   An enzyme is a specific body catalyst, usually a globular protein, that possesses active sites. The specificity of an enzyme is explained by the active sites at which substrates will bind. Substrates fit into the active sites as a key fits into a lock.

25.9   Cellulose forms the plant cell walls and is a linear polymer of β-D-glucopyranose units. Amylose is the major energy-storage substance of plants and is a linear polymer of α-D-glucopyranose units.

25.10   It is possible for glucose to exist in three forms in any solution such as blood: the straight-chain form, the α-D-glucopyranose form, and the β-D-glucopyranose form.

25.11   The complementary base pairs are the nucleotide bases that form strong hydrogen bonds with one another: adenine and thymine, adenine and uracil, and guanine and cytosine. A DNA molecule consists of two polynucleotide chains with base pairing along their entire lengths. The two chains are coiled about each other to form a double helix.

25.12   The only difference between ribonucleotides and deoxyribonucleotides is the sugar: Ribonucleotides contain β-D-ribose, and deoxyribonucleotides contain 2-deoxy-β-D-ribose, with both sugars having furanose rings. Both form polymers by condensation (loss of H_2O).

25.13   The genetic code is the relationship between the nucleotide sequence in DNA and the amino-acid sequence in proteins. Genetic information is coded into the linear sequence of nucleotides in the DNA molecule, and this coding then directs the synthesis of the specific proteins that make a cell unique.

25.14   A codon is a code structure in messenger RNA; each such structure has a particular sequence of three nucleotides and is usually denoted simply by its bases. An anticodon is a triplet sequence in transfer RNA complementary to the codon in DNA; the transfer RNA uses the anticodon to carry an amino acid to a ribosome. The messenger RNA and transfer RNA bond to each other through the codon and anticodon, respectively.

25.15   There are four RNA bases, and three of these are arranged in a specific order in each codon. Since there are four different possibilities for each base in the codon, there are $4 \times 4 \times 4 = 64$ different triplet codons.

25.16   A polypeptide is produced as follows: Imagine you have a ribosome with messenger RNA attached in the proper way for translation, and the first codon is in position to be read. (The messenger RNA has been synthesized with a sequence of bases complementary to that of the gene before attachment.) Transfer RNAs bring up various amino acids to be bonded to each other until a termination codon appears to signal the end of the chain, which is then released from the ribosome.

■ Solutions to Practice Problems

25.23 $nCF_2{=}CF_2 \rightarrow -CF_2{-}CF_2{-}CF_2{-}CF_2{-}CF_2{-}CF_2{-}$

25.25   The two monomer units for this polymer are

$$HOCH_2CH_2OH \quad \text{and} \quad HOC\overset{\displaystyle O}{\overset{\|}{}}CH_2CH_2C\overset{\displaystyle O}{\overset{\|}{}}OH$$

25.27 a. When this monomer undergoes an addition reaction, the triple bond is converted into a double bond, just as it is with acetylene. The resulting polymer contains alternating single and double bonds and, therefore, could produce a conducting polymer.

   b. When this monomer (propylene) undergoes an addition reaction, the double bond is converted into a single bond. Since the polymer does not contain any double bonds, this monomer cannot produce a conducting polymer.

25.29   The zwitterion for alanine is

$$CH_3-\overset{\overset{\displaystyle H}{|}}{\underset{\underset{\displaystyle NH_3^+}{|}}{C}}-COO^-$$

25.31   There are two possible dipeptides containing one molecule each of L-alanine and L-histidine, ala-his and his-ala.

ala-his   his-ala

25.33   DNA consists of two strands of polynucleotides with the adenine units in one strand paired to thymine units in the other. Similarly, guanine units are paired with cytosine. So, the adenine-thymine molar ratio is 1:1, and the molar ratio of guanine to cytosine is 1:1.

25.35   Adenosine consists of ribose and adenine.

25.37 Three hydrogen bonds link a guanine-cytosine base pair.

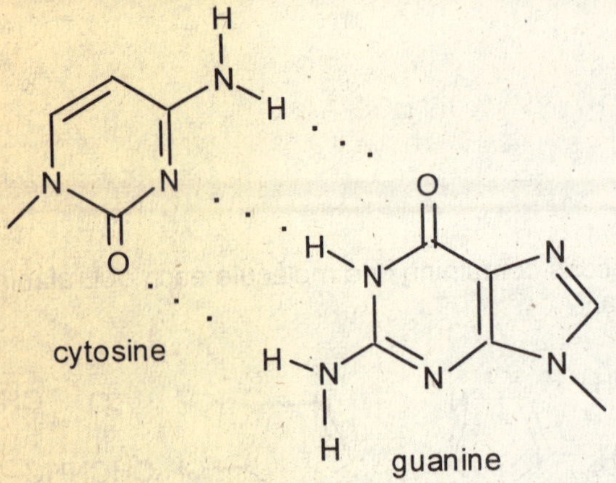

cytosine

guanine

Because only two hydrogen bonds link an adenine-uracil base pair, the bonding would be expected to be stronger in the guanine-cytosine pair.

25.39 If a codon consisted of two nucleotides, there would be 4 x 4 or sixteen possible codons using the four nucleotides. Because there are twenty amino acids that must be represented uniquely by a codon for protein synthesis, the codon must be longer than two nucleotides. A two-nucleotide codon would not be workable as an amino-acid code.

25.41 When DNA is denatured, the hydrogen bonds between base pairs are broken. DNA with a greater percent composition of guanine and cytosine would be denatured less readily than DNA with a greater percent composition of adenine and thymine because there are more hydrogen bonds between the former than between the latter.

25.43 Mark off the message into triplets beginning at the left.

GGA|UCC|CGC|UUU|GGG|CUG|AAA|UAG

Gly-Ser-Arg-Phe-Gly-Leu-Lys

Note that the UAG at the right codes for the end of the sequence.

25.45 The codons have bases that are complementary to those in the anticodon.

Anticodons	GAC	UGA	GGG	ACC
Codons	CUG	ACU	CCC	UGG

25.47 Consult Table 25.3 to find which nucleotides correspond to the amino acids in the sequence.

leu-ala-val-glu-asp-cys-met-trp-lys

CUU GCU GUU GAA GAU UGU AUG UGG AAA

25.49 The addition polymer that forms from *cis* -1,2-dichloroethene is

$$-CHCH-CHCH-CHCH-CHCH-$$
$$\;\;\;|\;\;|\;\;\;\;\;\;|\;\;|\;\;\;\;\;\;|\;\;|\;\;\;\;\;\;|\;\;|$$
$$\;\;\;Cl\;Cl\;\;\;\;Cl\;Cl\;\;\;\;Cl\;Cl\;\;\;\;Cl\;Cl$$

25.51 The monomer units that make up kevlar are

$$HOC-\bigcirc-COH$$

with $C=O$ groups, and $H_2N-\bigcirc-NH_2$

25.53 The amino acid with the nonpolar side chain is

$$\overset{NH_2}{\underset{|}{CH_3SCH_2CH_2CHCOOH}}$$

The other amino acid has a polar SH group in the side chain.

25.55 The zwitterion for serine is

$$\overset{NH_3^+}{\underset{|}{HOCH_2CHCOO^-}}$$

25.57 The two possibilities are

$$CH_3SCH_2CH_2\overset{NH_2}{\underset{|}{CH}}-\overset{O}{\overset{||}{C}}NHCHCOOH$$
$$\underset{|}{CH_2SH}$$

and

$$HSCH_2\overset{NH_2}{\underset{|}{CH}}-\overset{O}{\overset{||}{C}}NHCHCOOH$$
$$\underset{|}{CH_2CH_2SCH_3}$$

25.59 The number of possible sequences is 6 x 5 x 4 x 3 x 2 x 1, or 720.

25.61

APPENDIX A.   MATHEMATICAL SKILLS

■ Solutions to Exercises

1.   a. Either leave it as 4.38 or write it as 4.38×10^0.

    b. Shift the decimal point left, and count the number of positions shifted (3). The answer is 4.380×10^3, assuming the terminal zero is significant.

    c. Shift the decimal point right, and count the number of positions shifted (4). The answer is 4.83×10^{-4}.

2.   a. Shift the decimal point right three places. The answer is 7025.

    b. Shift the decimal point left four places. The answer is 0.000897.

3.   Express 2.8×10^{-6} as 0.028×10^{-4}. Then, the sum can be written

$$(3.142 \times 10^{-4}) + (0.028 \times 10^{-4}) \text{ or } (3.142 + 0.028) \times 10^{-4} = 3.170 \times 10^{-4}.$$

4.   a. $(5.4 \times 10^{-7}) \times (1.8 \times 10^8) = (5.4 \times 1.8) \times 10^{-7} \times 10^8 = 9.72 \times 10^1$. This rounds to 9.7×10^1.

    b. $\dfrac{5.4 \times 10^{-7}}{6.0 \times 10^{-5}} = \dfrac{5.4}{6.0} \times 10^{-7} \times 10^5 = 0.90 \times 10^{-2} = 9.0 \times 10^{-3}$

5.   a. $(3.56 \times 10^3)^4 = (3.56)^4 \times (10^3)^4 = 161 \times 10^{12} = 1.61 \times 10^{14}$

    b. $\sqrt[3]{4.81 \times 10^2} = \sqrt[3]{0.481 \times 10^3} = \sqrt[3]{0.481} \times \sqrt[3]{10^3} = 0.784 \times 10^1 = 7.84 \times 10^0$

6. a. log 0.00582 = -2.235

    b. log 689 = 2.838

7. a. antilog 5.728 = 5.35 x 10^5

    b. antilog (-5.728) = 1.87 x 10^{-6}

8. $$x = \frac{-0.850 \pm \sqrt{(0.850)^2 - 4(1.80)(-9.50)}}{2(1.80)} = \frac{-0.850 \pm 8.314}{3.60}$$

    The positive root is

    $$\frac{7.46}{3.60} = 2.07$$